# Locating Medical History

Locating Medical History

The Stories and Their Meanings

Edited by Frank Huisman and
John Harley Warner

The Johns Hopkins University Press
Baltimore and London

This book was brought to publication with the generous assistance of the Historia Medicinae Foundation, Delft; the Joannes Juda Groen Foundation, Amsterdam; and the Society Nederlands Tijdschrift voor Geneeskunde.

© 2004 The Johns Hopkins University Press
All rights reserved. Published 2004
Printed in the United States of America on acid-free paper
9 8 7 6 5 4 3 2 1

The Johns Hopkins University Press
2715 North Charles Street
Baltimore, Maryland 21218-4363
www.press.jhu.edu

Library of Congress Cataloging-in-Publication Data
Locating medical history : the stories and their meanings /
edited by Frank Huisman and John Harley Warner.
    p. cm.
Includes bibliographical references and index.
ISBN 0-8018-7861-6 (hardcover : alk. paper)
1. Medicine—Historiography. I. Huisman, Frank. II. Warner,
John Harley, 1953–
R133.L62 2004
610′.722—dc21
2003012875

A catalog record for this book is available from the British Library.

For
Jan Huisman
and
John Gustave Warner
and in memory of
Elly Walina Huisman-Hoogland
and
Dorothy Nies Warner

Contents

	Acknowledgments	ix
1	Medical Histories *Frank Huisman and John Harley Warner*	1

PART I: Traditions

2	To Whom Does Medical History Belong? Johann Moehsen, Kurt Sprengel, and the Problem of Origins in Collective Memory *Hans-Uwe Lammel*	33
3	Charles Daremberg, His Friend Émile Littré, and Positivist Medical History *Danielle Gourevitch*	53
4	*Bildung* in a Scientific Age: Julius Pagel, Max Neuburger, and the Cultural History of Medicine *Heinz-Peter Schmiedebach*	74
5	Karl Sudhoff and "the Fall" of German Medical History *Thomas Rütten*	95
6	Ancient Medicine: From Berlin to Baltimore *Vivian Nutton*	115
7	Using Medical History to Shape a Profession: The Ideals of William Osler and Henry E. Sigerist *Elizabeth Fee and Theodore M. Brown*	139

PART II: A Generation Reviewed

8	"Beyond the Great Doctors" Revisited: A Generation of the "New" Social History of Medicine *Susan M. Reverby and David Rosner*	167
9	The Historiography of Medicine in the United Kingdom *Roy Porter*	194
10	Social History of Medicine in Germany and France in the Late Twentieth Century: From the History of Medicine toward a History of Health *Martin Dinges*	209

11 Trading Zones or Citadels? Professionalization and
    Intellectual Change in the History of Medicine
    Olga Amsterdamska and Anja Hiddinga   237
12 The Power of Norms: Georges Canguilhem, Michel Foucault,
    and the History of Medicine   *Christiane Sinding*   262
13 Postcolonial Histories of Medicine   *Warwick Anderson*   285

PART III: After the Cultural Turn

14 "Framing" the End of the Social History of Medicine
    Roger Cooter   309
15 The Social Construction of Medical Knowledge
    Ludmilla Jordanova   338
16 Making Meaning from the Margins: The New Cultural History
    of Medicine   *Mary E. Fissell*   364
17 Cultural History and Social Activism: Scholarship, Identities,
    and the Intersex Rights Movement   *Alice Domurat Dreger*   390
18 Transcending the Two Cultures in Biomedicine: The History *of*
    Medicine and History *in* Medicine   *Alfons Labisch*   410
19 A Hippocratic Triangle: History, Clinician-Historians, and
    Future Doctors   *Jacalyn Duffin*   432
20 Medical History for the General Reader   *Sherwin B. Nuland*   450
21 From Analysis to Advocacy: Crossing Boundaries as a Historian
    of Health Policy   *Allan M. Brandt*   460

Notes on Contributors   *485*
Index   *491*

Acknowledgments

The idea for this volume grew from discussions at an international conference on the history of medical historiography held in Maastricht, The Netherlands, on 16–18 June 1999. The conference was supported by the Faculteit der Cultuurwetenschappen of the Universiteit Maastricht, the Nicolaus Mulerius Foundation in Groningen, the Netherlands Organization for Scientific Research, and the Huizinga Research Institute for Cultural History.

Although most of the contributions to this volume were commissioned after the Maastricht conference, five of the chapters, substantially reworked for the book, had as their springboard conference papers, those by Roger Cooter, Elizabeth Fee, Alfons Labisch, Hans-Uwe Lammel, and Vivian Nutton. Roy Porter's chapter is reprinted with permission from *Medicina nei secoli* 10 (1998): 253–269, and Ludmilla Jordanova's chapter is reprinted, with permission of the Society for the Social History of Medicine, from *Social History of Medicine* 8 (1995): 361–381 and contains a new afterword.

Of the many people who have helped shape this volume, we are especially indebted to Toby Appel, Wiebe Bijker, Lisa Boult, Thomas Broman, Marijke Gijswijt-Hofstra, Natsu Hattori, Godelieve van Heteren, Klasien Horstman, Manfred Horstmanshoff, Henk Huisjes, Gundolf Keil, Pieter Jan Kuijjer, Christopher Lawrence, Susan Lederer, Ramona Moore, Norbert Paul, Catrien Santing, Jenny Slatman, Manuel Stoffers, Janet Tighe, Janis Wethly, and Joseph Wachelder. To get this project going, Klaas van Berkel, Jo Tollebeek and Henk te Velde have been more important than they probably realize. We also greatly benefited from the insightful critique of an anonymous reviewer of the manuscript for the Johns Hopkins University Press, and from the thoughtful copyediting of Glenn Perkins.

Our editor at the Press, Jacqueline Wehmueller, has encouraged us in this project from the outset, and we are enormously grateful for her steadfast support, critiques, and patience. Naomi Rogers has been our most ruthless and caring critic and has pressed us at every turn to make the collection widely accessible to readers who are not history of medicine "insiders." Above all, we are grateful to the contributors to this volume for their feisty intellectual engagement, cheering humor, and willingness to explore new pathways, all of which have been indispensable ingredients in making the whole a truly collaborative venture.

The production of this volume has been aided by the generous support of the Historia Medicinae Foundation in Delft, the Joannes Juda Groen Foundation in Amsterdam, and the Society Nederlands Tijdschrift voor Geneeskunde. We are especially grateful to the Faculteit der Cultuurwetenschappen of the Universiteit Maastricht and to the Section of the History of Medicine at the Yale University School of Medicine, without whose support our transatlantic collaboration would not have been possible.

Locating Medical History

CHAPTER ONE

Medical Histories

Frank Huisman and John Harley Warner

Historians of medicine tend to be very ready to assert that the past gives important perspective on the present—that understanding the experience and management of illness in the past can aid patients, clinicians, policy makers, public health officials, ethicists, and voting citizens as they make difficult choices. Some historians are explicit about their conviction that tracing how earlier societies responded to epidemics, for example, or how they drew distinctions between the normal and the pathological, can help guide individuals, professions, and states today toward more effective and more just interpretations and interventions. Others write histories of medicine convinced that displaying a rich professional heritage offers intellectual satisfaction and reassurance to clinicians as they face daunting workaday demands. Still others see history as a vehicle for helping medical students make sense of the professional culture they are entering and realize their own agency as the pressures of socialization take their toll. Even medical historians openly wary about drawing lessons from the past are often quick to recognize the potential of historical perspective to aid a woman confronting childbirth, an adolescent coming to terms with her or his own body, or all of us in comprehending the ways in which culture shapes our sense of self in sickness and in health.

It is therefore surprising that a community so intent on calling attention to the salience of the past for the present has generally been roundly dismissive of the past of its own craft. During the last several decades, "traditional" medical history has been most frequently brought forward as a simplistic straw figure, cited only that it may be trounced. Authors who are decreasingly likely to have read work from the field's early years often assail that work in order to enhance the importance and novelty of their own contributions. Such historiographic posturing is a built-in rhetorical strategy in virtually all fields of history. Nevertheless, it remains ironic that, as historians of medicine, most of us reflect so little on the past of our own enterprise as we shape and reshape our historical practice.

Our aim in this volume is neither to defend nor to rehabilitate earlier approaches to medical history. However, we do want to challenge the now commonplace depiction of "traditional" medical history. In the new social history of medicine that took shape in the 1970s, the deployment of a flat caricature of older work helped to define the new program and to clarify a sense of mission for historians trying to consolidate their separate identity. Indeed, that movement gained in momentum partly by contrast with what it was *not*—the established, establishment medical history written exclusively by physicians for other physicians, consecrated to heroic celebration of great doctors and their achievements, Whiggish and triumphalist, unapologetically internalistic and naively positivist. By the early 1980s, this image of traditional history degenerated from an inspiring rallying call for historiographic revolution to an assumed but unexamined misrepresentation of the past.

During the 1980s and 1990s, invoking this stereotype had too often become a de facto substitute for fresh theoretical engagement and analysis. In book introductions, journal articles, and grant applications, parading an older medical history only to denounce it provided a readily available and intellectually undemanding way of asserting the importance of one's own work. This is one of the reasons why much work that is "intellectually flatfooted and theoretically unreflective" (as Roger Cooter puts it in chapter 14 of this volume) has managed to pass muster. Perhaps one of the best things we could do for our field is to make it disreputable to trot out "traditional" medical history as a simplistic foil, a maneuver that too long, too often has passed for theoretical and historiographic novelty.

More than this, invidious depictions of the past are often covert dismissals of other genres of medical history in the present, and precisely because these dismissals tend to be covert rather than open, they bypass any occasion for critical dialog. Dismissing approaches other than our own spares us the labor of spelling out our historiographic aims and allegiances while foreclosing opportunities for the kinds of debate that can keep a field vital. As individual writers we can perhaps dispense with such self-reflection and get on with our work, but our field and our students are the losers if we persistently shy away from explicit historiographic engagement.

The history of medicine, an eclectic field that accommodates remarkable diversity, is facing profound choices about its future. This is all the more urgently the case at a moment when the "new social history" program for understanding health culture and health care has not been *new* for a generation, and when the once new cultural history impulse has failed thus far to inspire a comparable sense of shared agenda and mission. Whether medical history after the cultural turn has been reinvigorated or derailed, the fact remains that we have yet to witness anything like the spirited debates of the 1970s and early 1980s that galvanized and ultimately transformed the field.

This volume explores the doing of medical history in the distant and recent past. We want to make it impossible, or at least unconscionable, to depict traditional medical history as unified in methods or aims. At the same time, we want to take a ruthlessly honest look at the practice of medical history in the present, recognizing diversity in historians' backgrounds, approaches, aspirations, and audiences. Disputes about who has the right credentials to pursue medical history seem, at the moment, a thing of the past. Yet, more than ever there is wide divergence in judgments on what stories merit telling, about what work those stories (told in a classroom or in a courtroom) can and should do, and about what constitutes meaningful medical history. We do not seek to privilege one answer, but we do insist that while the field is best served by a healthy pluralism, differing judgments about what medical history is worth doing should be the product of critical discussion about the variety of alternative choices at hand, and that vitality depends on such ongoing debate. The issues at stake in the history of medicine—how societies organize health care, how individuals or states relate to sickness, how we understand our own identity and agency as sufferers or healers—are simply too important for the practice of medical history not to be persistently subjected to vigorous reflection and reexamination.

The Trouble with Traditional Medical History

The invention of tradition long ago became a staple of historical fare.[1] Historians in general have recognized the extent to which nations, social groups, and professions create traditions to consolidate identity, to legitimate, and to inspire. In the process, historians have become acutely aware of "tradition" as a heavily freighted term, denoting clusters of practices and values fashioned at particular historical moments to serve specific purposes, constructions that gain in power as they come to be imbued with a sense of timelessness and to be regarded as inbuilt components of a culture rather than contingent creations.

Conversely, traditions—real or invented—have also been used to break with the past, to eschew continuities and inaugurate an assertively new tradition distinguished by its own dissenting values and distinctive practices. Particularly in eras that launched programs for change, what historical actors designated as "traditional" invites attention not as a descriptive category so much as a polemical construction that helped define, propel, and sanction reform.

It should give us pause when we read or hear of "traditional medical history" as an *invariant* entity—known, unified, timeless. Yet, for several decades now, the depiction of "traditional medical history" in precisely that way has littered our professional literature. Fairness to those who practiced medical history in the past is not what concerns us here. Rather, although the polemical service of this maneuver once was warranted as a tool in propelling the new medical history, it long ago became a crutch, holding the field back rather than helping it forward. Equally, by perpetuating a misrepresentation of the past, it also contributes to a misunderstanding of the present; a widespread sense that medical history is a field somehow in crisis looks rather different when viewed from the perspective of a longer past, which suggests that many of the tensions, doctrinal fissures, and institutional instabilities confronted today are not so new as most of us assume them to be.

From the start of the nineteenth century and the beginnings of medical history as a coherent field of scholarship with disciplinary aspirations, there was not one medical history but many. Rather than presenting medical history as a discipline with a self-evident identity and clear-cut boundaries, we would like to conceptualize it as a field. This spatial metaphor makes it possible to do justice to the family resemblances among widely disparate historical endeavors. The multiplicity of medical histories coexisting and sometimes competing within the wider field today can be seen not merely in disputes about what "medical history" should study but in the instability of the category itself. As early as the 1970s, some historians identified "history of medicine" as an excessively narrow and anachronistic label and asked whether the changing field might better be characterized as the history of health care. Similarly, prior to the historiographic revolution of the 1960s and 1970s, there was no single "medical history" transcendent over time but rather many different histories of medicine that were debated in specific times, places, and communities.

Instead of relating some single linear account of this earlier medical history, in this introduction we want to exemplify our point about diversity by selectively and schematically looking at the story of German medical history from the end of the eighteenth century through its transit to North America in the interwar period. It is not our aim to suggest that this account, with its narrow geographic focus, captures the plurality of medical histories pursued in the past. On the contrary, by limiting our focus to one of the tales about professional lineage that is most familiar to Anglo-American and Northern European historians in imagining "traditional medical history," we seek to reveal the extent to which change and diversity are to be found where many have assumed there to be stasis and homogeneity. We offer this not as a survey of medical historical approaches, much less endeavors, but as illustrative instances of the multiple and shifting programs of research and teaching, each characterized by its own narrative, audience, motives, method, and biases.[2]

Learned Western physicians before 1800 had long looked to history for much of the professional definition and authority that later physicians would derive from science. Historical knowledge was a requisite component of being a physician, and students were versed in Greek, Latin, philology, logic, and rhetoric in order to be able to read and understand the work of their predecessors, who in a way were also their colleagues. Well into the nineteenth century, history played a prominent role in transmitting and defending medical positions—physicians engaged in doxographic articulations of their own approaches by weighing the pros and cons of earlier medical positions—and the classical heritage served as a rich resource from which theories were selectively drawn, remolded, and rearranged according to contemporary needs. The best physician, one might argue, was not only a philosopher (as Galen would have it) but a historian as well.

At the end of the eighteenth century, however, the fundamental unity between historical and contemporary medicine began to be drawn into question, as part of a larger shift in German historical consciousness. This emerging historical sensibility made room for change and progress in medical knowledge.[3] The Halle physician and professor of botany Kurt Sprengel (1766–1833), who conventionally has been presented as the founding father of modern medical history, was both an embodiment and spokesman for this new historical consciousness.[4] The broader political context for Sprengel's program was the reform movement of the Prussian universities at Halle, Göttingen, and especially Berlin. It was no coincidence that the cradle of this new academic ideal was in Prussia—a country in ruin, its armies recently crushed by Napoleon. There was a widespread longing to compensate for material loss through a spiritual and intellectual renaissance, with the newly unified state as the embodiment of reason. Wilhelm von Humboldt's "new university" had the practical aim of training professionals (lawyers, ministers, and physicians) for Prussian society, while at the same time it was to be a refuge for the quest of knowledge and truth. The salvation of the German nation, as von Humboldt conceived it, was to come from a combination of teaching and research. Even if, as historians recently have argued, the Humboldtian university was only an ideal—a myth even—it nevertheless inspired people to think about the wider role of the university in society.[5]

In medical history, Sprengel and his "pragmatic" view of the field represented the transition to this new ideal of learning. The hallmark of German pragmatic history was the proposition that one can learn from the past—that history contains lessons for the present and can serve practical ends. In his five-volume *Versuch einer Pragmatischen Geschichte der Arzneikunde* (Essay on a pragmatic history of medicine, 1792–1803), Sprengel used medical history both to show the gradual development of the human mind and to promote a better understanding of contemporary medical knowledge.[6] The past would help medical students develop into physicians with a sense of civic responsibility. Because medical history would teach students to find something of value in theories that at first sight looked strange, he reasoned, history would instill in them modesty and tolerance. Sprengel openly broke with the doxographic method and with the presentation of the past through the biographies of great doctors to be emulated; history no longer entailed sharpening one's own medical judgment by entering into debate with colleagues who happened to be dead. Instead, Sprengel organized his narrative as an evolutionary process in which one school of thought succeeded another, presupposing links between medicine, philosophy, and general culture. Sprengel was at the inception of an educational *Handbuch* tradition that had *Bildung* as its most important goal and used medical history as a tool in socializing students into medicine. Later generations would canonize Sprengel for his pragmatic conception of the history of medicine, making him the most important point of reference for legitimating the institutionalization of medical history in medical faculties.

During the Romantic era, such writers of handbooks or compendia of the history of medicine as Heinrich Damerow (1798–1866) and Emil Isensee (1807–1845) went even further in their reliance on philosophy, drawing heavily on Schelling and Hegel.[7] Romantic historiography cherished an idealistic metaphysics in which ideas were considered to be the autonomous driving forces behind reality. The medical historian was not to lavish attention on the endless multiformity of life (individuals, events, acts, and so on) but to seek out the founding idea, the zeitgeist, of each era. Isensee claimed to be looking for "the charming goddess of general truth" and gave pride of place to periodization. "The eras have not been *constructed,* but are *real*," he insisted, in a flourish of romantic rhetoric; "they are the stations where medicine, on her European world voyage, changed the spiritual winged horses of her golden and never resting chariot."[8] Physicians who did not know the history of medicine, he suggested, were like strangers in their own house. Without historical perspective, they would have to discover everything anew, committing many unnecessary mistakes in the process. History would widen their mental horizon, and, Isensee argued, echoing Sprengel, foster both self-criticism and modesty.

Creating new medical knowledge, rather than teaching, informed another motive to engage medical history. This research program was not institutionalized as history of medicine but embodied in such individuals as the Berlin professor Justus Hecker (1795–1850), often named as the founder of historical pathology.[9] To Hecker, medical history constituted an epistemological tool, a research strategy to advance understanding of a *medical* problem, namely the etiology of epidemic disease. This was a research program rooted in eighteenth-century neo-Hippocratism.[10] Whereas the neo-Hippocrats had done the data-gathering on environment, meteorology, morbidity, and mortality, however, it fell the lot of nineteenth-century historical pathologists like Hecker to attempt the long-awaited synthesis by comparing and introducing a historical dimension. Historical pathology, Hecker argued, would serve general history, as major epidemics took part in shaping the thinking of epochs; contribute to pathology; serve the state and society by helping to prevent and combat large-scale disease; and further knowledge about individual diseases, such as bubonic plague, by studying epidemiological patterns from the Middle Ages onward.[11] In 1822, the same year in which Hecker published his *Geschichte der Heilkunde* (History of medicine), he was appointed professor extraordinarius (promoted to ordinarius in 1834), standing at the beginning of what can be called the Berlin research tradition in medical history.[12] Historical pathology seemed on its way to developing into a fully fledged medical subdiscipline.

From the 1840s, however, a younger generation of physicians, intent on infusing experimental methods into medicine, sought to jettison all vestiges of *Naturphilosophie.* Critics assailed historical pathology, pointing to the speculative foundations of what claimed to be a highly empirical discipline (such as the ontological character of its notion of "epidemic constitutions").[13] After Hecker's death in 1850, the Berlin chair remained vacant for thirteen years, until 1863, when the Danzig physician and sanitarian August Hirsch (1817–1894) was appointed.[14] By the time Hirsch published his own handbook in 1860, Hecker's program for historical pathology had all but disappeared.[15] In order to escape the specter of speculative natural philosophy, Hirsch stressed that his book was strictly based on notions derived from Rudolf Virchow's modern pathology and that Virchow approved of his book.[16] Virchow, in the 1840s, had himself based his public health activities in part on Hecker's historical epidemiology, an ingredient in developing his "sociological epidemiology" and radical call for democracy, education, liberty, and prosperity as the solutions to the 1848 epidemic of typhus in Upper Silesia.[17] Once historical pathology moved from being a clinical discipline to a branch of autonomous medical history, however, universities began to lose interest. Historical pathological research findings disappeared from clinical journals to find a new home in the short-lived medical history journal *Janus* (founded in 1846).

Midcentury reactions against medical history were even more pronounced in attitudes toward its value in the moral and intellectual shaping of future doctors. The teaching of medical history, far from offering practical lessons for the present, increasingly came to be dismissed as an indulgent exercise of little more than antiquarian value. With a growing faith in progress through experimental science, many physicians came more and more to regard medicine as an applied science. To the extent to which physicians sought to recast medicine in the mold of the experimental sciences, while the clinic endured, the library was being supplanted by the laboratory as the central institution of medical knowledge.

For the teaching of medical history, this epistemological break threatened a loss of relevance and legitimacy. With the symbiotic relationship between past and present challenged, what use could historical training be for medical practice? Indeed, this self-conscious break with the past was part of the creation and cultivation of a modern professional identity. As a sense of irrelevance developed into outright dismissal of medical history by medical students and physicians, universities began to doubt the necessity of appointing professors in the history of medicine. Just as the Berlin chair remained vacant for many years after Hecker died in 1850, the same pattern followed the deaths of incumbents at Halle in 1851, Bonn in 1853, and Breslau in 1856.[18]

Marking an early instance of how professionalization could alienate medical history from medicine, Carl Wunderlich (1815–1877)—best remembered as a founder of clinical thermometry—was among those who assailed the utility of history for medical students. Wunderlich also practiced medical history, but the only kind of history of medicine acceptable to him was one that legitimated modern, scientific medicine by documenting its rise, celebrating its achievements, and bolstering its self-confidence. In his eyes, scientific progress should be made manifest by displaying the sharp discontinuity between past and present.[19] In 1842 Wunderlich launched a vehement attack on Jena professor Heinrich Haeser (1811–1885) and his *Archiv für die Gesammte Medicin* (Archive for comprehensive medicine), a medical journal that included a historical section. Condemning its historical articles as "a heaping of historical material," Wunderlich charged that its "antiquarianism" could only estrange practicing physicians from historical study. Instead, history could—and indeed should—be a beacon to the present only if it dealt with the origin of the ideas that governed contemporary medicine.[20]

"Your judgment," Haeser replied to Wunderlich, "is a matter of complete indifference to me, since I am bound to assume your complete ignorance of the history of your science."[21] Yet Wunderlich was a better spokesman than Haeser for his profession and the attitude of most physicians toward medical history. This is clear in the fate of *Janus,* founded in 1846 as the first journal exclusively devoted to the history of medicine, only to fold in 1853. Edited by August Henschel, professor of clinical medicine at Breslau, the associate editors included leading German medical historians: Haeser, Hecker, Ludwig Choulant (1791–1861), and Julius Rosenbaum (1807–1874). They conceived of medical history as an autonomous research discipline, and the articles that filled the journal, with a strong emphasis on archival and philological investigations, were eclectic and had no overt didactic or pragmatic goal. To progressive German physicians, however, consecrated to experimental methods, the journal appeared not merely indulgent but reactionary.

A small minority of physicians used medical history as a tool of resistance against the scientific turn in medicine that seemed to be rendering history superfluous. Haeser, for example, who in 1845 had published his magnum opus, the *Lehrbuch der Geschichte der Medicin und der Volkskrankheiten* (Textbook of the history of medicine and epidemic diseases), wrote in 1859 to the Kultusministerium of Prussia warning against the "materialism" and skepticism that had been infecting the "new" scientific medicine and urged teaching in medical history as a means of counterbalancing this movement.[22] He deplored that philosophy had been traded for chemistry and the microscope. The dire state of medical history in Prussian universities was all the more striking, Haeser argued, because even the smallest faculties of law and theology both valued and taught the history of their disciplines. His letter clearly articulated one conception of the proper function of medical history and of the mission of the medical historian: to understand the present by looking at how it came to be and to create physicians who were more than technicians. To this extent he represented the pragmatic teaching tradition of Sprengel, whom he admired but criticized for being too presentist.[23] "We have become more humble, more just," Haeser asserted, "to us, history is no longer an enumeration of 'deviations' of the human mind, thus flattering our vanity. Rather, it is mirroring our weakness; it has grown into a shining torch to recognize the finger of God—the rule of an eternal law in the doing of mankind."[24] His aspiration was to write history without preconceptions based on an exhaustive study of the sources and with "a feeling for and a joy in the particular in and by itself," as Leopold von Ranke (1795–1886), a model for his approach, put it.[25]

At the same time that Haeser and Henschel were being assailed for the irrelevance of their medical history to medicine, a new philological impulse channeled the study of ancient medicine even farther away from using the history of medicine to shape medical students and aid the medical profession.[26] Before the mid-nineteenth century, editing and translating Greek and Roman medical texts had been chiefly a philological contribution to *medical* scholarship, drawing selectively on the past partly in an effort to create deontological ideals. The new approach in Germany, however, led by scholars whose primary professional allegiance was to philology, not medicine, was a historicist endeavor to reconstruct the past on the basis of newly translated and interpreted classical texts.[27] German fascination with the ancient Greek world grew into a science of antiquity, *Altertumswissenschaft*, with Berlin as its principal center and Theodor Mommsen (1817–1903), Ulrich von Wilamowitz-Moellendorff (1848–1931), and Hermann Diels (1848–1922) as its leading figures. In their hands, the study of ancient medicine increasingly became the domain more of classical philology than of medical history. It was this movement that, early in the twentieth century, was to culminate in the massive Corpus Medicorum Graecorum edition of ancient medical texts.[28]

The period between roughly 1850 and 1890, then, can be seen as an interlude during which the ideals of a new version of scientific medicine informed the judgment that medical history was largely irrelevant to preparing students for the emerging medical world. The final decade of the nineteenth century and start of the twentieth, however, witnessed an unmistakable renaissance in the history of medicine. The sheer success of the reductionist program in reshaping medical knowledge and medical culture prompted many leading physicians to worry that the transformation of medicine into a science, as well as the epistemological and technical successes of the new sciences, may have been bought at too great a price. In 1889, Theodor Puschmann (1844–1899), professor of medical history in Vienna, speaking at the annual conference of the Gesellschaft Deutscher Naturforscher und Ärzte (Society of German Naturalists and Physicians), called for a rehumanization of the physician in an age of scientistic ideals. Puschmann complained that most leading physicians were almost completely uninterested in history, a neglect that bordered on contempt. The history of medicine—as part of the history of civilization—could play a crucial role in a medical education. It would broaden future physicians, ennoble their character, prevent them from slipping into "superficial materialism," and lay a sturdy foundation for professional knowledge. His apology for medical history strongly resembled Sprengel's pragmatic ideal of Bildung, but with an important difference: Puschmann recognized that medicine had been transformed over the course of the nineteenth century and that in a sense two cultures—that of the laboratory and the clinic—had come into being. That cultural divergence within medicine, to his mind, made the need for the unifying influence of history greater than ever. The bounty of medical science could be preserved without losing sight of the demands of medical art.[29]

Puschmann's speech fell on fertile ground. It tapped into a movement of cultural reorientation among the German bourgeoisie aimed at combating the excesses of scientific materialism.[30] It also resonated with a growing sense that the research imperative of the universities had led to a measure of specialization and proliferation of subdisciplines that was counterproductive to authentic learning.[31] By the turn of the century the Humboldtian, neohumanistic ideal—including the unity of science—enjoyed an enormous revival as a way to counterbalance reductionist hubris.[32]

The year 1905 was to be a turning point for the field in Europe. Karl Sudhoff (1853–1938), who for thirty years had been a practicing physician, was appointed to the chair newly created in Leipzig, where he would direct the first research institute in medical history. In that same decade, chairs and societies for medical history were founded in Germany, Hungary, France, Italy, Switzerland, Austria, the Netherlands, and the United Kingdom. Viewed against the backdrop of Puschmann's pragmatic ideals, and particularly his emphasis on the *teaching* of medical history, it is ironic that Sudhoff's institute not only privileged research but also philological scrutiny and a historicist method.

The debates that surrounded the founding of this institute vividly display some of the strife at the start of the twentieth century over what the history of medicine should be and what ends it should serve. The Leipzig Institute was financed with money left by Marie Puschmann-Fälligen (1845–1901).[33] Her husband, Theodor Puschmann, had been a Privatdozent in medical history in Leipzig before he was appointed in 1879 to a professorship in Vienna. Having no children, the couple stipulated that after their deaths their estate should be used to support the "advancement of scholarly work on the history of medicine" in Leipzig, a place they remembered fondly. Shortly before his death in 1899, Puschmann changed his will, renaming the University of Vienna as beneficiary.[34] Puschmann-Fälligen, however, from whom he had just divorced in 1897, did not alter her will, and therefore when she died in 1901 Leipzig received the bequest of some half million reichsmarks.

But the question remained: How was "history of medicine" to be defined, and who was to profit from the money? The will had named the university as beneficiary, not a specific faculty, and both medicine and philology claimed the funds as their own. In 1904, once the legal dispute with Puschmann-Fälligen's relatives (who had contested the will) was out of the way, a debate erupted in the *Münchener Medizinische Wochenschrift* (Munich medical weekly) on the definition, relevance, and future of medical history.

The senior medical historian Johann Baas (1838–1909) instigated the battle over how the Puschmann money should be spent, arguing that the creation of an institute dedicated to philologically oriented cultural history of medicine was the best means of professionalizing a field populated too long by amateurs and by serious scholars forced to supplement their income by medical practice. It was time for the "Cinderella" of German medical sciences, as he designated medical history, to leave its station as subordinate stepsister and come into its own as a full-fledged discipline.[35] Max Seiffert (1865–1916), a Leipzig Privatdozent in pediatrics, countered by ridiculing antiquarian philologists, whose version of medical history served neither the education of future doctors nor the medical profession. He derided their work as nothing more than the retrieval of documents from the refuse dumps of history. He mocked the medicohistorical "guild" for compiling barren enumerations of names and dates, contrasting their approach to the field with that of Leipzig cultural historian Karl Lamprecht (1856–1915) and of the clinician-historian Wunderlich, who had used history to show the rise of contemporary scientific medicine. Stories about recent figures like Bichat, Mueller, Virchow, Helmholz, and Pasteur were of vastly greater relevance than those about distant figures from antiquity or the Middle Ages. The teacher of medical history, he concluded, should be a practicing physician, so as not to lose contact with medicine.[36]

Sudhoff's reaction was prompt, fierce, and ad hominem. As chairman of the recently founded Deutsche Gesellschaft für Geschichte der Medizin und Naturwissenschaften (German Society for the History of Medicine and the Natural Sciences), he cast himself as spokesman for the professional field.[37] He dismissed

Seiffert as an amateur who was after the post. The Puschmann money, moreover, was intended for "the advancement of scholarly work," which to Sudhoff meant principally research, not teaching. Both men took for granted the utility of medical history as a subdiscipline of medicine, although Sudhoff rejected Seiffert's contention that only the past hundred years held pedagogic value. However, Sudhoff insisted that to come of age as an academic discipline, the field needed the infrastructure to support basic research, just as in any other medical subdiscipline. Facts, gathered from meticulous research in libraries and archives, could free the field from the premature, speculative syntheses that still had hold of it. Although Sudhoff did not name the people who exemplified the medical history he envisioned, it is clear he had great admiration for the historicist medical historian Haeser and the French medical manuscript hunter and philologist Charles Daremberg (1817–1872).[38] The debate between Sudhoff and Seiffert thus echoed the clash between Haeser and Wunderlich of some sixty years earlier.

The Leipzig academic senate opted for the vision of Baas and Sudhoff, and it used the Puschmann funding to establish a full-fledged research institute that was not tied to any particular faculty but to the university in general. In his inaugural address, Sudhoff set out the conception of medical history as an autonomous discipline that he adhered to for the rest of his life.[39] The Leipzig Institute would provide the field with an infrastructure (in 1907 he founded his own journal, *Archiv für Geschichte der Medizin,* as a forum for original research, known after 1929 as *Sudhoffs Archiv*) and make source material available (which for him chiefly meant locating, editing, and publishing early texts).[40] Only the development of rigorous professional standards could put an end to the amateurism that had plagued medical history, Sudhoff asserted (a view shared by his successor, Henry Sigerist, when he became chair in 1925).[41] Teaching, though important, was not the core business of his institute.

In sharp contrast to the pragmatic medical historians, Sudhoff made no attempt to establish an explicit relationship between past and present by documenting medical *progress* and celebrating its heroes. It was his ambition to become the Leopold von Ranke of medical history, and, like the famous historicist, he sought to establish *wie es eigentlich gewesen.*[42] In 1929, looking back at his life, Sudhoff proudly reflected that he had refused to write a synthesis. He also edited Pagel's historical survey, but whereas the original 1898 edition contained an extended introduction on the pragmatic conception of history, that chapter was entirely missing in Sudhoff's 1922 edition.[43] As a historicist he wanted to banish speculative theory and let the sources speak for themselves.

There was at least one other major fault line at the turn of the century, a rift running *within* the ranks of professional German historians of medicine, with Sudhoff on the philological side and Julius Pagel (1851–1912) and Max Neuburger (1868–1955) on the philosophical side. Pagel and Neuburger, based in Berlin and Vienna, shared the pragmatic view of medical history that had been held by Puschmann, Neuburger's teacher, but they went on to proselytize for a distinctly cultural approach they believed would bridge the gap between science and the humanities. Pagel, in a 1904 programmatic essay, advocated what he called "medical cultural history" (*medizinische Kulturgeschichte*). He conceptualized medicine as a circle, overlapping, in Venn diagram fashion, with other circles representing science, philosophy, religion, the arts, theology, law, technology, industry, commerce, language—in short, all aspects of human life—which had to be studied together. His conclusion and credo was that "the true historian of medicine is a cultural historian."[44]

The cultural history program with its pragmatic goals for educating medical students did not gain much of a following in Germany.[45] Instead, it was chiefly the Sudhoff model that informed both professionalization and institutionalization in German medical history, and, with it, the estrangement of the field from medicine.[46] "Cultural history" did become something of an umbrella term, however, and Sudhoff, in his 1904 apology for medical history, used the phrase "we medical cultural historians" to help close ranks, drawing a line between professionals like himself, Pagel, Neuburger, and Baas and amateurs like Seiffert.[47] Sigerist, much later, at Neuburger's seventy-fifth birthday in 1943, would reminisce, "It was not easy to hold a leading position in medical history in a German-speaking university at the time of Karl Sudhoff. His domineering and aggressive personality tolerated disciples but not competitors."[48]

In the United States at the end of the nineteenth century, there was also uneasiness about the potential of the new scientific medicine to undercut physicians' appreciation of the art of healing, even as it brought them unprecedented technical expertise. Just as Puschmann had called for a rehumanization of medicine, physicians like John Shaw Billings (1838–1913) and William Osler (1849–1919)—both active members of the Johns Hopkins Hospital Historical Club, founded in 1890—pleaded much the same case. Representing medical history as a partial antidote to excessive reductionism, specialization, commercialism, and cultural disintegration in medicine, they cultivated an ideal of the "gentleman-physician" well versed in the classic liberal arts. Billings, head of the Surgeon General's Library who had an important voice in plans for the hospital and medical school at Hopkins, urged as early as 1877 that while the kind of "average" practitioner trained at most American medical schools could get along without formal instruction in medical history, for graduates of the Johns Hopkins Medical School as he envisaged them—the cadre of teachers and researchers in the new scientific medicine he expected to lead the profession—a course on medical history would be indispensable as "a means of culture."[49] From the moment he started working at the Johns Hopkins Hospital in 1889, Osler integrated medico-historical issues and problems into his clinical instruction. The "Oslerian method" used the example of past physicians as a tool to inspire.[50]

German and American discourses on the teaching of medical history as a contribution to Bildung and to the creation of gentleman-physicians were unmistakably similar, but there was an important difference: German medical history was firmly rooted in the university system. During the 1890s, a number of American medical schools began offering lecture courses on the history of medicine. Yet, however much leading medical educators may have been convinced of the worth of medical history in the curriculum, students tended to find lectures boring.[51] Although the Oslerian method was popular, there was a growing awareness of its amateurism and a sense that German academic standards should be introduced to the history of medicine in the United States.[52] It would be the deteriorating social and political climate of Europe in the 1930s and the intellectual diaspora it instigated that led such figures as Henry Sigerist (1891–1957), Owsei Temkin (1902–2002), and Ludwig Edelstein (1902–1965) to leave Germany for North America, moving the center of gravity in the field across the Atlantic.

Among the physicians who most energetically prepared American soil for German approaches to the history of medicine was Fielding H. Garrison (1870–1935). In 1891, after receiving his bachelor's degree from the Johns Hopkins University, Garrison (who would receive his M.D. from Georgetown in 1893 but never practiced medicine) joined Billings's staff at the Surgeon General's Library, where he spent most of his career. In 1913 he published *An Introduction to the History of Medicine,* which went through a number of editions, making him, in the appraisal of Genevieve Miller, the "first American medical historian to produce a major original work on the general history of medicine."[53] However much he admired the historical endeavors of Billings, Osler, and other contemporaries in the United States, however, Garrison was keen to professionalize the field, and held up Sudhoff as a model. Writing in 1923, in a reverential salute to Sudhoff on his seventieth birthday, Garrison asserted that Sudhoff was the only medical historian who merited the appellation "genius." At the same time, Garrison's remarks were an attempt to rehabilitate German culture in general, and Sudhoff's scholarship in particular, five years after the Great War. By drawing a sharp distinction between Prussian power and authentic German culture, Garrison suggested that the Wilhelminian epoch had been a barbaric and uncharacteristic interlude. Sudhoff, Garrison continued, was an embodiment of the best German culture had to offer and "easily the greatest and most accomplished of all medical historians," asserting that his methods represented a new departure and that the Leipzig Institute should be regarded as the center of medico-historical investigation.[54] Urging his American colleagues to take Sudhoff's example as their ideal, Garrison concluded, "Heil dem Meister!"[55] Starting in 1930, Garrison became medical librarian at Hopkins, where during the final years of his life he would be a colleague of Sigerist.

In 1926, William H. Welch (1850–1934), the Hopkins pathologist and disciple of the "German model" in medicine, accepted the endowed professorship in history of medicine at Hopkins—the first in the United States—reluctantly, because he did not regard himself as the right person for the job. He thought that German standards should be applied to American medical history as they had been to the rest of medicine, and to professionalize the discipline he wanted to create an institute comparable to the one at Leipzig. After Sigerist, Sudhoff's Swiss pupil and successor, came to Baltimore as a visiting lecturer, he was offered the position of head of the Johns Hopkins Institute of the History of Medicine, and Welch was pleased when Sigerist accepted in 1932. Sigerist, a Social Democrat with socialist ties and a supporter of the Weimar Republic, had strong incentives to leave Germany. Together with Temkin (who came with Sigerist) and Edelstein (who arrived in 1934), Sigerist introduced to the United States the ideal of history of medicine as a resource for Bildung. As a bridgehead of medical history in North America, the Hopkins Institute was to be a center where professional research standards were guarded, but also a place to train physicians who would teach medical history on the periphery at their local institutions.[56]

While Sigerist never quite understood Osler's popularity, both men used medical history to provide physicians with a sense of transcendent purpose. In the 1920s, the young Sigerist had joined in Sudhoff's philological project, but after a few years he had broken with the Germanic medievalism of his patron. Whereas Sudhoff wanted to build a stock of source material, Sigerist wanted answers to philosophical and ethical questions.[57] In the first volume of *Kyklos,* a journal he founded in 1928 while still in Leipzig, he had thrown out the challenge to medical history "whether in purely positivist fashion it wishes to add facts to facts, or whether it is capable of interpreting the past, of enlivening it, and of rendering it fruitful for a better future."[58] By the time he replaced Sudhoff at the Leipzig Institute, Sigerist had already begun exploring the history of medicine from a sociological perspective.[59]

More than this, Sigerist considered it to be his duty to educate physicians with a sense of civic responsibility, and he made the well-known dictum of Benedetto Croce—"every true history is contemporary history"—his own. In his 1931 *Einführung in die Medizin* (Introduction to medicine), which in 1933 was published in English as *Man and Medicine: An Introduction to Medical Knowledge*, Sigerist left the strict chronology that had characterized nineteenth-century handbooks and opted instead for thematic chapters with titles like "Causes of Disease," "Medical Aid," "The Physician," and "Sick Man."[60] In this work, one can also see a broadening of pedagogic aim that would intensify in American medical history over the next half century, from a focus on making responsible physicians to one aimed at developing reflective and responsible citizens.

Mirror, Mirror on the Wall

The Puschmann bequest compelled an answer to the question: What should the history of medicine be and what aims should it serve? And in the creation of Sudhoff's Leipzig Institute, one answer won out, but for a field, we would argue, under most circumstances a single answer is neither necessary nor desirable. Over the ensuing century, the array of replies has widened enormously. Yet in print and around seminar tables, the frequency with which differing answers are presented in an either/or format remains remarkable. In a field diverse in both practitioners and constituencies, however, a multiplicity of coexisting approaches, styles, and aims is not only inbuilt—as this volume shows—but to be desired.

History of medicine, broadly defined, is more fully integrated than ever into the larger field of history, where issues of health, the body, and the life cycle are pursued today by scholars who would never think to call themselves historians of medicine. Medical history as a distinct field even seems at some risk of dissolving, as a postmedical history of health cultures increasingly flourishes. Further, the more that work in medical history has joined the historical mainstream—driven by much the same methodological and interpretive currents and countercurrents that characterize any other historical subfield—the less its practitioners have needed to rely on separate institutions as places to discuss their ideas, publish their work, or secure jobs. As in so many realms, the successes of integration have tended to make separatism seem outmoded and undesirable. A scholar in another established historical subfield that, like medical history, is enjoying unprecedented prosperity, has recently suggested that the way to escape its self-imposed "ghetto status" might be to promote the field's "deinstitutionalization," and something of the same thinking can be heard in the hallways of history of medicine professional meetings.[61]

The specter of the deinstitutionalization of medical history has made the answers to questions about the field's identity and purpose more intense. Rumors about support being withdrawn by private foundations, states, or universities fuel rhetoric about disciplinary crisis. In Germany, an announcement by the Ministry of Health in 1997 that medical history would be removed from the medical curriculum—a move that, if implemented, would have led to the closure of nearly all institutes for the history of medicine—sent a shock wave through the field. In Britain two years later, concern that the Wellcome Trust was considering rechanneling funding from history of medicine into other areas excited similar anxieties. In North America, too, there has been perennial grumbling about merging medical history with one or another academic field, such as biomedical ethics, and about cutbacks in targeted federal or foundation support.[62] Warranted or not, such apprehensions would have less purchase if medical historians themselves were not so vocally troubled about the ghettoization of the field.

More often than not, strident, full-voiced answers to the question "What should history of medicine be?" are taken as signs of the fragility of its cohesive identity. When such answers take exclusionary, delegitimizing forms, there is reason for concern about the damage they might do. Yet, the lived reality of illness experiences and management, of health seeking and disease avoidance, gives our subject a universal intimacy and immediacy. If by changing caps (and voice) we can reach both professional historians and practicing clinicians, students in the health professions, policy makers, ethicists, economists, and a wider public—and do so through stories told not only in classrooms and in print but in museums and Websites—should that not be reason for celebration? Seen against the backdrop of the diverse approaches, audiences, and aims of medical history in the *past*, being confronted by divergent approaches does not necessarily present a dilemma we must resolve (save perhaps in our own individual writing and teaching) so much as the starting point for inciting a debate that has yet to take place in an engaged, constructive way. Doing history is about taking people seriously, and this includes fellow historians as well as historical actors. *Pace* science wars, if there is such a thing as a crisis in medical history, then one might well argue that it is not the fault of crude realism or postmodern relativism, or of any other such pairings of bipolar opposites, but a result of pluralism without tolerant yet critical exchange.

To that end, we have made a number of tradeoffs in designing this collection. We have elected to explore medical historical practices over a long period of time rather than to attempt anything like an encyclopedic mapping of recent work in the field. We decided early on not to structure contributions according to particular categories of analysis (gender, race, religion, age, geographic locale); we are aware that our contributors tend to draw their historical examples from the past two centuries; and, above all, our geographic default mode is the North Atlantic. We have not given explicit attention to the vigorous work that has emerged both from history and from area studies on medicine, health, and healing in Africa, Asia, and Latin America. Nor is our goal any exhaustive survey of historiographic approaches, much less their products, past or present. Finally, our aim is not overtly programmatic, at least in the sense of proselytizing for one or another tack we would like to see the field take, although we readily acknowledge an embedded argument in the way we have solicited and edited contributions that celebrates many of the historiographic changes of the past couple of decades while registering uneasiness about the risk of falling prey to complacency that dwelling overmuch on the success of our field can bring.

We have sought contributions from people who do not see eye to eye about how the history of medicine is best written, to what purposes, and for which audiences, and we have encouraged them to be bold in speaking their mind (using the first person singular pronoun whenever that might help to capture their own idiosyncratic voice). Therefore some of the essays strike counterpoints of dissonance, not harmony, and that is precisely what we hoped for. Why do historians tell the particular stories about medicine they do tell? How and why do they pose their questions? What work do they want their stories to do? What audiences do they envision? This volume sets out to rethink the missions and methods of medical history by looking at its history and its ambitions—legitimating and activist, didactic and emancipatory—exploring what the different kinds and styles have purported to offer. Our overarching objective in trying to capture some measure of the field's complexity and diversity is to provide a springboard for debate—for it is only through such exchanges that the field can remain vital and that, by a cultivated awareness of a common arena, its coherent identity can be sustained.

This volume is organized as a triptych, exploring an early history of the field, its transformations since the 1970s, and some of the divergent directions it has seemed to be headed after the cultural turn. The first section, "Traditions," looks at the work and aspirations of selected figures who practiced medical history during a long nineteenth century. The aim here is neither to be comprehensive nor to show some general developmental pattern but to give concrete texture to the backgrounds, preoccupations, national and local contexts, professional and political agendas, and relationship to larger changes both in medicine and in history that informed the pluralism that is characteristic of the field.

"To whom does medical history belong?" was an unresolved question at the end of the eighteenth century, as Hans-Uwe Lammel shows in the opening essay. He offers a double portrait of contemporaries Kurt Sprengel and Johann Moehsen, German physicians animated by different political and medical agendas who each sought to use his own version of medical history to foster stability in an insecure period. It was only the selective process of collective memory that established Sprengel as the "father" of modern medical history, as successors in medicine and in medical history sanctified the image of a predecessor whose legacy could help establish the identity of the discipline as they wished to see it. Danielle Gourevitch turns to two Comtean medical historians and the mid-nineteenth-century creation of a French positivist program in the history of medicine, focusing on the ambitious manuscript collecting of Charles Daremberg and his connection with the lexicographer, medical journalist, and utopian socialist Émile Littré.

A contest for medical history in Germany at the turn of the twentieth century is the focus of the next two essays. Heinz-Peter Schmiedebach, in his double portrait, explores in the cultural history approach of Max Neuburger and Julius Pagel, a formidable program that met only limited favor in Germany. Karl Sudhoff, on the other hand, investigated in Thomas Rütten's contribution, with his program to recover manuscripts and publish critical editions, was the clear winner. Sudhoff tightly controlled the key institutions of the field and the dominant program for research, all the way through "the fall" of the field in Germany when, in the 1930s, he and some other medical historians accommodated to the rise of National Socialism.

Vivian Nutton traces the transit of medical history from German-speaking centers to the United States. Instead of focusing on one or two individuals he presents a *tableau de la troupe,* organized around the study of Greek and Roman medicine from the 1870s through the early twentieth century. The periodization of this research field he suggests (from pragmatic medical history to philological classicist research to the history of ideas) conveys a clear sense of the fluidity of disciplinary boundaries, while it also underscores that the social history turn of the 1970s was not the first time medical history became estranged from medicine. In the final essay of this section, Elizabeth Fee and Theodore Brown offer a double portrait of two figures who, at first glance, shared Baltimore ties and a passion for medical history but little else—William Osler and Henry Sigerist. Here, however, their differences are registered mainly as backdrop for delineating their revealing similarities. Beyond their shared philhellenism and medical holism, Osler and Sigerist had a missionary zeal for the *utility* of medical history to inspire medical students, to create a professional self-image, and to foster a sense of professional responsibility. More overtly than the other chapters in this section, Fee and Brown examine not only two historians who went before us but also two legacies, embodied today historiographically and politically in the Osler Society and the Sigerist Circle.

"A Generation Revisited," the middle section, explores the impulses that have transformed the history of medicine since the 1970s. Three essays investigate the selective cultivation of "the new social history" and its interaction with other historiographic programs in North America, Britain, and continental Europe. In 1979, Susan Reverby and David Rosner published a programmatic essay titled "Beyond 'the Great Doctors'" that some have seen as a manifesto of the new social history movement in the history of American medicine.[63] Here, Reverby and Rosner look back at the aspirations and apprehensions of the generation of historians who at the time were taking their first steps into the history of medicine, and trace how and why the field has changed in the ensuing decades. As the editors urged, they use autobiographical voices to capture the processes that have led them from a single starting point to in some ways quite different historiographic places. Roy Porter, the late doyen of the social history of medicine in the United Kingdom, traces the course of the new medical history in Britain, and Martin Dinges, enlisting a comparative perspective, takes on the same task looking at Germany and France. The aim in these essays is not to provide a literature review or a roster of important contributions but rather to exemplify some of the major patterns of change that have transfigured the field.

A gloomier appraisal of change emerges from the two topical essays that follow. Olga Amsterdamska and Anja Hiddinga, analyzing publication and citation patterns in the four leading English-language medical history journals, question just how much history of medicine in recent decades has moved closer to general social and cultural history and away from medicine. Critiquing medical history from the perspective of the social studies of science, they conclude that the field is surprisingly traditional, conservative, theoretically adrift, and intellectually splintered. Following a very different course, Christiane Sinding explores the work of Michel Foucault and his mentor Georges Canguilhem and their creation of a critical countertradition in medical history. She explains how Foucauldian frameworks have acted as a force in the writing of medical history, with a keen comparative sensibility to French and Anglo-American contexts, but comes away with some dismay at how little she believes the more challenging of Foucault's theoretical propositions have been assimilated into history of medicine practice.

Geography—unlike gender, class, and race—was hardly singled out as a prominent and promising category of analysis in the programmatic calls of the 1970s that launched the new social history of medicine program. Yet, during the past two decades and especially in recent years, work on medicine and colonialism, turning in no small measure on the geographical and political construct of "tropical" medicine—on notions of both place and race—has proven to be one of the most vibrant arenas for carrying out the animating aspirations of that program. In a bridge to the final section, Warwick Anderson explores this growing current of the social history of colonial medicine, which came initially not from medical historians but from historians in area studies. He points to the promise that postcolonial studies offer as an analytical resource for historians studying medicine in any context, and he considers how this field of analysis draws medical history toward global history.

Where medical history is going and should go, often implicit in the essays of part 2, is a more explicit aim of the final section, which plunges into the methodological and political swamp of practices and controversies "After the Cultural Turn." It does not aim for a single, unified message but instead seeks to display some of the multiple meanings and uses of medical history. These essays speak frankly about the rhetorical battlefields of history, politics, and health care and passionately about the personal and social importance of doing medical history. Juxtaposing contributors with diverse points of view, from different national and disciplinary backgrounds, we have encouraged authors to get at the lived experience of medical historical practice by drawing freely on their own autobiographies.

Roger Cooter's opening salvo is the charge that the analytic program and activist power of the once new social history of medicine program have been derailed, not least of all by the cultural turn. The cultural studies impulse, he argues, the methodological relativism of postmodernism and the politics of global neoliberalism, has turned historians away from being relevant, hard-hitting, and intelligible social critics. As the key words—"social," "history," and "medicine"—lost their stable meaning, and as sociological categories were suborned by semiotic ones, the "social" came to be deprivileged, and with it the clarity of historical mission. His chapter is in dialog with the 1995 programmatic essay by Ludmilla Jordanova, reprinted here, which argues that the potential of social constructivism in medical history, far from being exhausted, remains a disappointingly underdeveloped avenue of social history analysis. Like Cooter, and resonating with the critical conclusions of Amsterdamska and Hiddinga, Jordanova calls for greater theoretical sophistication in the *social* history of medicine. In her new afterword, however, she finds ample ground for a productive rapprochement between the approaches of social and cultural history.

Mary Fissell provides another take on the cultural history impulse, exploring and explaining the new possibilities for historical insight that this current has brought. She offers a nuanced account—rooted partly in reflection on changes in the course of her own historical research programs—of what she sees as tradeoffs in a shift from a social to cultural history. Cultural historian as social activist is the central theme of the contribution by Alice Dreger. In her essay on the relationship between intersexuals and the medical profession, an account of how her historical scholarship on hermaphrodites drew her to the role of activist, she stresses that history is not able to supply easy lessons for present-day issues, but medical history can make a subtle and powerful difference in people's lives.

The physician as historian and the place of medical history in medicine today are investigated in the next two chapters. Alfons Labisch draws a telling distinction between history *of* medicine and history *in* medicine, without presenting them as either/or endeavors. Rather, he insists that history, as he has sought to implement it, can play a unique and important role in preparing students for medical practice and in the medical community. Jacalyn Duffin takes a very different tack. In keeping with the objective of this volume to engage rather than gloss over tensions that have come with the field's enriching diversity, she recounts her own experiences entering the field—confronting the bemusement of physicians who saw her squandering her professional training by taking up history and the suspicions of historians who considered her M.D. as an identifying badge of the historiographically naive Whig historian. She argues for important methodological and narrative similarities between history and medicine, and, like Labisch, feels that history offers something important to the medical students she teaches. In the end, however, she dismisses any dichotomy between historians who are and are not clinicians as a counterproductive source of antagonism within medical history and as meaningless to the practice of history. There are only the individual fascinations of individual historians—and we should not pretend otherwise—resulting in unique stories that inform and enlighten other people.

Audience, a central theme throughout this volume, is an explicit focus in the final two chapters. The general reading public—people who by and large have no professional tie either to history or to medicine—is a constituency that, Sherwin Nuland urges, is ready and waiting for the books on the history of medicine that most practitioners in our field seem unwilling or unable to write. He cautions, however, that the relativist methodology of the "new" medical history, as well as academic jargon, must be shed in order to write about health, disease, and healing in ways that have relevance and appeal in most people's lives. Allan Brandt, in the closing essay, turns to the equally promising, and equally (if differently) freighted audiences to be found in professionals engaged with policy, ethical, and legal decision-making. Relevance and responsibility go hand in hand in his treatment of the possibilities and problems of medical history storytelling. The more historians venture beyond speaking to each other, the more we must consider how our accounts of medical history, which we recognize to be interpretations rather than straightforward depictions of the past, might be put to use by our listeners—in the classroom, think tank, legislative chamber, or courtroom.

This volume is an invitation to explore and reflect on a "field"—one that can include widely disparate senses of what medical history is, should be, and should do. Whether "traditional" or "new," medical history is no monolith and never has been. It has been used as a didactic instrument to create responsible physicians; as a tool in researching the etiology of epidemic disease; as platform for legitimizing the medical profession; as a source of metaphors to analyze society; as a political device to expose and contest medical power; as a way to emancipate marginalized groups; as a means of taking science "to the people" in an attempt to promote participation and democratic debate; and as a vehicle of interdisciplinary discourse in the humanities and social sciences. While we hope that readers will be prompted to consider the merits as well as the shortfalls of approaches to medical history that are not their own, we are convinced that each historian can clarify his or her own ways and reasons for doing medical history by sizing up the alternatives.

NOTES

1. The classical works remain *The Invention of Tradition,* Eric Hobsbawm and Terence Ranger, eds. (Cambridge: Cambridge University Press, 1983), and *Les lieux de mémoire,* Pierre Nora, ed., 7 vols. (Paris: Gallimard, 1984–1992).

2. On the history of medical historiography, see Gert Brieger, "The Historiography of Medicine," in *Companion Encyclopedia of the History of Medicine,* W. F. Bynum and Roy Porter, eds., 2 vols. (London: Routledge, 1993), vol. 1:24–44; *Eine Wissenschaft Emanzipiert Sich. Die Medizinhistoriographie von der Aufklärung bis zur Postmoderne,* Ralf Bröer, ed. (Pfaffenweiler: Centaurus-Verlagsgesellschaft, 1999); John C. Burnham, *How the Idea of Profession Changed the Writing of Medical History* (*Medical History* supp. no. 18; London: The Wellcome Institute for the History of Medicine, 1998); Harold J. Cook, "The New Philosophy and Medicine in Seventeenth-Century England," in *Reappraisals of the Scientific Revolution,* David C. Lindberg and Robert S. Westman, eds. (Cambridge: Cambridge University Press, 1990), 397–436, 401–405; *Making Medical History: The Life and Times of Henry E. Sigerist,* Elizabeth Fee and

Theodore M. Brown, eds. (Baltimore: Johns Hopkins University Press, 1997); *Die Institutionalisierung der Medizinhistoriographie. Entwicklungslinien vom 19. ins 20. Jahrhundert,* Andreas Frewer and Volker Roelcke, eds. (Stuttgart: Franz Steiner Verlag 2001); *Médecins érudits, de Coray à Sigerist,* Danielle Gourevitch, ed. (Paris: De Boccard, 1995); Johann Gromer, *Julius Leopold Pagel (1851–1912). Medizinhistoriker und Arzt* (Cologne: Forschungsstelle Robert-Koch-Strasse, 1985); Edith Heischkel, "Die Geschichte der Medizingeschichtsschreibung," in *Einführung in die Medizinhistorik,* Walter Artelt, ed. (Stuttgart: Ferdinand Enke Verlag, 1949), 202–237; Susan Reverby and David Rosner, "Beyond 'the Great Doctors,' " in *Health Care in America: Essays in Social History,* Reverby and Rosner, eds. (Philadelphia: Temple University Press, 1979), 3–16; Dirk Rodekirchen and Heike Fleddermann, *Karl Sudhoff (1853–1938). Zwei Arbeiten zur Geschichte der Medizin und der Zahnheilkunde* (Feuchtwangen: Margrit Tenner, 1991); Volker Roelcke, "Die Entwicklung der Medizingeschichte Seit 1945," *Zeitschrift für Geschichte der Naturwissenschaften, Technik und Medizin* 2 (1994): 193–216; Henry Sigerist, "The History of Medical History," in *Milestones in Medicine* (New York: Appleton-Century Company, 1938), 165–184; John Harley Warner, "The History of Science and the Sciences of Medicine," *Osiris* 10 (1995): 164–193; Charles Webster, "The Historiography of Medicine," in *Information Sources in the History of Science and Medicine,* Pietro Corsi and Paul Weindling, eds. (London: Butterworth Scientific, 1983), 29–43.

3. On the idea of change in medicine, see Owsei Temkin, *Galenism: Rise and Decline of a Medical Philosophy* (Ithaca: Cornell University Press, 1973), and Wesley D. Smith, *The Hippocratic Tradition* (Ithaca: Cornell University Press, 1979). On this change in history writing, see Georg G. Iggers, *The German Conception of History: The National Tradition of Historical Thought from Herder to the Present* (Middletown: Wesleyan University Press, 1968).

4. On Sprengel, see Thomas Broman, *The Transformation of German Academic Medicine, 1750–1820* (Cambridge: Cambridge University Press 1996), esp. 139–142, and Hans-Uwe Lammel, *Klio und Hippokrates. Eine Liason Litteraire des 18. Jarhunderts und die Flogen für die Wissenschaftskultur bis 1850 in Deutschland* (Stuttgart: Franz Steiner, in press).

5. *Mythos Humboldt. Vergangenheit und Zukunft der deutschen Universitäten,* Mitchell G. Ash, ed. (Vienna: Böhlau, 1999); Sylvia Paletschek, "Verbreitete sich ein 'Humboldt'sches Modell' an den deutschen Universitäten im 19. Jahrhundert?" in *Humboldt International. Der Export des Deutschen Universitätsmodells im 19. und 20. Jahrhundert,* Rainer Christoph Schwinges, ed. (Basel: Schwabe, 2000), 75–104.

6. Kurt Sprengel, *Versuch einer Pragmatischen Geschichte der Arzneikunde* (Halle: J. J. Gebauer, 1792–1803).

7. See Guenter B. Risse, "Historicism in Medical History: Heinrich Damerow's 'Philosophical' Historiography in Romantic Germany," *Bulletin of the History of Medicine* 43 (1969): 201–211, and Dietrich von Engelhardt, "Medizinhistoriographie im Zeitalter der Romantik," in Bröer, *Eine Wissenschaft Emanzipiert Sich,* 31–47.

8. Emil Isensee, *Die Geschichte der Medicin und ihrer Hülfswissenschaften,* 6 vols. (Berlin: Liebmann, 1840–1844), vol. 1: *Ältere und Mittlere Geschichte,* xlvii, 9.

9. Johanna Bleker, "Die Historische Pathologie, Nosologie und Epidemiologie im 19. Jahrhundert," *Medizinhistorisches Journal* 19 (1984): 33–52; Richard Hildebrand, "Bildnis des Medizinhistorikers Justus Hecker," *Medizinhistorisches Journal* 25 (1990): 164–170.

10. Lammel, *Klio und Hippokrates.* See also James C. Riley, *The Eighteenth-Century Campaign to Avoid Disease* (Basingstoke: Macmillan, 1987).

11. See, for example, Justus Hecker, *Der Schwarze Tod im 14. Jahrhundert* (Berlin: Herbig, 1832).

12. Justus Hecker, *Geschichte der Heilkunde*, 2 vols. (Berlin: Theodor Enslin, 1822–1829). Like Hecker, Moritz Naumann, Johann Lukas Schönlein, Heinrich Haeser, Konrad Heinrich Fuchs, and Julius Rosenbaum looked to medical history in the hope of finding pathological patterns.

13. Bleker, "Die Historische Pathologie," 44–46. On Hecker as a Hegelian, "gothic" epidemiologist, see Faye Marie Getz, "Black Death and the Silver Lining: Meaning, Continuity, and Revolutionary Change in Histories of Medieval Plague," *Journal of the History of Biology* 24 (1991): 265–289.

14. See also Eugen Beck, "Die Historisch-Geographische Pathologie von August Hirsch," *Gesnerus* 18 (1961): 33–44.

15. August Hirsch, *Handbuch der Historisch-Geographischen Pathologie*, 2 vols. (Erlangen: Ferdinand Enke, 1860–1864).

16. Ibid., vol. 1: vii–viii. On Virchow's attitude toward medical history, see Erwin Ackerknecht, *Rudolf Virchow: Doctor, Statesman, Anthropologist* (Madison: University of Wisconsin Press, 1953), 146–155. Together, Hirsch and Virchow edited the *Jahresbericht über die Leistungen und Fortschritte der gesammten Medicin*.

17. Ackerknecht, *Virchow*, 125.

18. Peter Schneck, "'Über die Ursachen der Gegenwärtigen Vernachlässigung der Historisch-Medicinischen Studien in Deutschland': Eine Denkschrift Heinrich Haesers an das Preussische Kultusministerium aus dem Jahre 1859," in Frewer and Roelcke, *Die Institutionalisierung*, 39–56, 48–50.

19. Carl Wunderlich, *Geschichte der Medicin* (Stuttgart: Ebner & Seubert, 1859). See also Owsei Temkin, "Wunderlich, Schelling and the History of Medicine," in *The Double Face of Janus and Other Essays in the History of Medicine* (Baltimore: Johns Hopkins University Press, 1977), 246–251, and Richard Toellner, "Der Funktionswandel der Wissenschaftshistoriographie am Beispiel der Medizingeschichte des 19. und 20. Jahrhunderts," in Bröer, *Eine Wissenschaft Emanzipiert Sich*, 175–187.

20. Owsei Temkin and C. Lilian Temkin, "Wunderlich versus Haeser: A Controversy over Medical History," reprinted in Owsei Temkin, *On Second Thought and Other Essays in the History of Medicine and Science* (Baltimore: Johns Hopkins University Press, 2002), 241–249, 243 and 244. See also Toellner, "Der Funktionswandel."

21. Quoted in Temkin and Temkin, "Wunderlich versus Haeser," 247.

22. Schneck, "Über die Ursachen." The clinician Johann Lukas Schönlein (1793–1864) had petitioned along similar lines in 1850 (ibid., 39). See also Ludwig Edelstein, "Medical Historiography in 1847," *Bulletin of the History of Medicine* 21 (1947): 495–511, 507–511.

23. Sprengel's admiration of the Hippocratics and of vitalism, for example, informed a merciless treatment of the methodists and the iatrophysicists.

24. Heinrich Haeser, *Lehrbuch der Geschichte der Medicin und der epidemischen Krankheiten*, 2 vols. (Jena: Friedrich Mauke, 1853–1862), vol. 1: xviii–xix.

25. The quote is taken from Ranke's "On the Relations of History and Philosophy," in *The Theory and Practice of History*, Leopold von Ranke, ed., with an introduction by Georg G. Iggers and Konrad von Moltke (New York: Bobbs-Merrill, 1973), 29–32, 30.

26. Owsei Temkin, "The Usefulness of Medical History for Medicine," in *Double Face of Janus*, 68–100, 91–92; Anthony Grafton, "Polyhistor into Philolog: Notes on the Trans-

formation of German Classical Scholarship, 1790–1850," *History of Universities* 3 (1983): 159–192.

27. Although the German pattern in studies of ancient medicine was distinctive, other countries witnessed a similar transition in philology. For the positivist French physician-philologist Émile Littré (1801–1881), classical medical knowledge was to have direct relevance to contemporary medicine. His monumental edition of Hippocrates presented Hippocrates as a professional and moral model for practicing physicians; see Smith, *Hippocratic Tradition*, 31–36. This changed with his pupil, medical historian–philologist Charles Daremberg (1817–1872), who did not seek in the past guidance for contemporary physicians but instead conducted a kind of medical archaeology in the libraries and archives of Europe, retrieving, editing, and publishing texts as an end in itself (Temkin, "Usefulness of Medical History," 78).

28. See, for example, Stefan Rebenich, *Theodor Mommsen und Adolf Harnack: Wissenschaft und Politik im Berlin des Ausgehenden 19. Jahrhunderts* (Berlin: De Gruyter, 1997), and Jutta Kollesch, "Das Corpus Medicorum Graecorum. Konzeption und Durchführung," *Medizinhistorisches Journal* 3 (1968): 68–73.

29. Th. Puschmann, "Die Bedeutung der Geschichte für die Medicin und die Naturwissenschaften," *Deutsche Medicinische Wochenschrift* 15 (1889): 817–820. On Puschmann's plea for medical history as an academic discipline, see Schneck, "Über die Ursachen," and Gabriela Schmidt, "Theodor Puschmann und Seine Verdienste um die Einrichtung des Faches Medizingeschichte an der Wiener Medizinischen Fakultät," in Frewer and Roelcke, *Die Institutionalisierung*, 91–101.

30. On the relationship between this *Bildungsbürgertum* and the renaissance—and institutionalization—of medical history in Germany, see Volcker Roelcke and Andres Frewer, "Konzepte und Kontexte bei der Institutionalisierung der Medizinhistoriographie um die Wende vom 19. zum 20. Jahrhundert," in Frewer and Roelcke, *Die Institutionalisierung*, 9–25, and Ugo d'Orazio, "Ernst Schwalbe (1871–1920). Ein Kapitel aus der Geschichte 'Nicht Professioneller' Medizingeschichte," in Bröer, *Eine Wissenschaft Emanzipiert Sich*, 235–247.

31. See Björn Wittrock, "The Modern University: The Three Transformations," in *The European and American University since 1800: Historical and Sociological Essays*, Sheldon Rothblatt and Björn Wittrock, eds. (Cambridge: Cambridge University Press, 1993), 303–362.

32. See, for example, Friedrich Paulsen, *Die Deutschen Universitäten und das Universitätsstudium* (Berlin: Asher, 1902), 1–11, 63–65.

33. Andreas Frewer, "Biographie und Begründung der Akademischen Medizingeschichte: Karl Sudhoff und die Kernphase der Institutionalisierung 1896–1906," and Ortrun Riha, "Die Puschmann-Stiftung und die Diskussion zur Errichtung eines Ordinariats für Geschichte der Medizin an der Universität Leipzig," both in Frewer and Roelcke, *Die Institutionalisierung*, 103–126, 127–141.

34. On the Vienna institute and its competition with the institutes of Leipzig and Berlin, see Bernhard vom Brocke, "Die Institutionalisierung der Medizinhistoriographie im Kontext der Universitäts- und Wissenschaftsgeschichte," in Frewer and Roelcke, *Die Institutionalisierung*, 187–212, 194–200.

35. N. N. [Johann Hermann Baas], "Die Puschmann-Stiftung für Geschichte der Medizin," *Münchener Medizinische Wochenschrift* 51 (1904): 884–885. On Baas, see Burnham, *Idea of Profession*, 27–29.

36. M. Seiffert, "Aufgabe und Stellung der Geschichte im Medizinischen Unterricht," *Münchener Medizinische Wochenschrift* 51 (1904): 1159–1161.

37. Karl Sudhoff, "Zur Förderung Wissenschaftlicher Arbeiten auf dem Gebiete der Geschichte der Medizin," *Münchener Medizinische Wochenschrift* 51 (1904): 1350–1353. People like Max Neuburger, Julius Pagel, and Hugo Magnus seconded Sudhoff's defense of medical history; see "Protokoll der Dritten Ordentlichen Hauptversammlung," *Mitteilungen zur Geschichte der Medizin und der Naturwissenschaften* 3 (1904): 465–472, 468–471.

38. See Sudhoff's autobiographical essay, "Aus Meiner Arbeit. Eine Rückschau," *Sudhoffs Archiv* 21 (1929): 333–387, 339 and 364–368.

39. Karl Sudhoff, "Theodor Puschmann und die Aufgaben der Geschichte der Medizin," *Münchener Medizinische Wochenschrift* 53 (1906): 1669–1673.

40. Karl Sudhoff, "Richtungen und Strebungen in der Medizinischen Historik," *Archiv für Geschichte der Medizin* 1 (1907): 1–11. He also inaugurated in 1907 a series of monographs, *Studien zur Geschichte der Medizin Herausgegeben von der Puschmann-Stiftung an der Universität Leipzig*.

41. Henry Sigerist, "Forschungsinstitute für Geschichte der Medizin und der Naturwissenschaften," *Forschungsinstitute. Ihre Geschichte, Organisation und Ziele,* Ludolph Brauer, Albrecht Mendelssohn Bartholdy, and Adolf Meyer, eds. (Hamburg: Paul Hartung Verlag 1930), 391–405, 393.

42. Sudhoff, "Rückschau," 364, 366–368. Also see Iggers, *German Conception of History*.

43. Julius Pagel, *Einführung in die Geschichte der Medicin* (Berlin: Karger 1898), 1–22; Karl Sudhoff, *Kurzes Handbuch der Geschichte der Medizin* (Berlin: Karger, 1922).

44. Julius Pagel, "Medizinische Kulturgeschichte," *Janus* 9 (1904): 285–295, 287. He elaborated his ideas in *Grundriss Eines Systems der Medizinischen Kulturgeschichte* (Berlin: Karger, 1905). Also see Max Neuburger, "Die Geschichte der Medizin als Akademischer Lehrgegenstand," *Wiener Klinische Wochenschrift* 17 (1904): 1214–1219, and "Introduction," in *Handbuch der Geschichte der Medizin,* Max Neuburger and Julius Pagel, eds., 3 vols. (Jena: Verlag Gustav Fischer, 1902–1905), vol. 2:3–154.

45. Michael Hubenstorf, "Eine 'Wiener Schule' der Medizingeschichte?—Max Neuburger und die Vergessene Deutschsprachige Medizingeschichte," in *Medizingeschichte und Gesellschaftskritik,* Hubenstorf et al., eds. (Husum: Matthiesen Verlag, 1997), 246–289.

46. On the meanings of Sudhoff and his institute for contemporary and later medical historians, see *90 Jahre Karl-Sudhoff-Institut an der Universität Leipzig,* Ortrun Riha and Achim Thom, eds. (Freilassing: Muttenthaler, 1996), and Thomas Rütten, "Karl Sudhoff 'Patriarch' der Deutschen Medizingeschichte," in Gourevitch, *Médecins érudits,* 155–171.

47. Sudhoff, "Zur Förderung," 1351.

48. Henry Sigerist, "A Tribute to Max Neuburger on the Occasion of His 75th Birthday," *Bulletin of the History of Medicine* 14 (1943): 417–421, 420. See also Leopold Schönbauer and Marlene Jantsch, "Verbindungen Zwischen den Medizingeschichtlichen Instituten in Leipzig und Wien," *Wissenschaftliche Zeitschrift der Universität Leipzig* 5 (1955/1956): 27–31.

49. His lecture is reproduced in Alan M. Chesney, "Two Papers by John Shaw Billings on Medical Education, with a Foreword," *Bulletin of the Institute of the History of Medicine* 6 (1938): 285–359, 343. When in 1893 the School of Medicine at Hopkins opened, Billings was appointed a lecturer in history and literature of medicine, but he gave only a few lectures each year; see Genevieve Miller, "Medical History," in *The Education of American Physicians: Historical Essays*, Ronald L. Numbers, ed. (Berkeley: University of California

Press, 1980), 290–308, 296, and Sanford V. Larkey, "John Shaw Billings and the History of Medicine," *Bulletin of the Institute of the History of Medicine* 6 (1938): 360–376. For a particularly lucid example of medical history as a counter to other cultural forces in early-twentieth-century American medicine, see Lewis S. Pilcher, "An Antitoxin for Medical Commercialism," *Physician and Surgeon* 34 (1912): 145–158.

50. Michael Bliss, *William Osler: A Life in Medicine* (Oxford: Oxford University Press, 1999), 295–296, 350–353. On early medical history activity in America, see Genevieve Miller, "In Praise of Amateurs: Medical History in America before Garrison," *Bulletin of the History of Medicine* 47 (1973): 586–615.

51. Miller, "Medical History," 296–297.

52. Abraham Flexner, devoted to the German model in higher education and possibly unrivaled in his commitment to transforming American medicine into a bastion of the new sciences, welcomed the founding of the Institute of the History of Medicine at Johns Hopkins. Flexner liked the name "institute" because it implied "identity of aim with the great Institute founded by Professor Sudhoff at Leipzig" (*Universities: American, English, German* [New York: Oxford University Press, 1930], 111).

53. Miller, "In Praise of Amateurs," 586. On Garrison and his *Introduction,* see Erwin H. Ackerknecht, "Zum hundertsten Geburtstag von Fielding H. Garrison," *Gesnerus* 27 (1970): 229–230; Gert H. Brieger, "Fielding H. Garrison: The Man and His Book," *Transactions and Studies of the College of Physicians of Philadelphia* ser. 5, 3 (1981): 1–21; Burnham, *Idea of Profession,* 35; and Henry R. Viets, "Fielding H. Garrison and His Influence on American Medicine," *Bulletin of the Institute of the History of Medicine* 5 (1937): 347–352.

54. Fielding H. Garrison, "Professor Karl Sudhoff and the Institute of Medical History at Leipzig," *Bulletin of the Society of Medical History of Chicago* 3 (1923): 1–32, 2. Also see Fielding H. Garrison's "Foreword" and "Biographical Sketch" in his edition of Karl Sudhoff, *Essays in the History of Medicine* (New York: Medical Life Press, 1926).

55. Garrison, "Professor Karl Sudhoff," 32.

56. For a plea against amateurism and for a specialized research center in medical history, see Henry Sigerist, "The History of Medicine and the History of Science," *Bulletin of the Institute of the History of Medicine* 4 (1936): 1–13. Also see his survey of the state of medical history in the United States, "Medical History in the Medical Schools of the United States," *Bulletin of the History of Medicine* 7 (1939): 627–662. "On both sides of the Atlantic, the tendency is to leave historic medicine to the dilettante, but this should not be," Sudhoff asserted in 1929, in a benediction for the Baltimore enterprise offered at the dedication of the Department of History of Medicine. He urged that "without historic perception, the physician lapses into a mechanic." Medical history, Sudhoff concluded, "snubbed and rejected in the past, has labored diligently to perfect herself, and has now come into her own" (*Address of Professor Sudhoff at the Dedication of the Department of History of Medicine[,] Johns Hopkins University[,] October 18, 1929,* abstracted by Fielding H. Garrison, [n.p., n.d.], 3). For a poignant depiction of Temkin's journey, see Gert H. Brieger, "Temkin's Times and Our Own: An Appreciation of Owsei Temkin," *Bulletin of the History of Medicine* 77 (2003): 1–11, and Charles D. Rosenberg, "What Is Disease? In Memory of Owsei Temkin," *Bulletin of the History of Medicine* 77 (2003): 491–505.

57. To underscore the new direction he sought to lead the Leipzig Institute, in 1928 Sigerist founded a new journal, *Kyklos. Jahrbuch des Instituts für Geschichte der Medizin an der Universität Leipzig.*

58. Quoted in Owsei Temkin, "The Double Face of Janus," in *Double Face of Janus*, 3–37, 9.

59. Ingrid Kästner, "The Leipzig Period, 1925–1931," in Fee and Brown, *Making Medical History,* 42–62, 53.

60. Henry E. Sigerist, *Man and Medicine: An Introduction to Medical Knowledge* (New York: W. W. Norton, 1932). "Professor Sigerist," William H. Welch asserted in his preface to the American edition, "illustrates not merely the interest and cultural value of historical studies . . . but also the absolute necessity of such studies for an adequate understanding and interpretation of the present state and outlook of medical knowledge and practice" (vii). Also see Henry Sigerist, "Probleme der Medizinischen Historiographie," *Sudhoffs Archiv* 24 (1931): 1–18, 16.

61. Robert D. Johnston, "Beyond 'the West': Regionalism, Liberalism, and the Evasion of Politics in the New Western History," *Reconsiderations* 2 (1998): 239–277, 241. The discussion of "deinstitutionalization" is from an earlier version of the paper, presented as Jeffrey Ostler and Robert Johnston, "The Politics and Antipolitics of Western History," American Studies Faculty Colloquium, Yale University, New Haven, spring 1996.

62. In November 1997, Professor Johanna Bleker of the Institute for the History of Medicine of the Free University of Berlin issued a cry for help on the Internet on behalf of the field in Germany. By the following year, *Medizingeschichte: Aufgaben, Probleme, Perspektiven,* Norbert Paul and Thomas Schlich, eds. (Frankfurt: Campus Verlag, 1998) appeared, self-consciously defending the field. Two other volumes quickly followed, both animated in part by the sense that medical history needed legitimation in the face of threatened institutional dissolution; see Bröer, *Eine Wissenschaft Emanzipiert Sich,* and Frewer and Roelcke, *Die Institutionalisierung.* For Britain, see *Evaluation of the Wellcome Trust History of Medicine Programme* (London: The Wellcome Trust, 2000).

63. Elizabeth Fee and Theodore M. Brown, "Introduction," in Fee and Brown, *Making Medical History,* 1–11, 5.

Part I / Traditions

CHAPTER TWO

To Whom Does Medical History Belong?
Johann Moehsen, Kurt Sprengel, and the Problem of Origins in Collective Memory

Hans-Uwe Lammel

> When we use history to welcome each of them in turn as our guests, to take them in and observe how great and excellent they were, and to infer from their deeds the most important facets of their character, then we are involved in nothing less than an intimate relationship.... By studying history and constantly writing about them, we prepare ourselves always to receive into our souls the memory of these most noble and trustworthy men. And if inevitably in our dealings with the world we encounter wicked, malformed or ignoble things, we can push these things aside and banish them by quietly and resolutely turning our attention to these most noble and exemplary men.
> PLUTARCH, *Aemilius* 1

Though he was not exactly the "father of biography," well into the early modern era Plutarch (after 45–120 A.D.) was a much-read master of this genre. He founded a literary tradition by drafting the life stories of prominent personalities—war heroes and statesmen, famous poets, philosophers, and public speakers. At first, driven by pedagogic and philosophical considerations, he focused his attention on individuals, whose biographies have for the most part been lost. Plutarch's grand literary success and his lasting influence, however, came only after he decided to supplement his moral purpose—that is, the ethical cultivation of his readers through a depiction of worthy role models—with a further, eminently political aim: bringing the two peoples of the empire, Greeks and Romans, closer together. In his parallel biographies of individual Greeks and Romans, in whom he thought he could glimpse "comparable forms of thought and action," Plutarch tried to show how, in spite of their differences, they were both "children of the same spirit and worthy of one another."[1]

These remarks are intended to introduce the literary genre of parallel biographies to which this chapter is committed. It can hardly be claimed that the two historians of medicine that I will consider—the royal physician, district physician, and member of the Academy of Sciences in Berlin, Johann Carl Wilhelm Moehsen (1722–1795) and the professor of medicine Kurt Sprengel (1766–1833) from the University of Halle—were "children of the same spirit." Instead, this chapter will be concerned with the *différance* between the two forms of medical history these authors produced in an attempt to understand the aims, conditions, and contents of medical historiography in the latter half of the eighteenth century.[2] Sprengel and Moehsen exemplify two alternative engagements with the history of medicine that were developed in the eighteenth century.

In order to explain this period of German medical historiography, I have chosen to use the notions of "cultural space" and "changes in spaces." The double portrait that is presented here is intended to be chiefly a contribution to the history of medical historiography. In a sense, it could also be read as documenting the birth pains of the modern state, society, university, and medicine from the perspective of medical historiography. As Thomas Broman has argued, the modern medical profession did not arise from the ruins of the old regime. Instead, it developed out of the adaptation of an established elite to new circumstances.[3] The question is, what kind of circumstances were these? Johann Moehsen lived in an age that valued gentility and academic learning. He was imbued with humanist values and was rewarded by an important position at the court of Brandenburg. As a highly regarded member of the German *Gelehrtenstand,* he also represented the late Berlin Enlightenment with its pragmatic, local politics by little demarches.

It is correct to emphasize that Moehsen and Sprengel were of two different generations. Yet on the other hand, Kurt Sprengel does not fit totally in Broman's category of the (emerging) modern professional. He was a figure of the "great transition," as Roy Turner labeled this period. Of critical importance was that he worked at the university, with its special commitment to educating future physicians. His work in the history of medicine was subordinated to these commitments. At the same time, he argued for *Bildung,* through a mastery of everything Greek (with Hippocrates as an icon), as a way to acquire social status. Both Sprengel and Moehsen wanted to create a public domain, and both used medical history to do so. The difference between them is one of perspective and emphasis. Moehsen wrote an enlightened history of learning and erudition in Brandenburg in order to create a political domain by displaying the power and influence of reason in the development of science and the arts (including medicine); his potential audience was a circle of enlightened readers, including physicians and history writers. Sprengel on the other hand wrote a universal history of medicine (*Universalgeschichte der Medizin*), in which he tried to draw together, in a narrative format, the entirety of theoretical and practical knowledge of past medicine—knowledge that was relevant to future physicians as both healers and businessmen. Both men sought to provide a modern sketch of physicians' identity, but Moehsen situated that identity in the Enlightenment, while Sprengel insisted on the role played by the medical faculty itself and its didactic traditions in casting that identity.

Sprengel came to be remembered as a "founding father" for medical history precisely because his version of medical historiography was directed toward physicians, and—*malgré tout*—it was physicians who would dominate the field of medical history for nearly the next two centuries. This case study not only shows the diversity of medical history and the diverse uses to which it can (and has been) put. It also demonstrates the power of collective memory in cultivating one figure while letting another fade, as well as the need for clear stories of origins—founders' stories, myths to account for how things got started—both for professions (Hippocrates in medicine) and fields of scholarship (in this instance, the history of medicine) during periods of cultural transition.

Writing History in a Changing Academic Context

Kurt Sprengel was a founder of the field of scholarly medical historiography.[4] Whoever accepts this assertion has the satisfaction of knowing a starting point, a disciplinary origin. The generation from Sigerist to Grmek used this statement commonly. Yet to understand Sprengel's enterprise and the place he has come to hold as "founding father" of the discipline, it is essential to give close attention to the personal and institutional circumstances of Sprengel's work and life, his conception of learning, and his self-portrait as a botanist and as a scholar of medical history. Here, I am more interested in the *cultural setting* of a field of knowledge than in its internal development. Over the course of the eighteenth century, the influence of the University of Halle waned in favor of the reformed University of Göttingen. In Göttingen the university attempted to offer prospective state officials a modernized curriculum in jurisprudence that closely linked politics and history and thus enabled history writing to escape its auxiliary function and become more independent. Nevertheless, contemporaries in Halle could also point to an innovative didactic tradition. From its inception in 1693–1694, Halle had organized an attractive curriculum in theology and medicine. Bedside teaching and work with clinical case studies were the instruments with which its teachers sought to respond to calls, first made by Herman Boerhaave in Leiden, for autopsies and revitalized empiricism.

It is clear and hardly surprising that in the first half of the eighteenth century, given this practical orientation, historical studies of medicine's past either drew on customary forms of Baroque *Gelehrtengeschichte* or were limited to collecting biographical information. The impetus for renewed interest in the history of medicine and its sources in the second half of the century came from at least four factors characteristic of the eighteenth century: the Enlightenment discovery of the historical world, the growth of a critical historical consciousness, the development of a critical philological and philosophic method in the study of antiquity (for which Göttingen became renowned), and the discussion among naturalists about mechanism, organicism, and vitalism. Each of these shifts served to explicate fundamental questions about the relationship between the part and the whole, synchronicity and diachronicity, individuality and development.

My aim is to display *Aufklärungshistorie* as the last step in the development of historiography in the humanistic tradition and as the eighteenth century's special attempt to make historiography fit the enlightened literary public. From its beginnings as a self-confident, civil, rationalistic historiography, Aufklärungshistorie called into question, challenged, and ultimately reshaped historiography. Buoyed up by an emancipatory impulse to enlighten the burgher and evoke a sense of political agency, Aufklärungshistorie led to a new self-understanding of historical writing thanks to its historical critique of sources, its methodological reflections on the appropriateness of historical explanation, and its presentist moral and political judgments. More explicitly than ever before, past experiences were linked to their reflective interpretation. History was understood to be, in principle, open, and the narrator himself was an integral part of historical interpretation. This shift toward subjective agency was one of the greatest accomplishments of Aufklärungshistorie. My argument is that in Sprengel and Moehsen, we face two personalities who became deeply rooted in a medical historiographic tradition that had begun in ancient times and experienced an intellectual revival in the Renaissance, as well as in contemporary historiographic debate as stimulated by Aufklärungshistorie.[5]

The last decade of the eighteenth century witnessed an unprecedented boom of medical history on the German book market—a boom for which Sprengel was partly responsible. In dealing with his contributions to medical historiography, I will focus on Sprengel's well-known *Versuch einer pragmatischen Geschichte der Arzneikunde* (Essay on a pragmatic history of medicine), (which was translated into French and Italian. In comparing the prefaces published in the three German editions of the *Versuch* written between 1792 and 1828, it becomes evident that Sprengel imposed on himself the task of creating a timeless work, a great synthesis of a long European tradition of learned medicine. Before examining Sprengel and his magnum opus, however, we need to take a look at Johann Moehsen, who represented the tradition Sprengel was to replace.

Johann Moehsen: Royal Physician and Prussian Historiographer

Although accounts of how medical history as a field of knowledge was transformed into an academic discipline have long taken Sprengel as their starting point, it is possible to tell the story starting with Johann Carl Wilhelm Moehsen.[6] Born in 1722 in Berlin, Moehsen studied medicine in Jena and Halle, where he received his medical degree in 1741. A year later he returned to Berlin, and in 1744 he held the membership of the Academia Naturae Curiosorum. As a member of the Obermedizinalkollegium (from 1747) and the Obersanitätskollegium (from 1763), and as a physician to the Prussian king Friedrich II, he held important political positions with insights in medical administration of Brandenburg-Prussia. Later he was one of the founding members of an enlightened circle of bureaucratic elites known in Berlin as the Mittwochsgesellschaft and member of the Berlin Academy of Sciences (from 1786).[7] He died in Berlin in 1795.

Making medical history was a leisure job for Moehsen, and one of his special contributions to the literary public. As a fifth-generation member of a physician family and as member of two important medical administrations, he elaborated his insights not only into the problems of health and disease in Brandenburg but also into state history. Having started with a description of a coin collection on famous physicians (*Beschreibung einer Berlinischen Medaillen-Sammlung*) between 1773 and 1781, his historiographical interest shifted.[8] In part two of the *Beschreibung*, entitled *Geschichte der Wissenschaften in der Mark Brandenburg, besonders der Arzneiwissenschaft* (History of erudition and learning in Brandenburg, especially of medicine), Moehsen brought together interests in state history, erudition, and enlightenment.[9]

During the second half of the eighteenth century, a patriotic interest emerged in Brandenburg-Prussia that stressed the importance of local history writing.[10] Histories of states, regions, and towns proliferated. They emerged as strong competitors to other literary genres, such as tragedies, biographies, novels, and so forth. Both citizens at large and the court wanted to begin remembering the past in a form available to all who were literate; they sought to understand the present by knowing its relationship to the past. Moehsen is known as the author of several studies on the medical history of the German territorial states and on the history of learning as well, for example in Brandenburg.[11]

In his most important work, *Geschichte der Wissenschaften in der Mark Brandenburg*, Moehsen used the state history model of German Aufklärungshistorie in order to elaborate his main theme: the propagation and diffusion of reason throughout a particular territory from medieval times onward. Moehsen transferred Aufklärungshistorie into the historiographic reconsideration of a specific German territory, with particular emphasis on medical history. He sought to discover the institutions, people, and administrations that had improved or hindered conditions for the propagation and diffusion of reason. In doing so, he was forced to reconsider all aspects of the well-known political history of Brandenburg-Prussia and their influence on learned, erudite history. In his eyes, erudition and its history did not consist of a collection of book titles or lists of scholars' names.[12] Instead, he considered the political meaning of scholars, their social setting in society, and the influence that they attempted to garner. If his guiding historiographic principle was to regard history as a battle between reason and irrationality (*Irrtümer und Vorurtheile*), he also described the cultural settings of periods that concerned him.[13]

Moehsen differentiated four periods in the history of the Brandenburg state. Each period was characterized by certain relationships among the ruler, the clergy, and the other intellectual elites; by the wars in which Brandenburg was involved; and by its economic situation and trade relations. All these factors helped shape the peculiar situation of medicine in each period—the medical education at universities; the role of hospitals, barber-surgeons, and midwifes; the relationship between physicians, philosophers, and theologians; and the life and work of apothecaries, ordinary physicians, town physicians, and court physicians. For Moehsen, the framework of state history provided the opportunity to use the bricks of *Historia literaria* in order to reintegrate them into medical history, a unique aspect of the history of the Brandenburg state.[14] Historia literaria, the contemporary history of scholars, was to be the means for establishing a different form of recording knowledge. Knowledge as a whole was to be seized and made accessible and beneficial everywhere to all members of the *Republica literaria*, according to their various needs in making knowledge practicable. Republica literaria was a kind of virtual general library generated essentially by the Historia literaria. Scholars reported on what is going on in their institutions and in their subject fields. Extracts were produced from books, reworked and incorporated into handbooks, bibliographies, and encyclopedias. This form of compiling possessed a clear, antidogmatic orientation and was elevated to a critical method. In this way, eclecticism intended systematic compilation to be a comprehensive stocktaking in the field of learning.[15]

Kurt Sprengel: Botanist, Medical Professor, and Historiographer

Like Schiller and other German intellectuals, Sprengel attempted to transfer the older intellectual traditions into new public spheres by casting them as literature and text.[16] At the end of the century, this intellectual approach appeared to be the German way of getting a handle on a politically oriented literary public in an underdeveloped country—such as the Old Reich had been.[17] Previously, a real public sphere existed only in larger towns like Hamburg, Halle, Leipzig, Frankfurt, and Berlin.[18] The "social composition" of that public sphere was not bourgeois, but "educated." To be political meant being literary, nothing more or less. Although Germany lacked a real political consciousness, there did exist a rich awareness of intellectual traditions as cultural agents, whereby that awareness turned on people's inner emotions—inward toward Bildung rather than outward toward civic society.[19] Pre-Kantian theodicy as metaphysical optimism assumed that evil in this world had to be understood as an inevitable consequence of God's will to create the best possible world order, or as Leibniz put it, as "necessitas moralis ad optimum," a moral necessity in attaining the best possible circumstances.

Against the backdrop of this theodicy, which was so fundamental to the eighteenth century and so inspired by the notion of a divinely created world based on reason, the challenge historians faced was to show the workings of their providential rationality, in an attempt to make God's doing visible. In whatever he did—translating travel books, commenting on biographies of illustrious men, or studying natural history—Sprengel revitalized older intellectual traditions of collecting and compiling. There were two essential aspects of that eclecticism. As Martin Gierl put it, "First, it was opposed to elitist, minority-held and authority-linked knowledge—to learned schools and their constraints, especially to the theological-confessional restraints upon scholarly activity. Second, there was support for new forms of scholarly co-operation, exemplified by courteous dealings between scholars and by a new treatment of knowledge."[20] Everyday scholarly practice was to be converted from the old form of conflict-orientated learning into new cooperative forms. Joining forces in collecting and examining material meant compilation on a grand scale. In the process, Sprengel created a new type of medical historiography that was to appeal to many subsequent generations of physician-medical historians.

Sprengel was born as a son of a Pomeranian parson in Boldekow in 1766.[21]

According to his own curriculum vitae, his father taught him a wide range of subjects, including languages (Latin, French, Hebrew, Greek, Arabic, and Swedish), the natural sciences, and history. Before embarking on his studies in medicine he concentrated on theology at the University of Greifswald beginning in 1783. In 1785 he moved to Halle, changed his focus to medicine, and finished his studies in 1787. In Halle he received an academic position in medicine as doctor legens in 1788 and as professor extraordinarius one year later. In 1795, Sprengel became professor ordinarius without any indication of a special medical field. His historical interests shifted and broadened to include medical history and natural history. In 1797 he became the director of the Botanical Garden at Halle and was able to focus all his energies on the study of botany.

Sprengel was well versed in the literal explication of texts. By the 1780s Sprengel's literary productivity was astonishing. He began with some entirely traditional publications about ancient medicine: a commentary on a Hippocratic aphorism and Hippocrates' notion of eruption, a depiction of the Hippocratic doctrine of the pulse, and Galen's opinions on fever, as well as the two volumes of *Die Apologie des Hippokrates* (The apology of Hippocrates), published in 1789 and 1792.[22] In the *Apologie* he attempted to reanimate the grandfather of medicine by claiming his superiority in the art of observation, though not in conceptualization. This was a completely different interpretation from the current one, which had been introduced by Daniel LeClerc (1652–1728) in the so-called *querelle des anciens et des modernes,* and it met with violent opposition.[23]

Sprengel was obsessed with the figure of Hippocrates, in spite of debates during the first half of the century on the latter's putative atheism.[24] While conceding that the theories of the man from Kos were no longer up-to-date, Sprengel emphasized the ethical reliability of his vehement empiricism. This shift in focus made a new space of explanation accessible and secured a place for Hippocrates in the memory of the physician.[25] Whereas the discourse of early modern physicians had been of a general, humanistic nature, there was an explicit, discursive link between theory and practice in their modern successors. In this context, Sprengel did a very smart, yet subtle thing: he both historicized Hippocrates and made him a "contemporary colleague." Writing in an age that was convinced of the importance of case histories to medical practice, Sprengel made history the very basis for a new epistemology of medicine, with Hippocrates as its icon. At the time, the relation between theory and practice in medical teaching was a very uncomfortable one. In supposing that Hippocratic medical theory had been derived directly from bedside experience and had consisted of rules that guided practice, Sprengel presented Hippocrates as a role model.[26]

In writing history—gathering, accumulating, classifying, arranging, and reporting facts—Sprengel always behaved and acted as a botanist and critical compiler in the sense of the Historia literaria.[27] According to Sprengel, gathering and presenting data went to the core of historiography. For that reason, he also valued the travel story (*Erzählung*) "because, using vivid examples, it teaches us what humans are capable of and how an invisible eye watches over everything and often sees to the quick rescue of an unfortunate individual once he has been abandoned by human assistance."[28] Travel books were a part of the old natural history and of the new human science (*Wissenschaft vom Menschen*), as well as an instrument of learning by traveling intellectually in space and time. For Sprengel, they represented a new eclectic kind of empiricism.[29]

Sprengel made his intentions explicit by complaining about a lack of learning among physicians, most of whom were ignorant of their forerunners.[30] Progress in science, he argued, meant the accumulation of data. The latest accumulator had the task of gathering, summing up, and improving prior knowledge.[31] One should avoid hypothesizing and, instead, observe carefully and follow one's own experience. Sprengel's intellectual attitude as a translator was vital to his approach. He did not add his own observations, nor did he try to improve statements or annotations: "for that I lack the audacity."[32] Rather, he emended quotations, translated, and added authors and subject indices, including subjects in foreign languages. This was the expression of the critical philological viewpoint of a man of erudition. In the medical historiography of Sprengel, philology with all its facilities and techniques supplied the leading principles.[33] He sought to facilitate the acquisition of knowledge as a library does by its catalogs.

Sprengel's *Versuch* and the Meaning of "Pragmatism" in German Historiography

Compared with his contemporaries and predecessors, Sprengel did everything to maintain a high quality of writing in medical historiography. In the *Versuch*, Sprengel told the story of medicine from the beginnings (the childhood of mankind, "Kindheit des menschlichen Geschlechts") to the eighteenth century. In doing so, he divided medical history into eight epochs, which he classified according political or medical history. The first epoch concerned the origins of medicine and included chapters on Egyptian medicine, Greek medicine "from the centaur Chiron to Hippocrates," and so on. The eleventh and last section of the first edition concerned the period from Haller to the present, ordered "according to the individual fields of medicine."[34] Sprengel's accomplishment was to examine in a philologically critical manner information on medical history accumulated by the scholarly practice of Historia literaria and to transform that information into a continuous linear narrative. The result was a universal history of medical ideas, inventions and practices, written primarily for physicians.

During the three decades between the first (1792) and third (1828) editions, times changed. The Prussian state reorganized all institutions of higher education. A new university was born in Berlin, whose founders, among them Wilhelm von Humboldt, searched for another, modern pattern for academic life, research, and teaching. "Loneliness" (*Einsamkeit*) and "freedom" (*Freiheit*) were the new romantic buzzwords, the "magic wand," as Novalis put it, to solve earthly problems.[35] The changes Sprengel was forced to make in his historical approach, style, and method positioned him as a perfect candidate to the honorary title of the founding father of medical history—as it was conceived by physicians. He followed that "great transition" in a fascinating way; he outshone the entire next generation of medical historians, including Justus Friedrich Karl Hecker, Hermann Friedländer, Heinrich Damerow, Emil Isensee, and others down through the middle of the nineteenth century, to say nothing of other medical historians of his own epoch like Christian Gottfried Gruner, Johann Moehsen, Johann Christian Gottlieb Ackermann, and Philipp Ludwig Wittwer.[36]

When looking at the three editions of the *Versuch*, it becomes clear that the intellectual development of Sprengel as a medical historian was a function of the changes and reforms of the Prussian state and university.[37] As quotations from Lucian and William Hayley on the title-page of the first edition make clear, Sprengel was initially rooted in two historiographical traditions, one classical, the other modern.[38] On the one hand, he very much belonged to a tradition of Herodotus, Lucian, and, tacitly, Plinius. On the other hand, he had a great affinity for contemporary English historiography, particularly for Hayley.[39] The epigraphs both refer to the basic problem of the relationship between historical truth and falsehood. One can sum up the historiographic conflict as follows: given that the scholarly narrative representation of the past had become both a scientific and a literary process, what did an author need to take into account to write in an aesthetic and pleasing manner but to differentiate his work from writers of fiction? Sprengel's answer was that he never claimed to be a "history writer" (*Geschichtschreiber*); instead, he always used the term "history researcher" (*Geschichtsforscher*).

In the first and second editions Sprengel explained his notion of *Kulturgeschichte* (cultural history) by referring to Johann Christoph Adelung (1734–1806) and his book on the cultural history of mankind.[40] Sprengel shared Adelung's view that it was impossible to determine a direct relationship between cause and effect, "the careful development and demonstration of causes and inducements, from which events and actions emerge."⁴¹ In the third edition, however, the reference to Adelung is gone. Sprengel decided to replace the historiographic tool of cause and effect with the idea of divine providence. By claiming a providential cause, Sprengel was freed from the task of retroactively ascribing meaning to history and could dedicate himself entirely to the literary and narrative arrangement of the information. He considered it to be the task of the historian to use all source material available, in order to show how the worldly order was based on divine reason. However, every "history researcher" knew that sources were full of irrationality. One could have the wrong sources or be blinded to the rationality hidden in the sources by his own prejudice. The consequence of Sprengel's procedure was that in the second and third editions he virtually dismissed the secondary sources he had used in the first edition (authors such as Bayle, Gibbon, Muratori, Moehsen). Increasingly, he favored using primary sources.

It is no wonder that in the two subsequent editions of his *Versuch,* Sprengel excised everything that was linked to Enlightenment historiography. Similarly, he dismissed a more chronological representation of medical history in favor of a narrative form, and he opposed German Aufklärungshistorie by omitting former discussions about *fides historica* literature and debates emerging at midcentury.⁴² He succeeded in his efforts because he ignored questions of chronology, which had occupied much space in the *Apologie*. Instead, he chose a periodization for the history of medicine that was indebted to cultural history, a mixture of political and church history.⁴³ Finally, he dismissed the crucial practice of Historia literaria, the learned compilation of the most important works on medical history, and declared physicians' biographies and bibliographies of medical works to be nonessential for medical historiography.⁴⁴ Sprengel simply sought to examine the primary sources, staking everything on authenticity. The utility of such a scholarly undertaking—characteristic of the new learned elite—lay in the assumption that the history of medicine could be told as part of a temporal continuum from the beginnings up to the present, insofar as the sources allowed. A literary form had been created that allowed not only for renovation and extension but also for "moralizing" observations, an achievement of collective memory in literary form with historiographic pretensions.⁴⁵

The full title of Sprengel's magnum opus is *Essay on a Pragmatic History of Medicine,* but what, exactly, did "pragmatic" mean in this context? The term "pragmatic" had first been introduced by Thucydides,⁴⁶ but it reemerged on the eve of the eighteenth century in Göttingen in discussions about history, developing a five-fold meaning.⁴⁷ First, it referred to a new narrative form of writing history that aimed to enhance its scholarly status by searching for cause and effect. In this sense, pragmatism was directed toward increasing the credibility of a story. Although causality very much depended on the historian's point of view, it was considered an important weapon against historiographic skepticism, an issue that had been debated since the sixteenth century.[48] Second, historiography needed to be of use to contemporaries; it was intended to make them recollect the past and to be a tool of remembrance. In this respect, Sprengel used "pragmatic" in the sense of "a history *aimed precisely at useful instruction.*" Third, achieving these aims also meant attempting to improve the readability of history by engaging in a contest with other literary genres, such as biography and the novel.[49] Fourth, a pragmatic orientation was an attempt to escape the old pattern of Historia literaria, reconstructing the paradigm of older intellectual traditions into newer, modern ones.[50] Finally, it sought to entertain the ancients-versus-moderns debate.[51] In Sprengel's *Versuch,* we can find all five meanings of "pragmatic."

Kurt Sprengel seemed to have created a timeless work, but his *Versuch* was deeply rooted in the state of medicine in the late eighteenth and early nineteenth centuries. Nor is it surprising that this historiographic shift from Historia literaria to a historiography that pivoted on the notion of providence of God occurred immediately after the French Revolution. From the German point of view, violence and the reign of terror did not allow any other interpretation than the dismissal of rationalistic consideration of history in favor of the notion of divine providence. It is no small accomplishment that Sprengel succeeded in making his *Versuch* look like a universal history (*Universalgeschichte*).[52] He succeeded in doing so by using the concept of providence, a methodological tool used to mediate between past, present, and future.[53] It offered him the opportunity to reintegrate all aspects of medical history dating back to ancient times; at the same time, it stabilized the fragile situation of contemporary medicine by providing a reflexive tool (for instance, in the theory-practice debate around 1800).[54] By adopting his particular form of historiography, Sprengel contributed to the new intellectual and political situation in Germany, to neo-Humanism and Romanticism, and to a vital civic awareness among physicians.

Medical History: Tool of Remembrance and Shaper of Identities

Both Sprengel and Moehsen used teleological patterns to organize their stories. While Moehsen looked at history as a continuous process of propagation and diffusion of reason, Sprengel used the notion of providence to give his narrative coherence. Yet, there is an enormous gap between the way the work of these two medical historians has been remembered: whereas Moehsen has been all but forgotten, Sprengel is known as the founding father of modern medical history. The explanation for this huge difference in success of two men—who in their own days were both highly esteemed—can be found by looking at the cultural spaces in which they moved.

Moehsen's social setting allowed him to participate in the historiographic endeavors of the Göttingen historians. Those efforts were part of the enlightened search for historical roots and historical consciousness in order to increase civic influence on the state's internal formation, as well as to obtain and stabilize political discussion independent of (and in contradistinction to) official court discourse. However, the "space" of the Berlin Academy of Sciences did not allow Moehsen to pursue his work on medical history in Brandenburg once he had been elected as a fellow of this learned society. There are two possible explanations for this. First, the political historiographic tradition fostered by the academy was not strong enough for integrating other approaches, like medical historiography. After the demise of Friedrich II in 1786, the Academy of Sciences lost much power in the disputes between its French and German members (*Deutsche Deputation*) over whether communications were to be in French or in German.[55] Also, Moehsen merely became a fellow of the Philological Class of the Berlin Academy of Sciences, and at the forefront of erudite, academic endeavors stood the improvement of all auxiliary historical knowledge, not the enhancement of historiography in competition with fiction.[56] Friedrich Wilhelm II handed Moehsen the royal collection of coins to be put in order, a royal decision that illuminates the role of academies in eighteenth-century historiography. Collecting, compiling, and classifying as techniques of the scholarly practice mentioned above as Historia literaria and as a part of the self-image of this kind of scholarship made little distinction between the collecting and putting in order of coins and the collecting, revision, and translation of travel books.[57]

The university "space" in which Sprengel moved also offered opportunities and restraints. The role and function of academics was debated toward the end of the eighteenth century. Professors attempted to improve their reputations by increasing their literary activities beyond their own academic fields. The aim was to enlarge the number of students and to overcome a narrow understanding of teaching by improving the reputation of universities. It was necessary to have instructors who could depart from the older lecture style of reading from a rough notebook in favor of a more scientific point of view. Hence, new forms of literary presentation evolved. Intellectual independence and self-reliance were needed to overcome the old academic patterns and hierarchy. These new aims required an enlargement of the curriculum and a greater emphasis on the practical applicability of each course. Sprengel tried to accommodate this changing pedagogy in his writings. The large number of travel books that he translated represents just one of many attempts at increasing the social reputation of academic professors. Sprengel expanded this new field of academic duties with his *Versuch,* which became a manual (*Leitfaden*) "because of the chronology and the copious citations."[58] At the end of the eighteenth century, university reforms granted greater academic freedom, and as a result Sprengel produced medical historiography of unprecedented quality.

In short, while Moehsen believed his relationship to Aufklärungshistorie was enhanced by the Berlin Academy of Sciences and his attempt to integrate medicine into the history of the Brandenburg state, Sprengel's *Versuch* benefited from the promise medical history held for the university education of future physicians.[59] Moehsen could be characterized as an *Aufklärer,* whereas Sprengel was an academic professor, a man of Bildung and Wissenschaft. Both were enlightened eighteenth-century savants. What distinguished them was the space in which they worked. Throughout the Enlightenment, different types of medical education had been discussed and practiced, but what finally prevailed was the medical curriculum at universities that embraced reform, such as Halle, Göttingen, and Berlin.

It should come as no surprise, therefore, that Sprengel—and not Moehsen—held a prominent place in the collective memory of modern physicians. The work of Sprengel should be considered as the result of a highly self-conscious attempt to supply the medical profession with intellectual and moral points of reference that were dearly needed in an insecure age and an underdeveloped country like the Old Reich. After the fall of the ancien régime and the Holy Roman Empire and after the introduction of the Humboldtian university, his persona and his work were better adapted to the demands of the new era. Whereas in the "old world" of learning and patronage the place to be was the networks of German Enlightenment, in the "new world" of bureaucracy and professionalism the university was the center of culture. Sprengel created a new conceptual space and historiographical style. He did so first by creating a new image of Hippocrates, later on by building on the emerging civic ideal of the Humboldtian university. Like Moehsen, he suggested alternatives to traditional forms of historical writing and tried to create stability in a period of transition. Both of their narratives could be called medical history, but whereas Moehsen worked to create a new type of citizen, Sprengel wanted to educate a new kind of physician. While Moehsen attempted to show the propagation and diffusion of reason by writing a medical history of the state of Brandenburg, Sprengel was convinced that providence was the driving force of history. These differences, in both agendas and methods, were decisive for their legacies. Because of his providential idealism, Sprengel was canonized as the founding father of medical history by physician–medical historians, who were to dominate the field until the 1960s.

NOTES

I owe a debt of gratitude to Frank Huisman and John Warner for their lively encouragement to undertake a wholesale reworking of the original manuscript to make it a double portrait of Sprengel and Moehsen and for their numerous suggestions in relation to this problematic undertaking. Furthermore, I would like to express my appreciation to all those who generously assisted me with specific suggestions, especially Gabriele Dürbeck, Hamburg, Jutta Kollesch, Berlin, and Markus Völkel, Rostock/Berlin. Last but not least I would like to thank Vance Byrd, University of Georgia, Athens, and Eric J. Engstrom, Berlin, for their linguistic support and advice.

1. Konrat Ziegler, "Einführung," in *Plutarch, Fünf Doppelbiographien,* part 2 (Zurich: Artemis & Winkler, 1994), 1081–1110, 1089–1090.

2. Jacques Derrida's neologism *différance* comprises the two meanings of *différer*: "to differentiate" and "to defer." Différance itself is not a concept, as it has no clear semantic relation, but is instead simply a reference to that which, in poststructuralist understanding, makes a concept a concept: a reference to other concepts and the exclusion of contrary concepts. Ute Daniel, *Kompendium Kulturgeschichte. Theorie, Praxis, Schlüsselwörter* (Frankfurt: Suhrkamp-Taschenbuch, 2001), 140.

3. Thomas H. Broman, *The Transformation of German Academic Medicine, 1750–1820* (Cambridge: Cambridge University Press, 1996).

4. Henry Ernest Sigerist, "Die geschichtliche Betrachtungsweise der Medizin," in *Anfänge der Medizin. Von der primitiven und archaischen Medizin bis zum Goldenen Zeitalter in Griechenland* (Zurich: Europa Verlag, 1963), 1–34, 2; Mirko D. Grmek, "Einführung," in *Die Geschichte medizinischen Denkens. Antike und Mittelalter,* M. D. Grmek, ed. (Munich: C. H. Beck, 1996), 9–27, 13–15.

5. *Ancient Histories of Medicine: Essays in Medical Doxography and Historiography in Classical Antiquity,* Philip J. van der Eijk, ed. (Leiden: Brill, 1999). Nancy Siraisi, "Anatomizing the Past: Physicians and History in Renaissance Culture," *Renaissance Quarterly* 53 (2000): 1–30.

6. H.-U. Lammel, "Moehsen und die Lebensbeschreibung Thurneissers. Ein Beispiel aufgeklärter Medizinhistoriographie," in *Berliner Aufklärung: Kulturwissenschaftliche Studien,* Ursula Goldenbaum and Alexander Košenina, eds. (Hannover: Matthias Wehrhahn Verlag, 1999), 144–172.

7. J. K. W. Möhsen, "What Is to Be Done toward the Enlightenment of the Citizentry," in *What Is Enlightenment? Eighteenth-Century Answers and Twentieth-Century Questions,* James

Schmidt, ed. (Berkeley: University of California Press, 1996), 49–52. Günter Birtsch, "The Berlin Wednesday Society," in ibid., 235–252, 240–242. Jonathan Knudsen, "On Enlightenment for the Common Man," in ibid., 270–290, 271–272.

8. *Beschreibung einer Berlinischen Medaillen-Sammlung, die vorzüglich aus Gedächtnis-Münzen berühmter Aerzte bestehet; in welcher verschiedene Abhandlungen, zur Erklärung der alten und neuen Münzwissenschaft, imgleichen zur Geschichte der Arzneigelährtheit und der Literatur eingerücket sind,* part 1 (Berlin and Leipzig: George Jacob Decker, 1773).

9. *Beschreibung einer Berlinischen Medaillen-Sammlung,* part 2 (Berlin and Leipzig: George Jacob Decker, 1781; reprint, Hildesheim: Olms, 1976).

10. Hans Martin Blitz, *Aus Liebe zum Vaterland: Die deutsche Nation im 18. Jahrhundert* (Hamburg: Hamburger Edition, 2000).

11. *Beiträge zur Geschichte der Wissenschaften in der Mark Brandenburg von den ältesten Zeiten an bis zu Ende des sechszehnten Jahrhunderts* (Berlin and Leipzig: George Jacob Decker, 1783; reprint, Munich: Fritsch, 1976); "Rede . . . dem Andenken des Geheimen Raths Cothenius gewidmet," in *Sammlung der deutschen Abhandlungen, welche in der Königlichen Akademie der Wissenschaften zu Berlin vorgelesen worden in den Jahren 1788 und 1789* (Berlin: George Decker, 1793), i–viii; "Über die Brandenburgische Geschichte des Mittelalters und deren Erläuterung durch gleichzeitige Münzen," in *Mémoires de l'Académie Royale des Sciences et Belles-Lettres depuis l'avènement de Frédéric Guillaume II au thrône, 1786/1787* (Berlin: George Decker, 1792), 675–684.

12. Moehsen, *Beschreibung,* part 2, 4.

13. H.-U. Lammel, *Zwischen Klio und Hippokrates. Zu den kulturellen Ursprüngen eines medizinhistorischen Interesses und der Ausprägung einer historischen Mentalität unter Arzten zwischen 1750 und 1850 in Deutschland* (Habilitationsschrift, University of Rostock, 1999), 103–178.

14. Lammel, *Klio und Hippokrates,* 68–102; "Zum Verhältnis von kulturellem Gedächtnis und Geschichtsschreibung im 18. Jahrhundert. Medizinhistoriographie bei Johann Carl Wilhelm Moehsen (1722–1795)," in *Festschrift für Rolf Winau,* Johanna Bleker, ed. (Husum: Matthiesen, 2003); "Natur und Gesellschaft. Medizin als aufgeklärte Gesellschaftswissenschaft," paper read in October 2001 at the Potsdam Research Center for European Enlightenment.

15. Martin Gierl, "Compilation and the Production of Knowledge in the Early German Enlightenment," in *Wissenschaft als kulturelle Praxis, 1750–1900,* Hans Erich Bödeker, Peter Hanns Reill, and Jürgen Schlumbohm, eds. (Göttingen: Vandenhoeck & Ruprecht, 1999), 69–103, here 70–71.

16. Johannes Süßmann, *Geschichtsschreibung oder Geschichtsroman? Zur Konstitutionslogik von Geschichtserzählungen zwischen Schiller und Ranke (1740–1824)* (Stuttgart: Franz Steiner, 2000); Thomas H. Broman, "Rethinking Professionalization: Theory, Practice, and Professional Ideology in Eighteenth-Century German Medicine," *Journal of Modern History* 67 (1995): 835–872 and "The Habermasian Public Sphere and 'Science in the Enlightenment'" *History of Science* 36 (1998): 123–149.

17. Hajo Holborn, "Der deutsche Idealismus in sozialgeschichtlicher Beleuchtung," *Historische Zeitschrift* 174 (1952): 359–384.

18. Engelhard Weigl, *Schauplätze der deutschen Aufklärung. Ein Städterundgang* (Reinbek: Rowohlt-Taschenbuch, 1997).

19. Frank Baasner, *Der Begriff 'sensibilité' im 18. Jahrhundert. Aufstieg und Niedergang eines Ideals* (Heidelberg: Winter, 1988); Henri Brunschwig, *Enlightenment and Romanticism in*

Eighteenth-Century Prussia, F. Jellinek, trans. (Chicago: University of Chicago Press, 1974); Georg Bollenbeck, *Bildung und Kultur: Glanz und Elend eines deutschen Deutungsmusters* (Frankfurt: Insel, 1994).

20. Gierl, "Compilation," 70.

21. Earlier studies of Sprengel in context include Heinrich Rohlfs, "Kurt Sprengel, der Pragmatiker," in *Die medicinischen Classiker Deutschlands,* vol. 2 (Stuttgart: Ferdinand Enke, 1880), 212–279; S. Alleori, "Il sistema dottrinario medico di Curzio Sprengel, avversario dei sistema," *Collana di pagine di storia della medicina, Miscellanea* 19 (1968): 121–131; Fausto Bonora and Elio de Angelis, "La storiografia dell'illuminismo et la metodologia storiografica di K. Sprengel," *Medicina nei secoli* 20 (1983): 11–26; Wolfram Kaiser and Arina Völker, *Kurt Sprengel (1766–1833)* (Halle: Martin-Luther-Universität, 1982); Hans-Theodor Koch, "Curt Sprengel (1766–1833) und die Medizingeschichte," in *Die Entwicklung des medizinhistorischen Unterrichts,* Arina Völker and Burchard Thaler, eds. (Halle: Martin-Luther-Universität, 1982), 92–96; Alain Touwaide, "Botanique et philologie: L'édition de Dioscoride de Kurt Sprengel," in *Médecins érudits, de Coray à Sigerist,* Danielle Gourevitch, ed. (Paris: de Boccard, 1995), 25–44; Ugo D'Orazio, "Conoscenza come indisciplina: Kurt Sprengel e la moltiplicazione dei sistemi nell'età romantica," in *Per una storia critica della scienza* (Milan: Cisalpino, 1996), 245–282; and H.-U. Lammel, "Kurt Sprengel und die deutschsprachige Medizingeschichtsschreibung in der ersten Hälfte des 19. Jahrhunderts," in *Die Institutionalisierung der Medizinhistoriographie. Entwicklungslinien vom 19. ins 20. Jahrhundert,* Andreas Frewer and Volker Roelcke, eds. (Stuttgart: Franz Steiner Verlag, 2001), 27–37.

22. Kurt Sprengel, "Commentar zu Hippokr. Aphor. IV. 5," *Neues Magazin für Aerzte,* E. G. Baldinger, ed., vol. 8 (Leipzig: Friedrich Gotthold Jacobäer, 1786), 368–375; "Hippokrates Begriff vom Exanthem," in ibid., 375–378; and *Beyträge zur Geschichte des Pulses, nebst einer Probe seiner Commentarien über Hippokrates Aphorismen* (Leipzig and Breslau: Johann Ernst Meyer, 1787). Sprengel, *Galens Fieberlehre* (Breslau and Leipzig: Johann Ernst Meyer, 1788). Sprengel, *Apologie des Hippokrates und seiner Grundsätze,* parts 1 and 2 (Leipzig: Schwickertscher Verlag, 1789 and 1792).

23. On LeClerc, see Ian M. Lonie, "Cos versus Cnidus and the Historians," *History of Science* 16 (1978): 42–75, 77–92.

24. Sprengel, *Apologie*, part 1, 112–118.

25. Otto Gerhard Oexle, "Memoria in der Gesellschaft und in der Kultur des Mittelalters," in *Modernes Mittelalter: Neue Bilder einer populären Epoche,* Joachim Heinzle, ed. (Frankfurt: Insel Verlag, 1994), 297–323; Peter Burke, "Geschichte als soziales Gedächtnis," in *Mnemosyne: Formen und Funktionen der kulturellen Erinnerung,* Aleida Assmann and Dietrich Harth, eds. (Frankfurt: Fischer-Taschenbuch, 1993), 289–304.

26. Broman, *Transformation,* 139–143, 180–185.

27. Lisbet Koerner, "Goethe's Botany: Lessons of a Feminine Science," *Isis* 84 (1993): 470–495.

28. [John Hynes], *Die Schicksale der Mannschaft des Grosvenor, nach ihrem Schiffbruche an der Küste der Kaffern im Jahre 1782. Aus dem Englischen des Herrn Carter übersetzt von Kurt Sprengel* (Berlin: Vossische Buchhandlung, 1792), 3.

29. Justin Stagl, "Der wohl unterwiesene Passagier: Reisekunst und Gesellschaftsbeschreibung vom 16. bis zum 18. Jahrhundert," in *Reisen und Reisebeschreibungen im 18. und 19. Jahrhundert als Quellen der Kulturbeziehungsforschung,* Boris I. Krasnobaev, Gert Robel, and Herbert Zeman, eds. (Essen: Reimar Hobbing Verlag, 1987), 353–384.

30. [Kurt Sprengel], "Vorbericht des Uebersetzers" in P[aul] J[oseph] Barthez, *Neue Mechanik der willkürlichen Bewegungen des Menschen und der Thiere. Aus dem Französischen übersetzt von Kurt Sprengel* (Halle: Karl August Kümmel, 1800), n.p.

31. Gierl, "Compilation," 93–101.

32. [Sprengel], "Vorbericht": "The rich hypotheses of Georg Erh. Hamberger hardly deserve to be mentioned; and his great opponent, the most learned collector of all times and of all peoples, Albrecht von Haller, used only his forerunners without adding much new."

33. Herbert Jaumann, "Iatrophilologia. *Medicus philologus* and analoge Konzepte in der frühen Neuzeit," in *Philologie und Erkenntnis. Beiträge zu Begriff und Problem frühneuzeitlicher Philologie,* Ralf Häfner, ed. (Tübingen: Niemeyer, 2001), 151–176.

34. Kurt Sprengel, *Versuch einer pragmatischen Geschichte der Artzneikunde,* vol. 1 (Halle: Johann Jacob Gebauer, 1792), 17–18.

35. Liane Zeil, "Zur Neuorganisation der Berliner Wissenschaft im Rahmen der preußischen Reformen," *Jahrbuch für Geschichte* 35 (1987): 201–235. Gert Schubring, "Spezialschulmodell versus Universitätsmodell: Die Institutionalisierung von Forschung," in *'Einsamkeit und Freiheit' neu besichtigt: Universitätsreformen und Disziplinenbildung in Preussen als Modell für Wissenschaftspolitik im Europa des 19. Jahrhunderts,* G. Schubring, ed. (Stuttgart: Franz Steiner, 1991), 276–326, and *Mythos Humboldt: Vergangenheit und Zukunft der deutschen Universitäten,* Mitchell G. Ash, ed. (Vienna: Böhlau, 2001).

36. Lammel, *Klio und Hippokrates,* 299–324, 366–486.

37. Sprengel, *Versuch,* vol. 1 (1792), vol. 2 (1793), vol. 3 (1794), vol. 4 (1799), vol. 5 (1803); 2d ed. (Halle: Johann Jacob Gebauer): vols. 1–2 (1800) vols. 3–4 (1801), vol. 5 (1803); 3d ed. (Halle: Gebauersche Buchhandlung): vol. 1 (1821), vol. 2 (1823), vols. 3–4 (1827),vol. 5 (1828).

38. Lucian, *De historia conscribenda* 9, from Lukian, *Wie man Geschichte schreiben soll,* H. Homeyer, ed. (Munich: Fink, 1965), 106, 4–7; William Haylay, *An Essay on History in Three Epistles to Edward Gibbon* (London: J. Dodsley, 1780; reprint, New York: Garland, 1974), 70f., verse 215–220.

39. Vera Nünning, "'In speech an irony, in fact a fiction.' Funktionen englischer Historiographie im 18. Jahrhundert im Spannungsfeld zwischen Anspruch und Wirklichkeit," *Zeitschrift für historische Forschung* 21 (1994): 37–63.

40. Johann Christoph Adelung, *Versuch einer Geschichte der Kultur des menschlichen Geschlechts. Mit einem Anhang vermehrt,* 2d ed. (Leipzig: Christian Gottlieb Hertel, 1800; reprint, Königstein: Scriptor, 1979). See also Günter Mühlpfordt, "Der Leipziger Aufklärer Johann Christoph Adelung als Wegbereiter der Kulturgeschichtsschreibung," *Storia della storiografia* 11 (1987), 22–49.

41. Friedrich Rühs, *Entwurf einer Propädeutik des historischen Studiums* (Berlin: Realschulbuchhandlung, 1811), 252.

42. Johann August Ernesti, *De fide historica recta aestimanda disputatio* (Leipzig: Langenheim, 1746), and Johann Jacob Griesbach, *De fide historica ex ipsa natura rerum quae narrantur natura iudicanda* (Halle: Hundt, 1768), for example, were omitted in later editions.

43. Sprengel, *Versuch,* vol. 1: 16.

44. Gierl, "Compilation," 70–71.

45. Lammel, "Kurt Sprengel," 30f.

46. Rudolf Schottlaender, *Früheste Grundsätze der Wissenschaft bei den Griechen* (Berlin: Akademie-Verlag, 1964), 124–142.

47. Gudrun Kühne-Bertram, "Aspekte der Geschichte und der Bedeutungen des Begriffs 'pragmatisch' in den philosophischen Wissenschaften des ausgehenden 18. und des 19. Jahrhunderts," *Archiv für Begriffsgeschichte* 27 (1983), 158–186; also Lothar Kolmer, "C. Ch. Lichtenberg als Geschichtsschreiber: Pragmatische Geschichtsschreibung und ihre Kritik im 18. Jahrhundert," *Archiv für Kulturgeschichte* 65 (1983), 371–415, and Hans-Jürgen Pandel, "Pragmatisches Erzählen bei Kant: Zur Rehabilitierung einer historisch mißverstandenen Kategorie," in *Von der Aufklärung zum Historismus,* 133–151.

48. Markus Völkel, *"Pyrrhonismus historicus" und "fides historica": Die Entwicklung der deutschen historischen Methodologie unter dem Gesichtspunkt der historischen Skepsis* (Frankfurt: Lang, 1987).

49. Jürgen Jacobs, *Prosa der Aufklärung: Moralische Wochenschriften—Autobiographie—Satire—Roman. Kommentar zu einer Epoche* (Munich: Winkler, 1976), 40; Eckhard Kessler, "Das rhetorische Modell der Historiographie," in *Theorie der Geschichte: Beiträge zur Historik, vol. 4: Formen der Geschichtsschreibung,* Reinhart Koselleck, Heinrich Lutz and Jörn Rüsen, eds. (Munich: Deutscher Taschenbuch Verlag, 1982), 37–85.

50. Helmut Zedelmeier, *Bibliotheca universalis und Bibliotheca selecta: Das Problem der Ordnung des gelehrten Wissens in der frühen Neuzeit* (Cologne: Böhlau, 1992); Martin Gierl, "Bestandsaufnahme im gelehrten Bereich: Zur Entwicklung der 'Historia literaria' im 18. Jahrhundert," in *Denkhorizonte und Handlungsspielräume. Historische Studien für Rudolf Vierhaus zum 70. Geburtstag* (Göttingen: Wallstein, 1992), 53–80.

51. Peter K. Kapitza, *Ein bürgerlicher Krieg in der gelehrten Welt. Zur Geschichte der Querelle des Anciens et des Modernes in Deutschland* (Munich: Wilhelm Fink, 1981).

52. Geoffrey Barraclough, "Universal History," in *Approaches to History,* H. P. R. Finberg, ed. (London: Routledge & Kegan Paul, 1962), 83–109; *Universalgeschichte,* Ernst Schulin, ed. (Cologne: Kiepenheuer & Witsch, 1974).

53. See the article, heretofore attributed to Sprengel, in which direct reference is made to the events in France by Christian Gottfried Gruner, "Was ist Geschichte der Arzneykunde? Wozu nützt sie den Ärzten?," in *Almanach für Aerzte und Nichtärzte auf das Jahr 1794,* Christian Gottfried Gruner, ed. (Jena: Christian Heinrich Cuno's Erben, 1794), 3–18.

54. Broman, *Transformation,* 73–101.

55. Claudia Sedlarz, "Ruhm oder Reform? Der 'Sprachenstreit' um 1790 an der Königlichen Akademie der Wissenschaften in Berlin," in *Berliner Aufklärung. Kulturwissenschaftliche Studien,* Ursula Goldenbaum and Alexander Košenina, eds., vol. 2, (Hannover: Wehrhahn-Verlag, 2003), 245–276.

56. Andreas Kraus, "Die Geschichtswissenschaft an den deutschen Akademien des 18. Jahrhunderts," in *Historische Forschung im 18. Jahrhundert,* Karl Hammer and Jürgen Voss, eds. (Bonn: Ludwig Röhrscheid, 1976), 236–259.

57. All of this was to lead to a new order of knowledge anchored in the concept of visualization; see Francis Haskell, *Die Geschichte und ihre Bilder. Die Kunst und die Deutung der Vergangenheit* (Munich: C. H. Beck, 1995), 23–142.

58. Johann Christoph Nicolai, *Das Merkwürdigste aus der Geschichte der Medicin. Erster Theil,* (Rudolstadt, 1808), vi–vii. See also Kurt Sprengel, *Geschichte der Medicin im Auszuge,* vol. 1 (Halle, 1804), v–x. It should be remembered that he was also occupied with the foundation of his own journal of medical history "Beiträge zur Geschichte der Medicin," published in three parts (of which only the first volume was released between 1794 and 1796). Lammel, *Klio und Hippokrates,* 274–279.

59. Ludmilla Jordanova, "Writing Medical Identities, 1780–1820," in *The Third Culture: Literature and Science,* Elinor S. Shaffer, ed. (Berlin: Walter de Gruyter, 1998), 204–214. H.-U. Lammel, "Das Bad der Klio. Gelehrsamkeit und Historiographie," in *Bäder und Kuren in der Aufklärung. Medizinaldiskurs und Freizeitvergnügen,* Raingard Eßer and Thomas Fuchs, eds. (Berlin: Berliner Wissenschafts-Verlag, 2003), 129–159.

CHAPTER THREE

Charles Daremberg, His Friend Émile Littré, and Positivist Medical History

Danielle Gourevitch

In mid-nineteenth-century France, a strong positivist program developed in medical history, based on the philosophy of Auguste Comte (1795–1857)—which in turn had been inspired by that of Fourier, Saint-Simon, and Hegel. Comte's main book was *Cours de philosophie positive* (1839–1842), which brought him enthusiastic disciples, all of them convinced that the human race is perfectible and that this can be proved by historical studies. An extraordinary encounter between philosophy, philology, and medicine then occurred, and the positivist ideals informed several kinds of historical work—little teaching, some writing, and much editing of Greek and Latin texts, ancient and medieval.

Why was it so important? Every medical man was then aware that a new medicine was being born: with the French revolution, universities had been suppressed while nonacademic institutions, such as the Collège de France and the Muséum, were fostered, with their highly specialized teaching. Clinical medicine developed, with patients concentrated in the hospitals in the large cities giving students an enormous range of human material to observe. Experimental methods, specialization, figures, and facts became pivotal to the new scientific medicine. Daremberg's motto, "for history, texts; for science, facts," stressed the link between the new medicine and the history of its past, and the historical movement had a philosophical stance that encouraged collecting manuscripts, searching for unknown texts, and translating them in order to make them available to a medical audience unfamiliar with Greek and Latin.

The key figures of the movement were Auguste Comte and two medical historians, Émile Littré (1801–1881) and Charles Daremberg (1817–1872). Both had studied medicine but did not really practice; both had a passion for ancient Greek. For both, historical work expressed a positivist commitment, but, as I will show, not to the same extent: Littré was an activist, while Daremberg was far from being a militant. Littré is best known for his edition and translation of Hipppocrates,[1] but not everybody realized that he was also the author of the *Dictionnaire historique de la langue française*. Daremberg is best known for his translation of many Galenic works,[2] although many did not know that he also promoted the *Dictionnaire des antiquités grecques et romaines*.

In trying to understand the work of Daremberg and Littré, it is necessary to see them as members of a wider network of men across Europe. They shared common tastes and scientific ideals but varied widely in their personal background: a German Jew in Halle, Julius Rosenbaum; a respectable gentleman in Oxford, then in Hastings, William Alexander Greenhill; a nonconformist Neapolitan, Salvatore de Renzi; a bilingual Belgian doctor, responsible for an important hospital in Antwerp and the father of a large family, Corneille or Cornelius Broeckx; an abandoned child, born and reared in Dijon, then a doctor without a practice in Paris, Daremberg; and an ugly little man, worshipper of Comte, Littré. They did not all choose medical history for identical reasons, nor did they all write history in the same way.

I intend to focus on Daremberg because I know him well, almost as a friend; for years I studied his unpublished archives in Paris, London, and Oxford, and I organized a symposium about him and his friends.[3] I became interested in Daremberg because he exemplifies, but without excess as Littré does, one of the central currents in nineteenth-century medical history, one that even at that time competed with other visions of what medical history ought to be and what purposes it would serve. He embodied a positivist vision that had enduring consequences for medical historical understanding and for the conduct of medical history through the present day. This view pervaded even some scholars of the German romantic school of history, such as Heinrich Haeser (1811–1885); the third edition of his *Lehrbuch der Geschichte der Medizin*,[4] a perfect example of *Quellenforschung* (research based on primary sources), was deeply influenced by the recent experimental physiology and by French positivism, and, therefore was dedicated to Daremberg, Henschel, and De Renzi.[5]

Daremberg: Doctor, Librarian, Manuscript Hunter, and Translator

Charles Daremberg was born in secret in 1817, in Dijon, in the house of a midwife who declared him Charles Victor, that is to say, without a family name; not until 1865, by court trial, was he officially awarded the name that he already used, Daremberg.[6] As a schoolboy he was trained in the *petit séminaire*[7] Saint-Bernard in Plombières-lès-Dijon and as a young student at the school of medicine in Dijon. He then went to Paris where he took his medical degree in 1841, with a very unusual thesis dealing already with medical history, "Exposition des connaissances de Galien sur l'anatomie, la physiologie et la pathologie du système nerveux."[8] From that time onward, nothing would prevent him from dedicating his life to a better knowledge of ancient medicine. As soon as he could, he started missions collecting manuscripts all over Europe, the first in Rome in 1848–1849, then others in Italy, England, Germany, and Spain, paid or not paid by the Ministry de l'instruction publique et des cultes.

After receiving his medical degree, to earn his living, Daremberg took a position as hired doctor to a relief committee in a district of Paris. In addition, he was appointed as a librarian of the French (then Royal) Academy of Medicine and, from 1850 until his death, of the Bibliothèque Mazarine, housed on the premises,[9] with a salary of 2,000 (later 2,200) francs.[10] This was not enough to make daily life easy—although he was under the moral guidance of Dr. Descuret, a specialist of the human passions,[11] and under the protection of the rich and powerful de Broglie family[12]—since he had to support a wife and two children and he spent a lot of money on books. He was thus compelled to do extra work.

We do not know how or why Daremberg chose Galen on anatomy, physiology, and pathology of the nervous system as the subject of his medical dissertation.[13] He knew it would be a difficult undertaking, mainly because Galen was never clear about which animal he was actually dissecting and because he had a dangerous tendency to extrapolate his discoveries and conclusions from animals to humans. Daremberg worked terribly hard on his research, adding practical investigations to his literary knowledge, supported and helped at the Muséum d'histoire naturelle by such famous scientists as Blainville, Gratiolet, Duvernoy, Serres, Isidore Geoffroy-Saint-Hilaire, Emmanuel Rousseau, Jacquart, and Rouget. He was proud of this innovative study that was to decide of his whole life, although the professional press spoke very little of it, which was quite normal for a simple medical thesis.

His task at the Mazarine was not very demanding but neither was it a sinecure. His duties included being present once a month at the *consistoire,* a meeting of the staff. One day a week he had to spend in the reading room to help the readers, put a stop to the noisy talking between readers and attendants, and prevent the thefts and acts of vandalism.[14] He also took care of loans, a serious business, for the librarian was personally responsible if a volume was not returned. It was also embarrassing when he had to ask for books back from a former minister (as Abel Villemain) or members of the Académie Française (as Victor Cousin or Joseph Giraud).[15] Daremberg also prepared catalogs there: of the 600 books presented by Ampère (November 1851), of the manuscripts (January 1858), and of the incunabula (November 1858). He suggested purchases, including works by friends— "le Pline de M. Littré" (January 1851) and "la collection dite du Cardinal Maï" (February 1855), but he opposed the purchase of Voltaire's works (January 1860). Daremberg decided which books should be bound and which manuscripts and incunabula repaired, and he performed such sundry tasks as getting rid of a stoker who too often was drunk and helping to restore the whole system after the end of the siege of Paris (March 1871).[16] This left not much time for personal research, although his friends underestimated the amount of his work and sometimes envied his supposed freedom.

Daremberg had a real passion for manuscripts: all his life he enjoyed rummaging in libraries. Since he could not afford to buy them, nothing would distract him once he had decided to examine one, and when he could not collate it himself he would spend much money on copyists. His relationship with a private collector, Thomas Phillipps, baronet, illustrates how demanding and insolently insisting Daremberg could be. Part of their correspondence is kept at the Bodleian Library in Oxford. In August 1847, Daremberg wrote he would like to spend a few days in Middlehill with Phillipps in order to collate two manuscripts, one of Oribasius and one of Rufus.[17] He later changed the date, seeming oblivious to the trouble this caused his host. Years later, in 1861, he would send Bussemaker for the *Hippiatrica* and insist dryly that this had nothing to do with the French government and that he himself was paying the Dutchman for it.[18] When he eventually thanked the baronet, he did it so clumsily that his beautiful presents were not really welcomed. He accumulated numerous copies; although he used only a few, he cherished them all. When his private library was sold, Madame Daremberg, according to her late husband's will, would add to the books ninety bound volumes of copies, notices, and excerpts of unpublished ancient manuscripts he had collected in libraries all over Europe.[19] Some still await study.

One might ask why he had such a stubborn passion for collecting manuscripts and copying them, and why such an activity was so important to his vision of how best to advance medical history. The reason is that for a historian of literary texts, manuscripts are the facts so dear to positivists and can never be too numerous. But many of his fellow doctors were no longer able to read the original Greek and Latin texts; therefore, collecting and editing texts was not enough, and Daremberg thought that the good historian must also translate them.

Daremberg soon realized that his thesis was not so wonderful after all. In 1854, in the preface of the first volume of the *Oeuvres médicales et philosophiques de Galien,* he explained that it was difficult to understand ancient scientific texts, not to speak of translating them. He was fully aware that even in the Muséum he had sometimes been fascinated by mere words without understanding them and was therefore unable to write a clear translation.

In 1844, Daremberg and William Alexander Greenhill had gotten in touch, owing to Littré. Their friendship began when Daremberg presented Greenhill with a copy of his M.D. thesis, and a rich and frequent correspondence ensued, Daremberg writing in French, Greenhill in English.[20] They first met in 1846 and got along very well. A doctor of Trinity College (1842), Greenhill was supposed to practice at the Radcliff Infirmary, but he spent more time sitting at the Bodleian Library until he moved to Hastings in 1851. About that time Daremberg was undertaking the ambitious project of a Bibliothèque des médecins Grecs et Latins, a series of editions and translations of Greek and Latin medical texts, long before the Corpus Medicorum Graecorum and the Corpus Medicorum Latinorum were even considered.

At the time there was nothing in France resembling the present Centre national de la recherche scientifique, and therefore it was the duty of the academies and of the Ministry "de l'instruction publique et des cultes" to promote ambitious cultural projects. Some difficulties might arise between academicians, civil servants, and envoys sent on official missions to libraries for the purpose of scouting for new manuscripts, but in the main they would try to work seriously together, and an earnest scholar would receive money and help. The ministry asked for advice from the Royal Academy of Medicine and the Académie des inscriptions et belles lettres; both insisted that better manuscripts be used, that a better text be edited, and that a new translation be prepared but said not a word about the language to be chosen for that translation, probably because to them it went without saying that the language should be French.

In March 1847 Daremberg wrote to his British friend: "I sometimes thought of a Latin translation, but there are many disadvantages in so doing. And those who will use our Bibliothèque speak French and understand Greek. Also, translations into a living language are safer; difficult passages are more adequately taken care of than in a dead language, especially in Latin, which permits too easily to slip a word under another."

Yet the report prepared by the Academy of Medicine later that year was in favor of Latin, "still a universal language, or at least a language all scholars use." In November, the "prospectus," that is to say an attractive specimen, was printed and distributed, with a Latin translation of a few pages of Oribasius, by Bussemaker, the Dutch friend and employee.[21] Why such a change in the editorial policy? According to the prospectus, such a collection was aimed at scholars, not only of France, but of all learned nations. French was wider spread than any other living language, but Latin was more universal. Therefore medical practitioners and philologists who were supposed to partake in the enterprise and who were asked their choice all agreed that Greek texts should be translated into Latin, at least so the prospectus said.[22]

Masson and Baillière were the two publishing houses in charge of the project, and no better choice was possible: Victor Masson had learned the job with Louis Hachette, had just started a Bibliothèque Polytechnique, and was soon to create the *Gazette hebdomadaire de médecine et de chirurgie*. Jean-Baptiste Baillière had published scientific books since 1818; he was the *libraire de l'Académie de médecine, rue Hautefeuille, no. 19;* he would have a branch-shop in London, New York, Madrid, and Melbourne. They both realized that Latin books would be hard to sell and consented to publish the collection only if the translation were in French. When Daremberg informed Greenhill of this decision in a letter dated 17 March 1848, today in the French Academy of Medicine, Greenhill became absolutely frantic and refused to contribute further to the project. But both academies and the ministry were against Latin. "I maintain I am in favor of translations into living modern languages," Daremberg noted in the preface when in 1851 the first volume of Oribasius was finally published, "for only such languages prevent the translator from any compromise with the original text and are really helpful when a passage is difficult and perplexing."

Littré was of the same opinion when in 1839 he started his *Hippocrate* for Baillière, following Johann Friedrich Grimm (1737–1821) who had published a German *Hippocrates Werke; aus dem Griechischen übersetzt mit Erläuterungen*, in four volumes, in Altenburg, 1781–1791. In the preface to his first volume, Littré quoted his colleague who reminded the reader that a Latin translation was twice as difficult, being a translation into a dead language. In such cases the translator would inevitably make mistakes and impose his own idiomatic expressions so that the reader in order to understand his work ideally should know both common neo-Latin and the translator's native tongue![23]

The Bibliothèque des médecins had but a few volumes published, and many great projects contemplated by Daremberg were never completed at all, or were completed only after his death, partly because of his poor health, partly because of the death of a collaborator such as Bussemaker, partly because of the turbulent and sometimes dangerous political context,[24] but mostly because Daremberg started too many ambitious things at a time and extorted promises that could not be fulfilled.

Typical was the endeavor to produce for the Sydenham Society a translation of Choulant's *Handbuch der Bücherkunde für die ältere Medizin* (Leipzig: Voss, 1828). This learned society was an association of medical practitioners who wanted to promote the translation into English of standard books written in ancient or modern foreign languages, and the reprint of English books no longer available. The members quarreled about the books to choose and about the price to pay, and the society came to a quick end. But they left a legacy in their first publication, the translation of three essays by Justus Friedrich Hecker (1795–1850) entitled *The Epidemics of the Middle Ages* (1844); this German doctor was a follower of Schelling (1775–1864), but he mitigated the romantic enthusiasm and idealism of the school by a good amount of serious philology and precise analysis of sources.

Choulant's *Handbuch* had been a great success, and a second enlarged edition had appeared in 1841.[25] In the meantime, Rosenbaum had prepared *Additamenta ad Choulanti bibliothecam medico-historicam* (Leipzig: Engelmann,1842, then 1847).[26] Greenhill and Daremberg contemplated making additions to what the German scholar had already published, adding new chapters for the Middle Ages and Renaissance and translating the whole into English.[27] In December 1851, James Risdon Bennett—one of the founders of the Sydenham Society in 1843 and its secretary—wrote to Daremberg: "The Council of the Sydenham Society have directed me to say that they will be glad if you will undertake to complete the edition of the Choulant's book which Dr. Greenhill has commenced. They wish that you should fix a short account of each book, as brief as possible and that you should omit all particulars of the several editions except . . . title in full, size and no. of volumes, date, place of publication. . . . For this service the Council desire to offer you the sum of 1,500 frs."[28]

Under pretenses from Greenhill and Daremberg of bad health, overwork, and revolution, the society received nothing, and Bennett wrote Daremberg in November 1852: "The Council are anxious to know when they might expect to receive any portion of your ms. We doubt not that you are working hard and that the labor is considerable." Then he told Daremberg about the publications already in progress, by contemporary specialists such as Kölliker, Unzer, Prochashka, Rokitansky, and Romberg. The end was getting closer and closer; unable to accomplish its original intentions, the society would soon stop working,[29] and Greenhill and Daremberg did not keep to their promises. By December 1855, Bennett again wrote to Daremberg, "After this long delay and in the present circumstances of the Society, the Council find it difficult to contemplate the publication of such a work as the Bibliography."[30]

Littré the Mentor

Émile Littré had a very unusual career, which suffered no interruption under three political regimes.[31] Even during the Second Empire, he was never really persecuted by the police, although he was in political disfavor and was never invited with the rest of the intelligentsia to the castle of Compiègne, where the court resided during the imperial holidays.

Littré had followed a full medical course; however, he had not completed a thesis, and therefore was not formally a doctor and never officially practiced.[32] He earned his living as a journalist for the *Gazette de Paris* and the *Expérience,* which he had founded with Jean Dezeimeris (1799–1852), and he wrote for several medical dictionaries. He always knew he would use his medical knowledge for the sake of ancient and medieval medicine, and both his father Michel-François, and his brother Barthélemy, encouraged his intellectual ambition. His work would be useful, it would help the practicing doctors of the time who could not read Greek, it would make the invaluable Hippocratic experience available to them. To do this, Littré had to fulfill his threefold program: read all the manuscripts to fix a new text; translate that text carefully, accounting for semantic change; and, whenever possible, give a medical interpretation, enlightened by modern scientific discoveries.

Littré's image of Hippocrates as the father of Western medical thought, at the head of the school of Cos, and as a master of scientific observation proved appealing to several generations of philologists, medical historians, and physicians. He became a member of the Académie des Inscriptions et Belles Lettres in 1839, under King Louis-Philippe, on the merits of the first volume of his edition of Hippocrates.[33] The members of the academy had been impressed immediately by his punctiliousness in collecting and comparing all the manuscripts available, the accuracy of his translation, and his knowledge of the new sciences.[34] He would also become a member of the Académie Française in 1873, at the beginning of the Third Republic, for his monumental *Dictionnaire de la langue française,* which was to be used widely, in France and abroad, for at least a century.[35] Although positivism had never had any real political impact, Littré had become a living symbol to politicians who wanted to assert that their politics were based on science and directed toward progress.

Littré was an ascetic scholar,[36] a passionate philosopher, and a devoted friend, and he soon became a model for Daremberg.[37] He always helped his younger colleague, by reading his proofs, for example, and by putting him in touch with colleagues abroad. They often met in Paris and at Le Mesnil-le-Roy, in the suburb, where both had small country houses, they shared erudite talks, wrote critiques of each other's books and studies, and exchanged advice on gardening.[38] Both today might be considered politically left wing, although they differed in their level of commitment and creed: according to the present French standards, Daremberg would be called a *chrétien de gauche* and Littré a *gauchiste*.

Littré was a fervid proponent of Auguste Comte's doctrine, a fanatic positivist, the author of many books about his idol, and of a "little green book" (which came before a more famous red one), *Conservation, révolution, et positivisme,* published in 1852 in Paris. This was not actually a new book but the reprint of a few pamphlets, a project instigated by a M. Besnard, the mayor of the elegant Villers-Cotterets and a zealous positivist who paid for its publication. In the preface Littré proclaimed his thankfulness to the positivist philosophy, the inward consistency of which prevented him from aberrations and errors and the beneficial discipline of which kept him from losing heart or flaring up. Positivism, he then explained, became very successful after the nationalist European revolutions in 1848, as more and more people wished to ruin the previous order and to substitute a new one for it: no more revolution, no vote for all, but peace in Europe, disarmament, solidarity, and joint responsibility. Spiritual development and expression should be free: the state should not pay for the church and university, which delivered instruction that prevented a real reshaping of beliefs and customs; it should take care only of elementary instruction and then teach mathematics, astronomy, physics, and chemistry, without any theology or metaphysics. As Littré concluded, "The door is now wide open in front of us for great enterprises, endless labors, captivating and engrossing ideas. An unlimited career spreads out for us. This is a dogma, a political regime, and a cult we must expand, propagate, prove, and enlighten. . . . Among the workers who will be many, happy those whose name will become famous and who will receive gratefulness and acknowledgement just like the glorious founders of Christianity."[39]

Compared with such a utopian socialist, a political idealist like Daremberg was rather mellow, and sometimes Charles thought Émile was wasting his time, propagating positivist ideas or writing a biography of Auguste Comte just to please Comte's widow. By taste, by education, and in his desperate search for his parents, Daremberg needed respectability and peace; he was kind and generous, but he shared completely bourgeois values, developing a Christian sentimentalism, and a social and political conservatism.[40] He was very unhappy during the revolution of 1848 and thought he might flee to England. He never rebelled against any of the de Broglie's decisions and agreed, for example, that his wife should not be invited with him and his sons at their castle in Normandy for the holidays!

Daremberg loved Greenhill, a reserved and pious man with high moral standards, preoccupied with the welfare of the poor classes; he enjoyed his "scientific lectures" and his "talks about religion." Such "an excellent and respectable family" had "printed in [his] mind and heart indelible memories."[41] He even thought Greenhill might be the godfather of one of his sons and was sorry this was not possible, for theological reasons.[42] He liked the company of religious men and was on very friendly terms with Dom Pitra (1812–1889), the famous medievalist in Solesmes abbey and later a cardinal in Rome.

Upon Daremberg's death there was the immediate problem of disposing of his precious library, so large that Madame Daremberg was allowed to keep their apartment at the Mazarine until the books could find a new location. She and her two sons had no interest in the valuable collection and decided to sell it, in England if not in France. At this time, however, an overview of the collection was published in *L'Union médicale* by Félix-Alphonse Pauly,[43] who had been secretary to Daremberg, then curator at the Bibliothèque nationale and was to publish a *Bibliographie historique des sciences médicales,* for which a preface by Daremberg had been expected. In 1872–1873, Littré and Baillière senior were called as experts to settle the matter, and Littré explained that the collection comprised two different parts: one for letters, history, and archaeology, of no special interest; the other, a very precious gathering of books of medicine and history of medicine, probably unrivalled. All items were perfectly preserved and there were several incunabula. Littré insisted that the collection must not be broken up, but instead be deposited intact at a public library as a whole, and suggested the Academy of Medicine, for this place of great learning had but a sparse and inadequate library.

After some discussion, an agreement was signed; the books were purchased by the Academy of Medicine, and 45,000 francs were to be paid by the ministry and the academy in installments over three years. Two manuscript catalogs were prepared under the supervision of Georges, the beloved son and doctor,[44] but before an official one was established by the academy (11,981 items, from number 1,633 to 13,614), the books were kept in a special room at the Mazarine. Books and catalogs are still available at the academy, many marked with a special identification number with the letter *D*; a few are stamped with Daremberg's blue ex libris, which he had printed after he was elected as a professor at the faculty of medicine.

Daremberg in the Eyes of His Contemporaries

Daremberg worked extremely hard for his lessons at the Collège de France and at the faculty of medicine. In an approach quite characteristic of positivistic fact-finding, he accumulated documents and was attentive to details that might have been mere trifles for his audience, but which were precious facts to him. He entirely wrote his lessons, which were always strictly and obviously organized. The papers were soon published separately; then they were reused in his *Histoire des sciences médicales,* his magnum opus published at the very end of his life, when he still believed faithfully in the superiority of observation and experience.[45] Yet Daremberg was not a good teacher: his voice was low and hoarse, and his heart disease (probably angina pectoris) made him choke. At the beginning of his teaching in 1846 at the Collège de France,[46] he congratulated himself on the large audience and the applause they gave him.[47] This was not to last, however; students eventually wrote letters to him complaining that they could not hear him well, but he did not really care. As for the style of his lessons, it was judged too scholarly and clumsy. Regardless, his audience at the Collège was not real students but colleagues and amateurs. At the faculty his teaching was too short for us to know if it was really efficient. He was rather an *homme de cabinet,* and a note written at his death and kept by Baillière recorded that his personal tastes had committed him to a contemplative medicine, that he knew very little of the joys and sorrows of a practitioner, and that he never fought for a position in a hospital.[48] According to Achille Chéreau, Daremberg, after his death, was considered to have been more at ease and happy in his study than in an amphitheatre in front of a large audience, better suited to research and to translating and interpreting ancient medicine than to teaching.[49]

Among his many friends all over Europe, praise for his erudite expertise was almost compulsory; among his foes, criticism could be very unkind, whether it was adequate or inappropriate. In 1852, in the *Journal des savants,* after a long explanation about Oribasius and his work, Littré rejoiced because Daremberg and Bussemaker dedicated their book to him, which meant they agreed with the interpretative method he used systematically and earnestly in dealing with ancient medical texts. That is to say, they believed that the interpretation of such texts must be informed by the discoveries of modern anatomy, physiology, and pathology, but that, at the same time, a good knowledge of ancient medicine should lead to a better understanding of contemporary science. Littré approved of the painstaking care the authors had taken to collate all the manuscripts, noting each variant. He approved of their accurate and precise translation, especially of the technical terms for plants, animals, and ancient devices, although he disagreed with some details. Some emendations were wrong, for example, as were some translations; *stomachos* is not the "mouth of the stomach," but the stomach itself.[50]

In 1869 Daremberg dedicated his *État de la médecine entre Homère et Hippocrate* to his faithful Belgian friend Corneille, or Cornelius, Broeckx[51] in the following terms: "On many an occasion you made it clear yourself, that one can write history only by going to the original texts; I belong to that positivist school, and printing your name on the front-page of this book, I pay a tribute to the writer who follows the true historical method with as much erudition as persistence." This "Daremberg of Belgium," as Wachter called him,[52] often wrote reviews in the *Annales de la Société de médecine d'Anvers;* in 1854 he reviewed *Oeuvres d'Oribase* and judged the book perfectly appealing for philologists and useful for medical practitioners, as far as evacuations and bloodletting were concerned (although less so for purgatives, vomitives, and the theory of climates and places).[53] It traced the progress of science, scrupulously true to positivist ideals, without any vain and idle erudition.[54]

Broeckx mentioned the names of Drs. François Aran and Charles Robin, who had helped provide some interesting technical notes. Both were on the same ideological side as Daremberg. Robin was indeed the most competent observer and propagandist of positivism after Littré,[55] and they rewrote together the so-called Nysten's dictionary of medicine. "Nysten-Littré-Robin's," or just "Littré-Robin's," became the compulsory vade mecum of any materialistic doctor of the time and the aversion of the catholic party under the guidance of Cardinal de Bonnechose, who attacked it in the Senate on the 20 May 1868, *quod iuventutem corrumperet*.[56] Even in 1878 the preface to the eighteenth edition recalled Littré's method and philosophy.

An Austrian doctor who specialized in the history of Persian medicine, Franz Romeo Seligmann, for his part opposed Daremberg frequently and sometimes violently. He did not like him and disapproved of the way he behaved during his campaign for the chair of history of medicine at the faculty. His 1870–1871 reviews of *État de la médecine entre Homère et Hippocrate*[57] and *Histoire des sciences médicales* were both extremely critical and often unfair, using sometimes quite unpleasant words. As for the first book, Seligmann acknowledged Daremberg's diligence and persistence, but he should beware, he wrote, of fulsome commonplaces and grandiloquent tirades that might make him appear an obnoxious prig. Seligmann recognized the *Histoire* as the culmination of a life of hard work, Daremberg being often overtasked, not without detriment to his health.[58] Daremberg agreed with Littré that medicine should take its history into account, lest the discipline be lowered to a craft. Seligmann confessed that he approved of some excellent pages, rejoiced that the book had been available in Germany before the war started, and drew attention to the book's motto: "for history, texts; for science, facts," which indeed was the way Daremberg understood positivism. He thought, however, that even a positivist historian should not reduce medical history to a catalog of case histories. He paid tribute to Daremberg's positivistic researches but did not like the way they were put together. Sticking strictly to facts, opposing any theory or unifying principle, Daremberg, far from being an apostle of modern truth and new science, echoed the ancient times he tried to ridicule. Nor did Seligmann approve of Daremberg's bibliography. He ought to know, for example, Paolo Morello's *Istoria filosofica della medicina in Italia*. According to Seligmann, Daremberg understood nothing of Paracelsus, whom he considered a mad quack, or of Wunderlich, a mystic;[59] he felt no sympathy for Van Helmont, or even for Sprengel, although he had written a *pragmatische* history.[60]

Daremberg was so upset by such reproaches that he thought of answering Seligmann, as is attested in a letter from Haeser, kindly written in French, though a very clumsy French.[61] "I knew the article by Seligmann: his remarks about your book had embarrassed me, but I could not judge them because I have not read your book yet," Haeser told Daremberg. "Many of his assertions seem to me untrue and unfair. And since you ask me, I find your letter (which I am sending back without communicating it to anybody) so dignified and so discreet, that I think it should be sent to Mr. S.,[62] whom I know personally as you do, and whom I believe a good man. . . . I myself tried vainly to find Morello in Italy;[63] nobody has heard about him."

Was Seligmann right? About the details, he was mean, as mean indeed as Daremberg was sometimes. As for the bibliography, he was unjust, for Daremberg missed very little. Haeser, in contrast to Seligmann, was among those who wistfully anticipated the publication of Daremberg's *Histoire*. He was enthusiastic and contemplated a German translation even before its publication.[64] He even started reading it on the way up Righi Mountain during his holidays.[65] This is all the more striking because he did not share Daremberg's philosophical views, still influenced as he was by the Hegelian system. Haeser sometimes felt jealous of Daremberg, who was not a dean of the faculty and did not have to spend his time lecturing about general pathology and therapeutics. He was always bitter, confronted with apathy from the state,[66] although his *Lehrbuch der Geschichte der Medicin,* first published in 1845, was already a standard handbook.

Positivist History in Contemporary Context

The positivists were engrossed in the search for positivistic knowledge based on natural phenomena whose properties and relations could be verified by the empirical sciences. Although Daremberg lived in the romantic period, he had nothing to do with philhellenism, and he never contemplated fighting the Turks for Greek freedom or even visiting Greece. For him "positivistic" was the opposite of "romantic": no grandiose view, but precise and accurate details in order to offer good grounds for being believed.

We have a good example of his keen attention to details in the famous *Dictionnaire encyclopédique des sciences médicales,* published under the direction of Amédée Dechambre in one hundred volumes.[67] Daremberg (who had already started his *Dictionnaire des antiquités*) was in charge of the entries concerning Greek, Latin, Arab, and medieval doctors. Because Greenhill had done more or less the same job for the *Dictionary of Greek and Roman Biography and Mythology,*[68] Daremberg thought they might collaborate, with some articles being merely translated and signed *G.,* others rewritten and signed *D. G.,* and still others entirely new and signed *D.* Except for the translation itself, which was often rough,[69] the French articles were always more careful (for instance "Archagathos," translated and modified), sometimes richer and with references added ("Antyllus," co-signed), and sometimes entirely new (when new studies had completely changed the problems, as in "Ammonius" quoted by Celsus, a new edition of whose works had just been published).[70]

Further, Daremberg aspired to reform modern medical practice and ideology, and here he demonstrated a pragmatic view of the history of medicine, at a time when doctors, as far as philosophy was concerned, were adrift because so many changes had occurred over a short period. Daremberg was convinced a better knowledge of ancient medicine was useful for practitioners-to-be, and so was Broeckx. They were at the same time right and wrong: wrong because it is nonsense to look for a correct prescription in Galen or Van Helmont; right because in a time of intellectual and moral crisis, the knowledge of the past helps people maintain a certain distance from prevailing ideas and prevents them from behaving like a sheep in herd.

This period witnessed major epistemological changes, with medicine being transformed from humanist learning to a laboratory science, but it also saw many strange stages, such as Broussais's system. When Littré started his edition of Hippocrates, he thought he was working for his standard medical colleagues; by the time he finished it, he knew that in fact his contribution was chiefly for his fellow philologists.[71] According to Broeckx, doctors abandoned history for novelty; according to Daremberg, doctors would necessarily go back to history because there was too much novelty. The pathoanatomical viewpoint, Broeckx wisely and humorously noted when he reviewed Daremberg's *Oribasius,* had been raging, and many believed postmortem examination would discover not only the cause and seat of diseases but also the best way to cure them. Yet, a scalpel is not a deus ex machina. Moreover, a new medical system arose every day, and innovators maintained that no real medicine ever existed before them, calling themselves "homeopaths" (those who cured *similia similibus*), "hydrosudopaths" (those who dipped their patient in water) and "ontologophobes" (those who detested the ontological theory of disease, and used only antiphlogistics).

Around the middle of the nineteenth century, when modern scientific medicine was born, many a doctor felt that a scientific revolution had occurred, a "crisis" in the Hippocratic meaning of the word, which might be dangerous. Therefore, they maintained, letters and history should be added to the *curriculum studiorum* for medical students. Jean-Baptiste Parchappe de Vinay was an alienist who had taken care of the Maison de Saint-Yon for mental patients.[72] He was much involved in the organization of lunatic asylums, which, after passage of the 1838 law, were to be built in each French département, and wrote a book about it.[73] In 1858, he insisted on the usefulness, even the necessity, of a chair of medical history at the Collège de France precisely because it was the temple of the new science cultivated by such new priests as René Laennec, François Magendie, Claude Bernard, and Antoine Portal. History of medicine, he told the minister, should tell them the "right" doctrines and prevent them from new errors. But the minister would not listen to Parchappe.[74] Nor did Daremberg live apart from the workaday world. He knew what was going on at the Muséum d'histoire naturelle, at the Faculté de médecine and at the Collège de France. He was a friend of François Magendie, for whom he substituted for a few months at the college, and of Claude Bernard, whom he often visited in his private laboratory and tried, without success, to convert to his faith in the history of science.[75] His positivist creed was, in the way Littré had once written, enthusiastic for and never tired of great enterprises, endless labors, captivating and engrossing ideas. His bad health often impeded his legitimate ambitions, but he was happy to be at last a full professor at the faculty of medicine in Paris,[76] although he had lectured very little before death caught him in October 1872.[77]

Daremberg Today

For classicists,[78] Daremberg is today the well-known "author" of the *Dictionnaire des antiquités grecques et romaines* (alias "Daremberg et Saglio"),[79] with which Louis Hachette had entrusted him; in fact, he carefully promoted the work but wrote little of it.[80] This is especially striking because he was not interested at all in the historical and archaeological context of his beloved ancient and medieval doctors. He seemed to know nothing of the excavations then going on in the Halatte Forest, near Senlis, for example, or in Châtillon-sur-Seine, two Gallo-Roman sites where very important series of medical votive offerings were discovered. When he saw Christian votive offerings in an Italian church, he found them ugly and disgusting and made no medical commentary.[81] When he visited the Archaeological Museum in Naples, it was to criticize the royal government, and he did not really try to inspect the numerous surgical instruments collected there, although he had previously written an "Analyse de la notice de M. Vulpes sur les instruments de chirurgie trouvés à Herculanum et à Pompei."[82] Moreover in Châtillon, several oculists' stamps were already sheltered,[83] but Daremberg, as far as I know, never tried to see them, although he reviewed Dr. Sichel's *Nouveau recueil*,[84] the texts again being more attractive to him than the objects.

With all his delays, failures, and weaknesses, Daremberg was neither a genius nor a hero, but he was a virtuous man and a steady scholar. We are indebted to him for heaps of documents on medical history, which he had no time to use but carefully kept, and for a unique library in Paris, now open to the public in the Académie de médecine, on the rue Bonaparte. We are indebted to him for tidy researches on manuscripts; a huge reservoir of facts, sometimes obscured in his lengthy *Histoire* and his numerous notes in journals and papers;[85] and good, sometimes excellent, translations, such as that of Galen, not into Latin but into a living language. Because his Bibliothèque was to be "a memorial to the deserted manes of past medical princes,"[86] it behooves us to continue on that path. Above all, we are indebted to him for a convincing renewal of a humanistic faith, which he called positivism. An enduring lesson of Daremberg's approach may be that he believed that the best medical history should be philological, for this method teaches the student to read a text well and to understand the past, to know things deeply, to make distinctions and nuances, to reject the politically correct. In the end, the didactic project of such a history is to teach the medical student and the young doctor how to listen keenly to the patient.

NOTES

1. *Oeuvres complètes d'Hippocrate, tradution nouvelle avec le texte grec en regard,* 10 vols. (Paris: Baillière, 1839–1861).

2. *Oeuvres anatomiques, physiologiques et médicales de Galien, traduites sur les textes imprimés et manuscrits, accompagnées de sommaires, de notes, précédées d'une introduction ou étude biographique, littéraire et scientifique sur Galien,* 2 vols. (Paris: Baillière, 1854–1856).

3. See *Médecins érudits, de Coray à Sigerist,* Danielle Gourevitch, ed. (Paris: de Boccard, 1995), with an important bibliography, 197–212.

4. *Lehrbuch der Geschichte der Medicin und der epidemischen Krankheiten,* 3 vols. (Jena: Dufft, 1875–1882). The first edition had been published in 1845 as *Lehrbuch der Geschichte der Medicin und der Volkskrankheiten* (Jena: Muke, 1845).

5. Owsei Temkin and C. Lilian Temkin, "Wunderlich versus Haeser: A Controversy over Medical History," *Bulletin of the History of Medicine* 32 (1958): 97–104.

6. Although he sometimes spelt his name D'Aremberg, he had no official link with the famous family d'Arenberg.

7. A petit séminaire was a Roman Catholic secondary school, essentially but not exclusively training boys who wanted to go to the *grand séminaire* in order to take the holy orders.

8. "An outline of Galen's knowledge on anatomy, physiology, and pathology of the nervous system," dated 21 August 1841.

9. It was a rather small apartment for a family of four people, which he would nonetheless manage to enlarge for his books (without any right to do so).

10. For the sake of comparison, the doorkeeper earned 720 francs.

11. Some details are in *La mission de Charles Daremberg en Italie (1849–1850),* Mémoires et documents sur Rome et l'Italie méridionale, n.s. 5, Danielle Gourevitch, ed. (Naples: Centre Jean Bérard, 1994). Jean Baptiste Félix Descuret (1795–1872) had written *Médecine des passions, ou Les passions considérés dans leurs rapports avec les maladies, les lois et la religion* (Paris: Béchet jne & Labé, 1841); it was enlarged and reprinted at least four times, in France and in Belgium, and translated into Italian: *Medicina delle passioni: ovvero, Le passioni considerate nelle relazioni colla medicina, colle leggi e colla religione,* F. Zappert, trans., 4th ed. (Milan: Oliva, 1859).

12. The association with the de Broglies means certainly that his father or mother was of good stock.

13. I was amazed not to find this study quoted in a recent article by Julius Rocca, "Galen and Greek Neuroscience (Notes towards a Preliminary Survey)," *Early Science and Medicine* 3 (1998): 216–240.

14. When he was in charge of the Wednesday sessions, he took advantage of it to arrange for meeting his friends and collaborators that day.

15. The problem was the more complicated because there was a direct access from the institute to the Mazarine.

16. During that period, Daremberg and his son worked for a temporary hospital or *ambulance* and were very helpful; Charles then received a diploma from the Red Cross.

17. Daremberg to Phillipps, Oxford, 13 August 1847, Bodleian Library, University of Oxford, c 498 (120–124).

18. Daremberg to Phillipps, Oxford, n.d., Bodleian Library, b 150 (98–99).

19. See Archives nationales de France, Paris, F/17/3681.

20. Greenhill often complained about Daremberg's horrible handwriting.

21. See his "Notice nécrologique d'U. C. Bussemaker," *Journal des débats,* 29 January 1865.

22. N. N., *Prospectus* (Paris: Masson, 1847), 35–36.

23. Danielle Gourevitch, "La traduction des textes scientifiques grecs; la position de Daremberg et sa controverse avec Greenhill," *Bulletin de la Société des antiquaires de France* (1994): 296–307.

24. Danielle Gourevitch, "Un médecin antiquaire dans la tourmente de 1848 à Paris: Charles Daremberg," *Bulletin de la Société des antiquaires de France* (1992): 30–45, and "Une catastrophe dans les relations entre les érudits français et allemands: la guerre de 1870. L'exemple de Daremberg et de son ami Haeser," in Gourevitch *Médecins érudits,* 131–152.

25. The *Handbuch* is still very useful and available owing to a reprint in 1941 (Leipzig: Voss).

26. Samuel Kottek and Danielle Gourevitch, "Un correspondant (malchanceux) de Daremberg, Julius Rosenbaum (1807–1874)," in Gourevitch, *Médecins érudits,* 70–87.

27. Bennett to Daremberg, London, n.d., French Academy of Medicine, series 544, letters 48–72.

28. Bennett to Daremberg, London, 14 December 1855, French Academy of Medicine, series 544, letters 48–49.

29. Danielle Gourevitch, "Rapport," *Livret-annuaire de l'EPHE* 12 (1996–1997): 189 and 14 (1998–1999): 212.

30. Bennett to Daremberg, London, 14 December 1855, series 544, letter 49.

31. See Jean Hamburger, *Monsieur Littré* (Paris: Flammarion, 1989).

32. Roger Rullière and François Vial, "Émile Littré, étudiant en médecine," *Histoire des sciences médicales* 15 (1981): 215–220; R. Rullière, "Les études médicales d'Émile Littré," *Revue de syntèse* 103 (1982): 255–262.

33. He was also very active at the Academy of Medicine. See Jean-Charles Sournia, "Littré et l'Académie de médecine," *Bulletin de l'Académie nationale de médecine* 165 (1981): 941–947.

34. Jean-Charles Sournia, "Littré, historien de la médecine," *Revue de syntèse* 103 (1982) 263–269.

35. Alain Rey, *Littré, humaniste et les mots* (Paris: Gallimard, 1970); Jean-Charles Sournia, "Littré, lexicographe médical," *Histoire des sciences médicales* 15 (1981): 227–234.

36. Paul Ganière, "L'austère Monsieur Littré," *Bulletin de l'Académie nationale de médecine* 153 (1969): 169–174.

37. He was even his medical adviser sometimes (according to a letter of May 1864, now in private hands).

38. See Henry E. Sigerist, "Émile Littré über Charles Daremberg," *Sudhoffs Archiv für*

Geschichte der Medizin 23 (1930): 382–384, English translation in *Medical Life,* November 1932, 593–596. Pierre Paul Corsetti, "Quelques lettres inédites de Charles Daremberg à Émile Littré," in *Mélanges offerts en hommage au révérend père Rodrigue Larue,* Florent Tremblay, ed., *Cahier des études anciennes* 20 (1991): 251–279.

39. *Conservation, révolution, et positivisme,* xxxi; see also vi, xv, and xvi.

40. As an example of his generosity, at his suggestion, keepers Théophile and Maslon received a bonus for their good extra work (spring 1859), as well as the workmen who nicely repaired the building (February 1862). He also tried to help Dr. Jean Carolus; see Danielle Gourevitch and Simon Byl, "La triste vie du médecin anversois Jean Carolus, ami de C. Broeckx et de Ch. Daremberg: Vérité canonique et version intime, I. Les travaux historique," *Acta Belgica historiae medicinae* 5 (1992): 60–63, and "II . La biographie de Carolus," ibid., 139–145.

41. Greenhill to Daremberg, Oxford, 4 September 1846, French Academy of Medicine, series 541, letter 20.

42. Daremberg to Greenhill, Paris, 30 April 1850, French Academy of Medicine, series 541, letter 121.

43. *L'Union médicale* (1872): 824–826.

44. Marc, the other son, was a salesman in St. Louis in the United States.

45. See the interesting title of his *Cours sur l'histoire de la médecine et de la chirurgie. Leçon d'ouverture le 11 novembre 1871. Démonstration historique de la supériorité des méthodes d'observation et expérimentale sur les méthodes a priori* (Paris: Malteste, 1871) or in *L'Union médicale,* 3d ser., 12 (Nov. 1871): 733–740, 744–752, 769–776, 781–785.

46. Danielle Gourevitch, "Rapport," *Livret-annuaire de l'EPHE* 10 (1994–1995): 153–154 and 11 (1995–1996): 206.

47. Daremberg to Greenhill, Paris, 23 January 1846, French Academy of Medicine, series 541, letter 29.

48. The letter is now in private hands.

49. "Daremberg," in *Dictionnaire encyclopédique des sciences médicales,* vol. 25, (Paris: Masson-Asselin, 1880).

50. *Journal des savants* (August 1852): 509–522.

51. Danielle Gourevitch and Simon Byl, "Amitié et ambition: Broeckx, Daremberg et l'Académie royale de médecine de Belgique," *Acta Belgica historiae medicinae* 4 (1991): 12–19; and "Quelques aspects de la vie quotidienne du médecin anversois Cornelius Broeckx au milieu du XIXe siècle, d'après sa correspondance avec Charles Daremberg," ibid., 171–182. Simon Byl, "L'ami belge de toute une vie, Corneille Broeckx," in Gourevitch, *Médecins érudits,* 88–98.

52. P. F. de Wachter, "Notice sur la vie et les travaux de C. Broeckx," *Annales de la Société de médecine d'Anvers* (1870): 449–465, 505–520, 586–600.

53. He had written a *Discours sur l'utilité de l'histoire de la médecine* (Anvers: Heirstraeten, 1839).

54. *Annales de la Société de médecine d'Anvers* (1854): 519–526.

55. Georges Puchet, *Charles Robin (1821–1885). Sa vie et son oeuvre* (Paris: Félix Alcan, 1887).

56. Its success was enormous, even in Catholic Italy, where it was translated as *Dizionario di medicina e chirurgia, di terapeutica medico-chirurgica, farmacia, arte veterinaria e scienze affini* (Napoli: Dekten, 1879).

57. *La médecine dans Homère, ou études d'archéologie sur les médecins, l'anatomie, la phys-*

72   Traditions

iologie, la chirurgie et la médecine dans les poèmes homériques (Paris: Librairie académique, 1865) had been reviewed in *Canstatt's Jahresbericht über die Fortschritte in der gesammten Medicin in allen Ländern in Jahre 1865,* vol. 2 (Würzburg: Erlangen, 1866), as well as *La médecine, histoire et doctrines* (Paris: Didier & Cie, 1865).

58. According to a letter from Haeser to Daremberg, 9 August 1872, French Academy of Medicine, series 539, letter 166.

59. This is absolutely untrue, although "Wunderlich, like so many historians of this time, was under the influence of the philosophy of history which Schelling had helped to shape"; Owsei Temkin, "Wunderlich, Schelling, and the History of Medicine," *Gesnerus* 23 (1966): 188–195, 193.

60. *Versuch einer pragmatischen Geschichte der Arzneikunde,* 1st ed. (Halle: Gebauer, 1792–1803). Daremberg considered Sprengel's work as pragmatic but uncritical, and he felt sorry for Rosenbaum who spent too long a time trying to refurbish a decayed work, the 4th ed. (Leipzig: Gebauer, 1846); see *Revue des travaux relatifs à l'histoire et à la littérature de la médecine publiés en France et à l'étranger depuis le commencement de 1846* (Paris: Rignoux, 1848) 413–415.

61. This becomes clear from Daremberg to Haeser, Paris, 18 March 1872, French Academy of Medicine, series 540, letters 172–173.

62. This letter is presently in Cluj (Romania); see V. L. Bologa, "Une lettre de Charles Daremberg à Franz Romeo Seligmann," *Janus* (1930): 129–131.

63. I never found it either.

64. Haeser to Daremberg, n.d., French Academy of Medicine, series 539, letter 174.

65. Haeser to Daremberg, 9 August 1872, French Academy of Medicine, series 539, letter 166.

66. He had become a professor ordinarius in Breslau in 1862.

67. (Paris: Asselin, 1865–1889).

68. William Smith, *Dictionary of Greek and Roman Biography and Mythology,* 3 vols. (London: Taylor & Wallon, 1844–1849).

69. We do not know who helped him.

70. Only letter *A* was checked. See Danielle Gourevitch, "Charles Daremberg, William Alexander Greenhill et le Dictionnaire encyclopédique des sciences médicales en 100 volumes," *Histoire des sciences médicales* 26 (1992): 207–213.

71. Fernand Robert, "Littré et Hippocrate," *Histoire des sciences médicales* 15 (1981): 221–226.

72. See Lucien Deboutteville and Jean-Baptiste Parchappe de Vinay, *Notice statistique sur l'Asile des aliénés de la Seine-inférieure (Maison de Saint-Yon de Rouen), pour la période comprise entre le 11 juillet 1825 et le 31 décembre 1843* (Rouen: Péron, 1845).

73. *Des principes à suivre dans la fondation et la construction des asiles d'aliénés,* (Paris: Masson, 1853).

74. See my "Avant-propos" to the special issue of *La revue du praticien* (Nov. 2001), for its fiftieth anniversary.

75. See Danielle Gourevitch, "Claude Bernard lecteur de Galien?" *Transmission et ecdotique des textes médicaux grecs. Actes du IVe. colloque international Paris 2001,* A. Garzya and J. Jouanna, eds. (Naples: D'Auria, 2003), 173–185.

76. A rich politician, Salmon de Champotran, had given 150,000 francs and earmarked this chair for his friend Cusco; Cusco did not feel worthy and Daremberg won the position, although not without difficulties.

77. Haeser wrote a "Nekrolog" in *Berliner klinische Wochenschrift* 47 (1872): 570.

78. He was one of the founders of the Association pour l'encouragement des études Grecques en France.

79. *Dictionnaire des antiquités grecques et romaines* (Paris: Hachette, 1873–1923).

80. Danielle Gourevitch, "Un épisode de l'histoire du dictionnaire des antiquités connu sous le nom de 'Daremberg et Saglio': la publication du dictionnaire des antiquités chrétiennes de l'abbé Martigny," *Caesarodunum* 27 (1993): 79–95; "Un épisode de la longue histoire de 'Daremberg et Saglio': l'affaire Morel," *Caesarodunum* 28 (1994): 31–38; "Une lettre de Daremberg à Dom Pitra," *Bulletin de la Société des antiquaires de France* (1996): 125–128; and "Histoire du dictionnaire des antiquités chrétiennes de l'abbé Martigny, émule de J. B. de Rossi," *Acta XIII congressus internationalis archeologiae christianae,* Nenad Cami and Emilio Martin, eds. (Vatican City: Pontificio instituto di archaeologia christiana, 1998), 363–373.

81. Danielle Gourevitch, "La mission medico-historique de Daremberg et de Renan à Rome (octobre 1848–juillet 1850): le problème du rapport," *Bulletin de la Société des antiquaires de France* (1990): 232–242.

82. *L'Union médicale* 2 (1848): 126. This booklet (94 pp., ill.) by Benedetto Vulpes, *Illustrazione di tutti gli strumenti chirurgici scavati in Ercolaneo e in Pompei e che ora conservanso nel R. Museo Borbonico di Napoli; compresa in sette memorie lette all'Accademi ercolanese* (Napoli, 1847), was commented on by R. Felmann in *Antike Welt* 155 (1984): 69–79.

83. These are now typically called collyrium stamps, as they mark sticks and cakes of dry medicaments used especially for eye-diseases.

84. *Nouveau recueil de pierres sigillaires d'oculistes romains pour la plupart inédites, extrait d'une monographie inédite de ces monuments épigraphiques* (Paris: Masson, 1866), reviewed in *Revue critique d'histoire et de littérature* 32 (1867): 85–90.

85. These were mainly published in *Le Journal de l'instruction publique, Le Journal des missions, Le Journal des savants, Le Journal des débats, L'Union médicale,* and *La Gazette de Paris.*

86. Daremberg to Greenhill, Paris, 22 November 1846, French Academy of Medicine, series 541, letter 24.

CHAPTER FOUR

Bildung in a Scientific Age
Julius Pagel, Max Neuburger, and the Cultural History of Medicine

Heinz-Peter Schmiedebach

Around 1900, Julius Pagel (1851–1912) and Max Neuburger (1868–1955)—together with Karl Sudhoff (1853–1938)—laid the foundations for the modern development of medical history.[1] An increasing gap between medicine—newly defined by its grounding in scientific approaches and research methods of the modern experimental sciences—and the traditional methods and contents of medical history had become visible toward the end of the nineteenth century. During this transition, medical history lost its status as an integral part of contemporary medicine. Both Neuburger and Pagel took stock of this situation and attempted to create a medical historical discipline that, as an academically established discipline within medical faculties, needed to maintain strong ties to contemporary medicine while at the same time developing research methods that brought it close to other fields in the humanities.[2]

The medical history newly emerging at the end of the nineteenth century was characterized by several tensions, evident in the disputes between contemporary physicians and medical historians, as recently explored in historical examinations of the institutionalization of medical historiography at the German-language universities. Medical historians emphasized several key issues in the work of Pagel and Neuburger: utilitarian, pragmatic, and cultural-historical approaches to medical history, and the discipline's philosophical permeation. In particular, the utility of medical history was the object of considerable controversy among medical historians in the late nineteenth century. Recently published articles on Pagel and Neuburger focus on their delineation of the "usefulness" of medical history or on their inauguration of the cultural-historical approach.[3] Some authors have held that they adhered to a "positivistic-utilitarian" mode of thinking, which threatened medical history.[4] Others have reproached Neuburger for exploiting medical history to legitimate the emergence of a new scientific medicine grounded in the scientific approaches and research methods of the natural sciences.[5] In contrast, some older Anglo-American articles and a few German ones dating back to the 1950s assessed the "pragmatic" approach of Pagel in a far less critical tone. Walter Pagel and Paul Diepgen referred to the "pragmatic point of view" of Julius Pagel.[6] Diepgen wrote that Julius Pagel wanted to teach "pragmatic history" that was useful for contemporaries and that, by doing so, Pagel altered and reshaped eighteenth-century historiographic pragmatism.[7]

Another topic, mentioned by Walter Pagel as well as by Karl Sudhoff, was Pagel's attempt to construct a doctor-centered cultural history of medicine.[8] Sudhoff, who was one of the most influential German medical historians and founder in 1907 of the first medical historical journal (*Archiv für Geschichte der Medizin*), stated that Pagel's *Grundriss eines Systems der Medizinischen Kulturgeschichte* (Survey of the cultural history of medicine, 1905) was an original and distinguished piece of scholarship, yet constructed from an all too one-sided literary point of view.[9] Furthermore, a review in the *Historische Zeitschrift* from 1917 of the second edition of Pagel's textbook, as revised by Sudhoff, pointed out that Pagel had demonstrated that medical history was an important part of cultural history.[10] Yet, neither Max Neuburger in his obituary for Pagel in 1912 nor Henry E. Sigerist in his anniversary remarks of 1951 mentioned this aspect of Pagel's contribution.[11]

Finally, several assessments of Neuburger's work emphasized his "philosophical" approach to medical history. Sigerist juxtaposed the approach of Sudhoff with that of Neuburger. He saw Sudhoff as a philologist engaged in analytical work, whereas Neuburger was a philosopher bent on synthesis.[12] Other medical historians described Neuburger's work as permeated by a philosophical spirit sensitive to all the cultural relations of medicine.[13]

The above-mentioned "pragmatic" approach was rooted in a notion of "pragmatic history" that had emerged in the eighteenth century. Medical historian Kurt Sprengel (1766–1833) contributed to the concept of pragmatic history. Hans-Uwe Lammel has shown that in the eighteenth century the term "pragmatic" carried a four-fold meaning.[14] It bore the notion of a new narrative form of historiography, the scientific status of which was based on the search for cause and effect. Together, cause and effect would increase the credibility of a story. Historiography also needed to be useful for contemporaries. It needed to make them recollect the past, and it needed to function as a tool of remembrance. In doing these things, it was intended to improve the readability of historiographic texts by referring to other literary genres such as biography or the novel. Finally, the term "pragmatic" indicated an attempt to escape the old pattern of *Historia litteraria* by looking for a paradigm shift out of the older intellectual tradition into the modern one.

Owsei Temkin and Walter Pagel also considered pragmatic history in a critical sense. Pagel depicted the shortcomings of pragmatic history and held that its adherents did not accept the whole of historical truth but instead extracted from history what attracted their attention and promised reward in practical life. They set present-day standards of medicine as the gauge by which the past was measured, constructing a stepladder to progress.[15] Similarly, Temkin discussed the problematic implications of pragmatic history written as progressive history. The idea of progress could be taken to extremes, as when, for example, the Middle Ages were dismissed because they showed little progress or ridiculed because of their superstitions. In this extreme form, the idea of progress showed very clearly its relationship to the Enlightenment.[16]

In what follows, I discuss Pagel's and Neuburger's understanding not only of pragmatic medical history but also of medical cultural history. In the process, I explore Neuburger's oft-mentioned philosophical permeation of the past. I refer to several contextual fields, including the personal situation of Pagel and Neuburger and the contemporary methodological dispute within medical history. At the time when Pagel and Neuburger developed their concepts, the history of medicine became an independent discipline at the German-language universities. This process was characterized by several peculiarities that influenced the conceptualization of pragmatic medical history, or medical cultural history. One fact that has to be taken into consideration is their daily activities as physicians. Despite his appointment as adjunct professor of medical history in Berlin in 1898, Julius Pagel worked his whole life as a medical practitioner in the proletarian "Wedding-district" of Berlin.[17] Neuburger kept his private practice up until 1914; in 1904, he became professor extraordinarius and in 1917 full professor at the university in Vienna. Neuburger began his career in neurology, and, according to Sigerist, contributed to this field by writing a series of monographs and papers on the history of neurology "which inspired and directed neurological research."[18] Although in a slightly different way and time, both men were engaged in medical practice. This anchor in practical medicine turned the problems of contemporary medicine into guideposts, which partially determined at least the selection of their themes if not the intent of their arguments.

Apart from contemporary medicine, two other fields probably also influenced the works of Pagel and Neuburger. First, around 1900 medical historians discussed the role of the philological method embodied in the careful Latin and Greek editions with their huge bibliographic apparatus. This discussion also involved the relationship between physicians and philologists, both of whom were engaged in research in medical history. Moreover, it raised the question of how medical history wanted to represent itself to medical students and doctors; it was closely connected to questions of the legitimacy and "usefulness" of medical history.

Second, from the late 1880s, German academic history was involved in a discussion that opposed the "old" methods of "political" history to the "new" ones of "cultural" history. At the center of this quarrel stood Karl Lamprecht, who, as the leading figure of the "cultural" historians, was mainly criticized by the historians associated with the *Historische Zeitschrift*. Looking at this controversy within academic history, we have to ask to what extent this dispute influenced the almost simultaneously developing cultural historical approach of Neuburger and Pagel.

I discuss the works of Pagel and Neuburger with reference to the above-mentioned contextual fields and confine my considerations primarily to the period up to 1912, when Pagel died. I start from an internal perspective on medical history and sometimes cross the boundaries into medicine and history. I argue that Neuburger and Pagel attempted to use medical history as an instrument to elevate the quality of *Bildung* among medical students. They also wanted to contribute to the science and practice of contemporary medicine. By doing so, they underlined the usefulness of medical history but did not confine themselves to a positivistic-utilitarian approach. Rather, their cultural history was an attempt to create a particular profile of medical history that could stabilize medical history as a discipline of its own.

Pragmatic History of Medicine and Its Relation to Medicine

In the year 1904, German-speaking medical historians were confronted with two very polemical articles that stirred their community. The Leipzig pediatrician and Privatdozent Max Seiffert made fun of the "philological-antiquarian" methods of medical history and maintained that while the enumeration of heroes, books, and dates was commendable, it also had a "narcotic effect" on persons who did not belong to the community of medical historians. He called for a medical history that could delineate general problems of modern medicine and could render moral support to physicians who were obliged to struggle against the widespread hostility of quacks, social insurance companies, and other social and political forces.[19] The second attack came from the author of a historical account of Vesalius. Moritz Roth, professor for pathological anatomy in Basel, pointed out the difficulties of a critical assessment of the different books of the Hippocratic corpus. He condemned most previous studies and asserted that neither philologists nor a consortium of philologists, physicians, and cultural and other historians, as proposed by Pagel, would be able to solve the problems. Only a physician pursuing historical inquiry would be able to throw new light on the medicine of antiquity. Roth explicitly attacked both Neuburger and Pagel, as well as the pragmatic history of medicine, which he called "erudite rubbish" (*gelehrter Schutt*). Universities did not accept this kind of medical history.[20]

Both articles raised the question of the scholarly character and "usefulness" of medical history for medical students and physicians.[21] In 1904, a few months after the aggressive articles of Seiffert and Roth appeared, Neuburger read a paper on the history of medicine as an academic subject at Vienna's university. In this academic speech, Neuburger pointed out why the history of medicine was necessarily an integral part of the education of medical students. He referred to the knowledge of the old literature, which sensitized students to the interrelationship between past and present and thus enabled them to evaluate contemporary contexts critically. Into the hearts of students medical history implanted enthusiasm for the physician's art, for ethics, and for the corporate ethos of the profession; by enhancing philosophical and cultural historical awareness, medical history led the young physician out of intellectual isolation. Medical history fostered an understanding of the foundations of science by its study of past work, and because of that work the past supplied for the material to build future concepts. Medical history contained a large number of both remedies and therapeutic principles which had been scientifically confirmed by recent research. Knowledge of the medical past revealed that contemporary modes of homeopathy and natural healing were old medical methods; historical knowledge provided a shelter against a deceitful application of old remedies masquerading as new ones. Medical history elevated the general education of physicians, and in doing so it could bridge the gap between the two cultures of *Geisteswissenschaft* and science. Studying medical history trained the intellectual faculties. Going through the history of problems with their complicated controversial disputes prepared one for rigorous thinking about recent problems. Knowledge of the past provided a key to unlock the confusing abundance of isolated facts in contemporary medicine, an abundance that hid the aims and means of modern medicine. Without familiarity with the first elements of modern medicine, it was impossible to understand scientific research.[22]

Some points of Neuburger's statement touched on empirical-practical aspects or professional concerns. Others dealt with raising standards of education and training in critical thinking and the theory and practice of cognition. Neuburger referred to the attacks of Roth in a letter that was read at the Breslau meeting of the German Society for the History of Medicine and Natural Sciences in September 1904. He spoke of "utility as being beneath one's dignity" (*wahrhaft beschämende Utilitätsfrage*) and of the obligation to treat the history of medicine from different points of view. The existing literature had brought about some excellent achievements that—apart from their direct utility—were responsible for the reputation of the discipline as part of the *universitas literarum*. If a medical faculty excluded the history of medicine as a tool of general education, it would slip below the level of technical high schools and veterinary institutes, some of which even cultivated the history of veterinary medicine.[23] Neuburger did not neglect utility, yet the term "beneath one's dignity" expressed a certain distance to it. He emphasized the role of medical history for developing critical thinking and cognition, as well as for implanting enthusiasm in the hearts of young doctors. According to him, medical history was an indispensable part of university education and research. As long as medical education was located at universities and academic medicine maintained as a part of the universitas literarum, the disciplinary existence of medical history at medical faculties seemed to be secured.

In the context of this drawn-out dispute on the usefulness of medical history, Sudhoff made some general statements on the tasks of medical history. Like Neuburger, he too made favorable reference to the universitas literarum, to idealism, and to the general education (*Allgemeinbildung*) that medical history could enhance.[24] Moreover, he was convinced that the historical investigation of past epidemic diseases was a useful culmination of recent medicine. Despite this statement, he emphasized the importance of the philological method and maintained that it would be dangerous to view medical history as an applied science in any utilitarian sense. He underlined the independent and self-confident character of medical history, which had to go its own way and was valuable in its own right.[25] In 1908, Johannes Ilberg, a philologist and one of the best experts in ancient medicine, held that medical history needed to be "philologized." With this term he called for the acquisition of rules governing the study of historical sources, rules drawn either from Greek or Roman philology or from elsewhere. While Pagel objected to Ilberg's views, Sudhoff stressed that medical history would have to learn to engage in the study of historical sources.[26] In contrast to Pagel and Neuburger, Sudhoff obviously emphasized a philological method and maintained that it needed to become an important part of the methodology of medical history. Georg Harig, the former medical historian of the Humboldt University in Berlin, has persuasively argued that Sudhoff acted as an intermediary between classical philologists and medical historians, although for the most part he shared the aspirations of classical philology.[27] Still, these differences did not impede Neuburger's support for Sudhoff when he attempted to become full professor at Leipzig in 1918.[28]

Like Neuburger, Pagel refused to pander to modern utilitarianism.[29] He also maintained that the history of medicine sought to elevate Allgemeinbildung and to foster critical thinking.[30] Medical history revealed the crass changes and self-contradictions of theories, dogmas, and opinions; from history, one could learn how facts and laws in nature were discovered.[31] History also taught tolerance toward traditional and vulgar remedies; so-called people's medicine could only be understood in the light of history because, according to Pagel, its methods had been taught and recommended by scientific doctors at various times and were then adopted by the people.[32] Pagel put much emphasis on relations to practical medicine. For him, knowledge of history increased and confirmed the practical knowledge of medicine.

Like the French historian of medicine Charles Daremberg,[33] Pagel maintained that medical history embraced the whole of present-day medicine and summed up the stages of its development; thus, it enabled one to recapitulate the whole of human knowledge in a "genetic" way. History helped one not only understand the factual knowledge of the past but also discover new facts, in the same way that the embryological method had opened up many previously obscure fields.[34] With the help of history, one could find many remedies that had been unjustifiably forgotten and relegated to the therapeutic scrap heap, often in favor of bad innovations. Several discoveries and inventions of the past had been confirmed by the results of modern research.[35] Neuburger also used this argument. He referred to the successful introduction of hyperemia as a therapeutic means by the Berlin surgeon August Bier and held that this was an example for the revival of ancient ideas of cure, now based on modern biology.[36] Moreover, Pagel was eager to use history as a weapon against contemporary quackery, including homeopathy, and against the commercial behavior of doctors and their distasteful advertising. Historical knowledge was the best safeguard, a moral force in order to combat these dishonesties and untruths. History was the embodiment of medical ethics. It amounted to a religious cult in that it imparted fairness, modesty, and piety, and dispelled complacency.[37] Knowledge about the development of medicine enabled the physician to separate error from truth and to demonstrate how what was new in these ideas was untrue and what was true, not new.

Pagel's equation of medical history with a religious cult was rooted in the conviction that history embodied moral principles that the historian was obliged to detect and to teach. Besides that, it was evidence of science as substitute for religion in a secularized society. Pagel addressed important problems that preoccupied the medical profession in Germany around the turn of the century: the implementation of social insurance systems in the early 1880s; the growing popular opposition to university medicine, combined with criticism of materialistic mechanisms, and to human clinical trials;[38] and the rise of natural cures and other healing methods based on the nonprofessional skills of various healers.[39] All these phenomena were seen as a great threat to the autonomy and freedom of the medical profession. Pagel criticized the modern German state-run health insurance system,[40] and he wanted to use medical history in order to forge a powerful weapon against all such inconveniences. Thus, a particular aspect of the pragmatic approach became obvious: the usefulness of medical history for enhancing professional power. Pagel did much more than Neuburger to emphasize the contemporary professional battles he wanted to fight using weapons made from historical knowledge.

Their different assessments of homeopathy, for example, underline their differing degrees of engagement. Pagel labeled homeopathy as quackery,[41] as a theological dogma based on faith, as therapeutic fraud,[42] and as a product of suggestion;[43] thus, he subsumed homeopathy under the methods of nonscientific medicine, which all truthful physicians had to combat. Against such a statement, born as it was of the professional interests of physicians, Neuburger interpreted homeopathy as a result of the healing power of nature and saw Hahnemann working in the tradition of the so-called expectative therapy as invented by the Vienna school of medicine at the end of the eighteenth century.[44] Neuburger also reproached the homeopathic school for its arbitrariness and speculation, but he simultaneously related its aims and contents to the contemporary state of the art. Thus, he conceded that homeopathy had its merits: for the struggle against dogmatic therapies, the rejection of ontological concepts of diseases, the recommendation of individual and mild cures, and the suggestion of trials concerning the effect of remedies on healthy individuals.[45] Neuburger worked out a broader scope of contextual relations to both contemporary professional disputes and philosophical discussions.

Pagel, as Temkin has shown, adhered to the idea of historical progress.[46] Pagel criticized the Middle Ages because of its prevailing beliefs in religious miracles, superstition, orthodoxy, and its lack of scientifically educated physicians,[47] calling the period the most fruitless in the history of medical art and science.[48] He established a present-day standard as a gauge by which the past was measured and imagined a stepladder climbing from the dark beginnings up to the enlightened present, which represented the climax of a continuing process of development. Given this attitude, it might seem paradoxical that Pagel lavished so much time on researching medieval medical sources and editing medieval medical booklets. He conceded that from a pragmatic point of view the whole of medieval medicine was of little value. Yet, medicine in the Middle Ages had a certain importance because it provided evidence for the nonculture and for the sleeping spirit of medieval darkness (*Geistesschlummer mittelalterlicher Finsternis*). Moreover, only with a detour through medieval literature could one often find the way back to the ancient medicine.[49]

Another crucial point concerning the relation of medical history to medicine has to be mentioned. In 1897, Julius Pagel reviewed Neuburger's book on the historical development of experimental brain and spinal cord physiology. Pagel held that the book would legitimate Neuburger as physiologist as well as historian. Neuburger's book showed a very profound combination of the practical-experimental aspect with the historical exposition.[50] Similar to Sigerist, Pagel saw Neuburger's historical study as a real and worthy contribution to contemporary physiology. He viewed historical research as a practical contribution to contemporary medicine. Neuburger made a similar statement in 1904. Neuburger used the notion of *historische Medizin* (historical medicine), insisting that it had become a science that, apart from its literary and philosophical relevance, would prove to be an eminently practical discipline. He emphasized that a necessary prerequisite for this development would be the continuous, mutually fecundating interaction with the present and its needs and efforts.[51] The intensity of this interaction between the past and the present was not only a question of erudition, knowledge, and engagement of the persons involved; rather, there existed something like a law of nature that was applicable to all cultural and historical developments. In his work on brain and spinal cord physiology, he maintained that the shortcomings and failings of earlier research could be explained in part by its inferior quality and in part by the fact that it did not take adequate account of the auxiliary sciences. What was once mere philosophical speculation became clinical investigation and anatomy. Thus, each idea and each scientific fact could be resurrected, even though they may have long since receded into the past. Based on newly developed methods and experimental settings, old problems were considered anew. Neuburger insisted that now and again the study of the past was useful, as it held up to the present the mirror of its ideas and problems, its expectations and disappointments, its shortcomings and failings.[52]

This "law" said that there is a constancy of ideas and research questions in history, that these continuously existing ideas could remain hidden for a long time, yet that changes in auxiliary sciences and technical equipment led to a resurrection and to a "new" discovery, which was due only to a modern kind of technical permeation of an old problem. Because this law of the "historical constancy of ideas and problems" applied to the whole historical process up to the present, it was possible to bridge the gap between past and present. Therefore, medical historical research, when dealing with problems and questions derived from the needs of the present, had an inevitable relation to contemporary medicine.

Both Neuburger and Pagel connected past medicine to contemporary medicine, yet they did so by constructing two different kinds of connections. Whereas Pagel referred to historical progress in order to construct a line of development leading step by step up to the highest contemporary state of medicine, Neuburger did not see a line but a space within a historical continuum. Within this space, past and present interacted in a mutual and fecund way and were stimulated by recurrent ideas, problems, and needs. In other words, the relationship between past and present was defined not only in utilitarian terms but also in terms of laws of nature.

History of Medicine and Cultural History

Neuburger and Pagel often underlined the relationship between general cultural developments and the particular development of medicine.[53] In 1905, Julius Pagel published his *Grundriss eines Systems der medizinischen Kulturgeschichte*. Pagel decided to pursue a cultural historical approach because it offered the possibility of making medical history more attractive to students by addressing ways of living, social habits, and political attitudes.[54] Pagel in several places stated that medicine was an important part of general culture.[55] In his *Geschichte der Medizin* of 1898, he maintained that it was necessary to refer to all fields of civilization—housing, food, trade, art, pleasure, and politics, among others—in order to reconstruct a complete picture of a certain period, which, in turn, would lead to a deeper understanding of past medicine.[56] In his *Kulturgeschichte* Pagel delineated the positions of previous authors who dealt with *ärztliche Kulturgeschichte* (doctor-centered medical cultural history). He wanted to consider the whole cultural history of humankind from the perspective of medicine.[57] He preferred to consider the various branches of culture, assessing each one's contributions to the development of medicine and, vice versa, the physicians' contributions to each. Pagel endeavored to explain how, through this mutual influence, progressive development was initiated. In particular, his medical cultural history dealt with theology in medicine, homeopathy and mystic approaches, female physicians, medical religion, philosophy in medicine, law and medicine, medicine and natural science, social medicine, medicine and history of states, medicine and poetry, medicine and art, and a mixed chapter that discussed physicians as mathematicians, as pedagogues, ennobled physicians, physicians as husbands of princesses and of prominent actresses, and one-hundred-year-old doctors.

Pagel used his book to blame the Roman Catholic church of the thirteenth century for oppressing scientific thinking.[58] He accused Neoplatonism of having fostered mysticism.[59] He rejected women's right to study medicine and hoped the "strange movement" that received strong support in America and Russia would be soon subdued.[60] In accordance with a large majority of German doctors, Pagel attacked the German health insurance system. He rejected this system not because he was a friend of reactionary politics but because the system would foster indifference of the concerned workers to all measures which could improve health.[61]

Notwithstanding these negative statements, the major part of his book praised the contributions of physicians to culture and civilization. He followed the motto of the English liberal statesman Gladstone that now the time had come to render unto medicine a leading role in the affairs of mankind, not only with respect to healing procedures and health but to the whole of human life, including the prevention of diseases and the guarantee of a healthy life, happiness, and wealth. Governments and states began to accept medicine as an important factor of culture and civilization.[62] In another article, he maintained that a considerable number of academic teachers of philosophy belonged to the medical profession.[63]

Pagel placed his arguments within a tradition of medical cultural history, dating back to Sprengel.[64] Pagel referred to a few predecessors—Wilhelm Stricker, Heinrich Rohlfs, and Iwan Bloch[65]—yet did not mention his contemporary Karl Lamprecht who had initiated a drawn out dispute between 1891 and 1898 among German historians by proselytizing for the "new" approach of cultural history in contrast to the "old" one of political history. There is some evidence that this debate influenced medical history.

Karl Lamprecht had been appointed as professor for history in Leipzig in 1891.[66] Proponents of cultural history comprised a diverse assortment of historical economists, sociologists, and a few historians. Indeed, those who promoted cultural history seemed to make up such a diverse coalition that it remained difficult to identify even the common areas of their interest. Historian Roger Chickering has pointed out that the interest in Kulturgeschichte was due to the convergence of several groups on the historical profession's periphery, where sensitivity to contemporary social problems was evidently greater than in the profession's academic core. A great deal of what passed as cultural history was bereft of theoretical reflection. The only consensus of method among the cultural historians appeared to be that all phases of human activity belonged to the historian's purview and that the writing of history ought not to be confined to matters of the state.[67]

Yet the notion "cultural history" is a little bit misleading; Lamprecht wanted to place the problems and questions of social history at the center of his historical research. He emphasized the relevance of social and economic developments, denied the importance of ideas in historical development, and rejected the individualistic method of historicism, which he replaced with an approach designed to reveal the general laws in history. This change would transform history into an exact science based on recognizable laws. He called this approach, which was derived from the natural sciences, a causal genetic (*kausalgenetische*) approach.[68] Later on, Lamprecht emphasized the social psychological conditions that he wanted to see stressed by further historical research. He viewed psychology as a natural science of mental life and was therefore convinced that causal relations in the field of human intentions and mentality could be found. Finally, he maintained that there existed an analogous development between the human world and the biological processes of the natural world. This meant assigning biological growth and involution to historical conditions.[69]

Lamprecht's approach became a point of reference for medical historians cultivating what they labeled the medical history of culture. The terms "social" and "cultural" bridged the gap between biology and sociology and made it possible that the "laws" of biology could enter the realm of social conditions. When in 1904 Seiffert attacked the "philologic-antiquarian" methods in medical history, he referred to Lamprecht's "collectivistic" approach and to his successful "genetic" historiography (*genetisch entwickelte Geschichtsschreibung*). Seiffert viewed Lamprecht's success as a stimulus that could improve the quality of medical history.⁷⁰ In response to this attack, Sudhoff took up his pen as president of the Deutsche Gesellschaft für Geschichte der Medizin und der Naturwissenschaften.⁷¹ He too mentioned Lamprecht in a very positive way and spoke of the important and highly distinguished Leipzig historian. The medical historians valued his "genetic" approach and considered his method as a model for their work.⁷² When in August 1908 the Internationale Kongreß für historische Wissenschaften held its meeting in Berlin, it included a section on the *Kultur-und Geisteswissenschaften,* as well as a subsection on the history of science. In this section, which connected cultural history with the history of science, Lamprecht read a paper on the efforts of cultural and universal history in Leipzig.⁷³

At the end of the nineteenth century, then, medical and cultural historians were in close contact with one another. In some of his articles, Max Neuburger referred also to the positions of Lamprecht, yet without mentioning him by name. In his introduction to a short paper on Swedenborg, Neuburger, who considered himself a *Kulturhistoriker,* addressed the problem of "scientific collectivism." In using this term, Neuburger drew a direct line to Lamprecht's "collectivistic" approach, which rejected the accentuation of individualistic aspects in history. Neuburger stated that everywhere the pros and cons of this method would become obvious. He attempted to justify his considerations of Swedenborg by referring to a "law of contrary effects" (*Gesetz der Contrastwirkung*), which led him to the investigation of the work of this single person.⁷⁴ Obviously, Neuburger did not totally subsume his research under the rules of the collectivistic method, which considered the cultural power of a multitude to be a more important historical force than the achievement of an individual. He continued to study individuals, although he underlined elsewhere that changes in history flowed not from the activity of a single "reformer" but from sheer necessity that reigned over the whole course of history—meaning history is directed by an immanent purpose (*immanente Zweckmäßigkeit*).⁷⁵

Neuburger's reference to a "law of contrary effects" in this context seems to be more than simply a rhetorical phrase. He was eager to detect laws in the historical process that could help to understand the development of history. Pagel also presupposed historical regularities yet did not struggle to make them explicit.⁷⁶ In contrast to Pagel's reserve, Neuburger pointed out a "gloomy" (*düster*) law that would be valid for all cultural creations: when a period reached the highest degree of its particular characteristics, it was bound to perish. As examples he mentioned the excess of religious mysticism in the Middle Ages that had led to an anticlerical movement or the subtle scholastic thinking that provoked extreme skepticism. Thus, he formulated the law of highest determination that caused its own negation.⁷⁷ Neuburger derived this formulation of the law from Hegel's view of history, whereby a second period contained the negation of the first while a third contained a negation of the second.⁷⁸ Using Hegel, Neuburger provided an explanation for historical change from one period to another, and he tried to transform medical history into an exact science based on recognizable laws.

Throughout the whole work of Neuburger we find such attempts to detect laws in the process of scientific and medical development. From an analysis of the Viennese school of medicine he derived the thesis that "periods of eminence have always been coupled with liberal reforms in education."⁷⁹ In his book about the mechanisms of specific nutrition he set forth a basic law (*Grundgesetz*) of progress, which stated that in contradictions between opinions the divergence of principles become smaller in the same measure as positive experiences and observations grow larger.⁸⁰ Neuburger further tried—in accordance with the aims of Lamprecht—to assign laws from biology and the natural sciences to the historical development of medicine. With regard to the emergence of specific medical disciplines, he referred to the organically structured growth of scientific medicine as analogous to the growth of a living being that differentiates itself in the germ-layer and grows up by cellular division of labor.⁸¹ In the context of a new experimental elucidation of physiological problems, he referred to the law of conservation of energy. He said: "Even if certain characteristics determine the physiognomy of an era, the new and the representative appear less as recent strange features and more as a new arrangement that allows this or that particular feature to stand out more clearly. The sum of the elements, the expended intellectual endeavor—the law of conservation of energy holds here too—is more or less the same even if the net result is of variable usefulness."⁸²

Other elements of Lamprecht's approach of cultural history can also be found in Neuburger's work. At times he referred to psychology in order to explain the effects of new research results on the thinking of scientists.⁸³ Neuburger and Pagel dedicated a considerable amount of their work to the investigation of collective or social structures relevant for the development of medicine. Both worked on medical schools, primarily on those of Göttingen and Vienna, which were seen as crucial in raising the quality of medicine. Both considered professional problems. Neuburger reflected on the historical aspects of physicians' economic status and discussed the importance of scientific societies for the progress of medicine. Pagel wrote a monograph on the history of the Berlin society of district physicians for the poor; he also contributed to medical ethics. Both researched a group of Jewish physicians and the relation between religion and medicine. All these scientific and literary activities of Neuburger and Pagel met the contemporary criteria of cultural history. Beyond Sudhoff's general praise of Lamprecht's approach as a model for medical historical research, therefore, it is likely that the so-called Lamprecht dispute did influence the work of both Neuburger and Pagel.

Medical History and Academic Requirements

Neuburger and Pagel laid crucial foundations for the academic legitimacy of medical history as a discipline of its own at medical faculties in German-language universities. Medical history, at the time they approached the field, around 1900, was increasingly separate from modern experimental medicine. Likewise, the distance between it and mature modern philological disciplines became larger. Traditional history was primarily conceptualized as the history of ideas and as political history, without any relations to the history of science and medicine. Moreover, in accordance with the modern requirements of a university professor, they had to act as both researchers and teachers, which posed a particular problem. Because of the very special methods needed to conduct scientific research into the past, historical or medical historical education involved highly specialized training.[84] Yet, Neuburger and Pagel taught students of medicine, not students of medical history.

In order to advance the institutionalization of medical history within medical faculties, they had to address several problems. They had to teach in a way that would be accepted by medical students and physicians. They emphasized general education (Allgemeinbildung) for all medical students and not special education only for students interested in careers as medical historians. This affected how they presented the material. As Neuburger explicitly pointed out in his *Geschichte der Medizin,* he had dispensed with all philological-bibliographic and literary historical equipment.[85]

Because of the increasing gap between the methods and contents of medical history and medicine, they were obliged to find a way to bridge the emerging boundaries. Taking up the "old" concept of pragmatic medical history, they attempted to solve this problem and to secure strong links between medical history and recent medicine. Yet neither of them wholly subordinated their research and teaching activities to this narrow concept. They underlined the necessity of dealing with issues that did not have any usefulness for contemporary medicine in a pragmatic sense.

These steps beyond the concept of pragmatic medical history were necessary because medical history was obliged to develop and to present a disciplinary program of its own. This was difficult because scholars could only borrow ideas from other disciplines: history, philology, and medicine. But the newly emerging discussion on the notion of "culture"[86] offered a new opportunity, and the quarrel over cultural history at the end of the nineteenth century seemed to offer a solution. Apart from a tradition dating back to Sprengel that linked culture and medicine within medical history, this dispute between the outsider Lamprecht and the academic historians provided a chance for Pagel, Neuburger, and their contemporaries to discuss aims and methods anew, and to outline a particular medical historical perspective. Cultural history embraced a broad spectrum of different approaches. It contained social historical considerations, philosophical contextualizations, and profession-oriented political statements, and it was sometimes directed against a narrow "philological-antiquarian" method and the attention that method paid to arcane details. The widespread acceptance of the term "cultural history" was probably enhanced by its ambiguity, which allowed it to be applied in many different ways. The term did not signify a special research method.

In spite of all of his differences with Neuburger and Pagel concerning the role of philology in medical history, Sudhoff repeatedly referred to cultural history in very positive terms.[87] His affirmative remarks on Lamprecht, as well as the statement of Paul Diepgen in 1912 in which he endorsed Neuburger's connection between the development of general culture and medicine,[88] illustrate the broad support of contemporary medical historians. The term also fostered something like a corporate identity of medical historians who now could develop a program for their own discipline with ties to physicians and to certain groups of historians. The embryonic state of established academic medical history as a discipline in its own right at the beginning of the twentieth century made it necessary that medical historians demonstrated a united attitude toward the medical faculties—an attitude that cast doubt neither on the important role of physicians in history nor on their role in the newly emergent discipline itself. Cultural history functioned as an umbrella beneath which the community of medical historians could meet these requirements. Despite some severe differences within the community concerning the role of the philological method, cultural history as a commonly accepted point of reference enabled the medical historical community to avoid a "schism" that would have endangered its academic progress.

Pagel and Neuburger followed the same medical historical program, yet there existed several differences in the content and method of their works. Pagel produced detailed philological editions of the texts of medieval authors; he emphasized the professional interests of physicians; and he published on ethical and deontological problems. According to Pagel, the true historian would be a critic who introduced order into apparently chaotic material, found the right thread of continuity, and made himself a contemporary of the past by intuition.[89] Neuburger went beyond this definition and aimed at a conceptualization of medical history as an exact, scientifically based discipline in search of the laws of historical process. He did this on the basis of his philosophical erudition, which also helped him to consider the achievements of the past in the context of the contemporary intellectual conditions. This led not only to a thoroughgoing philosophical permeation of historical medical concepts but also to a consideration of topics concerned with the philosophical problems of medicine, such as mechanistic approaches and the healing power of nature, both of which were discussed by Neuburger's contemporaries. Despite these differences, both Neuburger and Pagel created important prerequisites for the professionalization and academic expansion of medical history at German-language universities.

NOTES

I would like to thank Hans-Uwe Lammel and Gabriele Moser for their suggestions.

1. F. N. L. Poynter, "Max Neuburger, 8 December 1868–15 March 1955: A Centenary Tribute," *Medical History* 13 (1969): 1.

2. See Bernhard vom Brocke, "Die Institutionalisierung der Medizinhistoriographie im Kontext der Universitäts-und Wissenschaftsgeschichte," in *Die Institutionalisierung der Medizinhistoriographie. Entwicklungslinien vom 19. ins 20. Jahrhundert,* Andreas Frewer and Volker Roelcke, eds. (Stuttgart: Steiner, 2001), 187–212.

3. Michael Hubenstorf, "Eine 'Wiener Schule' der Medizingeschichte? Max Neuburger und die vergessene deutschsprachige Medizingeschichte," in *Medizingeschichte und Gesellschaftskritik: Festschrift für Gerhard Baader,* Michael Hubenstorf, Hans-Uwe Lammel, Ragnhild Münch, Sabine Schleiermacher, Heinz-Peter Schmiedebach, and Sigrid Stöckel, eds. (Husum: Matthiesen, 1997), 246–289.

4. Werner Friedrich Kümmel, " 'Dem Arzt nötig oder nützlich?' Legitimierungsstrategien der Medizingeschichte im 19. Jahrhundert," in Frewer and Roelcke, *Institutionalisierung,* 75–89.

5. Richard Toellner, "Der Funktionswandel der Wissenschaftshistoriographie am Beispiel der Medizingeschichte des 19. und 20. Jahrhunderts," in *Eine Wissenschaft emanzipiert sich. Die Medizinhistoriographie von der Aufklärung bis zur Postmoderne,* Ralf Bröer, ed. (Pfaffenweiler: Centaurus, 1999), 175–187.

6. Walter Pagel, "Julius Pagel and the Significance of Medical History for Medicine," *Bulletin of the History of Medicine* 25 (1951): 207–225.

7. Paul Diepgen, "Julius Leopold Pagel und die deutsche Medizinhistorik seiner Zeit," *Berliner medizinische Zeitschrift* 2 (1951): 353–355.

8. Pagel, "Julius Pagel," 220–225.

9. Karl Sudhoff, "Julius Leopold Pagel. Ein Nachruf," *Münchener medizinische Wochenschrift* 59 (1912): 425–426.

10. K. Baas, "J. L. Pagels Einführung in die Geschichte der Medizin. In zweiter Auflage herausgegeben von Karl Sudhoff," *Historische Zeitschrift* 3d ser., 21 (1917): 146–148.

11. Max Neuburger, "Julius Leopold Pagel †," *Deutsche medizinische Wochenschrift* 38 (1912): 423; Henry E. Siegerist, "On the Hundredth Anniversary of Julius Pagel's Birth, 29 May 1851," *Bulletin of the History of Medicine* 25 (1951): 203–204.

12. Henry E. Sigerist, "A Tribute to Max Neuburger on the Occasion of his 75th Birthday, December 8, 1943," *Bulletin of the History of Medicine* 14 (1943): 417–421, 420.

13. See Robert Rosenthal, "Max Neuburger, December 8, 1868–March 15, 1955," *Bulletin of the History of Medicine* 29 (1955): 295–298, 297; F. N. L. Poynter, "Max Neuburger," 1; Owsei Temkin, "Professor Neuburger's Eightieth Birthday," *Bulletin of the History of Medicine* 22 (1948): 727–729; Lloyd G. Stevensen, "Max Neuburger's Centenary," *Bulletin of the History of Medicine* 42 (1968): 493–495; and Edwin Rosner, "Erinnerungen an Max Neuburger," *Medizinhistorisches Journal* 3 (1968): 328–332.

14. Hans-Uwe Lammel, "Kurt Sprengel und die deutschsprachige Medizingeschichtsschreibung in der ersten Hälfte des 19. Jahrhunderts," in Frewer and Roelcke, *Institutionalisierung*, 27–37, and "Interessen und Ansätze der deutschen Medizingeschichtsschreibung in der zweiten Hälfte des 18. Jahrhunderts," in Bröer, *Wissenschaft*, 19–29; see also chap. 2 in this volume.

15. Walter Pagel, "Julius Pagel," 215.

16. Owsei Temkin, "An Essay on the Usefulness of Medical History for Medicine," *Bulletin of the History of Medicine* 19 (1946): 9–47, 26.

17. Walter Pagel, "Julius Leopold Pagel (1851–1912)," in *Victor Robinson Memorial Volume: Essays on History of Medicine*, Solomon R. Kagan, ed. (New York: Froben Press, 1948), 273–297; see also Johann Gromer, *Julius Leopold Pagel (1851–1912)* (Cologne: Kohlhauer, 1985), 12–20.

18. Henry E. Sigerist, "Preface," in Emanuel Berghoff, *Max Neuburger. Werden und Wirken eines österreichischen Gelehrten* (Vindobona: Maudrich, 1948), ix–xi; on Neuburger, see also Gabriela Schmidt, "Der Medizinhistoriker Max Neuburger und die Wiener medizinische Fakultät," *Wiener klinische Wochenschrift* 105, no. 24 (1993): 737–739.

19. Max Seiffert, "Aufgabe und Stellung der Geschichte im medizinischen Unterricht," *Münchener medizinische Wochenschrift* 51 (1904): 1159–1161.

20. Moritz Roth, "Geschichte der Medizin und Hippokrates," *Münchener medizinische Wochenschrift* 51 (1904): 1396–1398.

21. On the function of history within medicine, see Alfons Labisch, "Von Sprengels 'pragmatischer Medizingeschichte' zu Kochs 'psychischem Apriori': Geschichte *der* Medizin und Geschichte *in der* Medizin," in Frewer and Roelcke, *Institutionalisierung*, 235–254.

22. Max Neuburger, "Die Geschichte der Medizin als akademischer Lehrgegenstand," *Wiener klinische Wochenschrift* 17 (1904): 1214–1219.

23. "Protokoll der dritten ordentlichen Hauptversammlung der Deutschen Gesellschaft für Geschichte der Medizin und der Naturwissenschaften zu Breslau," *Mitteilungen zur Geschichte der Medizin und der Naturwissenschaften* 3 (1904): 465–472.

24. I would like to thank Professor Gundolf Keil for making his recent paper on Sudhoff available to me. On Sudhoff, see also Gundolf Keil, "Sudhoffs Sicht vom deutschen Mittelalter," *Nachrichtenblatt der Deutschen Gesellschaft für Geschichte der Medizin, Naturwissenschaft und Technik* 31 (1981): 94–129, and Andreas Frewer, "Biographie und Begründung der akademischen Medizingeschichte: Karl Sudhoff und die Kernphase der Institutionalisie-

rung 1896–1906," in Frewer and Roelcke, *Institutionalisierung,* 103–126. Karl Sudhoff, "Theodor Puschmann und die Aufgaben der Geschichte der Medizin," *Münchener medizinische Wochenschrift* 53 (1906): 1669–1673, and "Richtungen und Strebungen in der medizinischen Historik. Ein Wort zur Einführung, Verständigung und Abwehr," *Archiv für Geschichte der Medizin* 1 (1907): 1–11.

25. Sudhoff, "Richtungen," 6.

26. On this conflict, see Georg Harig, "Sudhoffs Sicht der antiken Medizin," *Sudhoffs Archiv* 76 (1992): 97–105.

27. Ibid., 102.

28. See Ortrun Riha, "Die Puschmann-Stiftung und die Diskussion zur Errichtung eines Ordinariats für Geschichte der Medizin and der Universität Leipzig," in Frewer and Roelcke, *Institutionalisierung,* 127–141.

29. Julius Pagel, *Geschichte der Medizin,* 2 vols. (Berlin: Verlag von S. Karger, 1898), vol. 1:1.

30. Ibid., 2–3.

31. Ibid., 2.

32. Ibid., 9.

33. Charles Daremberg, *Histoire des sciences médicales,* 2 vols. (Paris: J. B. Baillière & fils, 1870), vol. 1:xiv–xvi, 7–9.

34. Pagel, *Geschichte der Medizin,* vol. 1:2.

35. Ibid., 5.

36. Max Neuburger, "Zur Geschichte des Problems der Naturheilkraft," in *Essays on the History of Medicine: Presented to Karl Sudhoff on the Occasion of His Seventieth Birthday, November 26th, 1923,* Charles Singer and Henry E. Sigerist, eds. (London: Oxford University Press; Zurich: Verlag Seldwyla, 1924), 325–348.

37. Pagel, *Geschichte der Medizin,* vol.1:6.

38. See Barbara Elkeles, *Der moralische Diskurs über das medizinische Menschenexperiment im 19. Jahrhundert* (Stuttgart: G. Fischer, 1996).

39. See Cornelia Regin, *Selbsthilfe und Gesundheitspolitik: Die Naturheilbewegung im Kaiserreich (1889–1914)* (Stuttgart: Steiner, 1995).

40. Julius Pagel, *Grundriss eines Systems der Medizinischen Kulturgeschichte* (Berlin: S. Karger, 1905), 72, and "Medizinische Kulturgeschichte," *Deutsche Geschichtsblätter. Monatsschrift zur Förderung der landesgeschichtlichen Forschung* 5 (1904): 145–156.

41. Pagel, *Geschichte der Medizin,* vol. 1:8.

42. Pagel, *Grundriss Kulturgeschichte,* 42.

43. Julius Pagel, "Homöopathie oder Suggestion ?" *Deutsche Medizinische Presse* 7 (1903): 98–101.

44. Neuburger, "Zur Geschichte des Problems," 343.

45. Max Neuburger, "Einleitung," in *Handbuch der Geschichte der Medizin,* Max Neuburger and Julius Pagel, eds., 3 vols. (Jena: Gustav Fischer, 1902–1905), vol. 2:122.

46. Julius Pagel, "Ueber die Geschichte der Göttinger medizinischen Schule im XVIII. Jahrhundert" (Med. diss., Friedrich Wilhelm University of Berlin, 1875), 8.

47. See, for example, Julius Pagel, *Die Entwickelung der Medicin in Berlin von den ältesten Zeiten bis auf die Gegenwart* (Wiesbaden: J. F.Bergmann, 1897), 5.

48. Julius Pagel, "Einleitung," in Neuburger and Pagel, *Handbuch* vol. 1:447.

49. Ibid., 450–451.

50. Julius Pagel, "Die historische Entwickelung der Gehirn- und Rückenmarksphysiologie vor Flourens—Von Max Neuburger," *Deutsche Medizinal-Zeitung* (1897): 675.

51. Max Neuburger, "Die Geschichte der Medizin als akademischer Lehrgegenstand," *Wiener klinische Wochenschrift* 17 (1904): 1214–1219.

52. Max Neuburger, *The Historical Development of Experimental Brain and Spinal Cord Physiology before Flourens*, Edwin Clarke, trans. and ed. (Baltimore: Johns Hopkins University Press, 1981), 290. The book was originally published as *Die historische Entwickelung der experimentellen Gehirn-und Rückenmarks Physiologie vor Flourens* (Stuttgart: Enke, 1897).

53. See, for example, Neuburger, "Einleitung," 4; Max Neuburger, *Geschichte der Medizin*, 2 vols. (Stuttgart: Enke, 1906), vol. 1:1–2.

54. Pagel, *Grundriss Kulturgeschichte*, 5.

55. See, for example, Pagel, "Medizinische Kulturgeschichte," 145–156.

56. Pagel, *Geschichte der Medizin*, 3.

57. Pagel, *Grundriss Kulturgeschichte*, 12.

58. Ibid., 33.

59. Ibid., 38.

60. Ibid., 45.

61. Ibid., 72.

62. Ibid., 68.

63. Julius Pagel, "Medizin und Philosophie. Ein Beitrag zur medizinischen Kulturgeschichte," *Reichs-Medizinal-Anzeiger* 32 (1907): 243–244.

64. On the historical use of "culture," see Jörg Fisch, "Zivilisation, Kultur," in *Geschichtliche Grundbegriffe*, Otto Brunner, Werner Conze, and Reinhart Koselleck, eds. (Stuttgart: Klett-Verlag, 1992), vol. 7:679–774; see also *Kultur und Kulturwissenschaften um 1900, Vol. 1: Krise der Moderne und Glaube an die Wissenschaft*, Rüdiger vom Bruch, Gangolf Hübinger, and Friedrich Wilhelm Graf, eds. (Stuttgart: Steiner, 1989).

65. Wilhelm Stricker, *Beiträge zur ärztlichen Kulturgeschichte. Fremdes und Eigenes gesammelt und herausgegeben* (Frankfurt: Auffahrt, 1865); Heinrich Rohlfs, "Die Aerzte als Culturhistoriker," *Deutsches Archiv für Geschichte der Medicin und medicinische Geographie* 7 (1884): 443–452; Pagel, *Grundriss Kulturgeschichte*, 11.

66. Roger Chickering, *Karl Lamprecht: A German Academic Life (1956–1915)* (Atlantic Heights, N.J.: Humanities Press, 1993).

67. Ibid., 155.

68. Friedrich Jaeger and Jörn Rüsen, *Geschichte des Historismus* (Munich: Beck, 1992), 141.

69. Ibid., 145.

70. Seiffert, "Aufgabe," 1159.

71. Karl Sudhoff, "Zur Förderung wissenschaftlicher Arbeiten auf dem Gebiete der Geschichte der Medizin," *Münchener medizinische Wochenschrift* 51 (1904): 1350–1353.

72. Ibid., 1352.

73. Paul Herre, "Bericht über den internationalen Kongreß für historische Wissenschaften zu Berlin, 6.–12. August 1908," *Historische Vierteljahresschrift 11, N.F. der Deutschen Zeitschrift für Geschichtswissenschaft* 19 (1908): 417–426.

74. Max Neuburger, "Swedenborg's Beziehungen zur Gehirnphysiologie," *Wiener medicinische Wochenschrift* 51 (1901): 2077–2081.

75. Neuburger, "Einleitung," 4.

76. See for example Pagel, *Entwickelung,* 32. Here he maintained that progress in anatomy is always combined with progress in surgery.

77. Ibid., 4–5.

78. Neuburger, *Brain Physiology,* 8.

79. Max Neuburger, "Development of Medical Science in Vienna," *The Lancet* 201 (1921): 536–538.

80. Max Neuburger, *Die Anschauungen über den Mechanismus der spezifischen Ernährung. (Das Problem der Wahlanziehung)* (Leipzig: Deuticke, 1900), 96.

81. Neuburger, "Einleitung," 72.

82. Neuburger, *Brain Physiology,* 4.

83. Neuburger, "Einleitung," 9.

84. Hans-Jürgen Pandel, "Was ist ein Historiker? Forschung und Lehre als Bestimmungsfaktoren in der Geschichtswissenschaft des 19. Jahrhunderts," in *Geschichtsdiskurs, vol. 1: Grundlagen und Methoden der Historiographiegeschichte,* Wolfgang Küttler, Jörn Rüsen, and Ernst Schulin, eds. (Frankfurt: Fischer Taschenbuch Verlag, 1993), 346–354.

85. Neuburger, *Geschichte,* v.

86. See Georg Bollenbeck, "Warum der Begriff 'Kultur' um 1900 reformulierungsbedürftig wird," in *Konkurrenten in der Fakultät. Kultur, Wissen und Universität um 1900,* Christoph König and Eberhard Lämmert, eds. (Frankfurt: Fischer Taschenbuch Verlag, 1999), 17–27.

87. See Sudhoff, "Richtungen und Strebungen," 9, and "Theodor Puschmann," 1672.

88. Paul Diepgen, "Literaturbericht. Geschichte der Medizin," *Archiv für Kulturgeschichte* 10 (1912): 465–480.

89. Julius Pagel, *Die Concordanciae des Johannes de Sancto Amando nach einer Berliner und zwei Erfurter Handschriften zum ersten Male herausgegeben* (Berlin: Reimer, 1894), xiii. The translation of Walter Pagel omits the words *gleichsam intuitiv;* see Walter Pagel, "Julius Pagel," 219.

CHAPTER FIVE

Karl Sudhoff and "the Fall" of German Medical History

Thomas Rütten

Karl Sudhoff (1853–1938) was born in Frankfurt am Main on 26 November 1853, the son of a protestant pastor. He attended primary school and two years of grammar school in Frankfurt, after which the family moved to Zweibrücken and, shortly thereafter, to Kreuznach, where he completed his secondary education and gained his *Abitur* in 1871. In the autumn of the same year, he began medical study at Erlangen University, then continued first at Tübingen and later in Berlin. On 2 August 1875, Sudhoff was awarded a doctorate in medicine. After brief stints as a hospital doctor in Augsburg and Vienna, he opened his own general practice in Bergen near Frankfurt in 1878, and married in August of the following year. In 1883, he moved his practice to Hochdahl near Düsseldorf, where he worked as a doctor to the local iron and steel works while using his spare time for extensive research in the history of medicine, notably on Paracelsus.

On 1 July 1905, Sudhoff was offered a professorship at Leipzig University, where he would subsequently be instrumental in establishing the Puschmann Foundation–funded Institute of Medical History. As a permanently appointed professor extraordinarius and, from 1922–1923 onward, as a full professor, Sudhoff taught and published exclusively in medical history until his retirement in

1925. After Henry E. Sigerist, who succeeded him to his chair, left for Baltimore in 1932, Sudhoff returned to his institute once more as its interim director, until Walter von Brunn finally took over from the eighty-year-old in 1934. Highly honored and a member of the National Socialist German Workers' Party (NSDAP), Sudhoff died on 8 October 1938 in Salzwedel, at the home of one of his sons.[1]

These, roughly, are the vital statistics of Karl Sudhoff's life. However, these rather conventional biographical facts, in their simplicity and innocence, do not begin to hint at the veritable Sudhoff cult that has characterized German medical history in particular from the first decades of the twentieth through the present. No other German medical historian ever received so many birthday honors, congratulatory messages, and Festschrift contributions,[2] not to mention honorary medals,[3] than Karl Sudhoff. Sudhoff was honored in Hochdahl in 1933 with a commemorative plaque outside his erstwhile residence and a "Professor Karl Sudhoffstraße."[4] In 1938, the Institute of Medical History at Leipzig was renamed "Karl-Sudhoff-Institut," while the scholarly journal he founded had already changed its title to *Sudhoffs Archiv* in 1922.[5] To this day, the Deutsche Gesellschaft für Geschichte der Medizin, Naturwissenschaft und Technik (DGGMNT, German Society for the History of Medicine and Natural Science) annually awards its Karl Sudhoff Commemorative Medal, with the Sudhoff Commemorative Lecture forming an integral part of the award ceremony.[6]

Such efforts, of course, do have a foundation: Sudhoff's role as the founding father of medical history as a properly institutionalized and professional discipline within the German academy, his unswerving commitment to strengthening the discipline in Europe and in North America, his impressive scholarly oeuvre, and his pioneering work in the field of research management are accomplishments that stand undisputed. Yet, the impartial observer cannot help but note a personality cult surrounding the man. Stereotypical epithets such as "doyen" (*Altmeister*), "master" (*Meister*), "leader" (*Führer*), "spiritus rector," "magister mundi," "Nestor," "patriarch," and "absolute king" are not meant simply to honor an eminent scholar but also to validate a small discipline that can see itself personified—and thus celebrate its past as well as anticipate future achievements—in one of its great figureheads.[7]

On the whole, the festive speakers, obituary writers, institutional successors, and pupils give the impression that Sudhoff was the pivot, the gravitational center of the entire discipline, in relation to which his colleagues were mere satellites. But Sudhoff's role as the arch progenitor of German medical history does not have its roots in his research and its subsequent reception alone. Rather, his international network of correspondents, in which Sudhoff invested a great deal of energy and which, to my mind, has not received adequate scholarly attention,[8] attests to his prominent role as a referee, initiator, organizer, and, most importantly, passionate writer of letters—in short, to his own efforts to guide and influence the reception of his works and legacy. If one agrees that the emergence and continued existence of an academic discipline are to no small degree dependent on a "homogenous communication network of scholars"[9] and their "strategies of self-promotion,"[10] letters and correspondences ought to be regarded as documents of vital importance to the history of any discipline. But before I put this hypothesis to the test, focusing on the fateful year 1933, let us first take a brief look at the structural preconditions for Sudhoff's networking activities.

Sudhoff had been the driving force behind the establishment of the DGGMN at the Assembly of Natural Scientists in 1901 as well as instrumental in establishing the association's in-house publication, the *Mitteilungen,* the following year.[11] He also held the first chair in medical history in the whole of the German-speaking world,[12] and he was, like no other, prepared for the "core phase"[13] of the institutionalization of medical history in Germany by virtue of his research on Paracelsus. Having thus risen to prominence, Sudhoff was in a position to shape and influence the young discipline—its methods, institutional expansion, and staffing strategies—in important ways. His professorship at Leipzig, which was funded by the Puschmann Foundation, offered freedom that no other medical historian of his day enjoyed.[14] Characteristic in this context was a discussion between Sudhoff and the dean of the Faculty of Medicine at Leipzig, the content of which had hitherto been unknown,[15] but about which Sudhoff told his Hungarian friend and colleague Tibor Győry (1869–1938) in a 4 May 1905 letter. Győry was a practicing doctor and scholar whom Sudhoff had first met at the 1898 convention of German Naturalists and Physicians and whose career (1902 habilitation in the history of medicine and, from 1919 onward, employed by the Ministry of Public Welfare and Industrial Medicine where he last held the rank of permanent undersecretary of interior affairs) Sudhoff had followed with interest. The friendship between the two men was characterized by mutual professional support and advancement and continued unabated until both their deaths in 1938. In the letter, Sudhoff wrote:

> On Easter Day, I received a dispatch from Geheimrat Curschmann, dean of the Medical Faculty at Leipzig, from Meran, in which he asked for a meeting "on an academic matter" on Wednesday at the "Hotel Rose" in Wiesbaden. I went, and C[urschmann] not only offered me the Leipzig chair in the History of Medicine, but also promised me more or less free hand as far as the financial resources of the Puschmann Founda-

tion are concerned. "We take an essentially idealistic view of the matter and would like to furnish you with a blank page—write on it what you will! If you manage to convince us, the committee [of the Puschmann Foundation], of which you would become a full voting member, will agree with all your proposals." If both faculty and ministry keep Curschmann's promises (made, mind you, "without official instruction"), I shall be on my way to Leipzig within a few weeks and [illegible] to complete the deal which, apparently, is secure even as far as pecuniary matters are concerned.[16]

"A blank page"—and Sudhoff, at the age of fifty-one, certainly knew how to fill it. He was ready for a professorship. On 1 January 1905, he received—"by COD letter (1.50) from the President of the Regional Assembly"—his "official certificate of appointment" to a titular professorship at Leipzig, which had been offered and accepted the previous year. On the evening of the same day, he also received "an inquiry from the University of Rostock asking whether I would be inclined to accept a full honorary professorship at their institution."[17] Sudhoff rejected the offer, since the dean, who had traveled to Düsseldorf especially for talks on the subject, could not assure Sudhoff of any "substantial funding" apart from income "from my general practice," "which I essentially would not wish to pursue once I took up a professorship."[18] After almost twenty years of having had to place his research into Paracelsus second to his general practice, Sudhoff knew that medical history was and ought to be a full-time job, and he undoubtedly was a "top candidate."[19]

Sudhoff used his prominent position to put his scholarly credo into practice, a credo that he had described to Győry in 1898 in the following terms: our wretched "earthly life" can only be imbued with any value "by way of our gathering the treasures of the soul and the intellect that continuously dwell among us as our outward possessions."[20] Very much in the spirit of such heuristic pronouncements, which attest to the absolute inseparability of life and work, Sudhoff proceeded to action. Over the years that followed, he developed a passion for gathering and collecting material that remains unrivalled in the field of medical history to this day. The library of his institute at Leipzig soon boasted the most extensive collection of scholarly material in the world; it put Leipzig firmly on the map as a center of information and specialist training and, generally speaking, as a model institution of the young discipline. During his first five years as a professor at Leipzig alone, Sudhoff's research missions took him all over Germany, as well as to Denmark, Sweden, Italy (thirty more research trips were to follow), and France; in search of medically relevant manuscripts, incunabula, early prints, illustrations, and archaeological finds, he rummaged through the libraries of Munich, Brunswick, Hildesheim, Göttingen, Hannover, Lüneburg, Hamburg, Lübeck, Berlin, Rostock, Schwerin, Kiel, Copenhagen, Lund, Uppsala, Stockholm, Florence, Bologna, and Paris; conferred with experts in his auxiliary disciplines; and brought all manner of treasured reproductions, excerpts, and bibliographical notes back to Leipzig. He even visited places like Wolfenbüttel's Herzog August Library, an all-but-forgotten institution at the time.[21] During World War II, the Leipzig collections would be moved into safe storage, but they were ultimately lost nonetheless. Today, it is only Sudhoff's own works that still attest to the wealth of material once gathered at the Leipzig institute.[22]

Methodologically, Sudhoff's oeuvre is characterized by a source-oriented philological approach; in terms of subject matter, the focus lies on the German-speaking regions and the Middle Ages. In addition to prints and manuscripts, Sudhoff also maintained a scholarly interest in medical realia, to which his exhibition catalogs pay impressive testimony—take, for example, an early one like that for the Rhineland Goethe Exhibition in 1899, which lists 2,402 separate exhibits,[23] or the catalog accompanying the major Hygiene Exhibition at Dresden in 1911,[24] where Sudhoff acted as chairman of the Historical Section and for which he and his eighty-five staff procured countless exhibits and commissioned over 600 reconstructed models to be displayed across 4,000 square meters of exhibition space.[25]

Ultimately, however, Sudhoff's success lay not so much in his methodological orientation, rooted in Ranke's historicism but also tolerant of a variety of other approaches,[26] as in his integrative powers, which enabled him to win scholars from all over the world as collaborators and suppliers of valuable information to the gigantic stock-taking exercise that he had in mind and the Baconian dimensions of which he alone was aware. Such an exercise, initially conceived to map out the field of medical history and gradually reveal the discipline's vast scholarly potential, had to be backed up by extensive bibliographic research. He used the *Mitteilungen* as an organ for reviews and the exchange of information about every conceivable medico-historically relevant activity. He even commissioned an array of international contributors whose task it was to sift through foreign-language publications that were largely unavailable to the German-speaking academic audience and to review them for the *Mitteilungen* in either German, English, or French.[27] Such "representatives" also proved useful in recruiting new members to the DGGMN. They were in a position to disseminate new research quickly and across linguistic boundaries, and they generally enhanced the reputation of an association that habitually organized its meetings in the spirit of "family reunions"[28] of members who rarely separated their professional from their private lives and who, through their membership, benefited not only as medical historians but also as keen international tourists.

Above all, however, Sudhoff, whose Leipzig institute served as the nerve center of all these networks, was able to put his "representatives" to good use when it came to appointments, internal politics, and scholarly disputes. Győry in particular entertained correspondences with almost every German speaking medical historian of the time and was generally very close to the DGGMN; he thus functioned as Sudhoff's loyal informer on morale, opinion, and political intrigue, and he repeatedly advised Sudhoff on ways to consolidate power. Thanks to the Leipzig institute's excellent research facilities, the association's annual meetings, and Sudhoff's attendance at numerous international conferences, Sudhoff was personally acquainted with most of his colleagues and had plenty of opportunity to weave his various webs. That countless medical historians all over the world, and in Germany in particular, seem to have had their gaze almost monomaniacally fixed on Sudhoff first of all illuminates Sudhoff's autocratic position of power, which he managed to build up over the years. This explains in large part how he was able to direct the fortunes of German medical history well into his old age and still set the course when it came to the response as a field to the challenge of National Socialism.

But let us begin at the beginning. On 28 April 1933, Sudhoff wrote to Győry: "I have taken things firmly into my own hands again and am now demanding decisions. Politically, I have decided resolutely to back Adolf Hitler, who alone promises a future, and have thus joined his party."[29] Sudhoff's membership in the NSDAP certainly caused something of a stir within medico-historical circles. In a letter to Győry dated 10 May 1933, Walter von Brunn expressed his regret over the fact that "Sudhoff has become a Nazi." One could, von Brunn continued, hardly spare Sudhoff the "accusation" of actually having been "quite pro-Semitic" in the past. "That Sudhoff of all people has now joined the Nazis is something I find extremely embarrassing, especially given the fact that, only a short while ago, when I very guardedly talked to him about my own leanings, he got all peeved and retorted in a rather condescending manner that he certainly wasn't a Nazi, while somehow simultaneously accusing me of being one." After all, it had, to von Brunn's mind, first and foremost been Sudhoff's "patronage of the Jews" that had run the DGGMN into the ground. As if he still struggled to believe it, von Brunn repeated in the same letter: "That Sudhoff of all people, Sudhoff, the intimate friend and protector of the Jews, should now have become a Nazi hurts me bitterly! But he wants to be part of the action whatever tune is playing at the moment, and will no doubt be cheered on by thousands of students! It must have been his insatiable vanity that drove him to take this, in my view rather embarrassing step!"[30] By July, Sigerist had also heard of Sudhoff's party membership and, like von Brunn, attributed Sudhoff's decision to vanity.[31] In the eyes of his contemporaries, then, it was not anti-Semitism as such that had motivated Sudhoff to join the NSDAP. Even if Sudhoff had, occasionally, made anti-Semitic comments in the past,[32] his reservations against the Jews never took the aggressive form of some of his colleagues' and correspondents' anti-Semitism,[33] or for that matter, that of his oldest friend and colleague Győry,[34] whose anti-Semitic tirades Sudhoff nevertheless never contradicted in his letters. A good deal of von Brunn's "accusation" against Sudhoff as a "protector of the Jews" had its roots in Sudhoff's close relationship with his pupil Henry E. Sigerist,[35] who was regarded in German nationalist and National Socialist circles not only as non-German, a socialist, capitalist, and a bolshevist, but also, erroneously, as a Jew. The envy and hatred directed at him as Sudhoff's "heir apparent" thus often appears indistinguishable from the more or less pronounced anti-Semitism of his colleagues.[36]

After Sudhoff had joined the NSDAP, he began reading Hitler's *Mein Kampf*, a book that seems, once again, to have been recommended to him by his Hungarian friend, who, for his part, had been alerted to it by Walter von Brunn.[37] While von Brunn underscored his recommendation with an emphatic exclamation mark and Győry called the two volumes a "most interesting and profitable" read, Sudhoff simply noted on 14 June 1933 that he, too, had "read the first volume of Adolf Hitler's *Mein Kampf* and shall get to part two as soon as I find the time."[38] Again, it appears that while Sudhoff had undoubtedly caught the National Socialist bug, he was not infected quite as vigorously by it as some of his colleagues were.[39]

All this took place against the backdrop of a period of radical change within academic medical history, and Hitler's impact was as drastic here as it was on other disciplines.[40] The Act for the Restoration of a Tenured Civil Service, which had been passed on 7 April 1933, deprived a whole host of mostly Jewish medical historians not only of their teaching posts but also of any alternative sources of income, their physical safety, and ultimately of their homes and homeland. One could mention Richard Koch (1882–1949),[41] Theodor Meyer-Steineg (1873–1936),[42] Walter Pagel (1898–1983),[43] Ludwig Edelstein (1902–1965),[44] and Ernst Hirschfeld, who was probably murdered by the Nazis.[45] Owsei Temkin (1902–2002)[46] and the non-Jewish Henry E. Sigerist (1891–1957)[47] had left the country in 1932, but Temkin was nevertheless—*post festum*—struck off the German register of university professors in 1933.[48] Among Győry's circle of correspondents, it was generally accepted that the new spirit of the times demanded certain sacrifices,[49] and it was occasionally even explicitly welcomed that medical history was now going to be a purely Christian discipline;[50] this way, the cultural-historical, sociomedical, and medicotheoretical research methods propagated largely by the Jewish members of the discipline would at last have to make way for philologically bibliographical or historicocritical approaches.[51] As far as we can glean from Sudhoff's correspondences, he himself did not, however, participate much in this debate, which either played down the fate of the persons affected[52] or sought to excuse the offenses committed against them.[53] At the same time, he did almost nothing to help those of his colleagues who had been dismissed or forced into exile.[54] A widely supported exception,[55] the only one, was made for Richard Hermann Koch, former professor of medical history, director of the Department of Medical History at Frankfurt (1926–1933), and a "domesticated pre-war Jew."[56]

While the Jewish colleagues who had thus been deprived of both their jobs and their rights barely scraped an existence, fled into exile, or groveled in the vain hope of gaining permission to stay, the *Gleichschaltung* (i.e., the appointment of Nazis to key posts in government agencies and other organizations) of the DGGMNT,[57] an organization over which Sudhoff still exerted a considerable amount of influence, got under way. At the annual meeting in Erfurt (8–10 September 1933), three members of the Berlin Institute of Medical History were elected to the centralized executive committee: Paul Diepgen (1878–1966)[58] became president, Walter Artelt (1906–1976)[59] secretary, and Ludwig Englert (1902–1981)[60] treasurer; previously independent local associations, meanwhile, were converted into affiliate *Ortsgruppen*. And even though Diepgen expressly regretted the "retirement of Richard Koch and Walter Pagel from their teaching posts," such regrets did not prevent the assembly from sending the following telegram to the German chancellor: "On 9 September, the individual German Associations for the History of Medicine, Science, and Engineering have merged to form one unified German Association for the History of Medicine and Science. The association equips our young people and our practicing German doctors, scientists, and engineers with the historical knowledge that arises from the German national health efforts and the historical deeds of German doctors, scientist, and engineers in their struggle to preserve our German national traditions and the German nation. The association fosters an appreciation of historical tradition as the eternal source of German culture. Finally, the association wishes to express its deeply felt respect to the Führer."[61] Last but not least, the twenty-six-strong general meeting at Erfurt adopted new statutes, in which the objectives of the association were defined as follows: "It thus also aims to further our knowledge of the character of the German peoples, German tradition, and German culture, and to contribute to putting our discipline into the service of a national political education." On the following day, Sudhoff was offered the association's honorary presidency, while Paul Diepgen, on behalf the German president von Hindenburg, presented Sudhoff with the Goethe Medal and announced the establishment of the commemorative Sudhoff Lecture, the first of which Diepgen delivered himself; it has, as we said above, remained an annual fixture to this day. "The reverent mood was gratifying in its sincerity," the thrice honored Sudhoff noted with some satisfaction.[62] To his mind, both the honorary presidency and the Goethe Medal had been long overdue.[63]

On his birthday on 26 November, Sudhoff devised a special treat for himself. He saw to it that the Sudhoff Medal was awarded jointly to Sigerist and Diepgen.[64] Diepgen, "who, in his current leading position, was suffering some pangs of conscience," was eventually "persuaded" to accept.[65] This surely suggests that the award of the medal to Sigerist had been felt within the association to be a somewhat awkward, if not explosive, issue. Overall, one gets the impression that this was Sudhoff's last and final attempt to halt the association's increasing international isolation while trying to strengthen the position of his own "camp" within Germany.

Finally, let us consider Sudhoff's motivation for actively participating in all this politicking in spite of his advanced years. One cannot help but wonder whether, at the age of eighty, his motives were not primarily personal rather than ideological or methodological ones. His letters from the year 1933 suggest that concerns for his chair and his institute, the question of his succession and his legacy, constituted the main motivation behind his various tactical and strategic moves. Before Hitler had come into power, rumors that medical history might lose the Leipzig chair altogether had been rife.[66] Indeed, after Sigerist's position had unexpectedly fallen vacant with his departure for Baltimore, the university had, albeit without clearance from the ministry, decided not to appoint a successor.[67] The academically controversial Nazi sympathizer Achelis, whom Sudhoff had cleverly installed as the institute's interim director,[68] left in March 1933 after only a brief stint at the Institute for the Prussian Ministry of Science, Art, and Public Education, where, following a short probationary period, he was appointed assistant secretary.[69] When, on 9 April—only eight days after the so-called *Judenboykott*—Sudhoff set off on an extended trip to Greece, everything still hung very much in the balance.

On his return at the end of April, Sudhoff may well have felt that events, particularly in politics, had come thick and fast during his absence. Terror, celebrations, and public marches had become everyday occurrences; the Gleichschaltung was well under way and concentration camps were set up; new laws and decrees passed every single day.[70] Perhaps he felt that the only way for him to "seize the reins" once again would be to follow the example of the Berlin institute and join the Nazi party,[71] thus consolidating his power and preserving his influence on appointments within the discipline. On 13 May, he wrote to Sigerist that he had "now gradually become used to the idea of having the chair offered to Prof[essor] Heinz Zeiss."[72] In the end, however, Zeiss chose to accept a position as a senior civil servant in the German Public Health Department instead. While both Haberling[73] and von Brunn[74] had entertained hopes for the Sigerist succession for a good year, Sudhoff, via Győry, gave them to understand that the ministry preferred the appointment of someone younger. On 24 August, 1933, Sudhoff asked Sigerist, whom he had planned but then failed to catch up with in Basel during one of his research trips, for a "frank and unvarnished" report on Fritz Lejeune, "who, as a Nazi, has a fair chance of coming to Leipzig if we do not take precautions."[75] Sigerist reacted with consternation and wrote a report that reflected his alarm at this prospect;[76] in the end, he proposed Martin Müller as a suitable candidate for the position.[77] It would take another full year before the institutional tug-of-war finally came to an end with the appointment of von Brunn. During this time, Sudhoff still reigned supreme at "his" institute. His decisionistic anti-Semitism, his joining of the NSDAP as an "April apostate," his eagerness to compromise, and his strategic maneuvering not only had saved his chair but also had, in Germany at least, consolidated his power, earning him a somewhat notorious posthumous reputation. Tellingly, neither Diepgen nor Sigerist attended the Sudhoff Medal award ceremony in 1933.

The spell of Sudhoff's integrative powers was broken, compromised by his political entanglements and the changing times. What Sudhoff had once advised Győry to write about the noble aims of the association simply was not true any more. Medical historians in Berlin and Baltimore had by now chosen to take very different and ultimately irreconcilable paths. Sudhoff had thrown in his lot with the wrong crowd and was unable, even at the cost of conformity and yesmanship, to halt the fall of German medical history. Perhaps power and fame at home ultimately were more important to him than the moral integrity of his academic achievements. Yet we cannot help but ask whether the price German medical history continues to pay for the "gains" from such an unholy alliance has, in the final analysis, not been rather high.

NOTES

I wish to thank Alexa Alfer, London, for her expert translation.
This chapter is dedicated to the memory of Owsei Temkin.

1. The only comprehensive biography is Dirk Rodekirchen, "Karl Sudhoff (1853–1938) und die Anfänge der Medizingeschichte in Deutschland" (Med. diss., University of Cologne, 1992).

2. See, for example, the Festschrift on occasion of Sudhoff's sixtieth birthday, which contains fifty-five individual contributions and was published as the sixth volume of the *Archiv für Geschichte der Naturwissenschaften*. On his seventieth birthday in 1923, von Brunn paid his respects in the *Deutsche Medizinische Wochenschrift*, while Henry E. Sigerist congratulated in the *Archivio di Storia della Scienza*, Paul Diepgen in the *Münchener Medizinische Wochenschrift*, Max Neuburger in the *Klinische Wochenschrift*, Arturo Castiglioni in the *Bolletino dell' Istituto storico italiano dell' arte sanitaria*, and M. A. van Andel in *Janus*. Karl Sudhoff's seventy-fifth birthday in 1928 was marked by the editors of the *Mitteilungen zur Geschichte der Medizin und der Naturwissenschaften*, as well as by a contribution by Ernst Hirschfeld to the same publication. On the occasion of his eightieth birthday, Sudhoff was honored by Ernst von Seckendorf in *Die Medizinische Welt* and by the Medical History Club at Baltimore's Johns Hopkins University. In 1938, obituaries were penned by Walter von Brunn for the *Deutsche Medizinische Wochenschrift*, by Henry E. Sigerist for the *Bulletin of the History of Medicine*, by Martin Müller for the *Klinische Wochenschrift*, by Pietro Capparoni for the *Bolletino dell' Istituto storico italiano dell' arte sanitaria*, by Walter Artelt for *Janus*, probably by Paul Diepgen for the *Münchener Medizinische Wochenschrift*, and once again by Walter von Brunn for *Sudhoffs Archiv* and the *Mitteilungen zur Geschichte der Medizin, der Naturwissenschaften und der Technik*. The centenary of Sudhoff's birth was marked by Johannes Hett in *Kosmos*. For detailed references see Thomas Rütten, "Karl Sudhoff, 'Patriarch' der deutschen Medizingeschichte. Zur Identitätspräsentation einer wissenschaftlichen Disziplin in der Biographik ihres Begründers," in *Médecins érudits, de Coray à Sigerist*, Danielle Gourevitch, ed. (Paris: Boccard, 1995), 154–171 and 208–211, 160–161, n. 22–39.

3. These include the Goethe Medal, 1938; Hungarian Order of Merit, 1938; German Red Cross Decoration First Class, 1938, to mention but a few.

4. See Wilhelm Haberling to Tibor Győry, Düsseldorf, 19 May 1933, Tibor Győry Papers, Archives of the Semmelweis University, Budapest (hereafter, Győry Papers): "On Wednesday, I went to Hochdahl and discussed all the necessary arrangements for the Sudhoff ceremony. The plaque for the house in which he first studied Paracelsus is going to be very dignified, and the good people of Hochdahl are wildly enthusiastic about this proposal. The street on which his house lies will be renamed Karl-Sudhoff-Straße—in short, I am very pleased that Sudhoff is still around to see himself honored in this way." See also Wilhelm Haberling to Tibor Győry, Göttingen, 7 September 1933, Győry Papers: "At the end of July, we unveiled a large commemorative plaque outside Sudhoff's house in Hochdahl. Diepgen made an excellent speech, and the large public turnout at the ceremony showed how popular S[udhoff] still is in the little town where he first started out."

5. Established in 1907, the journal changed its title in 1929 to *Sudhoffs Archiv für Ge-*

schichte der Medizin und der Naturwissenschaften. See Karl Sudhoff, "Aus meiner Arbeit," *Sudhoffs Archiv für Geschichte der Medizin* 21 (1929): 333–387, esp. 372–373; and Rodekirchen, "Karl Sudhoff," 17.

6. Initiated on Sigerist's suggestion, the DGGMNT has since awarded the medal annually to leading medical historians. The awards to Sigerist and Diepgen in 1933, to Győry in 1934, Haberling in 1935, and Zaunick in 1936, were, of course, centrally influenced by Sudhoff himself. On details of how the Sudhoff Medal came into being, see Henry Sigerist, "Erinnerungen an Karl Sudhoff," *Sudhoffs Archiv* 37 (1953): 97–103, 101. See also Rolf Winau, *Deutsche Gesellschaft für Geschichte der Medizin, Naturwissenschaft und Technik 1901–1976* (Wiesbaden: Steiner, 1978), 42.

7. "Altmeister": Max Neuburger, "Karl Sudhoff zum 70. Geburtstage," *Klinische Wochenschrift* 2 (1923): 2219; Martin Müller, "Karl Sudhoff †," *Klinische Wochenschrift* 17 (1938): 1639; Walter von Brunn, "Karl Sudhoff †," *Mitteilungen zur Geschichte der Medizin, der Naturwissenschaften und der Technik* 37 (1938): 297–302, 300; Gunter Mann, "Unser Bild," *Medizinhistorisches Journal* 17 (1982): 393. "Meister": Arturo Castiglione, "Carlo Sudhoff," *Bolletino dell' Istituto storico italiano dell' arte sanitaria* 3 (1923): 158; Paul Diepgen, "Karl Sudhoff als Medizinhistoriker. Zu seinem 70. Geburtstag am 26. November 1923," *Münchener Medizinische Wochenschrift* 70 (1923): 1414–1416, 1414; Henry E. Sigerist, "Karl Sudhoff. Sein siebzigster Geburtstag, 26. November 1923," *Archivio di storia della scienza* 5, no. 2 (1924): 139–147, 142 and 147; Ernst Seckendorf, "Karl Sudhoff, dem größten Paracelsusforscher, zum 80. Geburtstage," *Die Medizinische Welt* 7 (1933): 1697; Walter von Brunn, "Karl Sudhoff zum Gedächtnis," *Deutsche Medizinische Wochenschrift* 64 (1938): 1552–1553, 1553; Walter von Brunn, "† Karl Sudhoff," *Sudhoffs Archiv* 31 (1938): 338–342; Paul Diepgen, "Karl Sudhoff: Leben und Wirken eines großen Meisters," *Wissenschaftliche Zeitschrift der Karl-Marx-Universität Leipzig (Math.-Naturwiss. Reihe)* 5 (1955/1956): 23–25; [F.R.], "Karl Sudhoff †," *Beiträge zur Geschichte der Veterinärmedizin* 1, no. 4 (1938): 255. "Führer": Paul Diepgen, "Karl Sudhoff als Medizinhistoriker. Zu seinem 70. Geburtstag am 26. November 1923," *Münchener Medizinische Wochenschrift* 70 (1923): 1414–1416, 1414. "Spiritus rector": Max Neuburger, "Karl Sudhoff zum 70. Geburtstage," *Klinische Wochenschrift* 2 (1923): 2219. "Magister mundi": Paul Diepgen, "Zur hundertsten Wiederkehr des Geburtstages von Karl Sudhoff am 26. November 1953," *Archives Internationales d' histoire des sciences* 6 (1953): 260–265, 262. "Nestor": [Anon.], "Onoranze a Carlo Sudhoff, il nestore degli storici della medicina," *Bolletino dell' Istituto storico italiano dell' arte sanitaria* 13 (1933): 208; Pietro Capparoni, "Carlo Sudhoff," *Bolletino dell' Istituto storico italiano dell' arte sanitaria* 2, ser. 4 (1938): 339–341, 339. "Patriarch": Walter Artelt, "Karl Sudhoff †," *Janus* 43 (1939): 84–91, 90. "Absoluter König": Wilhelm Haberling to Tibor Győry, Coblenz, 28 March 1925, Győry Papers.

8. Even Andreas Frewer's ambitious attempt to reconstruct the "important steps and stages of the institutionalization as reflected in Sudhoff's biography and works" seems to work on the assumption that one can do without a detailed knowledge of the countless letters Sudhoff wrote during this "core period," notably to his oldest friend and colleague Tibor Győry in Budapest. Sudhoff's contribution to this correspondence, preserved in the archives of Berlin's Humboldt University and Budapest's Semmelweis University, undoubtedly represents the single most important source for an endeavor such as Frewer's. We are talking here about more than 350 letters by Sudhoff himself, plus a further 500 letters that Győry received from various other German-speaking colleagues between 1898 and 1937.

See Thomas Rütten, "Briefwechsel zwischen Tibor von Győry (1869–1938) und Karl Sudhoff (1853–1938)," *Wolfenbütteler Bibliotheks-Informationen* 21 (1996): 10–11, and Anne K. Halbach, *Briefe von Walter von Brunn (1876–1952) an Tibor Győry (1869–1938) aus den Jahren 1924–1937. Ein Beitrag zum Korrespondentennetz Tibor Győrys mit deutschen Medizinhistorikern* (Med. diss., University of Münster, 2003), 6–8. See also Andreas Frewer, "Biographie und Begründung der akademischen Medizingeschichte: Karl Sudhoff und die Kernphase der Institutionalisierung 1896–1906," in *Die Institutionalisierung der Medizinhistoriographie. Entwicklungslinien vom 19. ins 20. Jahrhundert,* Andreas Frewer and Volker Roelcke, eds. (Stuttgart: Franz Steiner Verlag, 2001), 103–126, 105.

9. See Rudolf Stichweh, *Wissenschaft, Universität, Professionen. Soziologische Analysen* (Frankfurt: Suhrkamp, 1994), 17.

10. Michael Stolberg, "Heilkundige: Professionalisierung und Medikalisierung," in *Medizingeschichte: Aufgaben, Probleme, Perspektiven,* Norbert Paul and Thomas Schlich, eds. (Frankfurt: Campus, 1998), 69–86, 70–71.

11. For details on the establishment of the DGGMN, which had emerged from the Historical Division of the Assembly of Natural Scientists (founded in 1886), see Yvonne Steif, "Die Entstehung der DGGMN aus den Versammlungen deutscher Naturforscher und Ärzte: Zur Institutionalisierung einer wissenschaftlichen Disziplin," in Frewer and Roelcke, *Institutionalisierung,* 143–161.

12. See most recently the figures provided in Bernhard vom Brocke, "Die Institutionalisierung der Medizinhistoriographie im Kontext der Universitäts- und Wissenschaftsgeschichte," in Frewer and Roelcke, *Institutionalisierung,* 187–212, 191.

13. The phrase comes from the subtitle of Frewer's "Biographie und Begründung."

14. For the Puschmann Foundation, see Ortrun Riha, "Die Puschmann-Stiftung und die Diskussion zur Errichtung eines Ordinariats für Geschichte der Medizin an der Universität Leipzig," in Frewer and Roelcke, *Institutionalisierung,* 127–141, 127–129.

15. Ibid., 130: "Details as to the content of these negotiations are not known."

16. Karl Sudhoff to Tibor Győry, Hochdahl, 4 May 1905, Győry Papers. Heinrich Curschmann (1846–1910) was a consultant internist. The meeting took place on 26 April 1905. Once the faculty had applied to the Royal Ministry of Culture and Public Education in Dresden on 23 June for the establishment of a chair in the history of medicine, ministerial approval was granted immediately and the faculty proceeded to nominate Sudhoff for the post; he was formally offered the chair on 1 July.

17. Karl Sudhoff to Tibor Győry, Hochdahl, 5 April 1905, Győry Papers.

18. Ibid. See also Frewer, "Biographie und Begründung," 120, n. 109, who puts the offer from Leipzig and Sudhoff's rejection of the Rostock offer in reverse chronological order.

19. Riha, "Die Puschmann-Stiftung," 130.

20. Karl Sudhoff to Tibor Győry, Hochdahl, 29 October 1898, Győry Papers.

21. Karl Sudhoff to Tibor Győry, Leipzig, 17 June 1907, Győry Papers

22. The Leipzig collections, packed up in 144 boxes and moved to a rock cellar underneath Castle Mutzschen on 3 May 1943, contained a large number of manuscripts and incunabula. Various documents attest to the fact that all of these boxes were confiscated and transported to the Soviet Union in May 1945. The material has since been regarded as lost without trace. See the list of contents of the collections of the Karl Sudhoff Institute moved to the cellars of Castle Mutzschen in 1943 (University Archive, Leipzig, Karl Sudhoff Institut, Schriftgutsammlung); see also Ingrid Kästner, "Walter von Brunn (1876–1952).

Direktor des Instituts 1934–1950," in *90 Jahre Karl-Sudhoff-Institut an der Universität Leipzig*, Achim Thom and Ortrun Riha, eds. (Leipzig: Karl-Sudhoff-Institut für Geschichte der Medizin und der Naturwissenschaften, 1996), 44–54, 51–52.

23. See *Rheinische Goethe-Ausstellung. Unter dem Protektorat seiner Königlichen Hoheit, des Prinzen Georg von Preussen, in der Aula der Königlichen Kunstakademie zu Düsseldorf. Juli bis October 1899* (Leipzig: Ed. Wartigs, 1899).

24. See Heike Fleddermann, *Karl Sudhoff und die Geschichte der Zahnheilkunde, die Beziehungen zu Karl August Lingner und die Dresdner Hygiene-Ausstellung 1911* (Med. dent. diss., University of Cologne, 1992), 22–25; Klaus Gilardon, "Karl Sudhoff und die medizinhistorischen Quellen—ein Beitrag zu seiner Sammel-und Ausstellungstätigkeit," *Nachrichtenblatt der Deutschen Gesellschaft für Geschichte der Medizin, Naturwissenschaften und Technik* 32 (1992): 106–109.

25. See Sudhoff's own account in his autobiographical essay "Aus meiner Arbeit. Eine Rückschau," *Sudhoffs Archiv* 21 (1929): 333–387, 376.

26. See Peter Schmiedebach's contribution to this volume, chap. 4.

27. In a letter dated 24 October 1901, Sudhoff "explicitly appoints" his friend Tibor Győry as "the Hungarian representative of our association." For a newspaper report Győry was going to write on the establishment of the association, Sudhoff advised him to emphasize the following points: "the universal character of the society, its complete lack of chauvinism of any sort, German, English, and French as our common languages, History of Medicine and Natural Sciences and Engineering! Research worldwide and not just Germany etc. etc. You have plein pouvoir!" Karl Sudhoff to Tibor Győry, Leipzig, 24 October 1901, Győry Papers.

28. See Anke Jobmann, *Familientreffen versus Professorenelite? Vergangenheitsbewältigung und Neustrukturierung der deutschen Wissenschaftsgeschichte der 60er Jahre* (Berlin: ERS-Verlag, 1998).

29. Karl Sudhoff to Tibor Győry, Leipzig, 28 April 1933, Győry Papers. A letter to Sudhoff from Richard Arthur Golf (1877–1941), then rector of the University of Leipzig, dated 19 June 1934 (University Archive, Leipzig, KSI 41a, 222), confirms Sudhoff's joining of the NSDAP: "I welcome you wholeheartedly as a card-carrying fellow party member. Heil Hitler!"

30. Walter von Brunn to Tibor Győry, Rostock, 15 May 1933, Győry Papers. See also Halbach, *Briefe*, 193–195. This assessment did not, however, stop von Brunn from enthusing in his obituary of Sudhoff that "He was a passionate follower of our Führer and his great works." See Walter von Brunn, "Karl Sudhoff †," *Mitteilungen zur Geschichte der Medizin, der Naturwissenschaften und der Technik* 37 (1938): 297–302, 301.

31. See Sigerist's journal entry dated 25 July 1933, Henry Ernest Sigerist Papers, Manuscripts and Archives, Sterling Library, Yale University, New Haven, Connecticut, box 1, folder 1: "A walk with Dr. Edelstein to Saret and to Flx-Fall, walking and discussing philological problems and again and again the German situation. He left after lunch. I was so glad to have him here in the peaceful surroundings of the Engadine. All the Germans ought to be able to go abroad for some time in order to realize that their life at home is mere hysteric, that the whole 'Hitlerei' is not more than an episode, that there are other values, real values. The intellectual isolation of Germany is complete. It gave me a shock to hear that Sudhoff has become a nazi. Why? Just in order to be 'dabei,' It is disgusting." The journal entry attests that, contrary to what has been assumed—not least on the strength of Sigerist's own

pronouncements (see *Henry E. Sigerist. Autobiographische Schriften,* Nora Sigerist-Beeson, ed. [Stuttgart: Thieme, 1970], 153, 238–239)—Sudhoff's obituary notice was not the first time Sigerist had heard that Sudhoff was a party member. See also Werner F. Kümmel, "Im Dienst nationalpolitischer Erziehung? Die Medizingeschichte im Dritten Reich," in *Medizin, Naturwissenschaft, Technik und Nationalsozialismus. Kontinuitäten und Diskontinuitäten,* Christoph Meinel and Peter Voswinckel, eds. (Stuttgart: Franz Steiner, 1994), 293–319, 295.

32. See Rütten, "Sudhoff," 166–171. Comments such as "Yes, Singer is an insolent Jew without the slightest trace of the charm his musical name might suggest, without the slightest trace!" (Karl Sudhoff to Tibor Győry, Leipzig, 15 December 1933, Győry Papers) are rare and are most probably rooted in personal animosity rather than in an anti-Semitic attitude per se, as Charles Singer, *From Magic to Science: Essays on the Scientific Twilight* (New York: Dover, 1958), xvi–xvii, suggests. For Singer, compare *Jüdisches Biographisches Archiv,* microfiche 596: 220–228, and the obituaries in *Isis* 51 (1960): 558–560, *Medizinische Klinik* 55 (1960): 1476–1477, *Gesnerus* 17 (1960): 73–74, *Medical History* 4 (1960): 353–358, and *Journal of the History of Medicine* 15 (1960): 420–421.

33. For the anti-Semitism of von Brunn, who stresses in a 10 May 1933 letter to Győry that he had been so "strongly nationalist" that the Nazis always "treated me as one of their followers, but only because I have always been nationalist and have always maintained a cool reserve toward the Jew," see Halbach, *Briefe,* no. 27, 193.

34. Győry even claimed an "innate anti-Semitism" for himself and confessed to "have always been a strong enemy of the Jews." See Tibor Győry to Karl Sudhoff, Budapest, 31 May 1933 and 23 September 1933, Herzog August Library, Wolfenbüttel, 99 Novissimi 4°, f. 164 and 167. Győry's letters show that he stereotypically equated Judaism with Bolshevism, Socialism, Communism, and Internationalism; his anti-Semitism thus seems to have been primarily politically motivated. This would also explain why Henry E. Sigerist was reputed to be Jewish. After Hitler's election victory, however, Győry seems to have toned down his own anti-Semitism somewhat in comparison to Hitler's: "I, too, am calling out Heil Hitler!, for he has put a stop to the Social Democrats and has got on at the Jews, even though that may have been a bit too radical. I once read Werner Sombart's book on economics and the Jews, and I will always remember how this German, nay, Teuton, speaks out against them yet hesitates to recommend we get rid of the Jews lock, stock, and barrel, for this would also harbor certain dangers for those of pure race. I would have made a clear distinction between the domesticated pre-war Jews such as Koch and some of the Nobel Prize winners etc., and those Jews who have come up in the world during and through the war." See Tibor Győry to Karl Sudhoff, Budapest, 31 May 1933, Herzog August Library, 99 Novissimi 4°, f. 164. See also Walter von Brunn to Tibor Győry, Rostock, 10 May 1933, Győry Papers, also in Halbach, *Briefe,* no. 57, 193–195,195. This quotation does, however, also show that Győry was no stranger to a eugenically motivated anti-Semitism.

35. For details on this relationship, see Ingrid Kästner, "The Relationship between Karl Sudhoff (1853–1938) and Henry E. Sigerist (1891–1957) as Shown in Their Correspondence," in *Actas del XXXIII Congreso Internacional de Historia de la Medicina. Granada, Sevilla, 1–6 septiembre, 1992,* Juan L. Carillo, ed. (Seville: San Fernando, 1994), 1073–1077.

36. Walter von Brunn to Tibor Győry, Rostock, 22 October 1933, Győry Papers: "Thank God you have for once shown your teeth to Sigerist the half-Jew (which, in my opinion, is exactly what he is)!" Georg Sticker to Tibor Győry, Würzburg, 21 January 1934, Győry Papers: "He [i.e. Sigerist] and Meyer-Steineg both show the same racial features." See also

Halbach, *Briefe*, 32, n. 202 and pp. 215–216. It thus seems that the feelings of resentment against Sigerist's academic successes, his wealth, and his prominent position as Sudhoff's heir apparent gave rise to his being labeled, with pejorative and discriminatory intentions, a Jew at the time; the remarkable persistence of this label to this day is illustrated by the following passage from a recent publication: "Naturally, Sudhoff was cooperative . . . and maintained close relations with Jews such as Temkin and Sigerist"; Gundolf Keil, "Paracelsus und die neuen Krankheiten," in *Paracelsus. Das Werk-die Rezeption. Beiträge des Symposiums zum 500. Geburtstag von Theophrastus Bombastus von Hohenheim, genannt Paracelsus (1493–1541) an der Universität Basel am 3. und 4. Dezember 1993,* Volker Zimmermann, ed. (Stuttgart: Steiner, 1995), 17–46, 20.

37. Walter von Brunn to Tibor Győry, Rostock, 7 May 1933, Győry Papers, also in Halbach, *Briefe,* no. 56, 192: "Please read Hitler, Mein Kampf!" Tibor Győry to Karl Sudhoff, Budapest, 31 May 1933, Herzog August Library, 99 Novissimi 4°, f. 163.

38. Karl Sudhoff to Tibor Győry, Leipzig, 14 June 1933, Győry Papers.

39. The extent to which enthusiasm for the national socialist cause was rife among Sudhoff's immediate colleagues is illustrated by one of Wilhelm Haberling's letters:

> These days, and in the bright glow of enthusiasm for our beloved German fatherland that has seized our entire nation, time simply seems to fly by. Every day brings something new that will eventually make it clear to everyone that Adolf Hitler has taken the right path toward uniting our nation. Gigantic things have been accomplished in a very short space of time, and the trust and confidence of the people will smooth the way for what lies ahead. I wish you could have been in Germany on the 1st of May and seen the happy and inspired faces everywhere—faith in Germany's future has at last been restored to our nation after all those long, difficult years during which many of us had almost begun to despair. I, for my part, never doubted, of course, that the day would come when our German nation would rise again. And now this day is indeed here, sooner and more splendid than we had ever expected! I firmly believe that such a lifting of our national soul will soon also have a stimulating effect on large sections of the national intellect, and that our discipline in particular will receive a new impetus for intensive research. My hopes for the History of Medicine are that our teaching efforts will at last be fully recognized and that, in the spirit of Hitler, our young doctors shall once again learn to respect and revere the great German physicians of the past and immerse themselves in their venerable works.

Wilhelm Haberling to Tibor Győry, Düsseldorf, 8 May 1933, Győry Papers.

40. See Paul Weindling, "Medical Refugees and the Renaissance of Medical History in Great-Britain, 1930s–60s," in *Eine Wissenschaft emanzipiert sich. Die Medizinhistoriographie von der Aufklärung bis zur Postmoderne,* Ralf Bröer, ed. (Pfaffenweiler: Centaurus Verl.-Ges., 1999) 139–151, 148. See also the emendation in Halbach, *Briefe,* 28, n. 172. Compare Marcel H. Bickel, "Medizinhistoriker im 19. und 20. Jahrhundert: Eine vergleichend-biographische Betrachtung," in Frewer and Roelcke, *Institutionalisierung,* 213–234.

41. For Koch, see *Jüdisches Biographisches Archiv,* microfiche 397: 117–120; Karl E. Rothschuh, "Richard Hermann Koch (1882–1949): Arzt, Medizinhistoriker, Medizinphilosoph," *Medizinhistorisches Journal* 15 (1980): 16–43 and 223–243; Hannelore Schwann, "Richard Kochs Beziehungen zum Karl-Sudhoff-Institut. Archivstudie anläßlich seines 100. Geburtstages," *NTM. Internationale Zeitschrift für Geschichte und Ethik der Naturwissenschaften, Tech-*

nik und Medizin 19 (1982): 94–103; and *Richard Koch und die ärztliche Diagnose,* Gert Preiser, ed., Frankfurter Beiträge zur Geschichte, Theorie und Ethik der Medizin, vol. 1 (Hildesheim: Olms Weidmann, 1988).

42. For Meyer-Steineg, see Susanne Zimmermann, "Theodor Meyer-Steineg (1873–1936) und die Medizingeschichte in Jena," in Bröer, *Eine Wissenschaft emanzipiert sich,* 261–269.

43. For Pagel, see Walter Pagel, "Erinnerungen und Forschungen," *Wege zur Wissenschaftsgescichte* II, Kurt Mauel, ed. (Wiesbaden: Steiner 1982), 45–66; Heinrich Buess, "Walter Pagel—12. November 1898–25. März 1983—zum Gedenken," *Clio Medica* 18 (1983): 233–239; and Isobel Hunter, "The Papers of Walter Pagel (1898–1983) in the Contemporary Medical Archives Centre at the Wellcome Institute for the History of Medicine, London," in Bröer, *Eine Wissenschaft emanzipiert sich,* 153–160.

44. For Edelstein, who in April 1933 had already been deprived by Paul Diepgen (1878–1966) of any further opportunity to work and who had emigrated in October of the same year, see Georg Harig, "Die antike Medizin in der Berliner medizinhistorischen Forschung," in *Tradition und Fortschritt in der medizinhistorischen Arbeit des Berliner Instituts für Geschichte der Medizin. Materialien des wissenschaftlichen Festkolloquiums anläßlich des 50. Jahrestages der Gründung des Instituts am 1. April 1980,* Dietrich Tutzke, ed. (Berlin: Institute für Geschichte der Medizin, 1980), 37–50; Owsei Temkin, "In Memory of Ludwig Edelstein," *Bulletin of the History of Medicine* 40 (1966): 1–13, reprinted in *"On Second Thought" and Other Essays in the History of Medicine and Science* (Baltimore: Johns Hopkins University Press, 2002), 250–263; John Scarborough, "Edelstein, Ludwig," in *Biographical Dictionary of North American Classicists,* Ward W. Briggs, ed. (Westport, Conn.: Greenwood Press, 1994), 153–156.

45. Elisabeth Berg-Schorn, *Henry E. Sigerist (1891–1957)* (Cologne: Hansen, 1978), 67–77.

46. For Temkin, see "Owsei Temkin: Historian of Medicine Who Insisted on the Importance of Ethics and Ideas in the Development of New Practices," *The Times* (London), 7 August 2002, 30.

47. For Sigerist, see, for example, *Making Medical History: The Life and Times of Henry E. Sigerist,* Elizabeth Fee and Theodore M. Brown, eds. (Baltimore: Johns Hopkins University Press, 1997).

48. On details of the revocation of Temkin's professorial qualifications, see "Hochschulnachrichten Leipzig," *Münchener Medizinische Wochenschrift* 80 (1933): 1458.

49. Walter von Brunn to Karl Sudhoff, Rostock, 4 May 1933, University Archive, Leipzig, KSI 41a, 210: "It is extremely regrettable, yet unavoidable, that this urgently needed new order will claim a number of casualties. I am and shall always remain critical, but am nevertheless committed to getting us back on our feet again at last! The fact that our History of Medicine, with its many Semitic colleagues, will suffer greatly is sad but has to be accepted; these days, the common interest demands sacrifices of all of us." See also Paul Diepgen, "Wie stehen wir in der Medizingeschichte?" *Deutsche Medizinische Wochenschrift* 59, *Praemedicus* 13 (1933): 834, who expresses very similar sentiments.

50. As early as 1925, Győry had written to Sudhoff: "Last semester, I had about 50 stud[iosi] med[icinae] registered for my classes, only a single 1 of whom was Jewish. Well, well—there isn't a lot of money to be made in our business. History of Medicine = a 'Christian Medicine.' " See Tibor Győry to Karl Sudhoff, Budapest, 18 December 1925, Herzog August Library, 99 Novissimi 4°, f. 65.

51. The so-called philosophical approach seems to have been particularly objectionable.

112 Traditions

Walter von Brunn to Tibor Győry, Rostock, 18 January 1933, Győry Papers, also in Halbach, *Briefe*, no. 53, 183–186, 184: "Pagel, whom I am not personally acquainted with but who, I am told, is a smug and unpleasant Jew with 'philosophical' ideas." Győry appears to be singing from a similar hymn sheet when he writes to Sudhoff: "Koch, a severe morphine addict /: as far as I know: / should be ruled out. He keeps forgetting and loosing all manner of things, a listless, perpetually tired fellow,—a good intellectual, but so terribly 'philosophical' that no one manages or cares to follow his line of argument. And then he always lectures in this awful 'Frankfurt-Jewish' accent of his." See Tibor Győry to Karl Sudhoff, Budapest, 6 March 1932, Herzog August Library, 99 Novissimi 4°, f. 135.

52. Walter von Brunn to Tibor Győry, Rostock, 7 May 1933, Győry Papers, also in Halbach, *Briefe*, no. 56, 192: "Richard Koch and Meyer-Steineg have been relieved of their duties—which I'm sure won't have any serious consequences for the History of Medicine, since they can continue with whatever research it is they are engaged in and didn't teach an awful lot anyway." See also Walter von Brunn to Tibor Győry Rostock, 10 May 1933, Győry Papers, also in Halbach, *Briefe*, no. 57, 193–194.

53. Tibor Győry to Karl Sudhoff, Budapest, 31 May 1933, Herzog August Library, 99 Novissimi 4°, f. 164.

54. This was demonstrably so in the case of Edelstein, who had written to Sudhoff from his Italian exile in a situation of dire existential need. See Thomas Rütten, "Zu Leben und Werk Ludwig Edelsteins," *Nachrichtenblatt der Deutschen Gesellschaft für Geschichte der Medizin, Naturwissenschaft und Technik e.V.* 48 (1998): 136–146.

55. See Halbach, *Briefe*, 29–30 (including references). According to this source, Sudhoff, who held Koch in very high regard, advised him in the spring of 1933 to resign from executive committee of the DGGMNT, to which Koch had been elected for the first time in the previous year. After this, Sudhoff continued to try in vain to help his esteemed colleague, with whom he maintained a regular correspondence until 1937. On Sudhoff's initiative, Koch was granted a meeting with Johann Daniel Achelis (1898–1963) at the Prussian Ministry of Science, Art, and Public Education in Berlin in May 1933. A resigned Koch, who had hoped for some sympathy and understanding for his precarious situation, summarized the meeting as follows: "Even such a benign man as Achelis now firmly believes in the Jewish poison." Next, Sudhoff, via a certain Mr. Wirz, a member of the Committee of Experts on National Health at the NSDAP Reichsleitung, contacted Gerhard Wagner, president of the National Socialist Medical Council, but Wagner, too, turned him down. In a letter to Sigerist, Sudhoff writes: "It is absolutely clear now that nothing can be done for Koch. Thank God that you have at least taken our good Temkin across the pond with you." Karl Sudhoff to Henry E. Sigerist, Leipzig, 13 May 1933, Sigerist Papers, Alan Mason Chesney Medical Archives of the John Hopkins Medical Institutions, Baltimore. On the general tolerance toward Sudhoff's efforts on Koch's behalf within the discipline, see also Tibor Győry to Karl Sudhoff, Budapest, 23 September 1933, Herzog August Library, 99 Novissimi 4°, f. 167: "I feel sorry for poor Koch with his lovely letter. True, I have always been a strong enemy of the Jews. But one ought to distinguish between the acclimatized and the wangling Jew."

56. Tibor Győry to Karl Sudhoff, Budapest, 31 May 1933, Herzog August Library, 99 Novissimi 4°, f. 164.

57. In 1931, the association first changed its name to Deutsche Gesellschaft für Geschichte der Medizin, der Naturwissenschaften und der Technik, followed in 1933 by a

further name change to Deutsche Gesellschaft für Geschichte der Medizin, Naturwissenschaft und Technik (DGGMNT).

58. For Diepgen, see Rainer Nabielek, "Anmerkungen zu Paul Diepgens Selbsteinschätzung seiner Tätigkeit an der Berliner Universität während des NS-Regimes," *Zeitschrift für die gesamte Hygiene* 31 (1985): 309–314, and Thomas Jaehn, "Der Medizinhistoriker Paul Diepgen (1878–1966). Eine Untersuchung zu methodologischen, historiographischen und zeitgeschichtlichen Problemen und Einflüssen im Werk Paul Diepgens unter besonderer Berücksichtigung seiner persönlichen Rolle in Lehre, Wissenschaftspolitik und Wissenschaftsorganisation während des Dritten Reiches" (Med. diss., Humboldt University, Berlin, 1991).

59. For Walter Artelt, see Gabriele Bruchelt, "Gründung und Aufbau des Berliner Instituts für Geschichte der Medizin und der Naturwissenschaften-eine archivalische Studie" (Med. diss., Humboldt University, Berlin, 1978), 56–58; Notker Hammerstein, *Die Johann-Wolfgang-Goethe-Universität Frankfurt am Main,* vol. 1 (Neuwied: Metzner, 1989), 355.

60. For Englert, see Jaehn, "Paul Diepgen," 17, 61; Harig, "Antike Medizin," 42, 49.

61. See Walter Artelt, "Bericht über die Verhandlungen der Deutschen Gesellschaft für Geschichte der Medizin, Naturwissenschaft und Technik zu Erfurt am 9. und 10. September 1933," *Mitteilungen zur Geschichte der Medizin, der Naturwissenschaften und der Technik* 32 (1933): 289–304; *Fortschritte der Medizin* 51 (1933): 938–939; *Medizinische Welt* 7 (1933): 1433–1434; *Der Chirurg* 5 (1933): 913–915; *Münchener Medizinische Wochenschrift* 81 (1934): 36–37. See also Winau, *Deutsche Gesellschaft,* 33, 43, 69–70.

62. Karl Sudhoff to Tibor Győry Budapest, 11 September 1933, Győry Papers.

63. See, for example, Wilhelm Haberling to Tibor Győry, Göttingen, 17 September 1933, Győry Papers: "He caused great merriment when he declared that he had been thinking for quite some time now that a Goethe Medal ought to be his by rights."

64. Karl Sudhoff to Henry E. Sigerist, Leipzig, 26 November 1933, Sigerist Papers: "A few weeks ago, I devised the plan, as a sort of birthday present to myself, that the Medal ought to be awarded to you, and I proposed a joint award to your good self and Diepgen, upon which the matter took its statutory course."

65. See also Karl Sudhoff to Tibor Győry, Leipzig, 26 September 1933, Győry Papers.

66. Walter von Brunn to Tibor Győry, Rostock, 18 January 1933, Győry Papers, also in Halbach, *Briefe,* no. 53, 183–186.

67. Karl Sudhoff to Henry E. Sigerist, Leipzig, 31 December 1932, Sigerist Papers.

68. See Walter von Brunn to Tibor Győry, Rostock, 22 April 1932, Győry Papers, also in Halbach, *Briefe,* no. 39, 150–151,150; Wilhelm Haberling to Tibor Győry, Düsseldorf, 9 May 1932, Győry Papers; Walter von Brunn to Tibor Győry, Rostock, 12 May 1932, Győry Papers, also in Halbach, *Briefe,* no. 40, 152–153, 153; Tibor Győry to Karl Sudhoff, Budapest, 25 June 1932, Herzog August Library, 99 Novissimi 4°; Tibor Győry to Karl Sudhoff, Budapest, 23 July 1933, Herzog August Library, 99 Novissimi 4°.

69. For Achelis, see *Der Nürnberger Ärzteprozeß 1946/47: Wortprotokolle, Anklage-und Verteidigungsmaterial, Quellen zum Umfeld. Commissioned by Stiftung für Sozialgeschichte des 20. Jahrhunderts,* Klaus Dörner, Angelika Ebbinghaus, and Karsten Linne, eds.(Munich: K. G. Saur, 2000), 74.

70. "The revolution had assumed the respectable face of bureaucracy." See Sebastian Haffner, *Geschichte eines Deutschen. Die Erinnerungen 1914–1933,* 6th ed. (Stuttgart: Deutsche Verlags-Anstalt, 2001), 95–240, 175.

114 Traditions

71. Karl Sudhoff to Paul Diepgen, Leipzig, 26 April 1933, Institut für Geschichte der Medizin, Archives of Humboldt University, Berlin, no. 13, f. 191. See also Jaehn, "Paul Diepgen," 62, n. 197.

72. Karl Sudhoff to Henry E. Sigerist, Leipzig, 13 May 1933, Sigerist Papers. Note the turn of phrase "*having* the chair offered." For Zeiss (1888–1949), who had joined the NSDAP as early as 1931, see *Deutsches Biographisches Archiv* 2, 1442:70–78; Paul-Julian Weindling, "German-Soviet Medical Co-operation and the Institute for Racial Research 1927–1935," *German History* 10 (1992): 177–206; and Ingrid Kästner and Natalija Decker, "Heinrich Zeiss (1888–1949) und die Versuche zur Institutionalisierung der Medizingeschichte in Rußland," *Acta medico-historica Rigensia* 5 (2000): 35–51.

73. Wilhelm Haberling to Karl Sudhoff, Düsseldorf, 9 May 1932, University Archive, Leipzig, KSI 41d, 28.

74. Walter von Brunn to Tibor Győry, Rostock, 12 May 1932, Győry Papers, also in Halbach, *Briefe,* no. 40, 152–153, 153.

75. Karl Sudhoff to Henry E. Sigerist, Leipzig, [24 August 1933], Sigerist Papers. While the letter itself is undated, its contents and contextual references establish the fact that it was indeed written on this particular day. See also Walter von Brunn to Tibor Győry, Rostock, 5 March 1933, Győry Papers, also in Halbach, *Briefe,* no. 54, 187–189,188: "That Lejeune should entertain hopes on Leipzig must surely be a bad joke." For Friedrich Josef August Lejeune (1892–1966), a member of the NSDAP since 1927, see *Deutsches Biographisches Archiv* 2, 800:396–400; Kümmel, "Im Dienst," 310.

76. Henry E. Sigerist to Karl Sudhoff, Florence, 3 September 1933, Sigerist Papers: "If he came to Leipzig, I should never again set foot in the institute for as long as I live."

77. For Martin Müller (1878–1960), see *Deutsches Biographisches Archiv* 2, 923:276; and Magnus Schmid, "Zum 75. Geburtstag von Prof. Martin Müller," *Münchener Medizinische Wochenschrift* 95 (1953): 586, and "Professor Dr. Martin Müller gestorben," *Münchener Medizinische Wochenschrift* 102 (1960): 870.

CHAPTER SIX

Ancient Medicine
From Berlin to Baltimore

Vivian Nutton

The historiography of medicine cannot be studied in isolation from other intellectual and social developments. How historians choose and treat their subjects is affected not only by what seems of interest to them but also by the ways in which others around them write about similar topics. Even within one and the same enterprise, there may be important differences in national styles and national priorities. This chapter focuses on the changing position of Greek and Roman medicine between 1870 and 1939, within both the history of medicine and the discipline of classics or classical philology, as it came to be called in Germany. From being central to the history of medicine, ancient medicine became largely the province of professional classicists, whose interests rarely coincided with those of medical historians. Germany's defeat in World War I and the subsequent collapse of the financial arrangements that had sustained much German research in ancient medicine almost ended the positivistic production of texts and editions of ancient authors. Younger scholars, like Owsei Temkin and Ludwig Edelstein, turned to the history of ideas, and took this approach with them across the Atlantic to Baltimore in the 1930s. Their immediate impact on historians of medicine and on American classicists was restricted, not least because of the different educational system they encountered. Paradoxically, they found a wider audience only after Edelstein had died and Temkin had retired from his post as director of the Johns Hopkins Institute, when new concerns and interests within classics, coming mainly from philosophers and from feminists, revived interest in ancient medicine.

This story can be interpreted in various ways. In one version, it shows the reduction in importance of the history of ancient medicine within the historiography of medicine in the face of the "historicizing" approach toward medicine's past that emerged in the middle of the nineteenth century. Before then, the outlines of the history of medicine were largely those laid down by three ancient writers: Cornelius Celsus, whose Preface to his *On Medicine* surveyed the history of medicine from Hippocrates to Hellenistic Alexandria;[1] the Elder Pliny, whose scabrous account of Greek doctors in Republican and early Imperial Rome exercised a powerful influence on all his readers;[2] and Galen of Pergamum, whose strident assurance that he alone knew what was true medicine effectively determined what past authors and what developments were considered significant.[3] Eighteenth-century antiquarians, like Conyers Middleton (1683–1750) and Johann Heinrich Schulze (1687–1744), supplemented this outline of the distant past with the evidence of archaeology, coins, and inscriptions, and with an increasingly sophisticated knowledge of ancient nonmedical sources.[4]

The "pragmatic" medical historians of the first third of the nineteenth century adopted a different perspective. Far from rejecting ancient medicine as outdated, they wished to show its relevance to present concerns. Particularly in Germany until the 1840s, the classical texts were still essential reading for the practical information of value that they contained, and the study of the history of Greek and Roman medicine was intimately linked to modern problems. To make available an ancient author, either in the original Greek or Latin or in an accurate translation, was a contribution to medicine and to medical scholarship. The massive program of reprints of ancient texts supervised by Karl Gottlob Kühn at Leipzig in the 1820s, the English translations of Greek texts by the Scottish physician Francis Adams, and (at least in its first volumes) the monumental edition of Hippocrates by Émile Littré, had a practical as well as a historical intent: to provide modern doctors with useful material that they might otherwise neglect or misunderstand.[5]

This approach to ancient medicine came to an end in the mid-nineteenth century, as developments within medicine made the diagnoses and remedies of the ancients superfluous. The claims of J. P. E. Petrequin (1809–1876) in the 1870s for a return to the true principles of surgery as laid down in his (still valuable) edition of the surgical texts in the Hippocratic corpus must have struck many of his medical readers as an old-fashioned and chauvinist attempt to turn back the clock and forget the achievements of German military surgery.[6] The new historicist approach, represented, for example, in the volumes of E. T. A. Henschel's new periodical *Janus*, set aside such immediately medical concerns in favor of an attempt to reconstruct the past on the basis of new texts and documents.[7] Scholars like Charles Daremberg, William A. Greenhill, and Theodor Puschmann devoted much of their energies to editing, translating, and interpreting classical texts for a historically minded public. The search for the genuine writings of Hippocrates was no longer intended to validate modern ideas by linking them to the Father of Medicine, as it had been for Littré, but seen as a form of intellectual archaeology, useful for what it revealed about a distant past.[8] In this, the history of ancient medicine was no longer alone. It faced strong competition for scholarly attention, not least from those more interested in the Middle Ages and in the earlier history of their own lands, but it remained a central theme in medical historiography certainly until 1914. Increasingly, however, research into ancient medicine was left to those with a professional interest in the ancient world, to philologists.

Ancient Medicine, Germany, and the New Science of Antiquity

This story can also be interpreted as an example of the German fascination with things classical and, above all, Greek. Beginning with Johann Joachim Winckelmann in the middle years of the eighteenth century, and continuing with Lessing and Goethe, there arose what E. M. Butler called the tyranny of Greece over Germany; *Bildung* and *Kultur,* education and culture, received in Germany a peculiarly Greek emphasis.[9] First in Prussia, under the influence of Wilhelm von Humboldt as minister for education in the early 1800s, and then in other German states, a university system that had primarily served to produce lawyers, doctors, and pastors was reoriented toward a new educational ideal, that of producing an intellectual elite whose practical skills were supplemented by an immersion in things Greek.[10] It was intended as both a spiritual and an intellectual process, bringing an enlargement of the mind that allowed a superior contemplation of the mundane practicalities of the world. The true philology (*Philologie,* as it was called) aimed to inculcate a higher understanding of the intellectual achievements of the Greeks through a precise understanding of linguistic style. In its claims for classics, it was more strident, more ostentatiously moral, and more Greek than what might be found in Britain or elsewhere. It appealed above all to the new middle classes, and, increasingly as the century wore on, it was backed up by a university system more extensive, better supported, technically more proficient, and with a higher public status than elsewhere in Europe.[11]

By 1890, in school and in university, the place of honor was claimed by the classics—even scientists and historians often conceded the role. All pupils in the Gymnasium had to study large amounts of Latin and Greek, whatever their future careers, and even in the less prestigious *Realschulen,* the ancient languages were not entirely absent. This devotion to the classics was not unique to Germany, but it was there pursued in a way that contrasted strongly with that elsewhere. In Britain, the emphasis was on language skills, on translation into and out of Greek and Latin, on a canon of "familiar authors," and on the moral values and social uplift gained from an acquaintance with Homer or Horace.[12] In Germany, although the moral element remained, there was a much wider conception of antiquity at work. The older, linguistically dominated Philologie, was supplemented, if not entirely replaced, in many universities by the idea of *Altertumswissenschaft,* a *science* of antiquity.[13] This change was further encouraged by the emphasis in the Humboldtian system on the union of teaching and research. In order to gain the coveted doctorate, students had to write a publishable thesis, and aspiring university professors had to write a second one, the *Habilitationschrift,* a thesis that was a major contribution to learning. The notion of research, of the steady accumulation of new knowledge, was thus built into the German system of education in the humanities at least as much as in the natural sciences. It was encouraged by all the German states, which established more and more posts to cope with an ever-expanding number of students. Between 1890 and 1910, the numbers of professors of the humanities (including law and theology) at the twenty-one German universities grew from 649 to 1051 (with 109 classicists and 185 historians).[14]

This overarching concept of Altertumswissenschaft was promoted by a series of remarkable scholars, based largely at the University of Berlin and at the Berlin Academy of Sciences, of whom the most famous today was the Roman historian Theodor Mommsen.[15] If it was Mommsen who laid the institutional foundations for the developments within classics in Berlin, it was his son-in-law, the patrician Ulrich von Wilamowitz-Moellendorff (1848–1931), and his friend, Hermann Diels (1848–1922), who exercised the greatest influence on the philological understanding of ancient medicine.

Wilamowitz, professor successively at Greifswald, Göttingen, and, for almost thirty years, Berlin, was the prophet of Altertumswissenschaft.[16] It was not enough, he argued, to be able to translate Latin and Greek; classical civilization needed to be approached in its entirety. The true scholar needed to be as familiar with the evidence of the papyri and inscriptions as with the tragedies of Sophocles or the epics of Homer, with the advice on diet of Diocles of Carystos as much as with Plato and Aristotle.[17] Only then could the scholar, a profession for Wilamowitz as important as that of priest or physician, begin to gain a knowledge of antiquity as a totality. It was a challenging ideal, fulfilled to an amazing degree by Wilamowitz himself in his long life, and corresponding in its optimism to other aspects of contemporary German medicine, science, history, and geography. Wilamowitz and his many supporters saw nothing strange in a science of antiquity comparable to a science of medicine or chemistry; all were founded on an ever increasing accumulation of factual data, produced according to the best scholarly criteria—and these, in classics as in science, were avowedly German.

In such circumstances, it is hardly surprising that the history of classical medicine continued to bulk large in German medical historiography. Sudhoff himself helped to bring to wider notice some of the earliest Greek medical papyri, and he wrote several useful little papers on antiquity.[18] The massive German *Lehrbücher* of the history of medicine, as well as the more specialized surveys of the history of anatomy or of ophthalmology, devoted many solid pages to the ancient history of their topic.[19] Modern discussions of the drugs and drug therapies used in antiquity depend heavily on the work done by German medical men in this period.[20] At Jena, the ophthalmologist Theodor Meyer-Steineg concentrated on classical antiquity in his lectures in the history of medicine and in building up his collection of historical instruments.[21] Increasingly, however, the classicists' agenda diverged from the doctors'. It was an agenda based solidly on texts: making editions, discussing manuscript affiliations, deciding on authenticity, collecting fragments, and *Quellenforschung* (the attribution of ideas in existing classical authors to more distant predecessors whose work has been lost).[22] Few doctors were able to master the ever more professional techniques of the new philology.[23] Robert Fuchs's edition of the so-called Anonymus Parisinus was a disaster from almost every point of view; he neither understood what he had found nor produced intelligible Greek.[24] Conversely, the philologists paid little or no attention to the medical content of their ancient treatises, occasionally investigating philosophical and historical questions but mainly concentrating on matters of language.

Particularly striking is the extent to which ancient medical texts were being edited anew using the modern methods of philology. Between 1870 and 1918, professors and doctoral students from almost every university in Germany published articles or theses on ancient medicine, many of them still not yet superseded.[25] An annual speech to mark the foundation of a university or an annual bulletin from a Gymnasium might contain an edition of an ancient medical text—

to the frustration of later scholars hunting for these bibliographical rarities.[26] A veritable tribe of teachers at classical Gymnasia showed their mastery of the ancient languages by producing editions of Galenic and Hippocratic texts, many still fundamental today, or by contributing articles on ancient medical authors to the massive and multivolume *Paulys Realenzyklopädie der klassischen Altertumswissenschaft*.[27] The acknowledged leader of studies in ancient medicine was Max Wellmann (1863–1933), the editor of Dioscorides, a Gymnasium teacher at Potsdam, and, from 1919 until his death, an honorary professor of classics in Berlin.[28]

Such productivity was (and is) without parallel, even in countries with a strong tradition in classics.[29] Britain, for example, with fewer universities, contained before 1939 in toto only two or three teachers interested in ancient medicine or science, and only one, W. H. S. (Malaria) Jones (1876–1963), who made it the main focus of his research.[30] But few now read his *Malaria in Greek History,* the book that made his name, and his edition and translation of selected works of Hippocrates in the Loeb series was based on decisions about the text that can most charitably be described as quirky.[31] Ancient medicine was firmly in the hands of gentlemen doctors, headed by Sir Clifford Allbutt (1836–1925).[32] The single volume of Galen in the Loeb series and the Loeb Celsus were simply translations by linguistically talented physicians made from texts edited by Germans.[33] The one great contribution made in Britain to ancient medicine, the edition of the so-called Anonymus Londinensis papyrus, was the result of collaboration between the London papyrologist Sir Frederic Kenyon (1875–1955) and Hermann Diels, professor in Berlin.[34]

Diels is the key figure in the philological study of ancient medicine. Unlike Wilamowitz, he came from relatively humble origins in the Rhineland and was primarily interested in ancient philosophy.[35] His doctoral thesis, submitted at Bonn in 1870, was devoted to an examination of the *History of Philosophy* ascribed to Galen, a treatise to which he returned at least three times in his life.[36] The work for which he is most renowned today, his *Fragments of the Presocratic Philosophers,* depended on a substantial acquaintance with the medical as well as the philosophical writers of antiquity.[37] He was also interested in ancient technology; his fundamental essays, *Antike Technik,* were reprinted in 1965, and his edition and translation of the *Mechanics* and *War Machines* of Hero and Philo was reprinted as late as 1970. His joint edition, with Kenyon, of the Anonymus Londinensis papyrus, with its precious information on early Greek medical theories, is still of considerable value to the specialist.[38]

As professor of Greek in Berlin, a member of the Berlin Institute for Antiquity, and, from 1895–1920, secretary of the philological-historical section of the Berlin

Academy of Sciences, Diels was at the very heart of Altertumswissenschaft and of the organs of the Prussian state concerned with academic matters. For him, and for Wilamowitz, the pursuit of science, *Wissenschaft,* was both patriotic and international: international in the sense that it transcended all boundaries, patriotic in that the monuments of scholarship brought honor to their country of origin.[39] Whether we are talking about history, theology, chemistry, medicine, or classics, there was a widespread belief in the progress of science, a progress that demanded both individual endeavor and national, indeed international, collaboration.[40] By discovering, publishing and organizing ever more data, one could secure as solid a foundation for the conclusions of historians as for those of chemists. This positivist history called for positive action, and action on a grand scale.[41] As another of the great scholars of Wilhelmine Berlin, the church historian Adolf von Harnack, put it somewhat scathingly, this was *Wissenschaft als Grossbetrieb,* "Scholarship as big business"—I am tempted almost to translate it as "heavy industry."[42]

Its monument was, and continues to be, the editions in the Berlin Corpus Medicorum. This series was originally planned for ancient medical texts in their original Greek or Latin, but it soon expanded to include others preserved only in later Oriental or medieval Latin translations.[43] It was the Danish historian of medicine and science J. L. Heiberg (1854–1928), who in 1901 first mooted the idea of a corpus of ancient medical writers, akin to those of the Aristotelian commentators, Greek and Latin inscriptions, or the early fathers of the Church, projects all carried out within the Berlin Academy. The suggestion was taken up with abundant energy by Diels, at first under the auspices of the International Union of Academies.[44] Diels knew what was required, having just brought to a successful conclusion the Berlin edition of all the ancient Greek commentaries on Aristotle. Within a month of Heiberg's initial suggestion, Diels had engaged his first collaborators and begun arranging for financial support from the Puschmann Stiftung. A plan of action was drawn up by Hermann Schöne, Privatdozent in Greek at Berlin, who spent much of 1901 and part of the next year visiting libraries in Italy, Paris, and Brussels.[45]

The first task was, at first sight, enormous: the preparation of a catalogue of surviving manuscripts of classical medical authors, a task that had not been attempted for over fifty years, and had not been even half fulfilled.[46] It is hard not to marvel at the ambitious timetable laid down by Diels, or the efficiency with which this aim was substantially achieved.[47] The bulk of the preliminary sifting of information was done by a young librarian, Dr. Rapaport, who was "introduced to the technique of organizing a corpus" by Professor Hirschfeld, the head of the Corpus Inscriptionum Latinarum in Berlin, and an expert on the Roman army and civil service. By 1903, emissaries had been sent to Britain and Spain; an Arabist and a Hebraist were also involved. The departure of Schöne to a chair at Königsberg in 1904 involved only a brief delay before the task of finally editing the catalog was handed to Johannes Mewaldt, who had just taken his doctorate with Diels that year.[48] By October 1905, Mewaldt had produced a first proof of a catalog of the manuscripts of Hippocrates and Galen, and copies had been sent to five senior professors for comments and corrections. As Mewaldt pointed out in apologetic tone, "the sheer difficulty of the text, the many corrections and additions that had to be made, and the heavy workload of the press" meant that the volume did not appear for a further six months, in May 1906.[49] Volume 2, the rest of the doctors preserved in Greek, took less time. Mewaldt began writing in October 1905; proofs were available by March 1906; and the catalog was published in October of that year. A further supplement appeared in 1907.

Errors and omissions there might be aplenty, but one is surprised at how relatively few there are, and still more at the speed of the whole operation. This was organized, Prussian science, done in a way and in a length of time that would have been unthinkable elsewhere. Thirty-one scholars from around the world lent their services, although the great bulk of the labor was undertaken in Berlin, with almost military efficiency.

Diels did not stop at the presentation of a catalog. He commissioned editors, many of them Gymnasium teachers as in his earlier corpus of the Aristotelian commentators, to work on producing texts.[50] In 1907 Mewaldt, now *Assistent* in Berlin, was appointed the supervisory editor of the corpus, a task he fulfilled adequately, despite worries about his somewhat arrogant personality and his move in 1909 to a junior post at Greifswald.[51] In 1908 the first volume in the series appeared, Max Wellmann's edition of Philumenus, *On Poisonous Animals,* a work hardly ever read or noticed since then.[52]

But Galen was the greatest prize, and Diels pushed ahead with the editing of Galen's commentaries on Hippocrates. He had three reasons for doing this. First, these commentaries stood in most need of even competent editing; second, it was thought that their tradition was almost entirely Greek and hence did not require expertise in oriental languages; and, finally, a proper edition of Hippocrates needed a proper edition of the earliest surviving major witness to the actual words of Hippocrates (i.e., Galen) before any progress could be made.[53] The *Abhandlungen* of the academy offered an appropriate forum for the publication of preliminary results and for discussions of manuscripts and authenticity. The first volume of Galenic commentary appeared in the corpus in 1914, followed the next year by a further large volume, part of which was edited by Diels himself.[54]

Weimar Idealism

But the clouds of war put an end to this optimistic progress of scholarship. Inflation almost wiped out the Puschmann Stiftung; death carried off many of the erstwhile collaborators, including Diels himself in 1922; and the political and social turmoil of the early years of the Weimar Republic only added to the general dismay and despondency. Since many of the professors of classics had been, like Wilamowitz, on the conservative, nationalist right, it is hardly surprising that they should find themselves out of sympathy with the new republic.[55] As Wilamowitz put it to the daughter of his Oxford counterpart, Gilbert Murray, in 1922, "The world that I knew is destroyed utterly."[56] Likewise, the young grew increasingly alienated from the imperialistic tendencies of positivist historiography, the piling up of fact upon fact, edition upon edition, document upon document.[57] Like Wilhelm's empire, the great collaborative enterprises, the Grossbetrieb of scholarship, seemed increasingly outmoded. The impulse to organize the world was no longer there, even when the institutions and collectivities continued. The younger scholars on whom Diels and Wilamowitz had placed their hopes, Schöne, Mewaldt, and Kalbfleisch, produced between them only a handful of articles on ancient medicine after the war.[58] Of their pupils, only Karl Deichgräber contributed substantially to the history of ancient medicine.[59]

This positivistic emphasis on texts and editions was replaced by a new emphasis, on ideas of eternal value.[60] Wilamowitz himself in his old age sought to discover the "inner depths of thought" in his book, *Der Glaube der Hellenen* (The creed of the Hellenes), in which *Glaube* is both more and less than religion and "Hellene" carefully distinguishes the pristine purity of the Athens of Pericles from the filthier Greek world of the Roman, let alone the Christian, period.[61] Younger scholars in classics and history turned to the world of ideas for their research—this is, of course, the period when the Warburg Institute and Library developed in Hamburg—or to an almost messianic superhero who embodied the qualities that would break with convention, as in Ernst Kantorowitz's 1927 study of Frederick II Hohenstaufen, *stupor mundi,* the wonder of the world.[62]

It is a process dramatically, indeed paradigmatically, illustrated in the history of ancient medicine.[63] The *editing* of texts of ancient medicine came to a near stop. The volumes in the Corpus Medicorum series bear melancholy footnotes to the death of an elderly editor, or to a twenty year gap between dissertation and book. There were new discoveries of texts, primarily made by Arabists like Gotthelf Bergsträsser, Franz Pfaff, and Richard Walzer, but they had little impact on classicists.[64]

The few professors who retained an active interest in ancient medicine, were no longer concerned with producing editions but with a search for the ideas that governed ancient Greek thinking about medicine.[65] Werner Jaeger (1888–1961), Wilamowitz's pupil and successor at Berlin, had investigated the interrelationship between ancient medicine and philosophy in his doctoral thesis of 1914, on Nemesius of Emesa, a fourth-century Christian bishop and theologian, whose treatise *On the Nature of Man* has been termed the first Christian anthropology.[66] It was an interest that remained with him to the end of his life, culminating in one book and one chapter.[67] In his *Diokles von Karystos* (1938), he argued that Diocles was a medical writer and investigator of the first rank, who derived his inspiration and method from his education at the hands of Aristotle and who provided the methodological bridge between him and the Alexandrians. Philosophy, in other words, guided the physician and anatomist.[68] The chapter formed part of the second volume of a large work called *Paideia* (a Greek word that, for Jaeger, involved education in its spiritual as well as its technical aspects) and is entitled "Greek Medicine as Paideia."[69] Hippocrates thus took his place alongside Homer and Plato as a spiritual guide in the formation of what some termed a "third humanism," a system of moral and ethical values deriving from Greece, which, he claimed, would give meaning to life. By studying the classics, one would become a better, and not only a wiser, man.[70]

A similar change can be seen among the medical historians. In Leipzig, Karl Sudhoff, titan, organizer, politically conservative and historically positivist, was succeeded by Henry Ernst Sigerist (1891–1957), a Swiss of broad sympathies.[71] In politics a Social Democrat, he was the only teacher in the Leipzig medical faculty who openly avowed his loyalty to the Weimar Republic.[72] In medical history he had begun with editions of a late Latin medical writer, published in the Corpus Medicorum series, and of early medieval *Antidotaries*. He continued this interest in the texts of late antique and early medieval medicine until the late 1930s, and his reverence for the classics remained with him throughout his life. His last work, his *History of Medicine* (1951–1961), a torso of one and a half volumes, is an unfinished attempt to recreate a Wilamowitzian totality of approach toward medicine in antiquity. It covers prehistory, Babylon, Egypt, Persia, and Early Greece, attempting to set the medicine of each civilization in its context. The tension between the older and newer approaches to the history of medicine is here palpable. Daring methodological insights are not always matched by sound practice; the sheer bulk of material obstructs a coherent narrative, and Sigerist's idealization of the Greeks is often at variance with his observations on other cultures. It is a glorious failure and is now rarely cited by historians of ancient medicine.[73]

But his lectures, books and articles from his Leipzig years show a very different approach to the writing of medical history. Whereas Sudhoff had laboriously edited and compiled for others to use, Sigerist raised questions. His courses bore such titles as "The Boundaries and Aims of Medicine," and "The Problem of Culture and Medical Psychology." He introduced his students to a whole range of new ideas and concepts, drawn from sociology, philosophy, and anthropology.[74] The awkward conjunction between old and new is illustrated in the contents of the first issue of *Kyklos. Jahrbuch des Instituts für Geschichte der Medizin an der Universität Leipzig* (1928), the new house journal founded by Sigerist. Alongside publications of medieval texts are essays on "Culture and Sickness," "Time-Implied Function," and "The Bases of Paracelsian Thought," as well as a report on a debate on the role of history of medicine in medical education. The change of emphasis can hardly be more marked.

Sigerist's most brilliant pupil, Owsei Temkin (1902–2002), shows the same development as his master. A brilliant linguist, his early work divides into two very different parts. One, linguistically, historically, and in a sense institutionally orientated, investigated the relations between the various commentaries on Hippocrates and Galen produced in late antiquity and surviving in Greek, Latin, and Arabic.[75] This was solid, Teutonic, and lasting scholarship that Sudhoff would have been proud of. The other was devoted to studying ideas within medicine—and in terms that a non-German, and a nonphilosopher, finds hard to translate and understand. "Studies on the Concept of 'Meaning' in Medicine," though full of insight, is not a text for the fainthearted or those unfamiliar with German philosophy, and the little essay on "The Humane Sciences (*Geisteswissenschaften*) in Medicine" is distinguished largely by the absence of any factual context and its high degree of abstraction.[76] What links the two sides of Temkin's early work is an almost Platonic belief in the superexistence of particular concepts. Temkin studies Hippocratism, not Hippocrates; ideas on epilepsy, not epileptic patients; how doctors think and thought, not necessarily what they thought or how this related to their actual practice. This emphasis on the crucial importance of ideas within the development of the history of medicine remained a feature of all Temkin's writings throughout his long life.[77] In his opinion, the ideas that the medical historian investigates should pass beyond the merely historical to inform the life of the individual reader or physician. One can, indeed perhaps one should, read *The Falling Sickness* as a guide to the ideal understanding of illness; *Galenism* as a manifesto arguing that the doctor needs to think, to be a philosopher, in order to heal; and *Hippocrates in a World of Pagans and Christians* as a demonstration of the value to the individual physician of possessing a medical, (supra-) religious ethic.⁷⁸ Even in his nineties, he wrestled with questions of the doctor's morality, drawing arguments from the past to enlighten present debate.

The same belief in the value of ideas to both the historian and the practitioner of medicine can be seen equally clearly in the work of Temkin's friend and Baltimore colleague, Ludwig Edelstein (1902–1965).⁷⁹ A student of Werner Jaeger in Berlin, Edelstein submitted his thesis on the Hippocratic treatise *Airs, Waters, and Places* at Heidelberg in 1929.⁸⁰ For more than thirty years following its publication in 1931, Edelstein produced a whole series of stimulating, and at times iconoclastic, books and essays. To such relatively familiar questions as "Who was Asklepius?," "When and why did the Greeks begin to dissect?," "What part was played in Greek medicine by philosophy?," "Who wrote the Hippocratic Oath?," he put forward vigorous new answers, which still provoke debate today.⁸¹ Although he did not neglect the practicalities of medicine, it was the ideas behind them that attracted his attention. In his view, philosophy supplied the guiding thread to Hellenism within medicine. As he later put it, man can live without philosophy as little as he can live without medicine.⁸² It was a message that resonated beyond the walls of a philological institute.

Its aims and its methodology were a direct challenge to the traditional philological way of doing things. How challenging can best be seen by a comparison with two contemporaries, Karl Deichgräber and Hans Diller, both of whom went on to distinguished careers as professors of Greek.⁸³ Edelstein was arguably their inferior in his feeling for Greek; he made occasional mistakes and was likely to draw major conclusions from a shaky textual basis. But his grasp of the big issues, and of what might prove of greater significance to nonphilologists, was far greater. Diller in his Hamburg dissertation and later in his Leipzig Habilitationschrift laid down extremely solid foundations for a future edition of *Airs, Waters, and Places*. In the former, he sorted out the manuscript tradition, including the various ancient Latin translations; in the latter, he produced a detailed commentary on the ethnographical and geographical ideas within that treatise.⁸⁴ Edelstein's study of the same work took the Greek text almost for granted. Instead, he tried to set it within the context of early Greek medical *practice*, seeking to establish what prognosis might mean to an ancient healer, and emphasizing the fragility of the life of an ancient medical craftsman.⁸⁵ Similarly, while Deichgräber in his 1931 Berlin Habilitationschrift sought to date the various sections in the Hippocratic *Epidemics* and to identify the genuine writings of Hippocrates, Edelstein pointed to the circular arguments used in such a process of identification, preferring instead to consider the whole Hippocratic corpus as a mass of material gradually being collected around a famous name.⁸⁶ Many of Deichgräber's individual points retain their value, but it is Edelstein's skepticism that has been more fruitful, opening up the whole of the Hippocratic corpus to scholarly investigation without the need to privilege one tract over another as being written by the father of medicine.[87]

Hippocrates in Baltimore

Edelstein was never an easy colleague—he trod on too many toes, and his moral rectitude made others less scrupulous uncomfortable—and he never obtained a paid post in a classics department in Germany.[88] Instead, he worked for a while at the new Institute for the History of Medicine in Berlin, under Paul Diepgen. He was also a Jew and was dismissed from his post and driven out of Germany in 1933. In 1934, he left Europe for America to join Sigerist and Temkin in the newly founded Institute of the History of Medicine at Baltimore. Here, at the leading medical school in the United States, at the university in which German teaching methods, organization, and programs had long been familiar; in a department created by William Henry Welch, perhaps the greatest of the American pathologists who had studied in Germany; and in an intellectual community from which sprang *The Journal of the History of Ideas,* it might be imagined that the approach to medicine as a history of philosophical ideas and questions would find fertile ground.[89] It did not—or at least not for decades. The list of Hopkins Ph.D.s in medical history reveals only one on a classical subject, and only a handful that could be described as dealing with the thought of physicians at any period.[90] Edelstein's later teaching in the Classics Department at Berkeley, again at Johns Hopkins, and at other U.S. universities produced few, if any direct, offspring.[91] Of those who held fellowships in ancient medicine at Johns Hopkins, only the name of I. E. Drabkin is at all familiar today.[92]

In part this was because of national differences in the education system. Temkin tells against himself the story of his first seminar, on the *Art of Medicine* by Galen, which he had given to an appreciative audience in Leipzig but which in Baltimore was almost entirely confined to the faculty, since he expected his audience to be as fluent in the classical languages as he. Almost in jest he ascribes part of the failure to the democratic style of the American university, in which a German sense of the dignity and mission of the professoriat fitted ill with a national utilitarian spirit or with one that did not allow an academic elite to impose its own values on the culture of society as a whole.[93] Temkin hints also at a more local reason, a preference among the Hopkins professoriat outside the institute for "positivist research," and hence a certain suspicion toward the historians of ideas.[94] Certainly, within Anglophone classics, and above all in 1930s Baltimore, the emphasis was on philology; on linguistics; on factual, political, and social history; and on archaeology.[95]

But there are other wider considerations. It is difficult to raise enthusiasm for the history of a tradition of ideas in an area that has not participated fully in that tradition.[96] The view from Chesapeake Bay is inevitably different from that from the Spree—or the Thames. German Romantic medicine, for example, is not at all easy to explain on the other side of the Channel, let alone of the Atlantic.[97] The Humboldtian civic and educational program embraced by a stable German bourgeoisie committed to notions of service to the state did not fit easily into the newer, more mobile and democratic, environment of the United States, with its long-standing suspicions of government involvement. Jaeger's claims for the prime values of humanism fell on deaf ears, even before they were laid open to derision by the advent of the Nazis. He and those like him, who attributed some moral value to the study of antiquity, were indeed prophets crying in a foreign land. The refugees from Nazi Germany, including Jaeger himself, undoubtedly brought with them new standards of scholarship, but their impact on the direction of academic research was rarely immediate.[98]

Nonetheless, what was achieved should not be underestimated. A steady stream of articles on aspects of ancient medicine appeared in the *Bulletin of the History of Medicine* and kept new developments before the (then relatively small) medicohistorical community. Indeed, far more was published on ancient medicine in this journal than in others aimed at a readership of classicists, like the *American Journal of Philology,* another Hopkins publication.[99] It was not until after Edelstein's death, with the appearance in 1967 of a selection of his articles, *Ancient Medicine,* many translated for the first time into English, that Anglophone classicists in general had easy access to his ideas.[100] Only then did his name become seen as representative of ancient medicine, and his opinions were taken to represent the *communis opinio* of scholars in the field.[101]

This volume, along with Temkin's *Galenism,* has contributed to a resurgence of interest in ancient medicine among classicists over the last thirty years that is without precedent since the days of Wilamowitz.[102] The revived Corpus Medicorum publishes a steady flow of editions produced by scholars around the world, and it now has rivals in France and Italy. Conferences proliferate, on topics from the uses of water to medical texts as literature.[103] This resurgence is partly the result of changes within classics, as it moves away from a philological to a more historical, literary, and interpretative discipline, indeed to a renewed idea of Altertumswissenschaft. Feminism is a more potent stimulus than fever.[104] One can talk about different schools of interpretation within the history of medicine, centered, for instance, on Paris, Cambridge, and Pisa.[105] But it is now the views of the social historian, the demographer and the historian of tantalizing and fleeting *mentalités* that dominate, not those of the believer in the eternal verities of Jaegerian humanism.[106] Even when scholars link ancient medicine with modern concerns or accept the notion of a medical tradition, they are also at pains to stress the great differences between ancient and modern and the malleability of that tradition.[107]

The results of this new enthusiasm for ancient medicine have still to penetrate deeply into the medical faculties and departments of the history of medicine. Certainly, fewer doctors today are capable of performing original research using the ancient languages than was the case even a generation ago, and the sense of the overriding importance of the European classical heritage is no longer as prominent in modern educational systems across Europe as it once was. The proliferation of academic journals means that specialists can spend their time talking to one another, with little concern for the wider impact of their ideas. But the increasing availability of good translations of many major, and even some minor, writings of the doctors of antiquity has also succeeded in broadening access, while the growing attention to the ancient world in the media, certainly in Britain, has allowed new discoveries to reach a wider audience. Conversely, some classical philologists have realized the need for a medical input into their interpretation of ancient sources, either through collaboration or by indicating to doctors where their contributions could be most valuable.[108] Other topics, most notably the Hippocratic Oath and the Thucydidean plague, continue to attract both medical and philological attention, although not always with satisfactory results. Recent surveys of the history of medicine have also begun to reflect these new developments, particularly in their treatment of the Hippocratic corpus, Galen, and ancient gynecology. The classicist, the doctor, and even the medical historian still have different priorities in their interests and skills, but there is an awareness of the importance of communicating these interests to more than a narrow few, and of convincing those eager for modernity that a look back into a very distant past is more than antiquarianism. This is not always easy, but a failure to make the attempt would be to betray the legacy of past scholars and to forget the history of the history of medicine.

NOTES

1. Philippe Mudry, *La Préface du De medicina de Celse* (Rome: Institut Suisse de Rome, 1982); Heinrich von Staden, "Celsus as Historian?" in *Ancient Histories of Medicine: Essays in Medical Doxography and Historiography in Classical Antiquity,* Philip van der Eijk, ed. (Leiden: E. J. Brill, 1999), 251–294.

2. Vivian Nutton, "The Perils of Patriotism: Pliny and Roman Medicine," in *Science in the*

Early Roman Empire: Pliny the Elder, His Sources and Influence, Roger K. French and Frank Greenaway, eds. (London: Croom Helm, 1986), 30–32.

3. Vivian Nutton, "Renaissance Biographies of Galen," in *Gattungen der Medizingeschichte,* Thomas Rütten, ed. (forthcoming); Mario Vegetti, "Tradition and Truth: Forms of Philosophical-Scientific Historiography in Galen's *De Placitis,*" in van der Eijk, *Ancient Histories,* 333–358, and "Historiographical Strategies in Galen's Physiology *(De Usu Partium, De Naturalibus Facultatibus),*" ibid., 383–396.

4. Vivian Nutton, "Murders and Miracles: Lay Attitudes towards Medicine in Classical Antiquity," in *Patients and Practitioners,* Roy Porter, ed. (Cambridge: Cambridge University Press, 1985), 23–24; Jürgen Helm, "'Der erste wahre Geschichtsforscher der Medicin.' Johann Heinrich Schulze und seine 'Historia medicinae' von 1728," in *Eine Wissenschaft emanzipiert sich. Die Medizinhistoriographie von der Aufklärung bis zur Postmoderne,* Rolf Bröer, ed. (Pfaffenweiler: Centaurus, 1999), 189–204.

5. Hans-Uwe Lammel, "To Whom Does Medical History Belong? Johann Moehsen, Kurt Sprengel, and the Problem of Origins in Collective Memory," chap. 2 in this volume; Vivian Nutton, "In Defence of Kühn," in *The Unknown Galen: Galen beyond Kühn, Bulletin of the Institute of Classical Studies* suppl. 77, Vivian Nutton, ed. (2002): 1–8.

6. Guy Sabbah and Sylvie Sabbah, "Joseph Pierre Éléonord Pétrequin (1809–1876), le 'correspondent lyonnais,'" in *Médecins érudits, de Coray à Sigerist,* Danielle Gourevitch, ed. (Paris: De Boccard, 1995), 113–128.

7. Ibid. The whole section in Gourevitch, *Médecins,* on "le cercle de Daremberg," 61–152, is relevant. Ludwig Edelstein, "Medical Historiography in 1847," *Bulletin of the History of Medicine* 21 (1947): 495–511, reprinted in Ludwig Edelstein, *Ancient Medicine: Selected Papers of Ludwig Edelstein* (Baltimore: Johns Hopkins Press, 1967), 463–478.

8. Wesley D. Smith, *The Hippocratic Tradition* (Ithaca: Cornell University Press, 1978), 31–44.

9. Elsie M. Butler, *The Tyranny of Greece over Germany* (Cambridge: Cambridge University Press, 1935).

10. Rudolf Pfeiffer, *A History of Classical Scholarship from 1300 to 1800* (Oxford: Clarendon Press, 1976), is exemplary both in its scholarship, and in its unconsciously German biases. Cornelia Wegeler, " . . . *wir sagen ab der internationalen Gelehrtenrepublik." Altertumswissenschaft und Nationalsozialismus. Das Göttinger Institut für Altertumskunde, 1921–1962* (Cologne: Böhlau Verlag, 1996), 26–28, surveys these changing trends in one major university.

11. Fritz K. Ringer, *The Decline of the German Mandarins: The German Academic Community, 1890–1933* (Cambridge, Mass.: Harvard University Press, 1969); Ulrich Sieg, "Im Zeichen der Beharrung. Althoffs Wissenschaftspolitik und die deutsche Universitätsphilosophie," in *Wissenschaftsgeschichte und Wissenschaftspolitik im Industriezeitalter: Das "System Althoff" in historischer Perspective,* Bernhard vom Brocke, ed. (Hildesheim: Edition Bildung und Wissenschaft, 1991), 287–306.

12. Christopher A. Stray, *Classics Transformed: Schools, University, and Society in England, 1830–1960* (Cambridge: Cambridge University Press, 1998); *Classics in 19th and 20th Century Cambridge: Curriculum, Culture, and Community,* Christopher A. Stray, ed. (Cambridge: The Cambridge Philological Society, 1999).

13. Ulrich von Wilamowitz-Moellendorff, *History of Classical Scholarship* (London: Duckworth, 1982), 155–178; Anthony T. Grafton, "Polyhistor into Philolog: Notes on the Trans-

formation of German Classical Scholarship, 1780–1859," *History of Universities* 3 (1983): 159–192.

14. Stefan Rebenich, *Theodor Mommsen und Adolf Harnack: Wissenschaft und Politik im Berlin des ausgehenden 19. Jahrhunderts* (Berlin: De Gruyter, 1997), 30–31. Student numbers quadrupled between 1865 and 1911, when there were 55,600.

15. Ibid., 55–79.

16. *Wilamowitz nach 50 Jahren,* William M. Calder III, Hellmut Flashar, Theodor Lindken, eds. (Darmstadt: Wissenschaftliche Buchgesellschaft, 1985).

17. His influential *Griechisches Lesebuch* (Berlin: Weidmann, 1910) includes selections from writings on mathematics, geography, and medicine, as well as the more familiar literary and philosophical authors. His three medical extracts, 269–286 (all supplied with a commentary), are the Hippocratic *Sacred Disease,* and two long extracts from Oribasius, citing Diocles of Carystos (fragment 182 in *Diodes of Carystos,* P. van der Eijk, ed. [Leiden: Brill, 2000]) and Athenaeus of Attaleia on diet. The last two are unfamiliar today even to professional medical historians.

18. Georg Harig, "Sudhoffs Sicht der antiken Medizin," *Sudhoffs Archiv* 76 (1992): 97–105.

19. Julius Pagel, *Geschichte der Medizin* (Berlin, 1898); *Handbuch der Geschichte der Medizin,* Max Neuburger, Julius Pagel, eds. (Jena: G. Fischer, 1902–1905); Max Neuburger [an Austrian], *Geschichte der Medizin* (Stuttgart: Enke, 1906–1911), translated as *History of Medicine* (London: Oxford University Press, 1910–1925); Ernst Julius Gurlt, *Geschichte der Chirurgie und ihrer Ausübung: Volkschirurgie, Alterthum, Mittelalter, Renaissance* (Berlin, 1898); Julius Hirschberg, *Geschichte der Augenheilkunde,* vol. 1 (Berlin, 1899).

20. Ludwig Israelson, "Die Materia Medica des Klaudios Galenos" (Diss., Dorpat, 1894); Rudolf von Grot, "Über die in der hippokratischen Schriftensammlung erhaltenen pharmakologischen Kenntnisse," *Historische Studien aus dem pharmakologischen Institute der kaiserlichen Universität Dorpat* 1 (1889): 58–133; Felix Rinne, "Das Receptbuch des Scribonius Largus, zum ersten Male theilweise ins Deutsche übersetzt und mit pharmakologischem Kommentar versehen," ibid. 5 (1896): 1–99. The majority of professors at the Russian university of Dorpat (Jurjev) at this period were Baltic Germans, and the *Studien* were published at Halle.

21. Susanne Zimmermann and Ernst Künzl, "Die Antiken der Sammlung Meyer-Steineg in Jena," in *Jahrbuch des römisch-germanischen Zentralmuseums Mainz* 38 (1991): 515–540, 41 (1994): 179–198; Susanne Zimmermann, "Theodor Meyer-Steineg (1873–1936) und die Medizingeschichte in Jena," in Bröer, *Wissenschaft,* 261–270.

22. Friedrich Münzer, *Beiträge zur Quellenkritik der Naturgeschichte des Plinius* (Berlin, 1897), and Max Wellmann, *Die pneumatische Schule bis auf Archigenes in ihrer Entwickelung dargestellt* (Berlin, 1895), are typical in their methodology.

23. There were exceptions. The edition of the works of Alexander of Tralles by Theodor Puschmann (Vienna, 1878–1879) is still valuable today, as is the *editio princeps* of the lost anatomical writings of Galen, edited by Max Simon, *Sieben Bücher Anatomie des Galen* (Leipzig: Teubner, 1906), but that, significantly, was based on the Arabic.

24. Hermann Diels, Hermann Usener, and Eduard Zeller, *Briefwechsel,* Dietrich Ehlers, ed. (Berlin: Akademie Verlag, 1992), vol. 1 (1892):447, 450. Cf. Ivan Garofalo, *Anonymi Medici De Morbis Acutis et Chroniis* (Leiden: Brill, 1997), xxxvi–xxviii.

25. The bibliographies to the Corpus Hippocraticum and the Corpus Galenicum assem-

bled by Gerhard Fichtner (Tübingen: Institut für Geschichte der Medizin, 1990) confirm the broad geographical spread of editors.

26. Two examples suffice: Georg Helmreich, ed., "Galenus De optima Corporis Constitutione. Idem De bono Habitu," *Programm des Kgl. humanistischen Gymnasiums in Hof für das Schuljahr 1900/1901* (Hof: n.p., 1901); August Brinkmann, ed., "Galeni de optimo Docendi Genere libellus," *Programm zur Feier des Gedächtnisses des Stifters der Universität König Friedrich Wilhelms II* (Bonn: Carl Georgi, 1914).

27. Among names still familiar to historians of ancient medicine are those of Johannes Ilberg (1860–1930), editor of Soranus, a teacher at Leipzig, and successively headmaster at Wurzen, Chemnitz, and Leipzig; Hugo Kühlewein (1870–ca. 1920), editor of the Teubner Hippocrates, who taught at Ilfeld and Kiel; Hans Gossen (1884–ca. 1944), a prolific Pauly contributor and a teacher in Rixdorf and Berlin; and Georg Helmreich (1849–1921), editor of Galen and Scribonius, who taught at Augsburg and became headmaster at Ansbach and Hof. Ernst Wenkebach (1875–1955) and Franz Pfaff (1886–1953), both teachers in Berlin, devoted almost fifty years to editing Galen's commentaries on the Hippocratic aphorisms.

28. Ilberg also held an honorary appointment in Leipzig, but these were rare prizes for a Gymnasium teacher. On Wellmann, see Georg Harig, "Die antike Medizin in der Berliner medizinhistorischen Forschung," in *Tradition und Fortschritt in der Berliner medizinhistorischen Arbeit des Berliner Instituts für Geschichte der Medizin,* Dieter Tutzke, ed. (Berlin: Humboldt Universität, 1980), 37–50.

29. The contemporary situation in historical studies was similar, see *British and German Historiography, 1750–1950: Traditions, Perceptions, and Transfers,* Benedikt Stuchtey and Peter Wende, eds. (Oxford: German Historical Institute London, 2000), and Pim Den Boer, *History as a Profession: The Study of History in France, 1818–1914* (Princeton: Princeton University Press, 1998).

30. Although not, I am told, of his Cambridge teaching. John F. Dobson (1875–1947, professor of Greek at Bristol), a friend of Charles Singer, translated a selection of the fragments of Herophilus and Erasistratus; Horace Harris Rackham (1868–1944, Cambridge) translated parts of Pliny for the Loeb edition; Arthur L. Peck (1902–1974), a doctoral student of Jones in Cambridge, worked later on Aristotelian biology, but all trace appears lost of his Cambridge Ph.D. thesis on Hippocrates, *On Regimen.*

31. William H. S. Jones, *Malaria in Greek History* (Cambridge: Cambridge University Press, 1907); *Hippocrates,* vols. 1, 2, and 4 (London: Heinemann, 1923–1931); cf. *Hippocrate, Tome V. Des Vents. De l'Art,* Jacques Jouanna, ed. (Paris: Les Belles Lettres, 1988), 95. Jones's translation of Pliny, *Natural History* 20–32, in the Loeb series (Cambridge, Mass.: Harvard University Press; London: Heinemann, 1951–1963) is more acceptable.

32. Sir Clifford (T. C.) Allbutt, *Greek Medicine in Rome, the Fitzpatrick Lectures on the History of Medicine delivered at the Royal College of Physicians of London in 1909–1910, with Other Historical Essays* (London: Macmillan, 1921).

33. *Galen, On the Natural Faculties,* Arthur J. Brock, trans. (London: Heinemann, 1916), was based on Helmreich's 1893 text. For Brock, whose translation of extracts in *Greek Medicine: Being Extracts Illustrative of Medical Writers from Hippocrates to Galen* (London: J. M. Dent, 1929) is still valuable, see David Cantor, "The Name and the Word: Neo-Hippocratism and Language in Interwar Britain," in *Reinventing Hippocrates,* David Cantor, ed. (Aldershot: Ashgate, 2002), 280–301. *Celsus, On Medicine,* William G. Spencer, trans. (London: Heinemann, 1935), was based on the Corpus Medicorum edition of Friedrich Marx (1915). Ed-

ward T. Withington's translation of the Hippocratic surgical texts in the Loeb edition, vol. 3 (London: Heinemann, 1928), was based on Petrequin's 1877–1878 edition.

34. Hermann Diels, *Anonymus Londinensis ex Aristotelis Iatricis Menoniis et Aliis Medicis Eclogae* (Berlin: Reimer, 1893). The English translation by Jones, *The Medical Writings of Anonymus Londinensis* (Cambridge; Cambridge University Press, 1947), was based on Diels. A new edition, by Daniela Manetti, based on a complete reexamination of the papyrus, will mark a considerable improvement.

35. Otto Kern, *Hermann Diels und Carl Robert. Ein biographischer Versuch* (Leipzig: O. R. Reisland, 1907), esp. 107–111, for contemporary views about his work on ancient science.

36. He joked in 1883 that he expected to get a chair in medicine for having written so much about Galen; Diels, Usener, and Zeller, *Briefwechsel*, vol. 1:283.

37. Hermann Diels, *Fragmente der Vorsokratiker,* 1st ed. (Berlin: Weidmann, 1903). *Briefwechsel*, vol. 1:44, 46, 50, 111, 120, 122, 138, 447, 450, shows his fundamental approach to the medical writers as sources for early Greek philosophy. In a letter of 1906, Diels claimed to have read through Galen while he was ill (!). See Maximilian Braun, William M. Calder, and Dietrich Ehlers *"Lieber Prinz." Der Briefwechsel zwischen Hermann Diels und Ulrich von Wilamowitz-Moellendorff (1869–1921)* (Hildesheim: Weidmann, 1995), 230.

38. Note Diels's assessment of its importance in *Briefwechsel*, vol. 1:451.

39. Kern, *Hermann Diels,* 116; Wolfhart Unte, "Wilamowitz als wissenschaftlicher Organisator," in Calder, Flashar, and Lindken, *Wilamowitz,* 720–770.

40. Kern, *Hermann Diels,* 113–118; Ulrich von Wilamowitz, *Kleine Schriften,* vol. 6 (Berlin: Akademie Verlag, 1972), 71–74.

41. Rebenich, *Mommsen,* 55–71; Unte, "Wilamowitz."

42. Theodor Mommsen, *Reden und Aufsätze* (Berlin: Weidmann, 1905), 209–210, vigorously denounced Harnack's criticisms. Harnack's doubts about this "waste of labour and waste of money" did not prevent him from participating fully in this "industry" as Royal Librarian and Diels's successor as academy secretary.

43. Jutta Kollesch, "Das Corpus Medicorum Graecorum-Konzeption und Durchführung," *Medzinhistorisches Journal* 3 (1968): 68–73, and "Das Berliner Corpus der antiken Ärzte: Zur Konzeption und zum Stand der Arbeiten," in *Tradizione e ecdotica dei testi medici tardoantichi e bizantini,* Antonio Garzya, ed. (Naples: D'Auria, 1992), 347–350. I merely add to her account material from the recently published correspondence of Diels.

44. Diels, *Briefwechsel,* vol. 2:289; cf. 429–434 for his experience with the Aristotelian commentators.

45. Schöne (1870–1941) took his doctorate at Bonn in 1893; he held chairs at Königsberg, 1903; Basle, 1906; Greifswald, 1909; and finally Münster, 1916–1935.

46. Danielle Gourevitch, *La mission de Charles Daremberg en Italie (1849–1850)* (Naples: Centre Jean Bérard, 1994).

47. Diels, *Briefwechsel,* vol. 2:289 (July 1901), proposed a terminus for the catalogue of October 1905, following three years of travel by Schöne (ibid., vol. 2:293).

48. Johannes Mewaldt (1880–1964), Redakteur of CMG, 1907, and Assistent in Berlin; he subsequently held posts at Greifswald, 1909; Marburg, 1914; Greifswald again, as full professor, 1916; Königsberg, 1923; Tübingen, 1927; and Vienna, 1931–1946.

49. Quotations are taken from Diels's introduction to Hermann Diels, "Die Handschriften der antiken Ärzte, Teil II, Die übrigen griechischen Ärzte ausser Hippokrates und Galenos," *Abhandlungen der königlichen Preussischen Akademie der Wissenschaften,* 1906, phi-

lologisch- historische Klasse, vols. 1–9 (Berlin: Georg Reimer). "Teil I, Hippokrates und Galenos," appeared in the *Abhandlungen,* for 1905; and the *Erster Nachtrag,* in those for 1908. All three were reprinted together in 1970 (Leipzig: Zentralantiquariat).

50. Braun, Calder, and Ehlers, *Lieber Prinz,* 258–259.

51. Ibid., 230, 235, 273. Diels's view of Mewaldt's personality (235) is hardly flattering, lacking almost totally in tact and utterly convinced of his own superiority. Whether, as Diels hoped, marriage cured him of his arrogance, I have been unable to discover.

52. *Philumenus, De animalibus venenatis,* Max Wellmann, ed. (Berlin: Teubner, 1908).

53. Wilamowitz, in Braun Calder, and Ehlers, *Lieber Prinz,* 231, doubted Mewaldt's competence, as yet, in Galen.

54. Johannes Mewaldt, *In Hippocrates De natura hominis Commentarii* (Leipzig: Teubner, 1914); Hermann Diels, *In Hippocratis Prorrheticum I Commentarii* (Berlin: Teubner, 1915). Diels had completed a first draft by 1910; see Braun, Calder, and Ehlers, *Lieber Prinz,* 267–272.

55. Rebenich, *Theodor Mommsen,* 243; Bernhard vom Brocke, "Wissenschaft und Militarismus," in Calder, Flashar, and Lindken, *Wilamowitz,* 649–719.

56. Gilbert Murray, "Memories of Wilamowitz," *Antike und Abendland* 4 (1954): 14.

57. There had already been complaints before 1914 about a study of classical antiquity that seemed to be unable to distinguish the trivial from the truly important; see the inspired satire by Ludwig [Lajos] Hatvany, *Die Wissenschaft des nicht Wissenswerten* (1908; reprint, Oxford: Blackwell, 1986). The title may be translated as "Knowing What's Not Worth Knowing."

58. Kalbfleisch's edition of the small work *On the Thinning Diet (De Victu Attenuante)* appeared in a composite volume of Galen's dietetic tracts in the corpus in 1923 (Berlin: Teubner). His work (over forty years) toward the editio princeps of Galen's *On My Own Opinions (De Propriis Placitis)* ended in tragedy when his house was destroyed by bombing during World War II.

59. Deichgräber (1903–1984) studied in Berlin with Jaeger before returning to his native Friesland to take his doctorate at Münster under Schöne. His thesis, *Die griechische Empirikerschule. Sammlung der Fragmente und Darstellung der Lehre* (1927; published, Berlin: Weidmann, 1930; 2d ed., Berlin: Weidmann, 1965), is a model of its kind. He became Redakteur of the Corpus Medicorum, before obtaining chairs at Marburg (1935) and Göttingen (1938). For his somewhat controversial career, see Wegeler, *wir sagen ab,* 234–235, and the obituary by Hans Gärtner in *Gnomon* 58 (1986): 475–480. His *Ausgewählte kleine Schriften* (Hildesheim: Weidmann, 1984) underplays the quantity and quality of his contributions to ancient medicine and reveals nothing about his importance as a teacher in directing dissertations on ancient medical texts in the 1950s and 1960s.

60. Wegeler, *wir sagen ab,* 53–59; *Altertumswissenschaft in den 20er Jahre. Neue Fragen und Impulse,* Hellmut Flashar, ed. (Stuttgart: Steiner, 1995), especially the articles by Joachim Latacz, "Reflexionen klassischer Philologen auf die Altertumswissenschaft der Jahre 1900–1930," 41–64; Peter Lebrecht Schmidt, "Die deutsche Latinistik vom Beginn bis in die 20er Jahre," 160–177 (especially, on 161, some scathing comments from contemporaries on the irrelevance [or triviality] of much of Wilamowitz's *Griechisches Lehrbuch*); and Albert Henrichs, "Philologie und Wissenschaftsgeschichte: Zur Krise einer Selbstverständnisses," 423–458.

61. Ulrich von Wilamowitz, *Der Glaube der Hellenen* (1931–1932; reprint, Darmstadt: Wissenschaftliche Bücherei, 1984); see Albert Henrichs, "Der Glaube der Hellenen: Reli-

gionsgeschichte als Glaubensbekenntnis und Kulturkritik," in Calder, Flashar, and Lindken, *Wilamowitz,* 263–305.

62. For the contemporary movements within historical studies, see Georg G. Iggers, *The German Conception of History. The National Tradition of Historical Thought from Herder to the Present,* 2d ed. (Middletown, Conn.: Wesleyan University Press, 1983), and Stuchtey and Wende, *British and German Historiography.* The old survey of George P. Gooch, *History and Historians in the Nineteenth Century,* 2d ed. (Boston: Beacon Press, 1959), remains invaluable for the sheer amount of its factual information. For history teaching, see Klaus Bergmann and Gerhard Schneider, *Gesellschaft, Staat, Geschichtsunterricht* (Düsseldorf: Schwann, 1982).

63. Some background is provided by Owsei Temkin, *"The Double Face of Janus" and Other Essays in the History of Medicine* (Baltimore: Johns Hopkins University Press, 1977), 7–10.

64. For the discussion about the propriety of including Bergsträsser's edition of the (Arabic) pseudo-Galenic commentary on Hippocrates' *On Sevens* in the Corpus Medicorum, see Braun, Calder, and Ehlers, *Lieber Prinz,* 283–284. The increased understanding of the complexity of the secondary tradition of the texts of Galen and Hippocrates acted as a further detriment to speedy progress.

65. I am thinking principally of Max Pohlenz at Göttingen; Bruno Snell at Hamburg, whose *The Discovery of Mind,* although not published until 1946, was conceived in the 1930s; Otto Regenbogen at Heidelberg, the teacher of Ludwig Edelstein and Hans Diller; and, above all, Werner Jaeger.

66. W. Jaeger, *Nemesios von Emesa. Quellenforschungen zum Neoplatonismus und seinen Anfängen bei Poseidonios* (Berlin: Reimer, 1914). For Nemesius, see William Telfer, "The Birth of Christian Anthropology," *Journal of Theological Studies,* n.s. 13 (1962): 347–354.

67. Jaeger for several years headed the Berlin Academy's Corpus Medicorum commission and directed Deichgräber's Habilitationschrift. For his interest in ancient medicine, see Heinrich von Staden, "Jaeger's 'Skandalon der historischen Vernunft': Diocles, Aristotle, and Theophrastus," in *Werner Jaeger Reconsidered,* William M. Calder III, ed. (Atlanta: Scholars Press, 1992), 229–265.

68. Werner Jaeger, *Diokles von Karystos. Die griechische Medizin und die Schule des Aristoteles* (Berlin: De Gruyter, 1938). Jaeger's views are no longer accepted; see Phillip J. van der Eijk, *Diocles of Carystus: A Collection of the Fragments with Translation and Commentary* (Leiden: Brill, 1999–2001).

69. Werner Jaeger, *Paideia,* vol. 2 (Berlin: De Gruyter, 1944), 11–58.

70. Donald O. White, "Werner Jaeger's 'Third Humanism' and the Crisis of Conservative Cultural Politics in Weimar Germany," in Calder, *Werner Jaeger,* 267–288.

71. Harig, "Sudhoffs Sicht"; Gundolf Keil, "Sudhoffs Sicht vom deutschen medizinischen Mittelalter," *Nachrichtenblatt der Deutschen Gesellschaft für die Geschichte der Medizin, Naturwissenschaft und Technik* 31 (1984): 94–129. Thomas Rütten, "Patriarch der deutschen Medizingeschichte. Zur Identitätspräsentation einer wissenschaftlichen Disziplin in der Biographik ihres Begründers," in Gourevitch, *Médecins,* 153–169, stresses Sudhoff's politics. Given his social class and nationalist ideals, it is hardly surprising that Sudhoff joined the NSDAP after 1933.

72. *575 Jahre medizinische Fakultät der Universität Leipzig,* Ingrid Kästner and Achim Thom, eds. (Leipzig: J. A. Barth, 1990), 147.

73. Heinrich von Staden, " 'Hard Realism' and 'A Few Romantic Moves': Henry Sigerist's Versions of Ancient Greece," in *Making Medical History: The Life and Times of Henry E. Sigerist,*

Elizabeth Fee and Theodore M. Brown, eds. (Baltimore: Johns Hopkins University Press, 1997), 136–161.

74. Temkin, *The Double Face,* 8–11.

75. Owsei Temkin, "Geschichte des Hippokratismus im ausgehenden Altertum," *Kyklos: Jahrbuch für Geschichte und Philosophie der Medizin* 4 (1932): 1–80, was Temkin's Habilitationschrift and was partly republished in *The Double Face,* 167–177. Note the change in the subtitle of the journal.

76. Owsei Temkin, "Studien zum 'Sinn'-Begriff in der Medizin," *Kyklos* 2 (1929): 21–105; "Die Geisteswissenschaften in der Medizin," *Vorträge des Instituts für Geschichte der Medizin an der Universität Leipzig* 3 (1930): 32–50 (delivered as part of a series on "Philosophische Grenzfragen der Medizin"). Note Temkin's own later criticism, *The Double Face,* 12–17.

77. Still enunciated in Owsei Temkin, *"On Second Thought" and Other Essays in the History of Medicine and Science* (Baltimore: Johns Hopkins University Press, 2002), 11–18; for earlier pronouncements to the same effect, contrast 231–240 (originally 1959) and 49–59 (originally 1981); *The Double Face,* 50–67 (originally 1949) and 14–16.

78. Owsei Temkin, *The Falling Sickness: A History of Epilepsy from the Greeks to the Beginnings of Modern Neurology,* 2d ed. (Baltimore: Johns Hopkins Press, 1971); *Galenism: The Rise and Decline of a Medical Philosophy* (Ithaca: Cornell University Press, 1973); *Hippocrates in a World of Pagans and Christians* (Baltimore: Johns Hopkins University Press, 1991).

79. Gary Ferngren, in the introduction to the reprint of Ludwig and Emma J. Edelstein, *Asclepius: A Collection and Interpretation of the Testimonies* (1945; reprint, Baltimore: Johns Hopkins University Press, 1998), xiii–xxii; Thomas Rütten, "Zu Leben und Werk Ludwig Edelsteins (1902–1965)," *Nachrichtenblatt der Deutschen Gesellschaft für die Geschichte der Medizin, Naturwissenschaft und Technik* 48 (1998): 136–146. Temkin's appreciation of his friend is reprinted in *On Second Thought* 250–263, and see also 1–5. A major study of Edelstein is expected from Rütten, whose "Zu Leben und Werk Ludwig Edelsteins" adds much to our knowledge of Edelstein in the 1930s.

80. Ludwig Edelstein, Περί ἀέρων *und die Sammlung der hippokratischen Schriften* (Berlin: Weidmann, 1931).

81. A bibliography is given by Temkin, *On Second Thought,* 259–263.

82. The quotation is from Ludwig Edelstein, "The Relation of Ancient Philosophy to Medicine," *Bulletin of the History of Medicine* 26 (1952): 299–316, 316, reprinted in *Ancient Medicine,* 349–366. See also "The Distinctive Hellenism of Greek Medicine," *Bulletin of the History of Medicine* 40 (1966): 197–255, reprinted in *Ancient Medicine,* 367–391. For the significance of philosophy, see James N. Longrigg, *Greek Rational Medicine: Philosophy and Medicine from Almaeon to the Alexandrians* (London: Routledge, 1993).

83. For Diller (1903–1977), professor at Kiel and a longstanding friend of Edelstein, see his *Kleine Schriften zur antiken Medizin* (Berlin: De Gruyter, 1973).

84. Hans Diller, *Die Überlieferung der hippokratischen Schrift* Περί ἀέρων ὑδάτων τόπων (Leipzig: Dietrich, 1932), and *Wanderarzt und Aitiologe. Studien zur hippokratischen Schrift* Περί ἀέρων ὑδάτων τόπων (Leipzig: Dietrich, 1934). Diller's edition finally appeared in 1970 as *Hippocratis De aere aquis locis edidit et in linguam Germanicam vertit* (Berlin: Akademie Verlag, 1970).

85. Edelstein, Περί ἀέρων, partially reprinted in *Ancient Medicine,* 65–110. See Diller's review, *Kleine Schriften,* 131–143, and his story (124) of his first meeting with Edelstein soon after hearing of his dissertation.

86. Karl Deichgräber, *Die Epidemien und das Corpus Hippocraticum* (Berlin: Weidmann, 1933; 2d ed., Berlin: Akademie Verlag, 1971). Cf. Ludwig Edelstein, "The Genuine Works of Hippocrates," *Bulletin of the History of Medicine* 4 (1936): 236–248, reprinted in *Ancient Medicine,* 133–144, a restatement in English of arguments published in German in 1935. I know of no ties of friendship between Deichgräber and Edelstein.

87. Geoffrey E. R. Lloyd, *Problems and Methods in Greek Science* (Cambridge: Cambridge University Press, 1991), 194–224. In his writings in the 1950s and 1960s Deichgräber showed that he too could see the wider picture.

88. In 1932 he was allowed to lecture (unpaid) at Berlin in the history of the exact sciences in the Classical Institute.

89. See Temkin, *On Second Thought,* 252, for his relationship with the History of Ideas Club.

90. *A Celebration of Medical History. The Fiftieth Anniversary of the Johns Hopkins Institute of the History of Medicine and the Welch Medical Library,* Lloyd G. Stevenson, ed. (Baltimore: Johns Hopkins University Press, 1982), 223–225, lists the degrees granted.

91. As far as I know, none of the historians of ancient medicine in the United States who came into prominence in the 1960s and 1970s (such as Wesley D. Smith, John M. Riddle, John Scarborough, Darrel Amundsen, Gary Ferngren, and Heinrich von Staden) spent time in Baltimore, and only von Staden (himself partly trained in Germany and Austria) retains a major interest in the history of medical concepts.

92. Drabkin (1905–1965), who taught classics at City College in New York, edited Caelius Aurelianus and, with Morris R. Cohen, compiled a well-known *Source Book in Greek Science* (Cambridge, Mass.: Harvard University Press, 1948). The historian of medieval pharmacology Jerry Stannard (1929–1988) also held a Hopkins Fellowship in the 1950s before moving to teach classics at the University of Kansas.

93. Temkin, *The Double Face,* 21–22. His account, as told to me, of an early visit to address the annual dinner of a Midwestern medical society revealed an even greater conflict of expectations and priorities.

94. Ibid., 22. By "positivist" is meant the empirical accumulation of data rather than any engagement with the underlying ideas revealed by this material.

95. Represented in classics by the epigraphist and archaeologist David M. Robinson and in the modern languages faculty by Henry Carrington Lancaster, whose studies of the French theater are invaluable sources of information.

96. See Stuchtey and Wende, *Historiography,* 2–4, for thoughts on the interchange of historical styles.

97. Sigerist's name is associated among American medical historians almost exclusively with an interest in the social history of medicine, and there have been relatively few American scholars who have written on the history of medical concepts. By contrast, this has long remained a major concern of many German institutes for the history of medicine, where the social history of medicine, influenced more from Britain than from the United States, did not take root until the 1990s.

98. Wegeler, *wir sagen ab,* 372–394, provides a detailed list of classical scholars driven out of their posts by the Nazis; for a similarly negative view of the immediate impact of German refugees in the United States, see William M. Calder, *Studies in the Modern History of Classical Scholarship* (Naples: Jovene Editore, 1982), 34–37.

99. Edelstein, for example, published only one article on ancient medicine and a handful

of reviews in American classical journals, and Temkin's name is, I suspect, unfamiliar to most students of classics and even, to judge from its frequent misspelling, their teachers.

100. Edelstein, *Ancient Medicine*. The equally important works of his contemporaries, Karl Deichgräber and Hans Diller, are largely neglected because they are not available in English. Deichgräber's, in particular, are scattered across a variety of journals, many accessible only in a major library.

101. Especially, *The Hippocratic Oath: Text, Translation, and Interpretation, Bulletin of the History of Medicine* suppl. 1 (1943), reprinted in *Ancient Medicine*, 3–64. Cf. Temkin, *On Second Thought*, 21–28; Pierre M. Bellemare, "The Hippocratic Oath: Edelstein Revisited," in *Healing in Religion and Society from Hippocrates to the Puritans: Selected Studies*, J. Kevin Coyle and Stephen C. Muir, eds. (Lewiston: Edwin Mellon Press, 1999), 1–64.

102. Vivian Nutton, "Ancient Medicine: Asclepius Transformed," in *Science and Mathematics in Ancient Greek Culture,* Christopher J. Tuplin and Tracey E. Rihll, eds. (Oxford: Oxford University Press, 2002), 242–255.

103. *L'Eau, la santé et la maladie dans le monde grec,* René Ginouvès, ed. (Paris: L'École française d'Athènes, 1994); *Les textes médicaux Latins comme littérature,* Anne and Jackie Pigeaud, eds. (Nantes: Université de Nantes, 2000).

104. E.g., Lesley Ann Dean-Jones, *Women's Bodies in Classical Greek Science* (Oxford: Clarendon Press, 1994); Helen King, *Hippocrates' Woman: Reading the Female Body in Ancient Greece* (London: Routledge, 1998); Rebecca Flemming, *Medicine and the Making of Roman Women: Gender, Nature and Authority from Celsus to Galen* (Oxford: Oxford University Press, 2000). Ann Ellis Hanson has also played a leading role in the study of ancient gynecology in the United States.

105. Centered respectively around Jacques Jouanna, Geoffrey Lloyd, and Vincenzo di Benedetti.

106. In Germany, Fridolf Kudlien (b. 1928) neatly shows this transition; he began as a textual editor with the Berlin corpus before investigating medical ideas and, most recently, the social status of doctors in antiquity. The essays in *Ancient Medicine in its Socio-Cultural Context,* 2 vols., Philip H. van der Eijk, Manfred F. J. Horstmanshoff and Piet J. Schrijvers, eds. (Amsterdam: Rodopi, 1995), illustrate well the new social approach. For mentalities, see Geoffrey E. R. Lloyd, *Demystifying Mentalities* (Cambridge: Cambridge University Press, 1990).

107. Geoffrey E. R. Lloyd, *In the Grip of Disease: Studies in the Greek Imagination* (Oxford: Oxford University Press, 2001); Smith, *Hippocratic Tradition;* Lawrence I. Conrad, Vivian Nutton, Roy Porter, and Andrew Wear, *The Western Medical Tradition, 800* BC to AD *1800* (Cambridge: Cambridge University Press, 1995).

108. The collaborative work of the late M. D. Grmek with Danielle Gourevitch, *Les maladies dans l'art antique* (Paris: Fayard, 1998) and with Jacques Jouanna, *Hippocrate. Épidémies V et VII* (Paris: Les Belles Lettres, 2000) is exemplary. John M. Riddle's *Dioscorides on Pharmacy and Medicine* (Austin: University of Texas Press, 1985) required a considerable input from ethnopharmacologists. For an example of a doctor using his specialist knowledge to raise questions about antiquity, see Peter Brain, *Galen on Bloodletting* (Cambridge: Cambridge University Press, 1986).

CHAPTER SEVEN

Using Medical History to Shape a Profession
The Ideals of William Osler and Henry E. Sigerist

Elizabeth Fee and Theodore M. Brown

The history of medicine has always been heterogeneous, reflecting the diverse stances, programs, and agendas of its practitioners. In the nineteenth century, it already had many different formulations.[1] As the field was reaching its first level of maturity in the late nineteenth and early twentieth centuries, different styles and priorities were still evident.

At Johns Hopkins, the two leading figures in the history of medicine, William Osler (1849–1919) in the 1890s and Henry E. Sigerist (1891–1957) in the 1930s, seem at first glance to occupy almost opposite positions along the spectrum of historiographic choices. Osler, the distinguished professor of clinical medicine, mined the history of medicine for inspirational messages and cultural enrichment. He wove historical examples and moral lessons into his regular clinical and bedside teaching, his formal and informal meetings with students, the lively gatherings of the Johns Hopkins Historical Club, and the innumerable commemorative lectures and orations he was invited to deliver. Sigerist, the European-trained scholar, pursued the history of medicine as a professional discipline, to which he applied his considerable literary, linguistic, and philological skills. He had been brought to Hopkins in 1932 to develop the Institute for the History of Medicine, started by William Henry Welch (1850–1934). Sigerist was determined to create a professional scholarly institute that would lay a sound basis for the development of the field in America along the lines of the German institutes with which he was familiar.[2] Later, he would become more politically engaged as he responded to the deepening crisis in Europe and his experiences in America.

These apparent differences in medical-historical orientation have taken on new meaning in recent decades because of the activities of two groups of medical historians affiliated with the American Association of the History of Medicine. One group, the older and longer-established Osler Society, is made up mainly of clinician-historians who meet to celebrate Osler's life and examine great moments in the history of medical practice. The other, more recently established, Sigerist Circle, consists largely of Ph.D.-trained historians and graduate students who meet to discuss topics in the social and political history of medicine and contemporary health policy issues. Each group has adopted its namesake as an iconic figure. The Osler Society members, remembering Osler's turn to bibliographic and historical interests, emphasize the contributions that knowledge of history can make to a broadened clinical understanding. The Sigerist Circle members, by emphasizing the later phases of Sigerist's career when he was known as an advocate of progressive social causes, promote the connections between historical scholarship and political advocacy. Because of the contemporary purposes to which their members put their respective inspirational figures, Osler and Sigerist have been converted into symbolic representations of themselves, exaggerating their differences.

These polarized differences between the remembered Osler and Sigerist hide some striking and perhaps unexpected similarities in the actual historical figures. In their Hopkins years, both men shared a fundamentally utilitarian view of medical history, as each sought to use it to influence and shape the profession, especially its youngest and most recently recruited members. Osler wanted to imbue physicians with a sense of their profession's overarching mission, with its timeless values and ideals.[3] Physicians, Osler frequently explained, needed to identify with the glorious past and even more glorious future of a profession that—to a greater degree than any other—combined intellectual preeminence with nobility of character.[4]

Sigerist wanted to motivate young physicians too, but with the possibilities of a socially and politically oriented medicine. Calling for the study of history from a wider social angle, he wanted to emphasize the evolution of physicians' societal responsibilities and not restrict its vision to the achievements of an ideal and heroic profession. Although he contributed to professionalizing the history of medicine, Sigerist, like Osler, worked to make sure that the subject would not become an isolated academic discipline but an instrument for shaping the medical profession.

There are other similarities and parallels between Osler and Sigerist. Both drew from the common intellectual tradition of romantic philhellenism, the conviction that all that was best in Western civilization derived from the achievements of Greek antiquity.[5] Neither one hesitated, when it suited his purposes, to employ a presentist and teleological framework. Both were interested in revealing progress in medicine and a purpose in history. Both had charismatic personalities and made an indelible impression on the generations of students who came under their influence. Although both had complicated careers that moved through several phases, during the period when they taught at Johns Hopkins both emphasized "pragmatic" goals for medical history (i.e., pragmatic not in the colloquial sense of the term, but in that of Sprengel, Neuburger, and Pagel).

Osler and Sigerist had arrived at Hopkins by different routes. As Michael Bliss explains in his recent biography, Osler did not turn seriously to the history of medicine until he had reached his forties and was already well established at the pinnacle of his career as Physician-in-Chief of the Johns Hopkins Hospital and Professor of the Theory and Practice of Medicine.[6] Until then, he had been fully engaged in developing his scientific reputation, clinical knowledge, medical practice, and teaching skills. Once settled at Hopkins, however, Osler became part of a new medical culture in which the demands of scientific rigor and achievement were tempered by cultural and humanistic sensibilities. Collecting vintage editions of medical classics, amassing a great library, and becoming conversant with the most instructive episodes of medical history were seen as the means for shaping the next generation of physicians as scientifically trained but cultured leaders of their profession.

Sigerist, by contrast, only practiced medicine during his brief early service in the Swiss army. He had begun studying classical and oriental languages before becoming interested in science—and selecting medicine on the grounds that it was the most inclusive of all the sciences. He then moved into the history of medicine as a way of combining his interests in medicine, history, languages, and literature. Sigerist was able to lead the life of an independent scholar (unlike Neuburger and Pagel) because of his family's wealth. Over time, he made several major changes in his approach to medical history. Without ever fully jettisoning earlier interests, he moved from the predominantly philological style of Karl Sudhoff, through the more philosophically oriented approach of Max Neuburger, to the socially and politically engaged mode of historical scholarship with which he is especially identified. In this third phase, begun at Leipzig but most fully developed at Johns Hopkins, he sought to influence contemporary health policy and to inspire physicians and medical students to social activism. Thus, although Osler and Sigerist might have been astonished to recognize it, they shared, during their Hopkins years, what in essence was a similar purpose; they both wanted to use the history of medicine as a tool for sculpting the future of the medical profession.

What can we learn from comparing the interests and commitments of these two Hopkins professors who both had—over the span of forty years—such a dramatic influence on their profession? Why was it that these men, who each aspired to use medical history to mold future generations of physicians, would themselves become emblematic role models for historians?

Sir William Osler: Medical History as Secular Religion

An important part of William Osler's formative intellectual and moral development lay in working through the apparent contradictions between religion and medicine. The son of an English Anglican clergyman posted to Canada, Osler had first intended to become a minister like his father. Following in the footsteps of one of his prep school mentors and his father's elder brother, however, he decided in college to pursue a career in medicine. A critical encounter with Sir Thomas Browne's *Religio Medici* (first published in 1642) helped pave the way.[7] This famous work by a seventeenth-century physician and naturalist provided a guide to the reconciliation of the scientific and religious spheres. Osler made the book his constant companion, regarding it as "a source of inspiration, wisdom, and solace."[8] He is said to have carried it everywhere, quoted it constantly, and was even buried with a copy of the book on his coffin.[9] This man who would have such an extraordinary influence on his students, and indeed on generations of English and North American doctors, thus worked to achieve some personal reconciliation of the sacred and the secular. He had grown up listening to the cadence of his father's sermons, and he in turn adopted a great deal of the language, structure of argument, and moral fervor of the sermon and turned it into a form of medico-historical inspirational rhetoric.

Osler turned to medical history in the 1890s in his early years at Hopkins. In 1889, at age forty, he began reordering his life and making—in addition to the new job—a series of important changes. He proposed marriage to Grace Linzee Revere Gross (1855–1928), the widow of Dr. Samuel W. Gross (1837–1889), one of the most prominent names in Philadelphia medicine. The marriage was a highly suitable match for both parties, but Grace refused to marry Osler until he had finished writing his textbook, *The Principles and Practice of Medicine*.

That textbook of some 1,050 pages was published in March 1892. A great synthesis, the book brought together the major strands of Osler's career to date: deep knowledge of clinical-pathological anatomy, keen appreciation of the correlation of post-mortem observations with careful bedside study of the manifestations of disease, and currency in the new and emerging fields of cellular pathology, bacteriology, and neurology. The book was an immediate success, selling 14,000 copies within the first two years and quickly becoming the dominant medical textbook in the English-speaking world.

The success of *Principles and Practice* marked a significant turning point in Osler's career. Writing the textbook had required a huge investment of energy and effort, and in its retrospective and synthetic orientation, it signaled the slowing down of his career as an active investigator. Although he was perfectly positioned to be the ideal clinical teacher and role model, Osler's move to Baltimore cut him off from the wellspring of his early research—post-mortem pathology—as this work at Hopkins was the territory of William Henry Welch and his staff. As Bliss suggests, Osler—a man of restless intellectual energy—now turned more of his attention to other interests, notably literary and historical pursuits.

With William Henry Welch, Howard Kelly (1858–1943), and some thirty other Hopkins doctors, Osler founded the Johns Hopkins Historical Club in 1889. The club soon became a focal point in the life of the medical institution. At early meetings, John Shaw Billings (1838–1913) spoke on "Rare Medical Books," Howard Kelly showed treasures from his private collection, and William Henry Welch lectured on Alexandrian medicine. For the winter of 1892, the club had decided to concentrate on the subject of Greek medicine. Osler's first major contribution to the club was a lecture on "Physic and Physicians as Depicted in Plato."[10] For his forty-third birthday in July 1892, Grace had given her husband a copy of Jowett's five-volume translation of the works of Plato, allowing him to study the *Dialogues* in depth and cull references to medicine from the Platonic writings.[11]

In the 1890s, Osler tried his hand at papers on a variety of literary and historical subjects. A number of essays—first read before the Historical Club and then published in the *Johns Hopkins Hospital Bulletin*—were biographical, with tones that could be playful and picaresque ("Thomas Dover: Physician and Buccaneer"), sober and erudite ("John Keats: The Apothecary Poet"), or inspirational and sermonic ("An Alabama Student").[12] This last essay evoked the life of the otherwise anonymous Dr. John Y. Bassett of Huntsville who "heard the call and forsook all and followed his ideal" by leaving his wife and children and going to Paris to study medicine.[13] Osler was here drawing on a tradition that John Harley Warner has characterized as "remembering medical Paris as a moral act."[14]

A more substantial essay in the same tradition, "The Influence of Louis on

American Medicine," read before the Stillé Society of the University of Pennsylvania, traced the shift in Parisian clinical medicine from theoretical speculation to statistical rigor and empirically grounded observation, primarily under the influence of Pierre Louis.[15] Osler then discussed the influence of Louis on two of his American students, James Jackson Jr. and W. W. Gerhard, quoting at length from correspondence that showed the great man's warm interest in his pupils and their enthusiasm for his teaching and fatherly mentoring.[16]

Osler further developed his style of biographical-historical exploration in a series of longer studies of Elisha Bartlett, John Locke, William Beaumont, and Thomas Browne. Each had a particular "claim to remembrance," was a "bright ornament" of the profession, and exemplified a "candid and truth-seeking spirit." Study of such exemplary lives promised to lend "stability to character and help to give a man a sane outlook on the complex problems of life."[17]

At the dedication of the Boston Medical Library in 1901, Osler explained that medical biography could inspire young members of the profession to envision bright possibilities for their lives. The past is a good nurse for the "weanlings of the fold," and the student's task is to become familiar with men on whose lives they can model their own. Through reading biographies, the young physician would experience "the silent influence of character on character . . . in the contemplation of the great and good of the past," and in "the touch divine of noble natures gone." Indeed, Osler went so far as to advocate that libraries set up altars for the worship of great men: "I should like to see in each library a select company of the Immortals set aside for special adoration. Each country might have its representatives in a special alcove of Fame."[18] Osler later got close to recruiting allies in his campaign for "alcoves of Fame" when in 1917 the high-toned and hero-worshipping *Annals of Medical History* was founded with several of Osler's well-placed American friends and students as associate editors—George Dock (1860–1951) and Harvey Cushing (1869–1939) among them.

Yet at the very moment when Osler was advocating the building of alcoves of Fame for the medical immortals, his own scientific reputation was being called into question. During his later years at Hopkins, Osler had become a prisoner of his own success. By the mid 1890s, he had more private patients than he could handle; he was consulting physician to the students, doctors, and nurses at Hopkins, as well as to the Baltimore elite, members of Congress, and denizens of the White House.

Unable or unwilling to cut back on his practice or to reduce his obligations at Hopkins, Osler felt increasingly beleaguered and began to worry about his own health. Yet the pace continued. In 1900, Osler's income, mainly based on his private practice, passed $30,000, grew to $40,000 in 1901, and reached $47,280 in 1903.[19] But at the same time, an undercurrent of critical talk was developing among some of his colleagues in Baltimore. Was Osler as good as he used to be? Bliss notes that some of the Hopkins medical scientists knew that Osler had drifted away from the leading edge of their disciplines; as an acute observer and classifier of disease, he published his clinical lectures and case reports but was not using his laboratory for active experimental work.[20]

Osler himself was conscious of a frustrating loss of scientific currency. Struggling to conform to his internal ideal of himself as an active investigator and as leader of the new specialty of internal medicine, he continued to publish in medical journals. He strongly supported the recruitment of Hopkins faculty such as John J. Abel (1857–1938), professor of pharmacology, who was deeply committed to the German model of laboratory-based research. He likewise supervised pioneering studies of the blood pressure of typhus and pneumonia patients, using the apparatus just developed by William Howell (1860–1945) and his students in physiology. Yet Osler was vulnerable to being stung by his waspish colleague, Franklin P. Mall (1862–1917), the professor of anatomy, who wanted the medical school to be a university-based research institute with its teachers devoted to science, not to clinical practice, making money, or even doing much teaching.[21] In 1905, Mall wrote that what was needed in Osler's department was a university professor who would "conduct his medical work along laboratory lines & will not continue publishing cases."[22]

Around this time, we see a marked shift in the nature of Osler's historical writing. He began focusing not on exemplary lives—men of extraordinary character and nobility—but on tracing ideas and trends in the march of medical science. In this new form of writing, Osler would emphasize the significance of laboratory-based scientific achievements culminating in the recent triumphal progress of bacteriology, biochemistry, and immunology. Although not sharp or complete— he did, after all, lead a delegation of eminent physicians to the tomb of Pierre Louis in 1905[23]—he began downplaying Hippocrates, the great clinical observer, as the reference point for measuring modern practitioners and began substituting Galen, portrayed as a pioneering experimentalist, as the model for scientifically grounded investigators.

This new approach to medical history is evident in "Medicine in the Nineteenth Century," a paper delivered at the Historical Club in January 1901 to mark the twenty-fifth anniversary of the founding of the Johns Hopkins University and the turn of a new century.[24] Perhaps in a not fully conscious effort to reinvent himself, Osler initiated his glorification of experimental investigations with a lineage that now went from Galen, Harvey, and Hunter to the mid-nineteenth-century laboratory researchers and concluded: "The study of physiology and pathology within the past half-century has done more to emancipate medicine from routine and the thralldom of authority than all the work of all the physicians from the days of Hippocrates to Jenner, and we are as yet but on the threshold."[25]

Much of this essay deals with bacteriology and the progressive conquest of infectious diseases, which Osler sees as the most remarkable achievement of the century. Bacteriological scientists with their laboratory methods "have in their possession a magic key to one of nature's secret doors." Those in the new pantheon of heroes—Pasteur, Koch, Cohn, and Von Behring, men who had achieved fame in the laboratory rather than the clinic—were the best hope of the future.[26]

In his Harveian Oration to the Royal College of Physicians in 1906, Osler focused on one great, experimentally based discovery, Harvey's early seventeenth-century recognition of the circulation of the blood.[27] Although even Harvey was liable to "catch the beliefs of the day," he was one of the rare few "who sees with his own eyes, and with an instinct or genius for the truth, escapes from the routine in which his fellows live." Harvey's discovery marked the passage from the "age of the hearer" and the "age of the eye" to the "age of the hand," that is, the age of modern experimental medicine.[28]

Osler combined his admiration for Greek medicine with his praise for experimental investigation by making Galen the father of experimental medicine. In an address to the Congress of American Physicians and Surgeons in 1907, Osler said that although the Greeks in general were limited to the level of fact and observation, Galen was "the one man alone among the ancients [who] could walk into the physiological laboratories today and feel at home."[29] Speaking directly to an audience of clinicians, Osler again traced Galen's legacy through Harvey, Hunter, and Jenner to the last half of the nineteenth century, which he termed "the era of experimental medicine." The tables had been turned; clinical medicine must now catch up with scientific progress.[30]

Osler reiterated these themes in his introduction to the multivolume compendium, *Modern Medicine: Its Theory and Practice* (1907), which had been created by leading specialists to bring busy practitioners up to date with developments in scientific medicine.[31] Osler again lauded Galen over Hippocrates, Harvey over Sydenham, and leapfrogged from French clinical medicine to the latter part of the nineteenth century, which saw "a complete revolution in our conception of the etiology of infectious diseases."[32] The primary audience for this work, Osler emphasized, was the man in general practice "who wishes to keep himself informed of the existing state of our knowledge in clinical medicine." However industrious, it was impossible for a young man unaided to keep abreast of all the latest scientific advances pouring out of the chemical and bacteriological laboratories; he needed a well-structured postgraduate education supervised by a small number of laboratory-oriented attending physicians.

Osler was responding to seismic shifts in the nature of medical education and practice, sympathetic to the difficulties of practicing physicians, and conscious that he himself was in danger of slipping from the pinnacle. When an offer came from Oxford University to become Regius Professor of Medicine, an honor dripping with English status and privilege, it gave Osler a pathway of escape from Hopkins, a chance to start over in a different, less encumbered setting. When his wife Grace read the invitation, she said she felt "a tremendous weight lifted from my shoulders" and wanted him to cable his acceptance right away. Moving to Oxford from Hopkins also gave Osler the chance to develop a new focus for his history of medicine, which now became much less tied to the needs of the profession and utilitarian educational purposes.

Osler left for Oxford in 1905, stating that he had opted for "a quiet easy berth for a man whose best work is done."[33] Once settled into his new situation, he avoided the various opportunities that came his way to become a busy consultant to Harley Street practitioners and the great London teaching hospitals. His limited professorial duties included a few weeks of examining medical students in the spring and a weekly clinic at the Radcliffe Infirmary for students and local doctors.

Osler now found time to spend some of his days in the laboratory. He collaborated with several prominent scientists: working with Archibald Garrod (1857–1936) on a rare urinary disorder and with Arthur Keith (1866–1955), using the new electrocardiograph machine for a study of Stokes-Adams disease.[34] For the most part, however, he devoted his time to the history of medicine. Interestingly, Osler's historical work increasingly emphasized the progressive march of science and presented the evolution of the experimental-laboratory-investigative ideal that had become ever more powerful to him even as the reality of it in his professional life faded away.

Intellectually, Osler was very much influenced by the classical revival of the latter half of the nineteenth century and particularly the wave of "philhellenism" that glorified the culture of ancient Greece.[35] He absorbed the atmosphere of Oxford, along with the assumptions of class-structured and imperial British society, as he served as a member of the Hebdomadal Council, the university's governing body, as a delegate to the university press, and as curator of the Bodleian Library. He was regularly reelected president of the Bibliographical Society; he turned down the presidency of the Royal Society of Medicine but accepted the presidency of the Classical Association. In 1911 he became William Osler, Bart. Osler gave expression to the values he imbibed in his 1910 "secular sermon" on "Man's Redemption of Man" to an audience of 2,500 at the University of Edinburgh. All traditions and ideas of any consequence came to us from the Greeks, he assured his audience, and the Greeks provided the germinal ideas for the entire development of modern science. "So far did unaided observation and brilliant generalization carry Greek thinkers, that there is scarcely a modern discovery which by anticipation cannot be found in their writings."[36]

Osler's intellectual approach was effectively captured in the metaphors he used recurrently in this period. He delighted in organic allusions that emphasized the natural, inevitable, even preordained growth of scientific and medical knowledge. Truth was a seed that might have been sown on stony ground but that would nonetheless, given the right conditions, emerge and flower: "Like a living organism, Truth grows, and its gradual evolution may be traced from the tiny germ to the mature product. . . . Or the germ may be dormant for centuries, awaiting the fullness of time."[37]

Another favorite metaphor was that of light, clear vision, and ability to perceive the truth in front of one's eyes. Those who fail to see the truth are "blinded," their eyes are "sealed," or they have "mental blindness." Riolan, who failed to acknowledge the circulation of the blood, was unable to see "the truth which was staring him in the face." Others saw "glimmerings" of the truth or the "brightness" of the image that is "clear as day to those with eyes to see."[38]

In his largely encyclopedic Silliman lectures, delivered at Yale in 1913 and later published posthumously as *The Evolution of Modern Medicine,* the lamp of reason was variously flickering, brightening, burning clearly, and being extinguished.[39] A few examples: "Following the glory that was Greece and the grandeur that was Rome, a desolation came upon the civilized world, in which the light of learning burned low, flickering almost to extinction." The fifteenth century, however, saw "a true dawn that brightened more and more unto the perfect day," whereas Pasteur would later bring "a light that brightens more and more as the years give us ever fuller knowledge."[40] Thus downplaying non-Western achievements and overstating the continuity between Greek thought and European "civilized" science, Osler again uses the metaphor of Truth as a living organism, sometimes of slow and difficult growth, but ultimately and inevitably flowering.[41]

Despite the continuity of metaphors, the Silliman lectures—his attempt at a grand survey and synthesis of the history of medical science from antiquity to the turn of the twentieth century—clearly marked a third phase in Osler's engagement with medical history. In the first phase, he had used the history of medicine as an integral part of his medical teaching at Hopkins to inspire the "weanlings of the fold" to enter the timeless profession with appropriately high ideals; in the second, as he got ready to leave Hopkins and settle in Oxford, he turned to medical history to validate the importance of experimental methods in the reconstruction of modern medical science, for practitioners and for himself; in the third phase, he would attempt to establish a new professional discipline of the history of science and medicine, not so much as part of medicine or for the benefit of physicians, but as a contribution to the intellectual and cultural history of Western civilization.

Osler organized a section for the history of medicine in the Royal Society of Medicine, persuading 160 of his colleagues to join and becoming its first president in 1912.[42] In 1914, he convinced Dr. Charles Singer (1876–1960) and his wife, Dorothea (1882–1964), both members of the historical section, to settle in Oxford and organize a history of science room at the Radcliffe Camera of the Bodleian Library. Although Osler's plans to establish the Singers more permanently at the Bodleian failed to gain support, he did succeed in offering Charles the title of Lecturer in the History of Biology.

Osler was also involved with the ambitious plans of George Sarton (1884–1956), the displaced Belgian scholar who, as an émigré in the United States, became a pioneering figure in the history of science.[43] While Osler and Sarton were in correspondence in 1917 and 1918, Osler alerted Sarton to the work of the Singers. Sarton contacted Singer and recruited him as a coeditor of *Isis,* which Sarton had founded in 1913 to serve as the field-shaping journal of the new discipline. Osler and Sarton exchanged further correspondence and reprints; in May 1919, in the last year of his life, Osler chose "The Old Humanities and the New Science" as the topic for his presidential address to the Classical Association and gave a ringing endorsement of Sarton's "new humanism." Sarton believed that the history of science captured the highest forms of human thought and civilization and thus provided the basis for a true philosophy: "The history of science is the history of mankind's unity, of its sublime purpose, of its gradual redemption."[44]

Such a conception fitted well with Osler's own progressive vision. In Sarton's spirit, he called for bridging the gap between the humanities and science by having students of the classics learn about classical science and students of the sciences combat the tendency to narrow specialization by studying the history of science. These were noble ambitions, and they had little to do with the practice of medicine or shaping the next generation of "weanling" practitioners. If anything, Osler had accepted Sarton's view that medicine was but one part of a comprehensive history of science as human knowledge. In leaving behind his Hopkins commitment to the history of medicine as a tool for shaping the medical profession, Osler at Oxford turned toward a grander intellectual mission.

Henry E. Sigerist: Medical History as Romantic Socialism

Henry Sigerist began his historiographical journey, in a sense, where William Osler's had ended. In his inaugural lecture as Privatdozent in medical history at Zurich in November 1921, Sigerist argued that one of the principal aims of the field was to help physicians realize that modern medicine had grown gradually, as the result of a long and troubled development in which "our grains of truth emerged from a sea of errors."[45] The history of medicine was "part of the general history of civilization," a bridge between the sciences and the humanities, and a facilitator in moving toward a new humanism. The field was thus especially valuable as a means to instill "idealism into the young students of medicine, an idealism more desirable than ever."

When Sigerist succeeded Karl Sudhoff as director of the Institute of the History of Medicine in Leipzig in 1925, he moved in a more pragmatic pedagogic direction while continuing his predecessor's development of the field as an autonomous academic discipline. Broadening Sudhoff's largely philological approach, Sigerist turned his scholarship in more "cultural" directions, first adapting Oswald Spengler's (1880–1936) general notions of "cultural morphology" to a broad periodization of medical history as successive cultural epochs, and then adopting art historian Heinrich Wölfflin's (1864–1945) more flexible cultural relativism as a way to understand the medical achievements of a given period as an expression of that period's general "style."[46] While in Leipzig, Sigerist also directed much of the institute's scholarly and teaching activities away from conventional historical subjects to a study of the pressing philosophical, ethical, social, and economic questions of the day as they related to medicine. With his characteristic energy, he added these new lines of inquiry to the older ones still being pursued at the institute.

In Leipzig, Sigerist also lectured on the broad scope of medical history as understood from philosophical and social perspectives (these lectures were published in 1931 as *Einführung in die Medizin*) and organized colloquia on contemporary political and economic issues. Students and younger faculty crowded into Sigerist's lectures and seminars, making the institute the vital center of the Leipzig medical faculty. They were attracted by his charismatic personality, his engaging teaching style, and by his vision of medicine made whole again by the galvanizing use of medical-historical scholarship.

Sigerist entered the Hopkins orbit when the institution was beginning to think seriously about the professionalization of medical history. Initially, William Henry Welch had in 1925 proposed the creation of a chair in the history of medicine and a central library within the Johns Hopkins Medical Institutions, both intended to counteract the unintended consequences of triumphal medical science and the centrifugal tendency of specialization. Welch was worried that, as an older cultured generation passed away and narrow scientific experts took its place, physicians could become mere technicians, specialists in a field but "devoid of the intellectual, cultural background so necessary for broadening the physician's influence."[47] Teaching the history of medicine would provide the "human note" needed to "make the physician a gentleman of culture."[48]

The General Education Board of the Rockefeller Foundation agreed to Welch's proposal but on the condition that Welch himself accept the professorship in the history of medicine.[49] Although uncomfortable and reluctant, Welch accepted the chair in order to secure the funds, but he soon set about trying to recruit an appropriate successor. He first offered the chair to Charles Singer, whose career William Osler had done so much to promote. When Singer declined, Welch offered the professorship to Harvey Cushing, the famous Hopkins product, noted neurosurgeon, and author of the then recently published Pulitzer Prize–winning biography of Osler. When Cushing declined, Welch, with Cushing's blessing, offered the position to Henry Sigerist, whom he had earlier met during a European tour.

When Sigerist took over the directorship of the Hopkins Institute of the History of Medicine in 1932, his first impulse was to import to Baltimore all the projects and activities with which he had been engaged in Leipzig. He thus arranged an appointment for his former student and assistant Owsei Temkin (1902–2002), and the two of them set out to create "a nucleus of German learning in foreign lands."[50] But Sigerist understood that he faced additional challenges, the most immediate being to elevate the work of the Hopkins Institute and of the larger American medical history community to a higher professional and scholarly level.

In his first full academic year in the United States (1932–1933), Sigerist launched a program of lectures and seminars at Hopkins similar to those he had previously offered at Leipzig. There were courses, for example, on problems of medical history, anatomy and anatomical illustration, the history of physiology, and the relationship between civilization and disease. Sigerist also pursued his own philological and analytical studies on medieval manuscripts and promoted Temkin's research on various Hippocratic texts. To help publicize this work, he launched a journal, the *Bulletin of the Institute of the History of Medicine*. He got involved with the activities of the American Association for the History of Medicine and in 1937 became its president. In short order, he revamped the constitution of the organization and in 1939 arranged for the adoption of the renamed *Bulletin of the History of Medicine* as the official organ of the association. He created "Graduate Weeks" at the institute as intensive workshops for the mostly amateur historians of medicine from around the country who came to Hopkins to study with Sigerist and his professional staff.

As in Leipzig, teaching was one of Sigerist's most central and dominant concerns. At Hopkins, influenced by the Great Depression and its consequences and by his need for supplemental funds, his teaching evolved quickly in a "sociological" direction. After an initial period of experimentation, by the 1934–1935 academic year, he had worked out an ambitious "vertical" curricular scheme: he would teach some aspect of medical history to all four years of Hopkins medical students and would key his teaching to their developmental stage. He would thus move from a first year survey of major historical periods, to a second-year outline of the history of pharmacology, to a third-year historical introduction to clinical medicine, and finally complete the series with a fourth-year capstone course on "the social aspects of medicine."[51] While his associates taught seminars on the history of anatomy and physiology and on topics in Greek medicine, Sigerist poured much of his creativity and energy into his fourth-year course and its burgeoning offshoots. By the 1935–1936 academic year, with the help of a grant secured from the Rockefeller Foundation, Sigerist's teaching included two "sociological" courses, "Social Aspects of Medicine" and a seminar whose primary purpose was to assist students with the preparation of original sociological papers. Sigerist also helped promote a student-initiated "Social Problems Forum" that included talks on contemporary issues and developments.

The momentum of his teaching helped pull Sigerist's scholarly work in a decidedly "sociological" direction. Already in his *Einführung*, he had examined relevant social problems, highlighted the history of public health and health insurance, and analyzed the social and economic conditions that shaped health services in the past. But as his sociological teaching expanded and drew more of his attention, Sigerist began to produce new historical essays on the social dimensions of medicine.[52]

As he was developing these ideas, Sigerist became embroiled in a debate with

George Sarton over the new direction in which he was taking the history of medicine. George Sarton had fired the first salvo in his editorial, "The History of Science versus the History of Medicine," in *Isis,* his pioneering history of science journal.[53] Sarton sharply criticized Sigerist for "overreaching" himself and implying that the history of medicine was "the best part of the history of science." On the contrary, Sarton claimed that the "core of the history of science should not be the history of medicine but the history of mathematics and the mathematical sciences, the other branches of science being dealt with in the order of their mathematical contents."[54] Sigerist responded that the history of medicine was a quite different scholarly enterprise from the history of science: "The history of medicine in a very large sense is the history of the relationships between physician and patient, between the medical profession . . . and society. . . . The morbidity of a given period, in other words, the task with which the physician is confronted, is the result of endless social and economic factors. If you want to understand the development of medicine in the 19th century, you will have to study the industrial revolution first."[55] Furthermore, because access to medical care depended on social class and most people derived little direct benefit from medical research, a comprehensive history of medicine had more in common with social and economic history than with the history of thought. As Sigerist put his point for special emphasis, "medicine is not a *branch of science* and never will be. If medicine is a science, then it is a social science."[56]

Sigerist's biggest sociological project was his 1937 book, *Socialized Medicine in the Soviet Union,* which presented the Soviet medical system as the endpoint of a long evolutionary process that began in Greece and culminated in the system the Russians had recently put into practice.[57] Sigerist strove to capture the dramatic historical novelty of the world's first full-scale socialist medical system: "All that has been achieved so far in five thousand years of medical history, represents the first epoch: the period of curative medicine. Now a new era, the period of preventive medicine, has begun in the Soviet Union."[58] The book, conceived as a primer on the Soviet medical system and as an introduction to socialism for young medical workers, presented an extravagantly optimistic endorsement of the Soviet system.[59] Sigerist explained that he had not "wasted time" in describing the inadequacies and inefficiencies of Soviet institutions; he had chosen to stress only "positive achievements," convinced that these would "enrich the world."[60]

Sigerist was now inescapably identified with the Soviet cause. One week after the publication of his book, *Socialized Medicine in the Soviet Union,* the Friends of the Soviet Union organized a "most impressive celebration" of the twentieth anniversary of the Russian Revolution in New York's Carnegie Hall. An orchestra played Russian music, Soviet ambassador Troyanovski was given an "endless ovation," and Sigerist spoke about Soviet achievements in public health.[61] Back at Hopkins, he found himself restless with the routine work of the institute and unable to concentrate on his usual lecture material.[62] When he now thought about "pure" medical history, which he had earlier referred to as his "real work," it seemed to him a kind of public relations exercise, a cover for his political activities. When he spoke of organizing the Graduate Week, for example, it was mainly in terms of staging a good show.[63]

Sigerist was increasingly beginning to think of himself as a Marxist, even if his Marxism seems at this stage rather postural and undigested. He was also starting to become acutely aware of the limits of his knowledge and of his need to spend time learning a radically different approach to history: "I have terrific gaps," he realized. "My knowledge of economic history is most superficial and I should be able to devote at least a year to studying it. I must ask for a leave of absence soon."[64]

He never did take time off to study economic history. Instead of acquiring a serious schooling in Marxism, he adopted what he termed a "Marxian attitude" that was "apparent to all."[65] He announced his intention to begin a course on medical economics in order to provide the students some instruction in Marxian economics.[66] He also allowed himself to become increasingly involved with the students who belonged to the local Hopkins unit of the Communist Party.[67] At this point, Sigerist recorded his admiration for the *Daily Worker*, "an excellent paper" with "a much higher standard" than the *Baltimore Sun*.[68]

Sigerist was having increasing difficulty focusing on his medical history lectures and writing commitments. In February, 1938, he accepted—with considerable misgivings—an invitation from Yale University to deliver three lectures, funded by the Terry Foundation.[69] He then fretted about the prescribed topic of "Science and Religion," trying to reassure himself that if Joseph Needham (1900–1995), biochemist, historian of Chinese science, and a Marxist, had previously given the lectures, it must be "possible after all." But the process of preparing the lectures proved miserable, attended by "insuperable inhibitions."[70] When he finally successfully completed his lectures in November 1938, and delivered them to an enthusiastic public reception, he still expressed grave disappointment because, as he admitted to himself, he was "uneasy . . . [that] what I said was not new to me. I have not done much research recently and I am repeating myself all the time. What I need is a whole year devoted to research—without any lectures."[71]

Sigerist's printed Terry lectures—completed in 1940 but not published until

1941 as *Medicine and Human Welfare*—are, in fact, mostly stale repetitions of earlier publications.[72] There are a few new elements, and these are largely politically inspired interpolations. In his lecture on "Disease," for example, Sigerist noted that despite the broad, general, historical decline of tuberculosis, the disease still persisted at high levels among the poor. It had become "a social disease, bred in slums, a disease of the low-income groups or unskilled workers' families."[73] In his lecture on "Health," Sigerist included an elaborate critique of the ancient Greek doctrine of hygiene, a doctrine and practice for the aristocratic, leisured, and privileged elite.[74] Finally, in his lecture on "The Physician," Sigerist called on doctors to lead the struggles for the general improvement of working conditions, the end of war, and the universal extension of medical care.[75]

Sigerist was in fact leading such struggles—especially for the universal extension of medical care—in his own, active political life. In addition to being in or lending his name to several Communist front and other progressive organizations, Sigerist was a charismatic inspirational force in the battle for universal health insurance and socialized medicine. He inspired a generation of students, interns, physicians, and public health professionals whose lives he touched and whose efforts to organize and advocate he encouraged and promoted.[76] He also reached out to the general public through talks at the "Peoples' Forum" in Philadelphia and other such venues, articles in the *Yale Review* and the *Atlantic Monthly,* interviews in *Time Magazine* and the *New York Daily News,* and radio talks and debates, including appearances on CBS's "Town Hall Meeting of the Air."[77] He was so well known for his political advocacy that he became a national symbol of socialized medicine. On May 2, 1940, he was startled but pleased to hear about a play in New York, *Medicine Show,* in which every night on stage an actor cried, "What we need now are men like Dr. Sigerist of Hopkins. . . . That's what we need!"[78]

Sigerist's political engagements, especially those in connection with medical students and young physicians, provided the energy for his main pedagogic passion at Hopkins, his "medical sociology" courses. By 1938, his teaching in this area included a formal course, "Current Events," led by Sigerist, and a seminar on "Medical Economics in the United States."[79] In 1939, he devoted the seminar to "socialized medicine" and assigned twenty-eight students from the Medical School and the School of Hygiene and Public Health to report in detail on the demographic, social, economic, and health service features of Baltimore and twenty-three Maryland counties.[80] The year 1940–1941 represented perhaps the peak of Sigerist's sociological teaching, with separate courses on the social aspects of medicine in historical perspective, medical economics, and current events.[81]

Inspired by the enthusiasm of his Hopkins students, Sigerist formulated plans for an ambitious and synthetic sociological treatise. In 1938 he announced his intention to publish a "two volume *Sociology of Medicine*."[82] In 1941 he returned to the idea and indicated that he would incorporate material on the economics of medicine and expand the work to four volumes.[83] In 1943 he wrote, "The plan of the *Sociology* . . . [has developed] rapidly. While it was rather hazy in the beginning, its outlines became clearer with every year. . . . The one-volume book soon developed into a four-volume plan."[84] A fragment found among Sigerist's papers hints broadly at how he planned to arrange the four volumes:

Sociology of Medicine: General Plan
1. Medicine as a Social Science
2. Health Insurance
3. State Medicine
4. Problems of Various Countries.[85]

Instead of writing this ambitiously projected four-volume book, however, Sigerist threw himself into another scholarly project that did materialize partially, a history of social security legislation. It began when, in April 1943, Martha May Eliot (1891–1978), head of the Children's Bureau in the Roosevelt administration, came to Baltimore to invite Sigerist to participate in a small, informal group that would plan for the organization of medical care in the United States after World War II.[86] Now full of creative energy, Sigerist discovered a topic that truly captured his interest: how the ostensibly liberal policies of the British government and the Beveridge report belied a conservative political agenda. This became the focus for his essay, "From Bismarck to Beveridge: Developments and Trends in Social Security Legislation."[87] Brimming with new historical insights and full of fire, the essay provided a fresh understanding of the economic, social, and political forces that Bismarck, the creator of modern social welfare policy, had been able to mobilize and manipulate.

Sigerist's fundamental insight was that Bismarck in the nineteenth century— just like Beveridge and Roosevelt in his own day—understood that to preserve their superior position, the ruling classes of society had to respond to the concerns of a working class made insecure by the economic realities of the Industrial Revolution.[88] Bismarck realized that to avoid both working-class impoverishment and revolution, the old feudal aristocrats and the new capitalist barons would have to use the mechanisms of the state to create systems of social insurance to give people security, "and in so doing he expected to take the wind out of the sails of the socialist movement and destroy it."[89] At first, both the Liberals and Social Democrats resisted Bismarck's legislative initiatives, but they worked out compromises as Bismarck moved relentlessly toward his long-term objective of antirevolutionary stabilization.[90]

Although Sigerist's book on the history of social security legislation never materialized, the Bismarck article, which Sigerist found "an absorbing piece of work," served as an important bridge from "sociology" back to "history." In the early 1940s, perhaps seeking refuge from his bruising political battles and bouts of depression, Sigerist returned to serious work on medieval manuscripts.[91] His abiding ambition, however, was to launch his *History of Medicine,* his grandly projected, synthetic, eight-volume work that he had been thinking about since 1936. He promised to begin writing in 1941, on his fiftieth birthday, a promise that he failed to keep.

Sigerist actually started writing his *History* in July 1945. Within two months, he had completed 225 pages of the first volume. Several important things had happened to Sigerist in 1944, but probably the most significant for his historical project were two trips late that year, one to the province of Saskatchewan in Canada, to serve as a fact-finding and health planning consultant to the impressively victorious socialist party, and the other to India, as an American representative on a special commission surveying health organization and administration.[92] He rationalized these trips as field excursions for his *Sociology,* but they were also energizers. All the while that Sigerist was in motion, and although he didn't realize it himself at the time, he was absorbing important lessons for his *History.*

The recognition now gradually dawned on Sigerist that his medical sociology and health care reform efforts were not the distractions from serious scholarship he sometimes thought them to be. They were, rather, keys to his creativity because they showed him the route he could travel on the way to integrating his scholarly and political selves. As he wrote in the foreword to the first volume of his *History of Medicine,* ultimately published in 1951, "Field work in social medicine . . . [at an earlier time] seemed to have no connection whatever with my historical studies; yet after every such tour I know that I had a deeper understanding of the workings of history."[93] Sigerist was now able to outline a new pattern of medical historiography derived from the insights and findings of his earlier-developed comparative medical sociology, but now used critically and systematically to explore the material foundations and social relations of medicine in past societies.

Sigerist argued that historians must first consider the economic and social structures and the prevailing health conditions in any past society by examining the dominant diseases to which rich and poor were subject. They should study the material conditions, starting with the geographical, physical, and socio-economic environment, then analyze the economic structure of the society, the means of production of food and commodities, and the conditions of work and recreation, housing, and nutrition. They needed to know what methods were used to maintain health and prevent illness and how such hygienic measures were distributed by class: did rich and poor have different possibilities for protecting their health? It was necessary for historians to understand the various kinds of health practitioners and the services they delivered to the different sectors of the population. They needed to explore the social history of the patient, the relationship of patients to doctors, and the relationship of illness to social structure. In addition, they had to understand the impact of sickness, medical care, and medical institutions on people's lives and to consider the level of development of a sense of collective social obligation, social welfare measures, and public health.[94]

Before finishing this volume, however, Sigerist had to come to terms with his circumstances at Johns Hopkins. He had been increasingly miserable for some time: sick, tired, overworked, politically embattled, and administratively overextended. The directorship of the institute felt like an increasing burden, and, from October 1942, he was also required to serve as acting director of the Welch Medical Library. As his friend Alan Gregg (1890–1957) of the Rockefeller Foundation sympathetically observed, "[Sigerist] is expected by Hopkins to be a routine library administrator. . . . The situation there gives me more or less the impression that I get from seeing a Steinway Grand used as a kitchen table."[95] Sigerist's onerous administrative responsibilities were partly the result of wartime exigencies but were also partly a punishment meted out by Hopkins president Isaiah Bowman (1878–1950) who found Sigerist's political activities and reputation an irritating liability. Sigerist's solution, however painful, was that he had to leave Hopkins, a decision that Gregg helped him with both practically and psychologically.[96]

What made Sigerist's decision particularly difficult was his realization that, despite his increasingly oppositional stance to Bowman and his growing alienation from Hopkins, he still derived great pleasure from teaching his students and interacting with a small but growing group of acolytes. He continued to offer his lectures and lead his seminars in the School of Medicine and in the School of Hygiene and Public Health. Many of those who studied with him in those years experienced his indelible imprint. With his charm, charisma, energy, and passion, he continued "to turn them on like so many light switches."[97] Several of Sigerist's most devoted students in this period went on to become key figures in the fields of public health, community and preventive medicine, and health care organization. In their subsequent careers, they translated into action, on the local, national, and international level, the lessons he had taught them.

Sigerist also put his stamp on a rising generation of historians of medicine, most notably Erwin Ackerknecht (1906–1988) and George Rosen (1910–1977). Both had long-standing relationships with Sigerist, Ackerknecht's going back to Leipzig in the late 1920s, and Rosen's beginning as a correspondent from Berlin in the mid-1930s. Sigerist had mentored each of them, helped shape their scholarly interests, and assisted at critical junctures in their careers. Now, in the late 1940s, as he got ready to leave Johns Hopkins and the United States for scholarly retirement in Switzerland, Sigerist consciously passed on the mantle of the new social history of medicine. He felt confident that Ackerknecht, with a new professorship in the history of medicine at the University of Wisconsin, and Rosen, as editor of the newly established *Journal of the History of Medicine and Allied Sciences*, launched in 1946, were well positioned to carry on his legacy in the United States. Each, he presumed, would use medical history to shape new generations of medical and public health practitioners.[98]

Biography—Legacy—Icon

We can now make sense of the puzzles and ironies alluded to in our introduction. Later generations of Ph.D. historians and graduate historians have tended to see Sigerist through the filter of Ackerknecht and especially Rosen, for it was Rosen who praised and to some extent exemplified Sigerist's integration of scholarship and advocacy and who called the first volume of his *History of Medicine* a masterful, commanding synthesis "on a grand scale."[99] These same historians tend to see Osler through the filter of Sigerist. The members of the Osler Society, for their part, generally distance themselves from the politically valorized Sigerist and turn to Osler apotheosized as saintly clinician-teacher at turn-of-the-twentieth-century Johns Hopkins. Neither group sees either their own iconic figure or his presumed opposite fully, each fixing only on a selected phase of their complicated careers. Both groups focus particularly on the Hopkins years, when for both Osler and Sigerist, although in different ways, medical history was most important in its utilitarian pedagogic application to the shaping of young medical professionals. This selective focus derives, in turn, from another iconic influence of great power: that of Johns Hopkins as the most salient twentieth-century symbol of American medical education. Hopkins stood Janus-faced, one aspect representing high clinical ideas and the other, haughty technocratic indifference. Which face you saw depended on where you stood.

NOTES

1. For a recent survey of the broad sweep of the history of medicine, see John C. Burnham, *How the Idea of Profession Changed the Writing of Medical History* (London: Wellcome Institute for the History of Medicine, 1998).

2. Sigerist knew he had his work cut out for him. He wrote—in private—of the Johns Hopkins Institute in early 1932: "The Institute is a superb instrument which has only to be tuned and played. Then the Old Ladies Home will amount to something." Henry E. Sigerist, Diary, 26 January 1932, Henry E. Sigerist Papers, Manuscripts and Archives, Yale University Library, New Haven, Addition (June 1987), Biographical Data and Memorabilia, group 788, box 1 (hereafter cited as Diary); *Henry E. Sigerist: Autobiographical Writings,* Nora Sigerist Beeson, ed. (Montreal: McGill University Press, 1966), 76.

3. See especially, Philip M. Teigen, "William Osler's Historiography: A Rhetorical Analysis," *Canadian Bulletin of the History of Medicine* 3 (1986): 31–49.

4. William Osler, "Books and Men," presented at the Boston Medical Library, 1901, reprinted in *The Collected Essays of Sir William Osler,* John P. McGovern and Charles G. Roland, eds. (New York: Gryphon Editions, Classics of Surgery Library, 1996), vol. 2:182–189.

5. Heinrich von Staden, "'Hard Realism' and 'A Few Romantic Moves': Henry Sigerist's Versions of Ancient Greece," in *Making Medical History: The Life and Times of Henry E. Sigerist,* Elizabeth Fee and Theodore M. Brown, eds. (Baltimore: Johns Hopkins University Press, 1997), 136–161. Michael Bliss, *William Osler: A Life in Medicine* (Oxford: Oxford University Press, 1999), 393.

6. Bliss, *William Osler,* 196.

7. Thomas Browne, *"Religio Medici," Together with a Letter to a Friends and Christian Morals,* H. Gardner, ed. (London: William Pickering, 1845).

8. William Osler, "An Address on Sir Thomas Browne," *British Medical Journal,* 21 October 1905, 993–998.

9. During his lifetime, Osler gathered the largest collection of Browne's work in existence. Sir Geoffrey Keynes, "The Oslerian Tradition," *British Medical Journal,* 7 December 1968, 599–604.

10. John Shaw Billings, "Rare Medical Books," *Johns Hopkins Hospital Bulletin* 9 (1890): 27–31. William Osler, "A Note on the Teaching of the History of Medicine," *British Medical Journal,* 12 July 1902, 93, reprinted in *Collected Essays,* vol. 3:568–570; "Physic and Physicians as Depicted in Plato," presented to the Johns Hopkins Historical Club, 1893, reprinted in *Collected Essays,* vol. 3:6–36.

11. *The Dialogues of Plato,* B. Jowett, trans. (Oxford: Clarendon Press, 1892).

12. William Osler, "Thomas Dover: Physician and Buccaneer," read to the Historical Club, January 1895, reprinted in *Collected Essays,* vol. 3:53–70; "John Keats: The Apothecary Poet," read to the Historical Club, October 1895, reprinted in *Collected Essays,* vol. 3:89–

106; "An Alabama Student," read to the Historical Club, January 1895; reprinted in *Collected Essays,* vol. 3:71–88.

13. Osler, "An Alabama Student," 72.

14. John Harley Warner, *Against the Spirit of System: The French Impulse in Nineteenth-Century American Medicine* (Princeton: Princeton University Press, 1998), 356.

15. William Osler, "The Influence of Louis on American Medicine," read before the Stillé Society of the Medical Department of the University of Pennsylvania, 1897, reprinted in *Collected Essays,* vol. 3:113–134.

16. Ibid., 134.

17. William Osler, "Elisha Bartlett: A Rhode Island Philosopher," an address to the Rhode Island Medical Society, 7 December 1899, reprinted in *Collected Essays,* vol. 1:100–150, 122; "John Locke as a Physician," an address to the Students' Societies of the Medical Department of the University of Pennsylvania, January 1900, reprinted in *Collected Essays,* vol. 3:186–225, 224; "A Backwoods Physiologist," an address before the St. Louis Medical Society, October 1902, reprinted in *Collected Essays,* vol. 3:277–306, 291; "Sir Thomas Browne," an address to the Physical Society, Guy's Hospital, London, October 1905, reprinted in *Collected Essays* vol. 3:350–385, 384.

18. Osler, "Books and Men," 187–188.

19. Bliss, *William Osler,* 296, 297, 300. Bliss estimates that these amounts would be equivalent to thirty times the sum in current dollars—and free of income tax.

20. Ibid., 303.

21. Ibid., 304.

22. Cited in ibid.

23. Warner, *Against the Spirit of System,* 364.

24. William Osler, "Medicine in the Nineteenth Century," an address to the Historical Club, January 1901, and published in the *New York Sun,* reprinted in *Collected Essays,* vol. 3:228–276.

25. Ibid., 234.

26. Ibid., 250–251.

27. William Osler, "Harvey and His Discovery," Harveian Oration to the Royal College of Physicians, October 1906, reprinted in *Collected Essays,* vol. 1:325–364.

28. Ibid., 359–360.

29. William Osler, "The Historical Development and Relative Value of Laboratory and Clinical Methods in Diagnosis: The Evolution of the Idea of Experiment in Medicine," in *Transactions of the Congress of American Physicians and Surgeons,* 1907; reprinted in *Collected Essays,* vol. 3:391–399.

30. Ibid., 398.

31. William Osler, "The Evolution of Internal Medicine," introduction to *Modern Medicine: Its Theory and Practice,* vol. 1 (Philadelphia: Lea Brothers & Co., 1907), reprinted in *Collected Essays,* vol. 2:303–322.

32. Ibid., 315.

33. Cited in Bliss, *William Osler,* 314.

34. William Osler and Arthur Keith, "Stokes-Adams Disease," in Clifford Allbutt and Humphrey Rolleston, eds., *A System of Medicine by Many Writers* (London: Macmillan, 1909), vol. 6:130–156.

35. Bliss, *William Osler,* 393. See also Heinrich von Staden, "Nietzsche and Marx on Greek Art and Literature," *Daedalus* (winter 1976): 79–96.

36. William Osler, *Man's Redemption of Man: A Lay Sermon* (New York: Paul B. Hoeber, 1915); reprinted in *Collected Essays,* vol. 1:390.

37. Osler, "Harvey and his Discovery," 326.

38. Ibid., 337, 339, 356, 357, 358.

39. William Osler, *The Evolution of Modern Medicine* (New Haven: Yale University Press, 1921).

40. Ibid., 84, 126, 208.

41. Ibid., 219–220.

42. Lewis Pyenson, "What Is the Good of the History of Science?" *History of Science* 27 (1989): 353–389, 370.

43. Arnold Thackray and Robert K. Merton, "On Discipline Building: The Paradoxes of George Sarton," *Isis* 63 (1972): 473–495.

44. Cited in ibid., 480.

45. Sigerist's inaugural lecture at Zurich as cited by Marcel Bickel, "Family Background and Early Years in Paris and Zurich, 1891–1925," in Fee and Brown, *Making Medical History,* 15–41, 30–32.

46. Owsei Temkin, "Henry E. Sigerist and Aspects of Medical Historiography," in Fee and Brown, *Making Medical History,* 121–135, 126–127.

47. "Proposal for the Establishment of an Institute of the History of Medicine at the Johns Hopkins University," n.d., 1, The Ferdinand Jr. Archives of the Johns Hopkins University, Baltimore, Records of the Office of the President, 1903–63, file 28.9 (Institute of the History of Medicine).

48. Ibid., 3.

49. Simon Flexner and James T. Flexner, *William Henry Welch and the Heroic Age of American Medicine* (New York: Viking Press, 1941), 418.

50. Sigerist, *Autobiographical Writings,* 81.

51. See Sigerist's report on the activities of the institute, *Bulletin of the Institute of the History of Medicine* 2 (1934): 123–139, 407–408, 512–513.

52. "The Physician's Profession through the Ages," *Bulletin of the New York Academy of Medicine* 2d ser., 9 (1933): 661–676; "Trends towards Socialized Medicine," *Problems of Health Conservation* (New York: Milbank Memorial Fund, 1934), 78–83; "An Outline of the Development of the Hospital," *Bulletin of the History of Medicine* 4 (1936): 573–581; "Historical Background of Industrial and Occupational Diseases," *Bulletin of the New York Academy of Medicine* 2d ser., 12 (1936): 597–609.

53. *Isis* 23 (1935): 313–320.

54. Ibid., 317.

55. "The History of Medicine *and* The History of Science," *Bulletin of the History of Medicine* 4 (1936): 1–13, 5.

56. Ibid., 4, 5.

57. *Socialized Medicine in the Soviet Union* (New York: W.W. Norton, 1937); "Socialized Medicine," *Yale Review* 27 (1938): 463–481, reprinted in *Henry E. Sigerist on the Sociology of Medicine,* Milton I. Roemer, ed. (New York: MD Publications, 1960), 39–53; "The Realities of Socialized Medicine," *Atlantic Monthly* 163 (1939): 794–804, reprinted in *Henry E. Sigerist on the Sociology of Medicine,* 180–196.

58. Ibid., 308.

59. Sigerist, Diary, 20 May 1937.
60. Sigerist, *Socialized Medicine,* p. 308.
61. Sigerist, Diary, 5 November 1937.
62. Ibid., 16 November 1937.
63. Ibid., 15 December 1937.
64. Ibid., 17 November 1937.
65. Ibid., 1 January 1938.
66. Ibid., 9 February 1938.
67. Ibid., 11 January 1938; 12 February 1938; 18 February 1938.
68. Ibid., 13 February 1938.
69. Ibid., 5 February 1938.
70. Ibid., 29 October 1938.
71. Ibid., 4 November 1938.
72. New Haven: Yale University Press, 1941. Early papers on which Sigerist drew extensively included "Die Sonderstellung des Kranken" (1929), translated and reprinted in *Sigerist on the Sociology of Medicine,* 9–22, and "Der Arzt and die Umwelt" (1931), translated and reprinted in *Sigerist on the Sociology of Medicine,* 3–8.
73. Henry E. Sigerist, *Man and Medicine: An Introduction to Medical Knowledge,* Margaret Galt Boise, trans. (New York: W. W. Norton, 1932), 46.
74. Ibid., 63.
75. Ibid., 133, 135, 139.
76. George A. Silver, "Social Medicine and Social Policy," *Yale Journal of Biology and Medicine* 57 (1984): 851–864.
77. Elizabeth Fee, "The Pleasures and Perils of Prophetic Advocacy: Socialized Medicine and the Politics of American Medical Reform," in Fee and Brown, *Making Medical History,* 197–228.
78. Sigerist, Diary, 2 May 1940.
79. *Bulletin of the History of Medicine* 6 (1938): 863–864.
80. Ibid., 7 (1939): 854.
81. Ibid., 8 (1940): 1131 and 10 (1941), 381–386.
82. Ibid., 6 (1938): 860.
83. Ibid., 10 (1941): 373 and 12 (1942): 446.
84. Ibid., 14 (1943): 253.
85. *Henry E. Sigerist on the Sociology of Medicine,* xii.
86. Sigerist, Diary, 14 April 1943.
87. Henry E. Sigerist, "From Bismarck to Beveridge: Developments and Trends in Social Security Legislation," *Bulletin of the History of Medicine* 13 (1943): 365–388. See also Sigerist, Diary, 20 December 1942; 15 March 1943; 16 April 1943.
88. Sigerist, "Bismarck to Beveridge," 368.
89. Ibid., 376.
90. Ibid., 386.
91. Michael R. McVaugh, "'I Always Wish I Could Go Back': Sigerist the Medievalist," in Fee and Brown, *Making Medical History,* 162–178.
92. Sigerist, Diary, 16 August 1944; 13 September 1944.
93. Henry E. Sigerist, *The History of Medicine,* vol. 1 (New York: Oxford University Press, 1951), xvii.
94. For a fuller discussion of Sigerist's accomplishment in volume 1 of the *History,* see

Elizabeth Fee and Theodore M. Brown, "Intellectual Legacy and Political Quest: The Shaping of a Historical Ambition," in Fee and Brown, *Making Medical History,* 188–189.

95. Alan Gregg to Raymond E. Fosdick, 27 April 1945, Rockefeller Foundation Archive, Rockefeller Archive Center, Sleepy Hollow, N.Y., Record Group 1.1, Series 200, Box 93.

96. Theodore M. Brown, "Friendship and Philanthropy: Henry Sigerist, Alan Gregg, and the Rockefeller Foundation," in Fee and Brown, *Making Medical History,* 288–312.

97. Leslie A. Falk, personal communication, 1995.

98. Theodore M. Brown and Elizabeth Fee, "'Anything but *Amabilis*': Henry Sigerist's Impact on the History of Medicine in America," in Fee and Brown, *Making Medical History,* 333–370.

99. George Rosen, "The New History of Medicine: A Review," *Journal of the History of Medicine and Allied Sciences* 6 (1951): 516–522.

Part II / A Generation Reviewed

CHAPTER EIGHT

"Beyond the Great Doctors" Revisited
A Generation of the "New" Social History of Medicine

Susan M. Reverby and David Rosner

A Bildungsroman for scholars within the field of the history of American medicine over the last three decades might be expected to take the traditional form of all coming-of-age stories: young whippersnappers question the wisdom of their elders, get sent into the wilderness to test their skills, come home wiser, if slightly bloodied, and ready to join the clan. Yet, when differences of race, class, gender, or politics keep adhering to the young (and the increasingly not so young), not everyone gets welcomed back so easily, or decides to stay. Our experiences within the field of what we wanted to be called the history of health care is particular to our biographies. Our story, however, tells a great deal about our generation that came of age in the 1960s and 1970s and how the field, and we, have changed.

In the late 1970s, when we were still graduate students, we wanted to make a statement with an edited book about the cutting edge work being done in the "new" social history of medicine. Editor Michael Ames was attempting to revive Temple University Press lists and garnered support of two senior medical historians who vetted the book (Gerald Grob and Charles Rosenberg). With the go-ahead, we set about contacting our friends, other graduate students, even David's adviser, most of whom were more than willing to send us a sample chapter. The result of this effort became *Health Care in America: Essays in Social History* (1979).

The book was our attempt to legitimize a type of history of medicine that, as we wrote, "both illuminates health policy concerns and explores the subtleties of medicine's past."[1] It contained thirteen essays with sections on medicine's boundaries, health care institutions, and professionals and workers.

To frame our efforts we struggled to write a short introduction we called, "Beyond 'the Great Doctors,' " an essay that has been likened to a "manifesto" by other historians. It appeared at a moment when the field itself was in turmoil, and our agenda for future work became a lightning rod for the ongoing debates in the new social history of medicine. We took the title from the physician-historian Henry Sigerist's line that "the history of medicine is infinitely more than the history of the great doctors and their books." We spent most of the essay seeking to establish continuity between our social historical interests and the work of some of our elders. Despite our attempt to situate this new agenda in ongoing traditions in the field, the book was received with a mix of reviews that saw it as the challenge we indeed meant it to be. Paul Starr's review in the *Journal of Social History* even likened it to the young impressionists' nineteenth-century "salon de refusés" that challenged the established traditions of classicism and romanticism in French art.[2]

Nearly a quarter of a century later, it seems appropriate for this volume to reexamine our efforts. Our aim is to reflect on the field's collective history and to discuss its problems and paradoxes. This is our renewed effort to stimulate a discussion to help all of us define more closely how all of our social and ethical views shape our scholarly work, define our lives, and shape our profession's collective ethical boundaries.

Why "Beyond the Great Doctors"

We realized it would be impossible in our edited book's introduction to trace out completely the intellectual history of the social history impulse within the history of medicine. It was our hope to validate a strand within the field we believed had been lost during the Cold War era. We argued that an engaged and useful history that was focused on the social relations of medicine and met rigid historical standards had existed and needed to be resurrected. We wanted written scholarship that would inspire doctors and also other health care professionals, workers, and consumers. Written before the push toward cultural and postmodernist history had taken hold, we sought to legitimize what we and others trained in social history were trying to do. Without consciously knowing this, we were indeed emulating Sigerist's sense of his own distance from what was seen as traditional history of medicine when he wrote in 1943: "They [Cushing, Welch, Klebs,

Fulton, et al.] all belong to the Osler school of *historia amabilis*. They 'had a good time' studying history. Their subjects were limited and never offensive. . . . My history is anything but *amabilis,* but is meant to be stirring, to drive people to action."[3] We had no way of understanding then what a threat *we* seemed to be as we were also being "anything but amabilis" too.

To begin, we think it important to reflect back on our journeys that got us to that essay and how this shaped what would become our intellectual passions.

Susan: I was raised in a medical family: my father was a physician and my mother a medical technologist who became a community college teacher. Despite my parents' love of the sciences, my high breakage fee in a chemistry class in a small upstate New York high school and my gender seemed to condemn me to some other future than medicine. With all the clarity of a seventeen-year-old, I settled upon personnel administration. However, after being politicized by my brief experiences in the civil rights movement, a year at the London School of Economics, and a longer effort in the antiwar movement, it was clear that this career choice was a serious mistake. I survived getting a B.S. degree in industrial and labor relations from Cornell because of a real love for labor history and the mentoring of an idiosyncratic social historian named Gerd Korman, who actually believed women could be intellectuals. Graduate school just did not seem acceptable at the time, however. I, as with many of my class in the late 1960s, went off to make war against my own government and then to a community-organizing-related job in New York City instead. My brief foray into graduate school in American Studies two years later was shortened when the women's movement and the invasion of Cambodia in 1970 intervened to transform my life, and cast me into the land of the dropouts. Searching for a job, I managed to parlay my burgeoning interest in women's labor history and my brief experiences in New York City's legal abortion clinics into a position as a "health policy analyst" at the Health Policy Advisory Center, or Health PAC as it was known.

Health PAC was formed out of the Institute for Policy Studies, the left-liberal Washington think tank. At Health PAC, we struggled to transform the left critique of health care from a doctor/AMA focus to one that explored what we labeled the "medical industrial complex."[4] For three years I learned to write and speak to and for an audience of health care providers, public health officials, and consumers, many of whom looked to Health PAC to provide an intellectual framework to understand the activism and varying crises in health care that swirled around.

My love of history, ambiguity, and my insistence on footnotes often put me at odds with the more journalistic bent of my colleagues. I left for a year in West

Virginia where I coedited a book on women's labor history.[5] But by the mid 1970s I was ready to return to graduate school. Even then, I thought my work would be more in women's labor history (my first edited book was on this topic) than health care, although I was accepted to work with the late George Rosen at Yale in the history of medicine. Personal commitments drew me instead to Boston and to the graduate program at Boston University in American Studies. A dissertation that was, I thought, going to focus on nineteenth-century domestic service became focused on "health" instead because that was where my experience had been and where funds to support me were available.

Serendipity as in all tales plays a large role in the rest of this story. Diana Long, the historian of science and medicine, was in the history department and quickly became both a mentor and friend. Trained in Yale's program, Diana's approach to history required us to learn in bio-bibliographic form our intellectual predecessors. Diana marched our seminar through the great men: Sudhoff, Sigerist, Temkin, Ackerknecht, Rosen, and Rosenberg. Feminist rhetoric aside, I was hooked by their intellectualism, if not their focus. At the same time, I continued to see myself as an "activist" historian. With two other colleagues I formed the Massachusetts History Workshop and continued to work on history in the working-class communities in and around Boston, participated in a health study group filled with health policy activists and practitioners, and wrote history pamphlets for health care workers.[6]

As I was beginning work on my dissertation on a social history of American nursing that explored the tensions between gender and class, I reconnected with David Rosner, then finishing up his dissertation across the river at Harvard. We had met briefly in New York while I was at Health PAC and he was in the New York City State Department of Mental Hygiene. As I had already coedited a book that provided the documents to redefine labor history in gendered terms, putting together a set of historical articles by others like us did not seem so impossible.

David: My own background certainly shaped the ways that I would later see the field of medical history. I had grown up in New York City in a working-class/lower-middle-class family. My father, an immigrant who edited a Hungarian language paper until the mid-1960s, became a linotype operator at the *New York Post,* where he stayed until his technologically forced retirement following the phase out of "hot type" in the early 1970s. My mother was a nursery school teacher in a small private school where the pay was miserable. Both of them had been deeply involved in various labor struggles over the decades, and I was raised in a world in which the dinner table conversation revolved around social and economic problems of working people. My family was certainly not middle class. But neither was it poor or "underprivileged."

In fact, my mother taught nursery school at Walden, a small private school that I was therefore able to attend on a staff scholarship. From the time I was eight I had attended school with very privileged kids, many of whom had parents who were physicians. In 1964, just as I was graduating, Andy Goodman, one of the school's recent grads who had been in the same class as my sister and was the brother of a classmate, vanished in Mississippi while trying to register voters during "Freedom Summer." I spent a good portion of that summer at the Goodman's home, awaiting word as the search went on for Andy, whose body was discovered in an earthen dam along with those of Mickey Schwerner and James Cheney. This was certainly a signature event in my life.

I went on to graduate from CCNY, a college that catered to, in the words of my mother, "the best of New York's working people" and had, as most CCNY students did at the time, immediately entered the world of work in anticipation of ultimately going to graduate school.[7] As a psychology major, the first job I took was with the New York State Department of Mental Hygiene. I joined a research unit involved in evaluating the impact of environmental damage on a group of overwhelmingly African American and Hispanic children. Deciding that most of the damage that we saw was a result of environmental exposures to lead, poor nutrition, and the like (rather than psychodynamics), I entered the University of Massachusetts in public health receiving my masters degree in public health two years later.

I returned to the New York State Department of Mental Health, nominally as the group leader for mental health services in lower Manhattan. This was a mammoth responsibility for a twenty-four-year-old, especially because of the "Willowbrook decree." This decision by the New York State courts, which mandated that developmentally disabled children in the state-run Willowbrook School on Staten Island were to be returned to their communities and that the facility itself be shut, sent the world of mental health and developmental services into a complete meltdown.[8] In a series of newspaper reports and television exposés, the horrifying conditions under which these children had been "warehoused" had led to public outrage and the court decree that called for the children's "deinstitutionalization." It was the nightmare of feeling partially responsible for sending kids back to ill-prepared communities and to families with few financial resources that led me to "retreat" back to graduate school. It was about then that serendipity took over and I happened to meet Barbara Rosenkrantz, who first suggested that I come to Harvard and try to join my training in public health with the history of science.

From the first, I was encouraged to see history as a tool that could help me understand the evolution of policy, and particularly, my own frustrations with the inadequacy of the mental health and public health system. Finding Susan in Cambridge and getting to know other younger historians like Harry Marks, Elizabeth Lunbeck, Martha Verbrugge and Marty Pernick (all who had gathered either as students and visiting scholars under the wing of Barbara Rosenkrantz in the History of Science Department at Harvard) and developing lifelong friendships with fellow graduate students like Elizabeth Blackmar, Roy Rosenzweig, and Jean-Christophe Agnew was immensely important particularly because it happened shortly after my first "formal" encounter with the world of historians of medicine, one that had left me quite shaken.

In June 1973, just before the beginning of my first year in graduate school, I received a thick envelope from Lloyd Stevenson, then editor of the *Bulletin of the History of Medicine*. I had submitted an article on the dispensary abuse controversy of the 1890s to the *Bulletin* three months before and I assumed that the thick envelope was filled with reviews and, perhaps, an acceptance of what I believed was a meticulously researched paper and hoped was to be my second publication.

Instead, when I opened the envelope I found a three-page rejection letter, detailing a deep antipathy toward the paper and, seemingly, toward me for being arrogant enough to question the motivations of physicians who were involved in the story I told. Stevenson wrote that he had expected "that a paper coming from Harvard should be better informed."[9] The letter was upsetting. The essay had not even been sent out for review, and a senior and powerful member of the field implied that I should not be in the field.[10]

What I did not understand was why Stevenson spent much of the letter discussing a paper I had published with Gerald Markowitz in *American Quarterly* the year before. That earlier paper, "Doctors in Crisis," had detailed the period before the Flexner report, during which major foundations had helped shape medical education. Stevenson claimed never to have read the article, but it was clear from his letter that he was deeply offended by its title. I felt that the article would have met a dismal fate had I sent it to the *Bulletin* instead.

I showed the Stevenson's letter to my adviser, Barbara Rosenkrantz, terrified that she might tacitly or openly agree with what Stevenson said. Her reaction to the letter was immediate: "Don't pay a moment's attention to this," was a polite paraphrase of her comment as I recall. She reassured me that I should remain in the department. She sent a copy to Charles Rosenberg who reinforced Barbara's view. It was really my first inkling that I had stepped into a deep schism in the field.

Social History and the Medicine Minefields

When we entered the field, we did not yet realize we were coming into a minefield of historical traditions and challenges. In the history of medicine, as within history in general, the social history tradition of the pre– and early post–World War II years was focused on a variety of social aspects of how medicine was received by the public and affected by social attitudes and values. By the 1960s, the work of a small group of American medical historians such as Charles Rosenberg, David Rothman, Barbara Rosenkrantz, James Cassedy, Diana Long, John Blake, and Gerald Grob had expanded the openings to explore social history that Sigerist, Richard Shryock, and others had created earlier. Few in number, they were often not seen as a threat to those trained in older historiographic tradition that had placed clinical practice and the physician as the center of the field.

The field of medical history was fairly insular early in the 1970s. Medical and biological questions were much less integrated into the historical mainstream. Medical historians worked more in isolation, generally associated with medical schools or history of science programs. Often historians of medicine were trained as physicians first and historians second. Sometimes, the lone medical historian at a medical center took up the field after receiving his or her medical degree or, at times, after retiring from the practice of medicine. Historians without medical degrees did make lasting contributions to the field. By and large, however, the field retained a parochialism that reflected the dominance of its membership's professional affiliations. In nursing history, a parallel story played out.

That was to change by the late 1970s as the number of doctoral students increased dramatically in the aftermath of the expansion of the university system as a whole. Many of us who saw ourselves as the new social historians were students of the older generation or found nonmedical social historians willing to support our work. While we certainly disagreed about numerous issues, we had a common faith that the field was ripe for social histories that delved into issues relating to race, gender, class, and politics. Those in or around our cohort sought to approach the history of medicine more as a social enterprise than as purely scientific or celebratory one.

Both groups of historians, the older physicians and the somewhat younger "professionals" (as we were called by some of the physician-historians), lived in relative harmony, each with its own set of questions and groups of interested scholars.[11] But, as more and more of us entered the field in the late 1970s from the periphery, the center of gravity shifted away from traditional centers of research such as Johns Hopkins, the home of the *Bulletin* and the Institute of the History of Medicine, and spread more widely throughout the historical landscape.

Following the antiwar, civil rights, and women's movements and other social upheavals, our own very naive view that we were creating a new field was undoubtedly seen by some our elders as ignorant at best and arrogant at worst. In the late 1970s and early 1980s a rather strong and, at times, vituperative debate broke out. At the annual meetings of the American Association for the History of Medicine (AAHM) as well as in the pages of the two major American journals, some of the young "professionals" faced a fairly bitter set of attacks by the editors and physicians writing book reviews and commentaries.[12]

The growing number of younger historians writing on nonclinical issues was deeply disturbing to some. Leonard Wilson, editor of the *Journal of the History of Medicine,* titled a January 1980 editorial "Medical History without Medicine." Wilson declared: "The study of the history of medicine by medical men [sic] derived from a deep interest in medicine itself, an interest that made them want to learn how medicine had arrived at its modern state through the course of history."[13] Medical historians, he argued, had previously seen themselves as "members of a long succession of physicians, scientists and teachers that extend back to antiquity in a continuous tradition of learning, teaching, and writing." He continued, "Medical historians had studied their historical predecessors" and "tended to look for those traits of medical character and quality of achievement which they respected and valued among their medical contemporaries." But the newer generations of younger non-M.D. historians were neglecting this tradition, he said. Despite the fact that he himself was not a physician, Wilson argued that younger historians were deficient in that they "focused on historical courses and seminars. They see little of the laboratory and less of the clinic," leaving them insensitive to traditional objects of historical inquiry, physicians and their activities.[14]

Part of Wilson's concern was certainly reasonable and at times prophetic. After all, by broadening out into a host of themes in gender, race, urban, political, institutional, demographic, and cultural history, there was a real danger that the history of medicine as a field would lose its master narrative. But larger forces were at work creating resentment.

Medicine's loss of control and status during the 1960s and 1970s (as well as the changes in the history profession) seemed to underline the sense of urgency that fueled the editorials linking attacks on critics of modern medicine, those promoting affirmative action for women and minorities within medical schools, and Ph.D.s who wrote medical history. The schism was never neatly doctors vs. social historians. The scholarship clearly demonstrated that there were "Ph.D.s" who paid close attention to clinical issues and medical research while there were other "M.D.s" who were closely attuned to the questions of social history. Nevertheless, it must have appeared that a growing number of us could not be counted on to see either medicine or medical history uncritically. It may also have been due to simmering fears that the professors who were sponsoring us were displacing their colleagues as the movers and shakers of the next generation. Control over the future of the field seemed at the time to be very much at stake.

Several other controversies illustrate the tensions that surfaced. In January 1979, two books on birthing and midwives were reviewed dismissively by obstetrician-historian Gordon Jones. Jones began his review of one book by saying that "the bias of this lay historian [sic] is obviously pro-midwife, pro-home delivery and against the obstetricians who, she believes, have for mercenary reasons obliterated midwifery in the United States." He dismissed the second book by stating it would be of interest only to "those who think socialized medicine is the ultimate and ideal solution to every imagined shortcoming of American medicine."[15]

In the *Bulletin of the History of Medicine,* the official organ of AAHM, the conflict escalated. Howard Berliner, a health policy/management Ph.D., was assigned by Lloyd Stevenson, the *Bulletin's* editor, to review our book, the first edition of *Sickness and Health,* another collection of social history of public health and medicine, edited by Judith Leavitt and Ronald Numbers, and a monograph critique by health educator/public health practitioner Richard E. Brown entitled *Rockefeller Medicine Men.* Berliner praised the new books, appreciating their differing attempts to stretch the traditional boundaries of medical history. The review so offended Stevenson that he took it upon himself to write an unprecedented five-and-a-half page response to Berliner's *review* as well as to the books themselves.[16]

In what Stevenson called his "second opinion," he accused the various writers of a number of professional crimes, some of which were perfectly valid, some of which were not. He found the writers not sufficiently respectful of physicians. According to Stevenson, the professionalization of the field by Ph.D.s had intimidated "amateur" [i.e., M.D.] historians, and he worried that "physicians intimidated by 'professionals' [i.e., Ph.D.s] should consider taking action."[17] Whether or not this meant purging the "professionals" or leaving the AAHM and starting another "amateur" association was never made entirely clear.

Underlying the controversy over the contours of medical history—who should do it, what it should address, what political or social content it should have—were more basic questions regarding the very definition of medicine itself. For "M.D."

historians, as Leonard Wilson explained, "in a strict sense the social history [of medicine] may not even be medical history. If such social history be considered medical history, it is medical history without basic medical science and clinical methods and concepts; that is, it is history of medicine without medicine."[18] Yet the work of our cohort, as demonstrated by the mainstream presses that published the reviewed books, tapped into the historical zeitgeist of the time. We too began to wonder whether our critics were right: did we have a place in this field?

In May 1980, the AAHM met in Boston.[19] At this point, the reviews had come out and tensions were relatively high. Half in jest in following Leonard Wilson's suggestions that we talk to physician-historian Gert Brieger, then the director of the Institute for the History of Medicine at Johns Hopkins, several of us did meet in a bar at the meeting hotel to consider bolting from the organization itself.[20] Brieger functioned as the go-between for the generations and training, assuring us that there was a place for us within the AAHM. Somewhat mollified, we stayed.

By then, many of us had started university positions or were close to attaining tenure. Our professional identifications ranged as widely as the jobs we were able to get: in public health, medical and nursing schools, health policy programs, women's studies, and traditional history departments. Throughout the 1980s, the threat that our kind of history posed lessened as we aged, took on positions of authority, and watched as the older generation retired or died.

> David: In 1980, my distinct sense that in a tight job market in traditional history of science and medicine departments and my continuing concerns about health care policy issues led me to take a position as an assistant professor in the Department of Health Care Administration at Baruch College in New York City.

> Susan: Two years later, in 1982, I searched for jobs in the Boston area because I was then married to a tenured academic and had a five-year-old daughter. I landed at Wellesley College, as their first hire in their Women's Studies Program (not the history department) with a position that was half time, one year at the start.

Many of our cohorts continued to see the AAHM as their primary source of professional identification outside their institutional appointments. Others attended meetings more sporadically and found homes in other public health or historical associations. As Stevenson may have feared, multiple professional identities were becoming more the norm.

The tensions within the AAHM simmered much below the boiling point through the 1980s. In 1990, through the efforts of Elizabeth Fee and Ted Brown, several historians (many in our cohort) created the Sigerist Circle, a separate section within the AAHM paralleling the Osler Society that met the day before the actual AAHM program began. The group's name reflected the identification with the activist and scholarly tradition that Sigerist himself represented. The circle would go on to present scholarly sessions each year, through the good works of Edward Morman to maintain a newsletter and bibliography, and to provide a home for those who needed an additional identity that made membership in the AAHM more than a scholastic endeavor.

The creation of this new section gave an institutional imprimatur to the more contemporary-oriented social historians while preserving their identity as medical historians. Thus, by the end of the twentieth century, social historians whose forays into medicine were more fleeting had left or never joined the AAHM. Others who continued to want to be seen as social historians of medicine and activist scholars found a home in the Sigerist Circle.

Reflecting the expanding acceptance of what counted as "medical history," the *Bulletin* began to publish a wider range of articles. In the early 1980s, about 50 percent of all articles published in the *Bulletin* focused on doctors. By the second half of the decade, this percentage dropped below 40 percent. By the end of the 1990s, it had declined below 30 percent. Significantly, articles focusing on gender, sexuality, race, and patients themselves increased from approximately 3 percent of the articles in the early 1980s to 10–15 percent by the end of the century.[21]

Given the growing emergence of gender, sexuality, race, and class as crucial categories in historical scholarship, it would have been nearly impossible for these arenas not to have grown within the history of medicine. It was becoming increasingly clear that what was once seen as peripheral to the core of medical history could be central. Further, newer work began to suggest that even considering that the very core of the history of medicine could exist without its relationship to the so-called periphery was exceedingly problematic.

Expanding the Social History Tradition

In 1979 we had argued that the social history impulse and the need to make our histories relevant were linked to the questions and concerns raised in the political movements of the 1960s and 1970s. In the next generation, historians influenced by social movements around women's health care, occupational health and safety, the AIDS epidemic, and racial disparities in health care delivery and health outcomes, and by the theoretical work that focused on the body, were being drawn into the field. Others, more influenced by the movements of postmodernism and poststructuralism, moved toward more theoretical considerations of multiple identities and cultural discourses on the body and health. In tracing out briefly the differing directions that this scholarship has grown, we will focus on issues of gender, class, and race. We will reflect on some of the processes that shaped broad changes in the field, dividing the categories up in the ways the field began at first to divide. We want to make clear, however, we think these analytic categories are *not separate,* and we share the theoretical position that they are "intersectional."[22]

Gender

As the field of women's history itself was expanding in the 1970s, explanations of the relationship between women and medicine played a central role. Any historical work that had to explain the social construction of womanhood came up against both the representations of the female body and the power of the institutions that defined womanhood. While some of the earliest books and articles took easy pot-shots at medicine with ludicrous quotes from nineteenth-century doctors, more thoughtful work attempted to put the beliefs about women's bodies within the context of medical theory more generally.[23] Interestingly, most of the literature on women's bodies was published in mainstream or women's studies journals rather than history of medicine publications.

Those of us who were also really taken with the internal workings of the health care system and for whom medicine and science themselves were of interest worked to integrate experiences of women and the concepts of gender into the history of medicine field. This scholarship initially took three forms: explorations of the experiences of women as workers and professionals in health care; critiques of medicine's ways of dealing with women's diseases, reproduction, sexuality, and health care needs; and deconstructions of notions of scientific neutrality and scientific discourses. All of these efforts were buffeted by debates in the larger historical community, first on the limitations of a narrow focus on white middle-class women as stand-in variables for all women. The analysis then moved to critiques of essentialist positions that assumed women's experiences could be knowable without a deconstruction of how the categories and representations were delineated.

Much of the earliest scholarship written in the 1970s was focused on the prescriptions and ideologies inherent in medical thinking. It assumed, however, a one-to-one correspondence between ideologies and women's internalized beliefs and experiences. As the theoretical frameworks became more sophisticated, efforts to understand how medical assumptions were internalized, acted on, and critiqued became more crucial. In the process of this development, there was much discussion between women's history specialists and medical historians with a focus on women's lives.

Underneath much of the scholarship was a desire to help the still growing women's health movement have a more reliable historical understanding of its own past and the institutions it was up against. As the AIDS and breast cancer epidemics spread and the attacks on hard won reproductive rights grew more violent, histories appeared to help make sense of these experiences. Historical accounts of how disease was defined, who became ill, and how their bodies were represented became essential.[24] New scholarship focused on the role of the state in monitoring women's bodies, defining the "normal" in sexuality, concerns with disease control in the past, the importance of gender to public health, and so forth. Others took their focus to be the wide range of reproductive experiences of women from abortion through birthing and menopause. In much of this work the focus moved from assuming the existence of all-powerful physicians to a search for various forms of women's agency as consumers, workers, and practitioners.

In the face of increasing contemporary demands for gender-based medicine, historians continued to provide more understanding of how gender and sex become biologized, under what conditions, and why. The link between the growing field of what became known as "science studies" and historians was forged. Historians of science provided complex historical narratives of the creation of understandings of the female body and the underlying gendered notions of science that supported such explanations. Those who worked on women as health care providers at first attempted to just chronicle the existence of such women (especially in medicine) and to understand the ways they (we) had coped with sexism and discrimination. Others took a more nuanced look at the disputes and differences among women physicians and created a less-conspiratorial and homogenous historical narrative. Nursing history's story paralleled medicine's, as scholarship in the 1980s explored themes of class, gender, and race. By the 1990s, other work focused on nursing-patient relationships, technology, and the importance of community to understanding nurses' and midwives' self-definitions and political organizations. Scholarship written by nurse historians and non-nurse historians together also expanded the audience for this work.[25]

By the mid-1980s and into the early 1990s, there were many criticisms being written that chided those who wrote as if womanhood was only a category that fit white and middle-class women. Much of the early work on women and their relationship to medicine and the state began to be reexamined as its universality was questioned. Many of us, especially those who came out of labor history or African American history, had never separated gender from other categories of analysis. In the face of wide-ranging political arguments over identity and who had the right to write about whom, however, concern over the very definition of womanhood spilled over into scholarly debates.

At the same time, the move within history from a focus on women to a focus on the concept of gender began to be felt in medical history. In her pathbreaking 1986 article in the *American Historical Review,* French historian Joan Wallach Scott made the argument for what she called "Gender: A Useful Category of Historical Analysis."[26] The concept that "women" were a stable historical subject came under attack as Scott and others influenced by poststructuralist arguments sought to undermine the assumptions that identity and the "authenticity of experience" had a one-to-one correspondence. As early American historian Kathleen Brown noted: "many of these approaches replace the search for stable, continuous and univocal meanings with analyses of contestation, discontinuity, and dissonance. . . . [this work] reflects a rejection of essentialism (the belief in a historical, transcendent core of experience and identity that is usually seen as derivative of the physical body), a project that many women's historians support in theory but find difficult to achieve in practice."[27]

The difficulty of achieving this within history of medicine was especially acute. Trying to parse out the links among internalized subjectivity, structures of nation, class, race, and empire, and then women's agency became much more complicated. The separation of such gender analysis from medical history continued. Those who wrote more theoretical pieces, such as Regina Morantz-Sanchez or Evelynn M. Hammonds, for example, tended to publish such work in history theory or feminist journals, rather than in the history of medicine journals.[28] In 1990, for example, Susan and Hammonds tried to lay out a research agenda that would link gender as a concept of power to class, race, ethnicity, and sexuality. In thinking back on this work, we did not attempt to give this paper at history of medicine meeting, but rather as part of a panel at the 8th Berkshire Conference on Women's History. We were arguing for a focus not just on actual bodies and what had happened to them, but the effects that the intersections among categories had as social formations in differing historical circumstances. We wanted, we wrote, "health care history in very particularistic ways to use race/class/gender/sexuality together as categories both when the 'bodies' are visible and when they are seemingly invisible, yet formative, in the creation of historical events."[29]

Such efforts at discourse or cultural studies history, as it became known, however, often came under attack from both the right and the left. On the right, such history was often seen as gibberish, with the loss of narrative power seeming to move this work out of history. On the left, the analysis of the power of discourse and representation often seemed to deprive individuals and groups of any agency or power to make change. In effect, this work seemed to erase politics as almost a possibility altogether. Influenced by postcolonial and subaltern cultural studies, however, more recent work in American medical history has begun to show new connections between central medical history concerns and class, race, and gender with the politics left in.[30] New analytic foci have also come from scholarship coming out of science studies and, in particular, understandings based on theories of embodiment that explore the ways gender, race, and class adhere to perceptions of the body.

Too often, however, the gender scholarship, and especially its emphasis on power, still has failed to influence the ways histories of hospitals, technology, or medicine are written. It was almost as if separate tracks existed.

Further, while the analytic depth of the historical scholarship grew, its clear relevance to health policy was often less obvious. If anything, various health policy analysts often selectively took from historical scholarship, turning "historical 'facts' into policy 'facts' that did not bear a close resemblance to one another."[31] Narratives that fit contemporary needs often overcame historical narratives that sought to provide a more nuanced past. While the "linguistic turn" and its emphasis on representing the multiple ways of seeing experiences and the relationship of seemingly binary opposites has often created a more sophisticated history, it has not always provided the kind of guide practitioners and consumers want.[32] The lessons of historians for contemporary policy remains much more problematic than any of us so innocently imagined several decade ago.

Class

"History from the bottom up" was the phrase that captured most directly the approach taken by social historians in the 1970s and 1980s. Attention to gender and race, and the intersection of both with issues of class, reframed the field and allowed for a flourishing of myriad local and national studies of the experiences of common people.

The creation of this new approach was profoundly influential in creating hybrid fields whose exact definitions were in high dispute as cultural, literary, postmodernist, and other approaches to the historical literature contended for control and space. For those interested in health, the growing attention to common laborers provided a new avenue for exploring the social production of disease and the impact of changing social environments on the health experience of Americans. During the mid-1980s and throughout the 1990s a plethora of new scholarship began to explore the ways in which the crucial importance of health issues was shaped by the experience of coal and hard rock miners, radium dial workers, workers in gasoline and chemical plants, and others in industrial work.

It is significant that much of the scholarship on labor and health appeared after some of the more basic investigations of the social history of the hospital and health care in general. Specifically, in the early and mid-1980s a series of books on hospitals began to call for greater and greater attention to the role of the patient as an object of medicine and as an agent of change in the organization of health care institutions. In some respects, the call for a history of "health care from the bottom up" was never accomplished, despite the efforts to focus attention on social class as a determinant in hospital organization.[33]

One clearly class-related issue that was largely avoided was the implications of the source of patients and the reasons for their entry into a facility: accidents and injuries on the job were major reasons for the felt need for the growth in the number of hospitals in the period between 1880 and 1920. While all the authors acknowledged that social and economic factors contributed to the "birth" of the institution in the late nineteenth and early decades of the twentieth century, we generally ignored the implication of these factors for the eventual evolution of the institution as a means of ameliorating the growing number of accidents and deaths related to work among working people that accompanied industrialization and urbanization.

Beginning in the mid- to late 1980s, a new type of literature on workers' illnesses and accidents began to appear that looked more closely both at the worker's experience on the job and the corresponding changes in the dangerous American work environment. Occupational safety and health history seemed to be a perfect merging of political, medical, and cultural history at a time when the boundaries between labor history and community history were becoming less and less distinct. Alan Derickson's work on the hospital system for hard rock miners in the West, for example, was a groundbreaking effort to blur the lines between the new institutional and social histories of hospitals that had recently appeared and the new labor history.[34]

The social creation of health and disease—central as they are to everyone's existence—were used as a kind of mirror on the social struggles and tensions in American society. They tied together a variety of historiographic traditions that were in danger of further fragmenting history as a field and isolating medical history as a sub-, sub-specialty.[35]

In the process of rewriting the history of occupational health, one of the central themes that emerged in a series of books in the 1990s, was the ways that medical science itself had incorporated a series of social assumptions into professional ideas about causation. Particularly glaring for us was the distinction that separated occupational medicine from environmental medicine in textbooks, professional associations, and etiological constructs and how completely the medical community and the science that was developed reified the growing social divides that separated laboring people from the rest of the community. "Occupational" medicine as a specialty served to distinguish the laborer from those who never stepped into a factory and further fragmented the professional and public understanding of the link between the "environmental" diseases that affected people outside the plant and "occupational" issues of the laborers themselves. A number of books were themselves a socially negotiated product, and the separation of occupational from other forms of illness had deep social meanings and implications for workers and their families.[36]

The history of working people has now really begun to transcend any narrow definition or parameter, linking varied groups inside and outside the factory gates.[37] What has been lost in terms of the clarity of subspecialization has been easily made up for in the ever-evolving richness of the questions that have developed and the breadth of issues pursued.

Race

In many ways the writing about race (usually translated into meaning the experiences of African Americans almost exclusively) in the history of American medicine parallels the developments in gender, except there has been much less scholarship. There are several explanations for the failure to take up race as fully, other than the effects of racism on academic scholarship. With several notable exceptions (Todd Savitt as the most prominent), in the 1970s and 1980s few white historians of medicine ventured into this topic area, and the number of African-American scholars could almost be counted on one hand. In turn, historians of the African American experience tended to focus on work, community studies, migration, sexuality, or gender, rather than medicine per se. The assumption that the history of scientific racism and eugenics had already been written left this topic somewhat understudied. The focus on the experience of particular people of color, rather than the concept of race itself as an indicator of power relationships and an underlying assumption inherent in medical thinking, limited understandings of why race was critical in the history of health care.[38]

As with the work on gender, much of the initial scholarship filled in the gaps, told the story of racial minorities (again primarily African Americans) in the professions, in the building of medical and nursing schools and creating hospitals, and in differential treatment of disease. The focus continued to be on the structures and experiences of racism in the delivery and organizing of care which often seem underproblematized and treated as a transhistorical experience.[39] Nevertheless, building up histories that provide the narratives of the racialized experience has proven useful. The beginning of a scholarship that moves beyond the African American experience to explore health care within other communities of color is a much more recent and welcome addition.[40] The introduction of understandings of discourse around the African American body, most visible in the works of medical historians Evelynn M. Hammonds and Keith Wailoo, has helped to bring the sophistication from African American, gender, and cultural studies to history of medicine.

Still, historians of medicine have often failed to understand, as a nonmedical historian noted, "the subject of race is at root a question of power and is, therefore, whether we like it or not, profoundly political."[41] Further, the continual saliency of the historically racialized experience in health care within communities of color makes it difficult for historians to historicize these experiences. Not only are there "facts" that are continually believed, there are standard narratives that are difficult to refute. This suggests that historians working on race need to consider, as historians working on memory and history have noted, the importance of understanding why certain truths and narratives continue to resonate and have power.[42]

Historical scholarship on race and medicine has begun to engage with the fast-paced sophisticated analyses that are coming out of recent African American scholarship, subaltern and postcolonial studies, science studies, and work on other racial and ethnic groups. The kinds of emphases that move away from simple binaries of resistance and accommodation, and that account for regional and time variability (although historians of Southern medicine have often done this) in more subtle form enrich our understandings. Whether the new work on whiteness studies will have any impact on the history of medicine remains to be seen.[43]

Of increasing importance in this area has been the work that begins to problematize the nature of the conceptions of race within medical and public health science, anthropology, and population genetics. Despite assumptions that the racial science and medicine of the nineteenth and early twentieth centuries had faded away, the search for a biological basis for race continues. In response to political demands made from within racial and ethnic communities and growing awareness of the health disparities between communities by public health practitioners, there is now a whole industry within medicine, nursing, and public health that starts with an assumption of differential outcomes based on race or ethnicity.⁴⁴ Historical understandings of the choices as to why and specifically how race becomes a particular kind of biological category and how this is used has increasing political and contemporary relevance. Historians have much to contribute here to the understanding of when race becomes a stand-in variable for other factors (class, nutrition, living conditions, access to medical care) rather than a category assumed to exist in "nature."

Recent specialty conferences have highlighted how much can be learned by cross-disciplinary and cross-cultural perspectives that suggest both the multiple roots of the contemporary scholarship and how much it has to offer to our understandings of medical thinking, disease, and medical institutions. Whether this affects mainstream history of medicine remains to be seen.

Beyond "Beyond"

When we wrote "Beyond 'the Great Doctors,'" it was in the hope that we, and those who thought about the history of health care as we did, would have a future in the field. That question has clearly been settled. Many of us are tenured, published, and respected. In a prescient way, however, we worried in our introduction about whether those who wrote an engaged social history could be in danger of becoming what we called "sophisticated antiquarians" in our own right. We wanted histories that would have meaning to a broader public that we felt responsible to speak to, but never for. It is not as easy as we thought then to tell what was antiquarian and what will become useful to other historians, practitioners, or consumers. Nor do we think that all history has to be directly applicable to a contemporary issue. We have come to appreciate in a way that we did not in the late 1970s the critical importance of understanding how medical and scientific ideas and practices are created. We no longer think the old "internalist vs. externalist" division can be made. The newest work on the intersections of race, gender, sexuality, class, and empire make this abundantly clear.

Although we have discussed briefly the developments of writing scholarship on gender, class and race separately for heuristic purposes, we are well aware that their integration is central to richer historical understandings. If these categories are seen as regimes of power, not just as characteristics of bodies, then they have much to offer even historians of medicine who want to concentrate on the most traditional historiographic foci of our field.

Our own paths to how we do this have of course been different, again reflecting both opportunities and differences within the field.

David: Perhaps the most interesting way that I have found to merge my ongoing interest in the uses of history in policy analysis has been in the creation of the Columbia's Program in the History and Ethics of Public Health and Medicine that I've helped establish with David Rothman as codirector and Ronald Bayer as associate director for ethics at Columbia University. I was first approached by Columbia to help them think through the ways that a history curriculum could be integrated into their school of public health and later to help establish a program in the history of public health that would join together the faculty of the History Department with the medical and public health schools to train students in the use of history in public health education, policy, and practice. Offering both an M.P.H. and a Ph.D., the program is unique in the country in that it provides both academic and public health credentials, and it is deeply gratifying to produce students who feel comfortable discussing Foucault's *Birth of the Clinic* as well as evaluating cohort designs for epidemiological and statistical research.

Recently, I was named director of a new Center for the History and Ethics of Public Health at Columbia's Mailman School of Public Health which broadens the educational and research agendas significantly. We are now seeking to help define a new type of public health ethics that will use history as its intellectual core. The center brings together an amazing array of scholars from Columbia's faculty and Gerald Markowitz and Gerald Oppenheimer from the City University of New York and Christian Warren from the New York Academy of Medicine to ask broad questions about the ways social issues, attitudes, and historical experiences shape the ways we address population health. Using history as the base discipline, we are seeking to avoid the pitfalls of understanding ethical dilemmas as rooted solely or even primarily in the individual doctor-patient relationship or questions of personal morality. Rather, as Susan and I said in our essay of twenty-five years ago, the new center will "provide [students with] essential tool[s] for analyzing current health . . . problems by providing a sense both of their origins and the possibilities to affect change." It will also do so with a sense of morality, ethics, and social responsibility.

I see my own experience as a historian involved in public health policy as extraordinarily rewarding. My work with Gerald Markowitz has played a part in reshaping the experience of workers and communities. Of special pride is the role that *Deadly Dust* and now *Deceit and Denial* are playing in ending certain types of dangerous practices and addressing a series of injustices to groups of workers, children, and communities who have been ravaged first, by silicosis, and second, by lead poisoning and exposure to chemical pollutants. In the case of *Deadly Dust,* it was quite remarkable to us that what we had initially seen as a highly scholarly account of the history of what we presumed to be a relatively obscure occupational disease came to play a role in court cases of workers currently suffering from the disease and in a national effort by three federal agencies—OSHA, MSHA and NIOSH—to eliminate the disease as a threat to workers. *Deceit and Denial* itself grew out of law cases we became involved in on the side of a variety of local and state governments and children who had been victimized by lead poisoning and communities polluted by petrochemicals. A Bill Moyers television special on the chemical industry, "Trade Secrets," an award-winning documentary called *Blue Vinyl,* and articles in *Newsweek* and other national publications have brought to public attention the importance of history in resolving questions of responsibility for past harms.[45]

Susan: After more than twenty years of teaching in a women's studies department it would be no surprise that my work would be more influenced by the theoretical developments in this field as in women's and African American history than history of medicine. My book on the history of nursing, *Ordered to Care: The Dilemma of American Nursing* (1986), melded debates on work relations coming out of labor and class studies to women's history concerns with the history of caring. By the late 1980s and early 1990s, however, debates about representation and discourse theory never seemed to me as separate from politics as others on the left and right had argued.

My scholarship has done this in several ways. When the debates with philosophy and science studies focused on an assumption of a gendered difference in the doing of science, I tried to test some of this theoretical more philosophical work within the field of nursing.[46] I have also continued to think about the history of women's activism within the health consumer movement, using work on the body and memory as theoretical touchstones.

For the last decade, I have been engaged in a multipronged effort to reconsider the infamous Tuskegee syphilis study, the United States' longest (1932–1972) nontherapeutic research "study." It provided me with an opportunity to mesh my understandings of class politics, race accommodation, and gender possibilities within the context of science and experimentation. It has also been an incredible experience of engagement with the larger health care community, from survivors of the study itself and their heirs in Tuskegee to officers of the U.S. Public Health Service.

I have become concerned that the multiple ways of understanding the study were not reaching historians and the wider health care community. I edited a collection of both primary and secondary articles on the study to provide actual documents for teaching and learning purposes.[47] This project is perhaps as traditional as work could get in history of medicine, except that I added poems, plays, and other forms of representation. Having supplied my "informants" book, as it were, I am now completing my own exploration of why the tales of the study are told in such differing ways. I am focused on what this teaches us and about power, views of science, and race, gender, sexuality, and class as integrated analytic concepts and lived experiences. In ways that I never expected, I have been caught up with both an internalist understanding of the medical views of syphilis and a more cultural political analysis of Tuskegee in the American imagination. It has also brought me back into intellectual conflict with differing views from medical practitioners and medical historians on how we understand the history of science and medicine in this story.

Teaching at an undergraduate college has left me without the pleasures (and difficulties) of having graduate students. Influence in a field can, I have learned, take other forms. My half time, one-year position in Women's Studies has turned into a four-person department that is growing. I am also building, along with other colleagues, a Health and Society major that will integrate on the undergraduate level the concern with ethics, history, and the social sciences that David is doing at the graduate level at Columbia.

The academy became more open to differences than we expected in the late 1970s. The methodological divides between historians are not as neat as we experienced then, as virtually all historians accept the "social history" approach as legitimate. On particular issues, however, we cannot predict who will take what side or another. The fault lines in the field do emerge again and again in the face of crises. When the AAHM went to meet in Charleston, South Carolina, in 2001, for example, tensions erupted within the association over whether the meeting should be moved to another state to honor an NAACP-called boycott of tourism over the continued flying of the confederate flag over the state house.[48] Perhaps because some of us have developed stature and place in the field, our roles in the AAHM as much as our politics led to differences on whether the meeting should be moved. In the end, the meeting stayed in Charleston, and a number of us made the decision not to attend.

Similarly, we have a profound sense of disappointment that our colleagues sometimes have few moral qualms about how they use their historical skills to cover up abuses by some of the industries that have caused Americans the most egregious harms. We stand by academic freedom, of course, for it can function to protect all of us. Yet thirty years after we first entered graduate school, we are troubled that leading medical historians have testified in lawsuits on behalf of the lead and tobacco industries, rather than consumers and communities harmed by their products.[49] It is heartening that others such as Robert Proctor at Pennsylvania State University and Allan Brandt at Harvard have served as experts on the behalf of states and those injured by tobacco company activities.

There will always, we suppose, be differences over how history is interpreted and in whose interests we should be producing our stories and providing our expertise. More than we understood two decades ago, we believe linking history to ethical understandings is crucial. We still feel that people's lives are at stake in what we write and say. Whether we are concerned about children poisoned by lead, workers whose lives were shortened by silicosis, African American consumers who will not trust health care providers because of deceits and inhuman treatment, or women who latch on to new technologies or drugs in hopes of cures, we believe we owe them the most truthful, nuanced, and carefully researched and argued history that we can write or testify to. We continue to believe that we and our students have much to offer in making the history we write more than an academic exercise, even as we meet the highest standards of the profession. We believe that there is much historians can do to be scholars as well as engaged and caring citizens, creating our own form of historical relevancy.

NOTES

We are grateful to John Harley Warner and Frank Huisman for their thorough comments. We also want to thank Elizabeth Robilotti of the Center for the History and Ethics of Public Health for her data on articles in the *Bulletin*. We also thank Nitanya Nedd for helping us with the very difficult technical aspects of getting Macs and Windows to speak to each other. This was our first joint writing effort since 1979. It reminded us of how enjoyable (and how much work) it was the first time around!

1. Susan Reverby and David Rosner, "Beyond 'the Great Doctors,'" in *Health Care in America: Essays in Social History,* Reverby and Rosner, eds. (Philadelphia: Temple University Press, 1979), 3–16, 3.

2. Elizabeth Fee and Theodore Brown, "Introduction: The Renaissance of a Reputation," in *Making Medical History: The Life and Times of Henry E. Sigerist,* Fee and Brown, eds. (Baltimore: Johns Hopkins University Press, 1997), 1–11, 5. See Paul Starr, Review of *Health Care in America: Essays in Social History, Journal of Social History* 14 (1980): 142–143.

3. Quoted in Theodore M. Brown and Elizabeth Fee, "'Anything but *Amabilis*': Henry Sigerist's Impact on the History of Medicine in America," in Fee and Brown, *Making Medical History,* 333–370, 333.

4. See the *Health PAC Bulletins* (New York: Health PAC, 1968–1992); Lily M. Hoffman, *The Politics of Knowledge: Activist Movements in Medicine and Planning* (New York: SUNY Press, 1989).

5. Rosalyn Baxandall, Linda Gordon, and Susan Reverby, *America's Working Women: A Documentary History* (New York: Random House, 1976).

6. See James R. Green, *Taking History to Heart* (Amherst: University of Massachusetts Press, 2002) for the details on the Massachusetts History Workshop and its connections to a similar historical effort in Great Britain and their still ongoing *History Workshop Journal*.

7. At CCNY, David had been active in a variety of civil rights and antiwar activities and had taken a psychology course with Kenneth Clark, whose work on racism and children had been critical in the 1954 Supreme Court decision *Brown v. Board of Education*. These activities led him to coauthor with Gerald Markowitz, *Children, Race, and Power: Kenneth and Mamie Clarks' Northside Center* (Charlottesville: University Press of Virginia, 1996; reprint, Routledge, 2001).

8. See David and Sheila Rothman, *The Willowbrook Wars* (New York: Harper & Row, 1974).

9. The letter is available online on David's syllabus website, www.columbia.edu/itc/hs/pubhealth/rosner/g8965.

10. The paper ultimately ended up as a chapter in David's book, *A Once Charitable Enterprise: Hospitals and Health Care in Brooklyn, New York 1885–1915* (New York: Cambridge University Press, 1982; reprint, Princeton University Press, 1986).

11. At AAHM meetings in the 1970s, for example, it was not unusual to see an older doctor-historian taking the arm of a younger woman historian to escort her to the bar. As we often quipped, behavior that the women's movement taught us to object to with our contemporaries was sometimes tolerated as a quaint chivalrous practice not worth criticizing.

12. Some of this was briefly discussed in David Rosner, "Tempest in a Test Tube: Medical History and the Historian," *Radical History Review* 26 (1982): 166–171.

13. [Leonard Wilson], "Medical History without Medicine," *Journal of the History of Medicine* 35 (1980): 5.

14. [Leonard Wilson], "History vs. the Historian," *Journal of the History of Medicine* 33 (1978): 127–128. See also "Schizophrenia in Learned Societies: Professionalism vs. Scholarship," *Journal of the History of Medicine* 36 (1981): 5–8.

15. Gordon Jones, M.D., book review in *Journal of the History of Medicine* 34 (1979): 112–114.

16. Howard Berliner and Lloyd Stevenson, book reviews in *Bulletin of the History of Medicine* 54 (1980): 131–141.

17. Lloyd Stevenson, "Second Opinion," *Bulletin of the History of Medicine* 54 (1980): 135–136.

18. [Wilson], "Medical History without Medicine," 7.

19. Susan also chose to show the association's senior members that you could write about a physician's ideas in a social historical context. At the 1980 meeting, with Lloyd Stevenson in the audience, she presented a paper that would become "Stealing the Golden Eggs: Ernest Amory Codman and the Science and Management of Medicine," *Bulletin of the History of Medicine* 55 (1981): 156–171. By the time the paper was published, Caroline Hannaway had become the *Bulletin*'s editor.

20. Our memory is that among those present with us were Harry Marks, Marty Pernick, Judy Leavitt, and Ron Numbers.

21. Elizabeth Robilotti, a graduate student in Columbia's Program in the History and Ethics of Public Health, did the research on topics in the *Bulletin* and compiled these statistics.

22. Kimberlé Crenshaw, "Demarginalizing the Intersection of Race and Sex," *The Univer-

sity of Chicago Legal Forum (1989): 139–167. See also Valerie Smith, *Not Just Race, Not Just Gender* (New York: Routledge, 1998).

23. The influential 1978 book *For Her Own Good* by Deirdre English and Barbara Ehrenreich (New York: Anchor) was one of the earliest historical critiques (written by two nonhistorians) of medicine's paternalism toward women. Written in the tones of the antipatriarchy argument of early second-wave feminism, the book did little to differentiate between ideology and practice or to put medical ideas on women within the context of medical theory in general. Nor did it allow for any sense of agency on the part of women. It came under almost immediate attack by most feminist historians. The book did have a powerful influence on nonhistorians looking for explanations of medical power.

24. Susan M. Reverby, "Thinking through the Body and the Body Politic: Feminism, History, and Health Care Policy in the United States," in *Women, Health and Nation: Canada and the United States since 1945,* Georgina Feldberg, Molly Ladd-Taylor, Alison Li, and Kathryn McPherson, eds. (Toronto: McGill-Queen's University Press; Ithaca: Cornell University Press, 2003), 404–420.

25. The politics within history of nursing seemed to suffer a somewhat parallel fate in the beginning. In 1984, the first meeting of what was to be called the American Association for the History of Nursing, was held. Under the same scholarly umbrella were retired nursing practitioners, nurses trained in the education schools to do history, nurse historians trained by Ph.D. historians, and the social historians with no nursing experience. In nursing history, the difficulty was more a genteel tradition that tried to deny, rather than elucidate, the historical divisions (especially by race and class) within the profession. Over time, the Ph.D. non-nurse historians maintained an infrequent presence, and the associations' annual meetings became much more a home to the increasingly professionalized nurse historians. But it was clear that this organization, especially as led by the nurse historians with Ph.D.s in history, that high historical standards were to be established as a new journal, prizes, and lectureships were used to define the field's parameters. Although tensions existed among the groups for a short while, the field was too small to sustain large divisions. Turf battles also ended when it became clear that non-nurse historians could not be hired by nursing schools since they could not obviously teach public health or medical-surgical nursing and few schools could afford full time historians.

26. *American Historical Review* 91 (1986): 1053–1075. Two years later Scott published a collection of her essays that further expanded her position in *Gender and the Politics of History* (New York: Columbia University Press, 1988). For a critique of Scott, see Laura Lee Downs, "If Woman Is Just an Empty Category, Then Why Am I Afraid to Walk Alone at Night? Identity Politics Meets the Postmodern Subject," *Comparatives Studies in Society and History* 35 (1993): 414–437.

27. Kathleen M. Brown, "Brave New Worlds: Women's and Gender History," *William and Mary Quarterly* 50 (1993): 311–328, 312.

28. Morantz-Sanchez, for example, published her comparative paper on Elisabeth Blackwell and Mary Putnam Jacobi first in *American Quarterly* in 1982. Her rereading of Blackwell to take into consideration gender theory appeared in *History and Theory* ten years later. Hammonds, in turn, tended to publish her more theoretical papers on sexuality, AIDS, and black womanhood in such journals as *Radical America* and *differences,* and in collections on cultural studies and feminism.

29. Evelynn M. Hammonds and Susan M. Reverby, "Playing Clue: Do We Need the

Bodies to Name the Crime? Race, Gender and Class in American Health Care History," paper delivered to the 8th Berkshire Conference on the History of Women, New Brunswick, N.J., 9 June 1990, 4. Susan used this occasion to critique the analytic limits of her book *Ordered to Care* because it did not consider race but focused on gender and class.

30. See, for example, Laura Briggs, "The Race of Hysteria: 'Overcivilization' and the 'Savage' Woman in Late Nineteenth-Century Obstetrics and Gynecology," *American Quarterly* 52 (2000): 246–274.

31. S. Ryan Johansson, "Food for Thought: Rhetoric and Reality in Modern Mortality History," *Historical Methods* 27 (1994): 101–125, 101.

32. For an example of a struggle with how to present this viewpoint to a consumer audience and to think about how much "experience" could not be transparent, see Susan M. Reverby, "What Does It Mean to Be an Expert? A Health Activist at the FDA," *Advancing the Consumer Interest* 9 (1997): 34–36. Between 1993 and 1997, Susan served as the consumer representative on the U.S. Food and Drug Administration's OB-GYN Devices Expert Panel.

33. For the most part, these early attempts were aimed at examining the ways the institution incorporated social and class distinctions into its organizational structure and social relationships. See Morris Vogel, *The Invention of the Modern Hospital* (Chicago: University of Chicago Press, 1981); Charles Rosenberg, *The Care of Strangers* (New York: Basic Books, 1987); and Rosner, *A Once Charitable Enterprise.*

34. Alan Derickson, *Workers' Health, Workers' Democracy: The Western Miners' Struggle, 1891–1925* (Ithaca: Cornell University Press, 1988).

35. David Rosner and Gerald Markowitz, eds., *Dying for Work: Workers' Safety and Health in 20th-Century America* (Bloomington: Indiana University Press, 1987) and *"Slaves of the Depression": Workers' Letters about Life on the Job* (Ithaca: Cornell University Press, 1987). David Rosner and Gerald Markowitz, *Deadly Dust: Silicosis and the History of Occupational Disease in Twentieth-Century America* (Princeton: Princeton University Press, 1991) and *Deceit and Denial: The Deadly Politics of Industrial Pollution* (Berkeley: University of California Press/Milbank Memorial Fund, 2002).

36. Christopher Sellers, *Hazards of the Job: From Industrial Disease to Environmental Health Science* (Chapel Hill: University of North Carolina Press, 1997); Christian Warren, *Brush with Death: A History of Childhood Lead Poisoning* (Baltimore: Johns Hopkins University Press, 2000); and Claudia Clark, *Radium Girls: Women and Industrial Health Reform, 1910–1935* (Chapel Hill: University of North Carolina Press, 1997) are among the best recent works. David's most recent book with Gerald Markowitz, *Deceit and Denial,* builds on this theme and looks at the ways that industry has shaped our understanding of danger and risk.

37. Among the most challenging new efforts are Samuel Roberts, "Infectious Fear: Tuberculosis, Public Health, and the Logic of Race and Illness in Baltimore, Maryland, 1880–1930" (Ph.D. diss., Princeton University, 2002), and Amy Fairchild, *Science at the Borders: Immigrant Medical Inspection and the Shaping of the Modern Industrial Labor Force* (Baltimore: Johns Hopkins University Press, 2003).

38. For more on this, see Evelynn M. Hammonds, *The Logic of Difference* (Chapel Hill: University of North Carolina Press, forthcoming).

39. See W. Michael Byrd and Linda A. Clayton, *An American Dilemma: Race, Medicine and Health Care in the United States* 2 vols. (New York: Routledge, 2000, 2002). These authors are physicians and public health professionals. As they say clearly, "though history is uti-

lized as a major organizational and analytic tool, the book is not written as pure history or medical history."

40. For examples, see Nayan Shah, *Contagious Divides: Epidemics and Race in San Francisco's Chinatown* (Berkeley: University of California Press, 2001), and Laura Briggs, *Reproducing Empire: Race, Sex, Science and U.S. Imperialism in Puerto Rico* (Berkeley: University of California Press, 2002).

41. Douglas A. Lorimer, "Race, Science and Culture: Historical Continuities and Discontinuities, 1850–1914," in *The Victorians and Race,* Shearer West, ed. (Aldershot: Scolar Press, 1996), 12–33, 12.

42. See, for examples, Stephen B. Thomas and Sandra Crouse Quinn, "The Tuskegee Syphilis Study, 1932–1972: Implications for HIV Education"; Vanessa Northington Gamble, "Under the Shadow of Tuskegee: African Americans and Health Care"; and Amy L. Fairchild and Ronald Bayer, "Uses and Abuses of Tuskegee," reprinted in *Tuskegee's Truths: Rethinking the Tuskegee Syphilis Study,* Susan M. Reverby, ed. (Chapel Hill: University of North Carolina Press, 2000), 404–417, 431–442, 589–603; and Spencie Love, *One Blood: The Death and Resurrection of Charles R. Drew* (Chapel Hill: University of North Carolina Press, 1996).

43. Peter Kolchin, "Whiteness Studies: The New History of Race in America," *Journal of American History* 89 (2002): 154–173.

44. Waltraud Ernst, "Introduction: Historical and Contemporary Perspectives on Race, Science and Medicine," in *Race, Science, and Medicine, 1700–1960,* Waltraud Ernst and Bernard Harris, eds. (London: Routledge, 1999), 1–28.

45. "Trade Secrets" aired on PBS stations in March 2002. *Blue Vinyl,* by Judith Helfand and Dan Gold, also used materials we uncovered. See also Geoffrey Cowley, "Getting the Lead Out," *Newsweek,* 17 February 2003, 54–56.

46. Susan M. Reverby, "A Legitimate Relationship: Nursing, Hospitals and Science in the 20th Century," in *The American General Hospital,* Janet Golden and Diana Long, eds. (Ithaca: Cornell University Press, 1989), 135–156.

47. Reverby, *Tuskegee's Truths.*

48. The year before, the Organization of American Historians, at a cost of over $100,000, boycotted the Adams-Mark Hotel in St. Louis over a racial legal dispute and moved its entire convention meetings to other places in that city. But questions of possible organizational bankruptcy hung over both decisions.

49. See Laura Maggi, "Bearing Witness for Tobacco," *American Prospect* 11 (27 March–10 April 2000) and available on the Web at www.prospect.org/print/V11/10/maggi-l.html. For the actual transcript of one historian's testimony for the tobacco industry, see www.tobaco.neu.edu/box/BOEKENBox/transcripts.html (accessed 28 August 2002). For a somewhat different view of the role of historians as experts, see David J. Rothman, "Serving Clio and Client: The Historian as Expert Witness," *Bulletin of the History of Medicine* 77 (2003): 25–44. It is a testament to the dynamism of the program at Columbia that divergent views exist among our faculty. A cautionary tale about the dangers of the effect of litigation based historical research is related in Ellen Silbergeld's review essay, "The Unbearable Heaviness of Lead," *Bulletin of the History of Medicine* 77 (2003): 164–171, a review of Peter English's book on lead poisoning, which outlines some of the historical distortions that can occur when a historian is hired by industry. The one substantive correction we would make is that she unfortunately included Christian Warren's fine book in her larger critique of English's work.

CHAPTER NINE

The Historiography of Medicine in the United Kingdom

Roy Porter

We've all endured those "Twenty Countries in Seven Days" package holidays from which the wretched tourist emerges dazed and dizzy, remembering nothing at all about anywhere he's been. If I attempted to visit all the main trends in British history of medicine in this occasion it would induce a similar sort of academic travel sickness. In the interests of mental health—yours and mine—I shall impose a strict regimen.

First, I propose to say nothing about movements common to Western scholarship at large.[1] Thus, I shall not rehearse yet again the rejection of Whiggish triumphalism or revisit the impact of feminist history, of structuralism, of Foucauldian *savoir-pouvoir*, postmodernism, Derridean textual analysis, and the wider linguistic turn. These tendencies have been felt from San Diego to St. Petersburg and even in Sheffield and Southampton.[2]

Second, I shall keep silent about fields and periods beyond my competence, for example medieval studies, or the politics of modern health care.[3]

Third, I shall restrict myself to British-based historians writing about British medicine—while reminding you that many of the finest works produced by British scholars in recent years have been on foreign topics, for instance Lawrence Brockliss and Cohn Jones's magisterial *The Medical World of Early Modern France*.[4]

Fourth, I hope to steer my historiographical ship between two reefs. On the one hand, I shall not speak abstractly of -isms and -ologies. To my mind, not only would that be tedious, it would also be misleading because the practice of the history of medicine in Britain is not, in fact, ideologically polarized into doctrinaire sects but is characterized by a healthy pluralism and diversity: among historians English individualism still rules OK.[5] On the other hand, I shall refrain from bombarding you with fleeting and instantly forgettable references to hundreds of names, topics, and titles. My plan, rather, is to address in some detail a mere handful of books that seem to me indicative of new trends and influential as rethinkings of the field. That way, I trust, we will at least avoid intellectual indigestion.

By way of prologue, I must say something about the institutional developments underpinning such scholarly tendencies. When I started out in the mid-1960s, history of medicine in Britain was generally thought an intellectually mediocre pursuit, holding no fascination for brash and bumptious apprentice historians of science like myself: it had no big issues, no clashes of the kind provoked by Popper, Lakatos, Kuhn, or Feyerabend. It seemed to be the unproblematic chronicle of how dreadful diseases had been conquered by great doctors.

All this was to change. Over the last thirty years, new diseases like AIDS have challenged the progress saga, while public attitudes toward scientific medicine and the medical profession have grown critical. As an inevitable consequence, the history of medicine has itself been problematized.[6]

British scholars have been well placed to take advantage of such new ferments thanks to two developments. The discipline has been energized during the last quarter-century thanks to the founding and flourishing of the Society for the Social History of Medicine, a radical outfit that brought together younger historians, social scientists, and left-leaning health professionals. Its thrice-yearly journal, *Social History of Medicine,* is now ten years old.[7]

A comparable stimulus has come from the Wellcome Trust. By supporting the Wellcome Institute in London; units in Oxford and Cambridge,[8] Manchester and Glasgow; and lectureships in almost thirty universities, the Trust has set study of the history of medicine—once largely conducted by retired or Sunday doctors—onto a proper academic footing. Most Wellcome appointees are trained historians working in or alongside history departments. That has its pros and cons—arguably certain research topics really do require professional medical expertise and experience. But it has ensured that the history of medicine has been exposed to the trade winds of history and is now undertaken with due historiographical sophistication.

The most influential scholarship during the last generation has not been history of medicine in the traditional, narrow sense at all—that is, top-down accounts of doctors, by doctors, for doctors. It has been about health, in many cases the healthiness of populations at large. And in this regard, there can have been no more influential contribution to our understanding of how healthy people were, how long they lived, and what killed them than that of the Cambridge Group for the History of Population and Social Structure. Hence, I would first like to pay tribute to the work of these and other historical demographers, in establishing the population history of England—a topic especially relevant just now since this is the two hundredth anniversary of the publication of Malthus's *An Essay on the Principle of Population,* a work that emphasized the positive check of epidemic disease and stirred considerable controversy among doctors.[9]

Malthus's portrayal of Nature as ceaseless struggle long dominated scholarly approaches to the population history of preindustrial Europe. Malthusian orthodoxy taught that preindustrial societies sustained extremely high birth rates. Hence, they must also have suffered a correspondingly high death rate. And did not the facts bear this out? After all, even relatively advanced France had undergone decimating famines well into the eighteenth century; from the Black Death onward, Europe at large had been pestilence-ridden while war, too, had been endemic. Nevertheless, the Malthusian trap had evidently finally been sprung, since from around 1800 the major Western societies had supported the rising populations essential for industrialization.

So how had that great escape come about? Explanations traditionally looked to a relaxation in the regime of death. Exactly how or why people had stopped dying at such a shocking rate no one knew, but the cause was bound to lie there—in the death rate—since the Malthusian model presumed that the birth rate was always near its ceiling.

For the last thirty years, however, this received population model has been under fire, and authoritative documentation of such revisionist thinking came in 1981 with Tony Wrigley and Roger Schofield's *The Population History of England.*[10] Their achievement was twofold. Making national projections grounded on scrutiny of the registers of over four hundred parishes, they established for the first time reliable population aggregates of deaths, births, and the total numbers of inhabitants alive at any time between 1541, when parish registers began, and the coming of civil registration at the dawn of the Victorian era.

Moreover, they proposed an interpretation of the dynamics of change which has since won acceptance. *Pace* the Malthusian model, early modern English society (and, to some degree, other Western European nations too) was marked by only a moderately high birth rate—one far lower than the possible biological maximum or that now commonly found in the Third World. It also had a correspondingly moderate mortality rate. This equilibrium was maintained less through catastrophic Malthusian positive checks than through the other kind of check that he was to forefront in subsequent editions of his *Essay*—the preventive check which stopped too many surplus mouths being born in the first place.

Above all, Wrigley and Schofield held that the principal population regulator in preindustrial England had been deferred marriage. By world standards, the English customarily married very late; in 1700, women commonly did not wed until they were around twenty-five, and men some years later still. This delay served as an effective fertility curb. Focusing on nuptiality, they demonstrated that the source of the dramatic population rise from around 1750 lay in changes in marital habits. Couples began marrying earlier, having children earlier, and having them over a longer overall span; in other words, the explanation for what Professor Thomas McKeown called the "modern rise in population" lay more in fertility than in mortality.[11]

The Population History of England was a colossal and magisterial counting exercise. A follow-up work, *English Population History from Family Reconstitution 1580–1837,* published last year, supplements it by recourse to a further method, *family reconstitution,* pioneered by the French scholar Louis Henry.[12] Simply put, family reconstitution aims to exploit the fact that parish registers record baptisms, marriages, and burials. Where registers have been conscientiously kept, and if a sufficient percentage of parishioners passed all their days in their native parish, it should be possible, Henry concluded, to plot precisely when in their lives identifiable individuals got married, when their offspring came along, and when they died; in other words, one could proceed from mere aggregates to the reconstruction of the demographically significant acts and rhythms in the lives of actual individuals and groups. Disaggregating trends, one would be able to document whether particular cohorts of individuals were actually marrying earlier or later, were having their children more bunched up or more spaced out, were giving birth more frequently, and so forth.

So what does this follow-up volume demonstrate? It comes more as a relief than a disappointment that, with a few minor exceptions, its findings bear out the conclusions of the earlier volume. That can hardly be a surprise, given that both are products of the Pop Group, drawing on much the same raw data. Between them, these two volumes provide the indispensable factual foundations for all future study of mortality patterns and epidemiology.

In many ways a parallel work, drawing on statistical and demographic expertise and developing methods of huge potential importance to medical history, is *Height, Health, and History: Nutritional Status in the United Kingdom, 1750–1980,* by Roderick Floud, Kenneth Wachter, and Annabel Gregory, a challenging essay in anthropometry.[13]

The old question—did the Industrial Revolution make life better or worse?—sparked a long-running *standard of living debate,* in which the main kind of evidence traditionally used was wage rates. The shortcomings of such data are, however, all too familiar. The authors of *Height, Health and History,* by contrast, come up with fresh answers using data about physiques.

Biologists are confident that, ceteris paribus, variations in height reflect distinctions in well-being, tallness being a proxy for *nutritional status.* So is it possible to reconstruct how the physical stature of the British changed? Floud and colleagues attempt this for army recruits, who have been assiduously measured since the late eighteenth century, and then seek to extrapolate on that basis.

What is revealed? The eighteenth-century base level for the laboring man was low, perhaps under 5' 4" (161 cm). There was then a slow rise until around the 1840s; these gains were then lost during the next generation, but from the 1870s, heights began to rise in a continuous curve up to the present. Privates were once twelve or thirteen centimeters shorter than their officers: the upper (or, better perhaps, taller) classes really did look down on the lower.

If we may make inferences from height to healthiness and so to quality of life, these are challenging findings. They contradict *pessimists* who have interpreted the advent of industrialization as eroding working-class living standards. They suggest a period in the mid-nineteenth century when, despite improving wages, survival prospects may have gotten worse—thanks, presumably, to the worsening sanitary condition of the early Victorian *shock town.* They undercut the scaremongering claims of fin-de-siècle eugenists about national deterioration and racial suicide. And they may give some indirect support to Thomas McKeown's belief that improvements in health were mainly due to better nutrition—though medical historians like Anne Hardy counter that public health and urban improvements played the chief role.[14]

Thanks to works such as these, medical history has been forced to engage with the wider history of the body, considered simultaneously as a biological entity and as a social actor. Another link-up between the history of populations and the history of medicine is study of sexuality and sexual behavior, an inquiry stimulated by feminism, by Foucault, and by the *bottom-up* history of everyday life.[15] Here I shall select for discussion a fine work which falls within the time-period covered by Wrigley and Schofield and by Floud and Co. It is Tim Hitchcock's

English Sexualities 1700–1800.[16] How did sex change, and how did such changes mesh with medical history?

Dismissing as *Whiggish* the interpretation advanced by Edward Shorter—the view that *modernization* overcame traditional taboos and led to sexual emancipation, more and more pleasurable sex[17]—Hitchcock contextualizes sex in terms of shifting male-female relations within local communities and new readings of the sexual body. Avoiding Lawrence Stone's excessive emphasis on the unrepresentative upper crust,[18] Hitchcock draws a distinction between the *public* sexual culture formerly predominant—sexuality as a kind of public play, legitimate so long as it was properly handled within the community—and the new private sexual milieu (associated with such phenomena as pornography and antimasturbation literature) that he traces as emerging during the eighteenth century.

The earlier model squared with a broadly *humoral* view of the sexual body and with what Thomas Laqueur has called the *onesex* model.[19] It was a nest of beliefs that stressed the power of female sexuality, and it prescribed courting practices in which the successive granting of sexual favors led along the road to marriage. Casual premarital sexual activity was permitted, so long as it stopped short of penetrative intercourse.

This erotic world we have lost gradually gave way to a new system of sexual expectations and gender relations. Biomedical teachings began to stress the essentially passive nature of female sexuality—female orgasm was no longer reckoned necessary for pregnancy. And the delineation of the active, phallocentric male and the vulnerable virgin ushered the young into sexual role models that accented what Hitchcock calls *compulsory heterosexuality* among *opposite sexes,* with penetrative intercourse becoming the norm. Hitchcock's model is not only highly suggestive in itself, it explicitly integrates the best recent medical and social histories of sexuality.

If our understanding of bodily health in its full biosocial richness is enhanced by the teaming up of demographic history with the history of sexuality, another fruitful alliance forged in recent studies has been that between demographic and epidemiological history on the one hand, and environmental history on the other. The outstanding work of this kind is Mary Dobson's *Contours of Death and Disease in Early Modern England,* which makes full use of the demographic researches of the Cambridge Population Group, while also drawing on another significant tradition, the *Annales* school.[20] A generation ago, Annaliste history stimulated some fine regional studies of the health of populations, notably Jean-Pierre Goubert's 1974 account of Brittany.[21] The Cambridge and the Paris traditions have now been expertly combined in Dobson's pioneering essay in historical medical topography. Taking the counties of Sussex, Kent and Essex, Dobson—a geographer turned medical historian—exploits demographic data to explain dramatic geographical differentials in health.

Using burial/baptism discrepancies and other such indices, she reveals that certain environments in the South-East were far less salubrious than others. It was largely a matter of contours: high ground had the lowest morbidity, low ground the highest mortality. Particularly unhealthy were the salt marshes and creeks typical of Romney Marsh, the Thames and Medway estuaries, and the Essex coast—a fact well known to the pioneering eighteenth-century student of population, Dr. Thomas Short.

Today's historical demographers have principally dwelt upon the *urban graveyard* phenomenon; Dobson shows it would be a mistake to associate the countryside unequivocally with healthiness, for there were *rural graveyards* as well. She has also discovered what an excellent *health record* could be enjoyed in early modern times by certain upland areas, far off the main highways, with a dry soil, running streams, and ample wood for fuel. The biological *ancient regime* was not everywhere inevitably unhealthy.

Making use of medical records as well as parochial data, Dobson analyzes the fevers that decimated coastal fringes, paying attention above all to *marsh fever* or *ague,* which she confidently identifies as benign tertian malaria. She also shows that bitter winters produced severe mortality among the old, while humid summers bred the enteric fevers that worsened infant deaths.

While stressing how topography had a profound influence on mortality variations, Dobson avoids the trap of geographical determinism. She underscores *social* factors as well—migration patterns and the roles of wealth, class, and occupation in the gradients of sickness. She also provides an illuminating account of the medical resources of the region, while not claiming that these weighed very heavily in the ultimate mortality scales, at least before the widespread adoption of quinine against malaria in the nineteenth century. Rather, she submits that what was important in reducing mortality levels was that *civilization* was changing *nature.* Especially after 1750, the once-fatal marshy areas were growing less hazardous, thanks to agricultural improvement in its widest sense—marsh reclamation, fen drainage, and new field systems.

These findings raise challenging questions respecting the interpretation of demographic change. Wrigley and Schofield have argued, as we have seen, that the population increase after 1740 owed more to a rising birth rate than to a declining death rate, and they have attributed this change to earlier marriage. Emphasizing the topography of mortality, Dobson, by contrast, naturally focuses more on deaths than on births and nuptiality. The two approaches are not necessarily in conflict, for even Wrigley and Schofield, while primarily concerned with the national picture, do not rule out regional variations.

My discussion so far has suggested that crucial to the new medical history have been inquiries into the dialectics of disease and society. It will be no surprise that numerous works have appeared during the last decades examining the social development of specific health practices and medical provisions, against such a backdrop of epidemiological rhythms and demographic change. I shall single one out in particular because of the exemplary manner in which it ranges all the way from social change, through medical provision, to the theory and practice of medicine itself: it is Mary Fissell's *Patients, Power, and the Poor in Eighteenth-Century Bristol*.[22] Concentrating on modernizing processes in an age of industrialization, Fissell pans, somewhat like Hitchcock, from the *traditional* medical milieu of the mid-seventeenth century through to the *modern* world of the New Poor Law of 1834, exploring transformations in the medical beliefs and practices of the people at large in context of tensions between high and low cultures.

The typical late-seventeenth-century lower-class Bristolian was likely to be a participant in various overlapping health care systems. These included magical, astrological, and faith healing, often practiced by *wise women;* home-brew herbal medicine and other forms of self-help kitchen physic; and, not least, regular medicine that, though beyond the pockets of the poor, could be available through charity. Patients picked the forms of therapy they preferred; everyone was, in a sense, his or her own physician.

The eighteenth century was to bring remarkable changes. In an emergent consumer society, medicine became commercialized.[23] There were swelling numbers of itinerants and regulars jostling in the medical marketplace, and drug stores proliferated to meet the new preference for pills. This commodification of medicine impacted most upon the middle classes.

For the poor, by contrast, the turning points were the foundation of the Bristol Workhouse in 1696, and of the Bristol Infirmary forty years later. Fissell maintains that the traditional reputation of Georgian hospitals as *gateways to death* is undeserved: records suggest that the Bristol Infirmary played a positive, if minor, role in nurturing the health of ordinary people.[24]

The true significance of the hospital, however, lay not in its cures but in its role in the reformation of popular health care. The infirmary's day-to-day running soon fell into the hands of the medical staff, above all the surgeons. Trained in Edinburgh's *medical factory,* such men brought an aggressive professionalism to their job. Out went the vestiges of magical and folk medicine; out went the old-style diagnosis dependent upon the sick person recounting his or her *complaint* to the doctor. All this was replaced by the practitioner inspecting the sick person for diagnostic signs, which could, in turn, be expressed in the technical, Latinate jargon of scientific medicine. Through the infirmary and similar institutions, the people were deprived, in the name of progress, of the medical belief systems that had given meaning to their sufferings. Popular medicine, too, was thus *reformed*. No longer was it every man his own physician; medicine, doctors now emphasized, was too complex, too serious, to be left to the sick. Patients, Fissell claims, were thus *deskilled*, and a patient-oriented system was replaced by a doctor-driven medical economy. Hence, to deploy Ivan Illich's rhetoric, the hospital *expropriated* the health of the poor by medicalizing them.[25] More dramatically, Fissell maintains that a significant function of the Bristol Infirmary lay in dissecting patients who had died in its beds. The corpses of the indigent become teaching fodder, and surgical operations themselves excited fear as perhaps heartless medical experimentation. In the popular mind, hospital and jail, medicine and punishment tended to be elided.

Mention of dissection is a further reminder of how the new history of medicine has been stimulated and strengthened by the development of body history, especially analysis of what might be called the people's two bodies, the physical and the cultural. Here a focal point has been the history of death and the corpse. Attention has particularly been directed to the rise of anatomy as the final meeting-point of doctors and the people, as brilliantly shown by Ruth Richardson's *Death, Dissection, and the Destitute: A Political History of the Human Corpse*.[26]

I wish instead to concentrate, however, on a similar work, *The Body Emblazoned: Dissection and the Human Body in Renaissance Culture*, by Jonathan Sawday, a literary historian,[27] because his study affords strong confirmation of the feasibility and desirability of ambitious interdisciplinarity in the history of medicine.

Addressed to the transformation of understanding of the body since the Renaissance, *The Body Emblazoned* seeks to establish intimate interactions between changing medical practices and intellectual and artistic images. The key new activity impacting on the body was anatomy: from Vesalius through Harvey, the knife cut into corpses as never before. Anatomy became the cutting edge of medical investigation and of a doctor's training; moreover, with the erection of magnificent anatomy theatres, it also became a public display of the alliance between medical and civic power.

One consequence of the anatomical revolution was a discrediting of traditional thinking about the body and its relations to mind, soul, and self, which had dominated medieval Christendom. Ancient taboos about bodily sanctity could no longer stand once dissecting was routinized. In certain ways the body became degraded—an object exposed to violation by the gaze, to be cut, dismembered, and experimented upon. Yet in the eyes of others it could equally be ennobled, praised as a masterpiece of beauty or mechanical design, proof of Divine Wisdom. The polysemicity of the anatomized body is a point of emphasis with Sawday.

The newly exposed body did stout service as a metaphor and marker. Because the corpses surgeons cut were those of criminals, dissection assumed a penal character. But the cruel invasiveness of the knife could also suggest other modes of mastery, not least the bloody colonization of the New World, or the misogynistic conquest of women as envisaged by Restoration courtly poetry. *Anatomizing* also became a popular literary and philosophical genre, as in Robert Burton's *Anatomy of Melancholy* (1621).

The most daring aspect of *The Body Emblazoned* lies in its exploration of the symbiosis of the medical, the philosophical, and the artistic. Especially in the Dutch Republic in the 1630s, painters were incorporating dissection scenes into their repertoire, audaciously alluding in their renderings of the corpse on the anatomist's slab to the Pietà tradition of the crucified Christ. To paint the corpse was to dissect it with the artist's not the surgeon's knife, as in Rembrandt's *The Anatomy Lesson of Dr. Nicolaes Tulp.* Can it be purely coincidental, Sawday asks, that Descartes, too, was living close to the butchers' quarter of Amsterdam at roughly the same time and was himself performing dissections?

The anatomical tradition culminated in William Harvey. His demonstration, in *De motu cordis* (1628), that the heart was but a pump undermined age-old correspondences and archetypes (the heart as monarch) and, for all Harvey's personal conservatism, corroborated the dualistic Cartesian separation of mind from body—now available at last to science but treated as a thing apart. If anatomy was to be destiny, consciousness had to be incorporeal, or at most a ghost in the machine. Between them, Sawday argues, Harvey, Descartes, and Rembrandt—or rather their shared *mentalité*—created the new mind/body dualism.[28] Through such suggestions, the salience of history of medicine for the history of philosophy and the history of the arts is established.

Finally and briefly, I wish to refer to a branch of the discipline flourishing in Britain in recent years: the history of mental disorder. For long, the history of psychiatry was undertaken mainly by psychiatrists, with the predictable admixture of strengths and weaknesses associated with in-house approaches. Perhaps because Britain had no Pinel, Kraepelin, or Freud, the history of psychiatry excited rather little attention.

Since the mid-1970s, the scene has changed remarkably. The history of psychiatry has become contested, partly because psychiatry itself has, thanks initially to the antipsychiatry movement. The last fifteen years have seen extensive research, above all into the history of the asylum. We now possess, for the first time, full and critical studies of particular institutions. Patient records have been analyzed by computer to create in-depth profiles, decade by decade, of developments in diagnostics and treatment, admissions policies, length of stay, and so forth.[29] Charlotte Mackenzie and Trevor Turner have separately studied Ticehurst House, the plushest private asylum for the rich, looking respectively at the institution's management and its psychiatric categories.[30] Anne Digby has surveyed the chief charitable enterprise, the York Retreat.[31] Recent histories question stereotypes about conditions at Bethlem.[32]

The outcome, not surprisingly, is a more complex picture than the benign vision of progress painted by older in-house histories or that of callous exploitation and social control assumed by antipsychiatry. If some institutions were scandals, others were well run; if asylums never fulfilled the great curative expectations of optimists, they did not deteriorate into mere workhouses, jails, or dumps. Two volumes, entitled *150 Years of British Psychiatry, 1841–1991*, and edited by German Berrios and Hugh Freeman, form an extremely welcome foray into recent history and give the lie to the historians' sneer that insiders invariably write self-serving, Whiggish whitewashes of their profession.[33]

The history of madness is another instance—like that of the history of the body—in which medical history and cultural history can interact, through discourse analysis and history from below. A fine example of this is Allan Ingram's *The Madhouse of Language: Writing and Reading Madness in the Eighteenth Century*, a study of mad people's writings.[34] It has become something of a historiographical orthodoxy that madness was *silenced* in the Classical Age. In coining that expression, Michel Foucault meant that the discourse of the insane ceased to be regarded as possessing any meaning and hence stopped being heeded by public authorities.[35] It is within this interpretative framework that Allan Ingram proceeds.

Like Sawday, Ingram brings to his study the linguistic skills of the literary historian. Scrutinizing both the explicit messages and the latent meanings of a spectrum of texts—psychiatric writings, fiction, autobiographies, and so forth—he finds Foucault correct to a degree. Yet his investigation chiefly emphasizes the continuing vitality of traditions that privileged the words of the insane, whose punning, freewheeling, dislocated discourse attracted attention. Yet its signification underwent a transformation. In the Renaissance, mad language had been judged revelatory about the body politic or the cosmos, transmitting divine or diabolic messages. By contrast, Enlightenment auditors evaluated such talk as imparting messages about the individual psyche and personality. This psychologization of *inner voices,* Ingram maintains, owed much to Lockean notions of the association of ideas—chains of thought that were twisted and tangled in cases of delusion. The mad were not so much silenced as subjected to new modes of interpretation.

Looking back, I am appalled at all the important developments in British medical history upon which I have not touched at all.[36] But if I have emphasized certain developments over others, it is because these corroborate the chief theme of this essay. Thirty years ago, the history of medicine seemed a back alley, of no special relevance to history at large. Today the history of medicine is everywhere, it commands widespread scholarly attention; healing—the care and discipline of the body in the largest sense—is the thread that links together recent investigations into population, sexuality, gender, the disciplines of power, institutional history, the history of representations and so forth. From the wings the subject has moved stage center.

NOTES

1. For bibliographical surveys, see Gert Brieger, "History of Medicine," in *A Guide to the Culture of Science Technology and Medicine,* Paul T. Durbin, ed. (New York: Free Press, 1980), and "The Historiography of Medicine," in *Companion Encyclopedia,* W. F. Bynum and Roy Porter, eds. (London: Routledge, 1993), 24–44; L. J. Jordanova, "The Social Sciences and History of Science and Medicine," in *Information Sources in the History of Science and Medicine,* Pietro Corsi and Paul Weindling, eds. (London: Butterworth Scientific, 1983), 81–98; and Margaret Pelling, "Medicine since 1500," in ibid., 379–407. *Companion Encyclopedia* contains many up-to-date surveys of recent scholarship.

2. For aspects of these, see Johanna Geyer-Kordesch, "Women and Medicine," in Bynum and Porter, *Companion Encyclopedia,* 884–910, and *Reassessing Foucault: Power, Medicine, and the Body,* Colin Jones and Roy Porter, eds. (London: Routledge, 1994). For such elements in general historiography, see Richard Evans, *In Defence of History* (London: Granta Books, 1997).

3. It will thus be obvious that many key dimensions of medical history are not mentioned in this essay. One is medicine and the state. Here a major achievement is Charles Webster, *The Health Services since the War,* vol. 1: *Problems of Health Care. The National Health Service Before 1957* (London: Her Majesty's Stationery Office, 1988) and vol. 2: *Government and Healthcare: the National Health Service, 1958–1979* (London: The Stationery Office, 1996). For conceptual discussion and a literature review, see *The History of Public Health and*

the Modern State, Dorothy Porter, ed. (Amsterdam: Rodopi, 1994). It will also be apparent that, in a broad-brush survey such as this, I have chosen to omit references to the periodical literature and have concentrated on books.

4. Lawrence W. B. Brockliss and Colin Jones, *The Medical World of Early Modern France* (Oxford: Clarendon Press, 1997); see also Paul Weindling, *Health, Race, and German Politics between National Unification and Nazism, 1870–1945* (Cambridge: Cambridge University Press, 1989). I shall similarly exclude works on the history of medicine in the British empire, such as David Arnold, *Colonizing the Body: State Medicine and Epidemic Disease in Nineteenth-Century India* (Berkeley: University of California Press, 1993), and Mark Harrison, *Public Health in British India: Anglo-Indian Preventive Medicine, 1859–1914* (Cambridge: Cambridge University Press, 1994).

5. For some insight, see Alan Macfarlane, *The Origins of English Individualism: The Family, Property, and Social Transition* (Oxford: Basil Blackwell, 1978).

6. See Ivan Illich, *Limits to Medicine: The Expropriation of Health* (Harmondsworth: Penguin, 1977).

7. Dorothy Porter, "The Mission of Social History of Medicine: An Historical Overview," *Social History of Medicine* 8 (1995): 345–360; Ludmilla Jordanova, "The Social Construction of Medical Knowledge," *Social History of Medicine* 8 (1995): 361–382. [The journal was founded in 1988. *Eds.*]

8. As of 1998, the Cambridge Wellcome unit will close as a result of a deplorable decision by the Wellcome Trust, but it will be replaced by a comparable unit at the University of East Anglia.

9. For the influence of Malthus on historical demography, see Richard M. Smith, "Demography and Medicine," in Bynum and Porter, *Companion Encyclopedia,* vol. 2:1663–1692.

10. E. A. Wrigley and R. S. Schofield, *The Population History of England, 1541–1981: A Reconstruction* (London: Edward Arnold, 1981). Some of the thinking underpinning this may be found in Peter Laslett, *The World We Have Lost,* 3d ed. (London, Routledge, 1983).

11. Thomas McKeown, *The Modern Rise of Population* (London: Edward Arnold; New York: Academic Press, 1976).

12. E. A. Wrigley, R. S. Davies , J. E. Oeppen, and R. S. Schofield, *English Population History from Family Reconstitution, 1580–1837* (Cambridge: Cambridge University Press, 1997).

13. Richard Floud, Kenneth Wachter, and Annabel Gregory, *Height Health and History: Nutritional Status in the United Kingdom, 1750–1980* (Cambridge: Cambridge University Press, 1996). For further studies that stress the interplay of medicine with demographic change, see S. Szreter, *Fertility: Class and Gender in Britain, 1860–1940* (Cambridge: Cambridge University Press, 1995), and John Landers, *Death and the Metropolis: Studies in the Demographic History of London 1670–1830* (Cambridge: Cambridge University Press, 1993). For the role of food, see *Famine, Disease, and the Social Order in Early Modern Society,* John Walter and Rogers Schofield, eds. (Cambridge: Cambridge University Press, 1989). For nutrition and medical history, see *The Science and Culture of Nutrition, 1840–1940,* Harmke Kamminga and Andrew Cunningham, eds. (Amsterdam: Rodopi, 1995).

14. Anne Hardy, *The Epidemic Streets: Infectious Disease and the Rise of Preventive Medicine, 1856–1900* (Oxford: Oxford University Press, 1993). See also James C. Riley, *Sickness, Recovery, and Death: A History and Forecast of Ill Health* (London: Macmillan, 1989).

15. For some thoughts on this, see Roy Porter, "History of the Body," in *New Perspectives on Historical Writing,* Peter Burke, ed. (Cambridge: Polity Press, 1991), 206–232.

16. See Tim Hitchcock, *English Sexualities, 1700–1800* (Basingstoke: Macmillan, 1997).

17. Edward Shorter, *The Making of the Modern Family* (New York: Basic Books, 1975).

18. Lawrence Stone, *The Family: Sex and Marriage in England, 1500–1800* (London: Weidenfeld and Nicolson, 1977).

19. Thomas W. Laqueur, *Making Sex: Gender and the Body from Aristotle to Freud* (Cambridge, Mass.: Harvard University Press, 1990). In this area, see also Roy Porter and Lesley A. Hall, *The Facts of Life: The Creation of Sexual Knowledge in Britain, 1650–1950* (New Haven: Yale University Press, 1995), which examines the roles of doctors and medical beliefs in regulating sexuality.

20. Mary J. Dobson, *Contours of Death and Disease in Early Modern England* (Cambridge: Cambridge University Press, 1997).

21. Jean-Pierre Goubert, *Malades et médecins en Bretagne, 1770–1790* (Rennes: Institut Armoricain de Recherches Historiques, 1974).

22. Mary E. Fissell, *Patients, Power, and the Poor in Eighteenth-Century Bristol* (Cambridge: Cambridge University Press, 1991). See also her "The Sick and Drooping Poor in Eighteenth Century Bristol and Its Region," *Social History of Medicine* 2 (1989): 35–58, and "The Disappearance of the Patient's Narrative and the Invention of Hospital Medicine," in *British Medicine in an Age of Reform,* Roger French and Andrew Wear, eds. (London: Routledge, 1992), 92–109. For other regional studies, see John V. Pickstone, *Medicine and Industrial Society: A History of Hospital Development in Manchester and Its Region, 1752–1946* (Manchester: Manchester University Press, 1985). Issues underlying such works are astutely discussed in Christopher Lawrence, *Medicine in the Making of Modern Britain, 1700–1920* (London: Routledge, 1994).

23. For studies that make use of the concept of the eighteenth-century medical marketplace, see Roy Porter and Dorothy Porter, *In Sickness and in Health: The British Experience 1650–1850* (London: Fourth Estate, 1988); Dorothy Porter and Roy Porter, *Patient's Progress: Doctors and Doctoring in Eighteenth-Century England* (Cambridge: Polity Press, 1989); and Roy Porter, *Health for Sale: Quackery in England 1650–1850* (Manchester: Manchester University Press, 1989).

24. On hospitals, see Lindsay Granshaw and Roy Porter, eds., *The Hospital in History* (London: Routledge, 1989).

25. Illich, *Limits to Medicine*.

26. Ruth Richardson, *Death, Dissection, and the Destitute: A Political History of the Human Corpse* (London: Routledge & Kegan Paul, 1987). For the wider history of death, see Philippe Ariès, *The Hour of Our Death,* trans. Helen Weaver (London: Allen Lane, 1981), and Nigel Llewellyn, *The Art of Death: Visual Culture in the English Death Ritual c.1500–c.1800* (London: Victoria & Albert Museum, 1991).

27. Jonathan Sawday, *The Body Emblazoned: Dissection and the Human Body in Renaissance Culture* (London: Routledge, 1995).

28. For an earlier study from a literary viewpoint, see Francis Barker, *The Tremulous Private Body: Essays on Subjection* (London: Methuen, 1984). For a comparable work exploring the body, see Joanna Bourke, *Dismembering the Male: Men's Bodies, Britain, and the Great War* (London: Reaktion Books, 1996).

29. For conceptual discussion, see Anne Digby, "Quantitative and Qualitative Perspectives on the Asylum," in *Problems and Methods in the History of Medicine,* Roy Porter and Andrew Wear, eds. (London: Croom Helm, 1987), 153–174. For instances of asylum histories, see *Let There Be Light Again: A History of Gartnavel Royal Hospital from Its Beginnings to the Present Day,* J. Andrews and I. Smith, eds. (Glasgow: Gartnavel Royal Hospital, 1993).

30. Charlotte MacKenzie, *Psychiatry for the Rich: A History of Ticehurst Private Asylum, 1792–1917* (London: Routledge, 1993); T. Turner, "A Diagnostic Analysis of the Casebooks of Ticehurst Asylum 1845–1890," *Psychological Medicine,* Monograph Supplement 21 (Cambridge: Cambridge University Press, 1992).

31. Anne Digby, *Madness, Morality, and Medicine* (Cambridge: Cambridge University Press, 1985)

32. Jonathan Andrews, Asa Briggs, Roy Porter, Penny Tucker, and Keir Waddington, *The History of Bethlem* (London: Routledge, 1997); P. Allderidge, *Bethlem, 1247–1997: A Pictorial Record* (London: Phillimore, 1996).

33. *150 Years of British Psychiatry, 1841–1941,* German E. Berrios and Hugh Freeman, eds. (London: Gaskell, 1991); *150 Years of British Psychiatry,* vol. 2: *The Aftermath,* Hugh Freeman and German E. Berrios, eds. (London: Athlone, 1996). Other major works dealing with psychiatry include Roger Smith, *Trial by Medicine: Insanity and Responsibility in Victorian Trials* (Edinburgh: Edinburgh University Press, 1981). Much new research appeared in W. F. Bynum, Roy Porter, and Michael Shepherd, eds., *The Anatomy of Madness,* vol. 1: *People and Ideas* (London: Tavistock, 1985); vol. 2: *Institutions and Society* (London: Tavistock, 1985); and vol. 3: *The Asylum and Its Psychiatry* (London: Routledge, 1988). For a fuller historiographical discussion, see Roy Porter, "History of Psychiatry in the U.K.," *History of Psychiatry* 2 (1991): 271–280. Much recent scholarship has been published in this new journal, set up in 1990. A further recent development is study of mental deficiency. See *From Idiocy to Mental Deficiency: Historical Perspectives on People with Learning Disabilities,* David Wright and Anne Digby, eds. (London: Routledge, 1996).

34. Allan Ingram, *The Madhouse of Language: Writing and Reading Madness in the Eighteenth Century* (London: Routledge, 1991). Compare Ingram's other study, *Boswell's Creative Gloom* (London: Macmillan, 1982).

35. Michel Foucault, *La folie et la déraison: Histoire de la folie à l'âge classique* (Paris: Librairie Plon, 1961), trans. and abridged by Richard Howard as *Madness and Civilization: A History of Insanity in the Age of Reason* (New York: Random House, 1965; London: Tavistock Publications, 1967). For evaluation, see Arthur Still and Irving Velody, eds., *Rewriting the History of Madness: Studies in Foucault's Histoire de la Folie* (London: Routledge, 1992).

36. Among the many fields about which all too little has been said in this survey are history of pediatrics; clinical medicine; laboratory medicine; medical education; pediatrics; hospitals; medicine and empire; medicine and war; women's medicine; medical economics; history of surgery. The reason is not a lack of good work in these fields but a lack of space and competence on the author's part.

CHAPTER TEN

Social History of Medicine in Germany and France in the Late Twentieth Century

From the History of Medicine toward a History of Health

Martin Dinges

What has distinguished the social history of medicine might be explained by contrasting it with other, earlier work. By and large, before 1970 medical history had been written by physicians, mainly for physicians, and about physicians and their world view. The social history of medicine, however, has tended to widen the focus to encompass all medical personnel, including, for example, non-academically trained healers and nurses; to consider the economic aspects of medicine; and to give the patient—all but unknown in earlier medical history—his or her own place as an actor. A sufferer is interested in health matters long before any physician becomes interested; people experienced pain long before the emergence of a medical discipline like anesthesiology.

At the core of my idea of the social history of medicine are questions of power and the inequality of chances. The inequality concerns the access to life chances and medical care, as well as the chance to become a historiographical subject: patients' voices are weak in the bulk of sources used for medical history because this material originally was produced to allow physicians to work and health care administrations to function.

No doubt one might also define the field by the object of research—medicine and society and their relations—or the specificity of theoretical or methodological approaches.[1] Different combinations of these three have also been discussed. There is no doubt that medium-range theories—such as the patient's role or professionalization—have helped to shape the field, and the same holds true for the use of larger concepts, ranging from historic materialism to theories of social differentiation, modernization, and the more recent theories of knowledge. Nevertheless, I prefer here not to restrict my focus to specific approaches to the exclusion of others but instead to concentrate on a field of research and, as a general approach, the social construction of illness.[2] In this conception the different actors—patients, healers of all sorts, authorities (such as the state), and interested third parties—all have an influence on what is socially considered to be a disease.[3] Their intellectual, symbolic, economic, and political resources varied from century to century. This "equipment" determines their chances to impose their view, but there is no need to presuppose that a specific—for example, the medical—vision of illness always and necessarily imposed itself over the other conceptions.

In this chapter, I explore several interrelated shifts in historiographic attention over the past generation. I first look at the move toward a more differentiated view of actors, institutions, and experiences in the epidemiological transition and from general attitudes toward patients' experiences with epidemics. I then turn to the shift from professionalization as a concentration of (monopolistic) power toward a better understanding of the variety of health care services offered in a pluralized medical marketplace, and from medicalization from above toward the power of demand. Finally, I examine the movement from ideas and ideologies toward discourses informing practices. My aim is to explain these shifts by limiting myself to some key examples, not to offer anything approaching an exhaustive literature review.[4]

Trained as a social historian of early modern Europe (with research in the 1980s on poverty and deviant behavior in France), I will concentrate on the modern era, chiefly up until World War I.[5] Due to limited space, I will bypass the history of psychiatry, a large field that would merit a study of its own.[6] I shall also emphasize the German-speaking countries, referring to France as a point of comparison. This focus might suggest that research in these two linguistic spaces interact strongly, which—aside two conferences in the 1980s and two recent monographs[7]—definitely is not the case. Because of linguistic competences on both sides of the Rhine, German and French research has been much more influenced by English-language publications. Nevertheless, I will concentrate on research conducted *in* Germany and *in* France, generally excluding research *on* Germany and *on* France by Anglophone historians such as Richard Evans, Colin Jones, Mary Lindemann,

Mathew Ramsey, and many others who were instrumental in developing the social history of medicine of France and Germany.

To understand the development of the field, some introductory remarks on institutions, persons, and debates are necessary. Medical history in Germany—as in Austria and Switzerland—was traditionally institutionalized at the medical faculties. Medical historians were nearly exclusively physicians, and only a few scholars (for example, Christian Probst and Gunter Mann) were interested in the social history of medicine. In the late 1960s, Ernst Klee, a journalist writing for the leading weekly of West Germany, *Die Zeit*, and the psychiatrist Klaus Dörner reported on the problematic role of physicians during the Nazi period.[8] These newspaper articles, mainly moral incentives, were the first efforts to reconsider the social and political influences on medicine. They had a slowly developing impact on professional societies and medical faculties and led—often only during the 1980s—to research projects reconsidering their role during the Nazi period. Debates on less politicized periods were inaugurated during the 1970s inside the academy by the sociologist and physician Alfons Labisch (b. 1946).[9] He advocated the rediscovery of the "progressive traditions of health politics," for example, of the labor movement during the Weimar Republic. In the 1980s, he broadened his approach toward the social history of medicine as a sort of historical sociology of medicine, considering English-language research an example to follow.

In that decade, an economic historian publishing on the demographic transition (Reinhard Spree),[10] social historians doing research on professionalization (Ute Frevert, Claudia Huerkamp, Anette Drees),[11] and sociologists working on the nineteenth-century state's health politics or on issues of insurance (Gert Göckenjan)[12] joined the field, as did the body historian Barbara Duden.[13] Historical demography was reintroduced to Germany from abroad. The Swiss historian Arthur Imhof published on this subject after some academic years in Norway.[14] The subjects of these monographs characterize well the understanding of the social history of medicine that was guiding these scholars. Several conferences in Bielefeld, beginning in 1982, aimed in vain at cooperation between traditional historians of medicine and representatives of the new approaches. In 1987, Frevert (b. 1953), like Labisch a scholar at the beginning of a university career, identified the consideration of the social and political conditions of medicine as the central difference between the new social history of medicine and traditional medical history.[15] She suggested that the traditional approach was too exclusively interested in the history of the medical disciplines, thus misinterpreting the entire history of medicine as progress to a brilliant present. She also stressed the backwardness of German compared to English and French research. In response, Gunter Mann tried to show that traditional medical history had already realized a lot of what Frevert asked for and regretted the social historians' insufficient perception of these achievements.[16]

Since these beginnings in the 1980s, the bulk of social historical work has been produced by scholars trained or working in history departments. Recently, medical historians at some medical faculties have also introduced innovative work on topics closer to their traditional focus, such as the history of science, medical practice and scientific disciplines, and the reception of medical concepts.[17] Concerning work presented as "new intellectual history," I do not to see how the German species of this approach might not contribute to a new social history of ideas.[18]

In France, medical history as a separate field is institutionalized neither in medical nor in history faculties.[19] Since the 1970s it has been practiced instead in some departments for the history of science and in general history departments. One might call Jacques Léonard the "founding father" of the field, given his expansive body of work on physicians, medical knowledge, and medical culture mainly during the nineteenth century. In the 1970s, traditional historians of medicine and of science were reluctant to accept his social historical approach. French social history of medicine, much more than its German counterpart, has another root in historical demography, which since the 1960s played a crucial role in general historiography of the *Annales* school.[20] The 1970s work of François Lebrun, Maurice Garden, Jean Pierre Goubert, Alain Croix, and Pierre Guillaume indicate links between this demographic perspective and later research programs focused on, for example, epidemics (Jean-Noël Biraben, Bartolomé Benassar), institutions such as hospitals and foundlings' asylums and their demographic effects, and health care politics (Patrice Bourdelais). The more integrated role of anthropological and historical research in France might explain an earlier trend toward a more cultural approach to the entire field, as in studies of popular medical culture (Françoise Loux, Philippe Richard) during the 1970s.[21]

The French social history of medicine of the 1960s and 1970s had specific French roots, Michel Foucault being another specific source of inspiration. English-language publications played a very secondary role until the 1980s; nevertheless, the major synthetic contribution on French medical history during the early modern period stems from two English historians.[22]

In what follows I show how the two historiographies developed from a massive discourse about the relation between medicine and social change toward a more localized one that takes the individual sufferer seriously. Whereas systems of health care were at the center during the 1970s and 1980s, a praxeological turn has taken the stage since the 1990s.

The Epidemiological Transition and Experiences of Illness

In his groundbreaking 1981 publication on social inequalities of death and health, Spree analyzed the general improvement in health conditions in imperial Germany.[23] Affirming the McKeown thesis, he insisted on the importance of general improvements of life conditions and urban sanitation over any direct impact of either physicians or hospital care, without denying the indirect impact of bacteriology as a convincing scientific idiom that physicians deployed in propagating sanitation.[24] He found that the significant increase in life expectancy was mainly connected with infant care and proposed a model in which socioeconomic and cultural factors came together in explaining this change. Stöckel, with a case study on Berlin, recently refined these conclusions, emphasizing the importance of decreasing birth rates and stressing the socially differentiated outcome of measures taken first by private groups, then by municipal services, and finally by the Reich up to 1933.[25]

Case studies on the public health infrastructures instituted in individual cities during the last third of the nineteenth century transcended the problem of the highly aggregated data Thomas McKeown and his followers had used. Beate Witzler, as one example, in her *Großstadt und Hygiene* (1995), compared the politics of health care in six German cities.[26] In all these cities, concern about mortality was dominated by the risk of infectious diseases, such as cholera, and therefore until the end of the nineteenth century the more important question of infant care attracted little political notice.[27] Witzler insisted on the complementary roles played by urban politicians, specialists, and populations. Various problems of implementation characterize the path toward hygienic ameliorations. There was a major shift from health politics as a regulatory activity of the state toward municipal administration and technological intervention. Only after 1900 was there a social hygiene turn toward such issues as the containment of alcoholism and venereal disease.[28] Witzler concluded that competition among cities was a major motivation, evident after 1900 in the rapid growth of investments in hospitals as an emblem of civic pride.

Vögele in his 1998 book on urban mortality change between 1870 and 1913 compared the ten largest German and English cities.[29] Focusing on measurable health effects, he found German health conditions were worse at the beginning of the period but the "urban penalty" (higher mortality in the city than in the countryside) was always less evident than in England.[30] Health improvement started earlier in England but was implemented slower there than in Germany. It was England that first introduced health legislation and infrastructures, whereas what mattered more in Germany was not general legislation so much as the local initiative of cities. Ironically, the Prussian three-class electoral system facilitated costly infrastructural reform—partly because the ruling class in German cities directly profited economically—whereas English voters were more reluctant to support expensive investments. Because digestive disease played a much larger role in Germany than in England across the entire period, with diphtheria as a major killer, the same improvements in infrastructures were more effective in Germany. The rise of living conditions had a strong—but not linear[31]—impact on the mortality decline. Central water supply and sewerage systems allowed the populace to put into practice the precepts of personal hygiene that physicians propagated, behavioral changes that were the main contribution of the medical profession to mortality change.[32] Finding that different local and national paths led to comparable improvements in the health of the population, Vögele concluded that only a combination of multinational comparison with microhistorical studies might enable historians to isolate the decisive factors that explain the change.[33] Comparison of conceptualizations of the urban environment by two different professions—physicians and engineers—opens up particularly promising avenues for research.[34]

Since the 1990s, studies of urban and rural environments have proven to be a powerful context for linking the social history of medicine to other fields. Early in the nineteenth century, health figured as an important argument in the resistance to the development of factories.[35] Olivier Faure for Lyon, and Ulrike Gilhaus for Westphalia showed that economic interest was more important than social class in determining postures toward polluting enterprises.[36] The changing attitude toward animals in the city is another health-related issue receiving increased attention.[37]

Hospitals and clinics, emblematic of the infrastructure of health changes in the nineteenth century, have been studied in France and Germany with a partly different historiographic impetus. In France, Michel Foucault's 1963 *Birth of the Clinic* raised a renewed interest in such institutions. As early as 1982, Faure, analyzing the case of Lyon, reevaluated Foucault's ideas, which had been based too exclusively on the Parisian example.[38] Faure underlined the belated medical specialization of these institutions in Lyon. Von Bueltzingsloewen considered the German case comparing Göttingen, the core of her study, with other cities.[39] The important initiatives in such large cities as Vienna and Berlin came from political or municipal authorities; only in minor cities did the universities play a crucial role. Local interest groups and students asking for clinical instruction also fostered the development of these clinics. Outside the university cities, the medicalizing impact on populations, as seen in the number of treated patients, was very limited.

Among German historians, by contrast, hospitals became an important subject of research chiefly because of their role in exploring and explaining the epidemiological transition. Johanna Bleker completely ignored the *Birth of the Clinic* in her important 1995 study on patients of the Juliusspital in Wuertzburg.[40] Her leading interest in this innovative study is in the social composition of patients and in diagnostics in the hospital, which was a forerunner of clinical medicine in Germany.[41] She found that the mortality rate in the Juliusspital was 6 percent, much lower than the overcrowded Charité in Berlin. Bleker's volume is exemplary of the recent trend in hospital studies.[42] By and large, historians have been less interested in assessing to what extent the early clinic was a medical institution and instead have concentrated on detailed evaluations of the inmates, their illnesses, and the treatment they underwent.

Historical investigations of the reasons why the sick entered hospitals further reflect this trend. Faure and Dominique Dessertine, comparing a sample of hospitals in the region of Lyon, had in 1991 analyzed long-term trends (1866–1936) in hospitalization.[43] They revealed that youth and professional and local instability—not misery and destitution—were key factors leading the sick to the hospitals, and that entering a hospital was much more a matter of choice and a sign of social integration than of marginalization, old age, or widowhood. The population adopted this institution before it became medically effective in a modern therapeutic perspective, a pattern that equally could be demonstrated for German hospitals during the first half of the nineteenth century.[44]

Faure and Dessertine emphasized the methodological difficulties involved in comparing hospital and global populations.[45] This might be the reason why recent German studies insist on tracing the interior differentiation of hospitals, as in a recent study on the long-term medical and administrative development of the general hospital through the late twentieth century.[46] The most sophisticated comparative analysis we have, published in 2001 by Gunnar Stollberg and Ingo Tamm, concerns scientific, medical, status-centered, economic, and social differentiation of the inmates of hospitals in six cities from the early nineteenth century to 1914, showing how different groups of patients were included or excluded, in ways that varied from city to city, according to the statutes of each institution.[47] Still very little is known about patients' attitudes to the institution.

Only one contribution to the 1996 volume on hospitals draws on such sources as autobiographies of working-class inmates to stress their generally positive reception of the hospital.[48] Jens Lachmund and Gunnar Stollberg used mainly bourgeois autobiographies and the concept of the patient role in their 1995 historical-anthropological monograph, sketching the somewhat different attitudes of this class toward the institution and the problems that arose as hospitalization transformed the patient's social identity.[49]

Aside from what can be gleaned from studies cited above, we still know much too little about morbidity as distinct from mortality.[50] The only monograph that has centered on morbidity—apart from epidemiological change—is Marlene Ellerkamp's 1991 study of female textile workers in Bremen.[51] Comparing working women in different companies, she shows the illness risks women incurred from the working place and from living conditions at home. Ingrid von Stumm has used insurance data to reconstruct specific morbidity in Leipzig from 1887 to 1905.[52] While where one worked seems to have been a more important determinant of morbidity for women than for men, it is not possible to say with certainty whether gender or the specific conditions of employment were the determinant for gender-specific morbidity, findings that invite further research.

Ever since the emergence of HIV/AIDS, epidemics have again been at the core of the social history of medicine—already the focus of earlier work, mainly French studies on the history of diseases like plague and leprosy, which had been inspired by demographic interests.[53] Medievalists in the 1990s, for example, produced studies that transcended more traditional investigations of institutional care.[54] In 1996, François-Olivier Touati traced how the meanings of leprosy changed between the twelfth and the fourteenth centuries. Although there was no sharp break or simple linear trajectory of change, earlier associations of illness with saintliness shifted toward the marginalization of the lepers, expressed both in discourse and in the topography of leper houses. In Kay-Peter Jankrift's comparative study on how eleven medieval cities in northwest Germany reacted to different kinds of epidemics, he shows how cities developed locally variable and durable strategies for responding to the challenge of epidemics.[55]

Mental attitudes toward plague had already been treated by Jean Delumeau in his 1978 history of fear.[56] Studies on responses to threats of disease increasingly have focused on the past two centuries, examining the social construction of such diseases as cholera, tuberculosis and syphilis.[57] In a case study on smallpox vaccination in Württemberg during the first half of the nineteenth century, for example, Eberhard Wolff criticized the concept of the "traditionality of popular medical culture" as insufficient and revealed the rationalities and the irrationalities of both "traditional" and "modern" behavior.⁵⁸ At the same time, he questioned reigning historical assumptions about people's resistance to vaccination. Taking seriously the socioeconomic background and voice of patients, he showed that the refusal to vaccinate their very young fifth child might have been a "rational" strategy of very poor parents in the countryside. The realistic aim was not survival of all family members under all conditions, and economic survival of the rest of the family might prevail over the life of a single member.

Tuberculosis, for German and French historians, as for their counterparts elsewhere, has proven to be a particularly appealing subject for privileging patients' conceptions of and experiences with disease. Guillaume, in his 1986 study on France, tested the public image of this disease—framed by literary sources—against actual illness experiences of inmates in tuberculosis clinics.⁵⁹ His refreshing approach to individual experience put the patient visibly at the center of his study, while he also included a more conventional social historical analysis of knowledge, institutions, and politics. Exploring a different dimension of patient experience, Sylvelyn Hähner-Rombach, in a regional long-term study on Württemberg up until World War II, mapped the options available to the sick during peacetime, World War I, and the National Socialist period, thus underlining the political economy of illness experience.⁶⁰ She showed that while most employed men had access to the clinic, women, without the same health insurance benefits, had much more limited access. They either had to be employed and insured themselves (still rare around 1900) or had to be privileged as an insured family member, which varied widely around the country. Mothers, moreover, rarely felt that they could leave their families in order to undergo a "cure" of two or more months. Fathers seemed less bound by domestic obligations, but taking time away from work to pursue a "cure" meant that their families had to live on a fourth of the regular income. Therefore, married men used the possibility of the cure less than could be expected. Around 5 percent of all male inmates interrupted treatment before the end of the cure, usually in order to return to work, compelled by fear of the destitution of their family. Another 5 percent of the male inmates were expelled for disciplinary reasons.

Still other studies have used cross-national comparison as a way of illuminating variations in illness experiences. Limitations of access to the expensive "cure" were different in England as Flurin Condrau showed in 2000, in a keen comparative analysis of England with the German Reich from the late nineteenth century through World War I.⁶¹ Financed by poor law, the English system focused less on clinics, provided shorter stays in institutions, and promoted less expensive forms of care, often given nearer to the patient's residence. What is most interesting about this study is its skilled combination of quantitative and qualitative methods to understand the patient during the cure. In English clinics, women always represented a third or more of the inmates. This different pattern shows how much the wider German gender gap can be attributed to purely financial considerations and had little to do with the prevalence of the disease or the relative inclination of men to accept medical aid. Using quantitative results on the continuing coming and going of patients in the clinic, Condrau concluded that literary images of the disease experience, for example, considering it as a "common voyage" of the inmates, are completely wrong. Condrau is best when using autobiographical material and a diary of a group of German inmates to explore the patient role. He also looks at self-organization of patients' groups within hospitals, which shaped everyday life in the institution, partly conforming to the hospital's disciplinary requirements and partly diverging from them.[62] Especially interesting are the English sources on individual personal and social problems after discharge, which sometimes permit a reconstruction of inmates' biographies across ensuing decades. The lady almoner kept former patients in contact with the institution and also gave advice that went well beyond questions of medical care. Life with tuberculosis continued to be very difficult, and death seemed always to come as completely unexpected following periods of hope. Former inmates continued to look to the institution because they saw it as helpful. Medicalization does fit these specific clinic studies, but a conception of medicalization modified by taking the patients' view seriously into account.

Professionalization, Medicalization, and the Medical Marketplace

Historical writing in the 1970s and 1980s tended to place physicians center stage in the medical marketplace of the eighteenth century onward, giving early modern physicians an anachronistically pivotal role in the provision of health care. More characteristic of later scholarship is the shift from physician-centered studies toward a more global perspective on the variety of sources of medical care available. Earlier studies relied on the number of physicians as an indicator of what some historians regarded as "deficient" provision of healers during the eighteenth and nineteenth centuries, demonstrating the overestimation of the importance of physicians, who in fact were a minority among health care providers.

Attention to professionalization, a sociological concept widely taken up in the 1970s and 1980s, actually reinforced historians' narrow focus on physicians. Professionalization, understood as a process through which the medical profession gained ever more professional independence and with it an ever-growing quasi-monopolistic market position, enjoyed a historiographic heyday, exemplified by works such as Léonard's *Les médecins de l'Ouest au XIXe siècle* (1978).[63] In some less nuanced early studies, physicians' claims of competence—from the 1770s onward—as advisers not only on public health policies but also on almost everything related to the health of states and individuals alike, tended to be taken at face value, thus overestimating their importance.[64]

Seen from the perspective of the twentieth century, the ascension of the physicians' power was institutionally, ideologically, and economically impressive.[65] Under the German system of statutory sickness insurance, for example, physicians had the discretionary power to certify illness, the precondition for state financial aid to a sick individual. The professionalization paradigm recently prompted study of the various pathways toward the creation of a professional corps in France, Germany, and Switzerland, which sometimes led to the integration of second-class practitioners.[66] Studies on professionalization of subgroups (for example, dentists) continue.[67] More innovative is recent French research scrutinizing the profession from the inside, stressing the internal economic, political, and cultural differentiation that was often masked by its collective fight for professional upliftment.[68]

Concomitant with this interest in physicians were considerations about the importance of other healers. Faure, in 1993, stressed the challenges of an overcrowded market for France during most of the nineteenth century, during which the physicians were only one group of many offering medical services. Politicians did not back a particular side during the professional battles but generally preferred any (even low qualified) provision of medical services to none.[69] Quantity won out over quality, and it is imaginable that the quality of the academically trained physicians' care was—until the 1920s—not always preferable to the services of lay practitioners.

Lay practitioners were still important players in the medical market during the 1930s, representing a fifth of the registered healing personnel in the German Reich.[70] Sander's 1990 study on nonacademically trained practitioners and their corporation in Württemberg, which analyzed their unsuccessful battle to professionalize during the eighteenth and nineteenth centuries, was groundbreaking in that it provided a better social, economic, and cultural understanding of this group.[71] Offering illustrative individual examples, she succeeded in transcending the kind of historiography only interested in systemic developments. A decade later, Faltin studied the ideological background of a lay practitioner and his practice in various villages and cities of the German southwest during the first half of the twentieth century.[72] As the only registered health care practitioner in a village around 1900, a lay healer could function as the general provider of all medical services and was consulted by people from all age groups and of all classes. Under the more competitive conditions of a larger city, however, his market share narrowed to the middle-aged. It would seem that in order to subsist he had to become more specialized. From these recent studies it becomes evident that the power of physicians was more limited and the medical market never as monopolistic as had been supposed.

As early as 1980, Léonard stressed the importance of these healers and especially of nurses and their (partial) complementarity with the interests of physicians.[73] Studies on the old regime gave historical depth to this view.[74] Now French research took up again the still underestimated role of other medical and paramedical personnel in medicalization, taking a closer look at their practice.[75] In a 1996 comparative paper on Germany and France, Isabelle von Bueltzingsloewen underscored the crucial role of nurses during the nineteenth century. Not only were they teachers of cleanliness in hospitals—long before a professional practice in a modern sense could be theirs—but they also developed a great capacity to adapt to the more technical standards of "modern" medicine.[76] This approach also reconsidered the different roles churches played in the two countries in the process of medicalization.[77] To consider medicalization automatically as a process of secularization is certainly too simple.

Recently, scholars have stressed the contribution of veterinarians in showing the effectiveness of inoculation and other methods on animals. During the first half of the nineteenth century, Ronald Hubscher argued in a 1999 study, veterinarians, like "charlatans" and others, should perhaps be considered as agents of a "soft" medicalization. They had a propaedeutic role in medicalizing the population, involving enculturation more than a top to bottom diffusion. They may have been allies rather than competitors of physicians, a challenge to the conclusions of earlier research.[78]

These studies have led to changes in what historians mean when they speak of "medicalization." It has come to be conceptualized as encompassing the growth of medical services over time, mainly since the nineteenth century; the inclusion of ever more people into an expanding medical market; and the increasing role and power of physicians—allied to the state—along with the shrinking of the role and power of the patient.[79] Yet, during the 1980s, studies of medicalization, like those of professionalization, still tended to focus on the rhetoric of physicians and of the state, reinforcing the vision of medicalization as a top-down process.[80] Studies on the institutional arrangements of statutory sickness insurance (earlier in Germany, in France only in 1930–1945) also suggested the notion of an "enforced medicalization of the lower classes."[81] The large majority of patients, historians supposed, were culturally distant from physicians and used medical services only in extreme necessity.[82]

According to such a view, medical enlightenment could be analyzed largely as a process of legislative innovation; thus, Bettina Wischhöfer's 1991 study emphasized how, during the second half of eighteenth century, the small state of Lippe reconstructed itself by means of "modern" health politics.[83] At the same time, this innovative study showed that demand from the poor for free medications grew much faster than what was planned for in the state's budget. Moreover, other studies on the medical market of early modern towns revealed that the assumptions that physicians and patients maintained a cultural distance and that use of the physicians' services by large parts of the population was nonexistent were simply wrong.[84] Health was a central issue for all classes, and the sick not only knew about the differences between licensed and other healers but selectively chose among them on the basis of their own opinions about the practitioners' specific competences.[85] Francisca Loetz's 1993 case study on Baden was the first in Germany to question systematically the top-to-bottom assumptions about medicalization.[86] Critiquing the entire concept, Loetz insisted that social demand was the crucial force driving change.

Simultaneously, Faure stressed that whenever access to health care provision was liberated from economic restrictions, demand tended to grow very fast.[87] The change from poor relief via mutual assurance to a system of statutory insurance greatly enlarged the number of people integrated into insurance systems.[88] Research on these systems has just begun to reveal attempts to limit expenditure;[89] the diversity of these institutions;[90] and the persisting role of private intiative.[91] Specific forms of funding even had an impact on the social construction of illnesses and medical specializations.[92] Further research on the funding of health care will provide additional insight as the interests of patients are taken into serious consideration. In any case, the demand for cures seems to be much stronger than earlier assumptions about the cultural gap between physicians and sufferers had assumed.

The implementation of medical enlightenment involved a piecemeal negotiation between the government and various local forces, interested more in health care provision by a variety of sources than in the introduction of licensed healers.[93] In an enormous study on the health politics of the French occupational administration of the Rhine departments between 1794 and 1814, Calixte Hudemann-Simon showed the specific problems of implementation of a new centralized order under the adverse conditions of war and in the face of a different medicopolitical heritage. On the institutional and legislative level, the study opens multiple comparative issues that cannot be discussed here. Hudemann-Simon also has stressed the "locality of administration" as a precondition for efficient political implementation.[94]

Gender also remains a burning issue in studies of the medicalization process. Midwives have been—and for certain authors still are—a major focus for understanding power and gender in the medical market. Earlier studies represented the midwife—in Germany and France—as a victim to an unfriendly medical takeover by male obstetricians.[95] Physicians were supposed to have had an unlimited will to dominate women as birth givers and as midwives, an assumption that ontologized and essentialized gender in ways remarkably similar to nineteenth-century anthropologies of "the" woman's nature.[96] This earlier vision certainly sharpened the gender issue, but it underestimated women's role in demand for medical care—demand that could not be mapped simply onto the gender of the healer—a fact that Gélis in 1988 was the first to stress.[97] Recent research has shown how eclectic villages were in taking up the proposal by authorities that they accept the services of midwives trained in new courses around 1800; the decision depended on whether the "state" or the village itself had to pay for this new service.[98]

In a feminist interpretation of this historical change, Eva Labouvie has recently studied early modern midwives in the Franco-German border region of the Saar.[99] Discovering the countryside as an important site of investigation and broadening the chronological scope to include the entire early modern period, Labouvie has expanded our approach to professionalization and medicalization alike. As Gélis had done earlier, she underlined the cultural importance of women's sociability surrounding childbirth and infant care, placing the medical aspects of birth into a much broader cultural field.[100] Nevertheless, while Labouvie recognizes how medical history stands to be enriched by seeing childbirth as more than just a medical issue, she interprets the growing role of male obstetricians too simply as a male takeover of power and destruction of female self-help culture, replaced by professionalized midwives' care.[101] This approach ignores women's capacity to choose.

Feminist studies of clinical obstetrics have also depicted clinics as machines of male oppression, using the female patients as "training material" for students.[102] No doubt free clinics were intended to serve the training needs of midwives and obstetricians and the moral and biopolitical aims of the state more than the needs of the "patient."[103] At the same time, social analysis shows that patients were mainly young women of low social status without sufficient social capital to give birth at home. This obliged them to disciplinary restrictions that were sometimes scandalous. Yet, during the nineteenth century the demand for birth clinics was ever growing, clearly suggesting that these young women had no better alternatives.

Recent comparative research on clinics in Göttingen, Vienna, and Paris has revealed the very different approach that midwives and obstetricians took in using instruments. In Paris, where midwives had a more important position than in Göttingen and Vienna, instruments were used with more reluctance, but the more invasive practice of Friedrich B. Osiander and his followers in Göttingen had much more to do with the individual than with his gender, as their use of these instruments was a hundredfold greater than in the practice of the Vienna obstetricians.[104] Quantitative comparison also confirms the limited impact of this type of medicalization of birth during the nineteenth century. On the basis of statistics from several German territories, Hans-Christoph Seidel showed in 1998 that well past 1900, most childbirth did not take place in the clinic.[105] Until the 1880s, moreover, the growing involvement of better trained midwives and of obstetricians led to neither an increase nor a decrease in the incidence of stillbirth. In general, in the first half of the century, the role of obstetricians was only to supervise midwives and to manage problematic births; in the latter half, in large cities, physicians showed a slowly growing interest in the midwives' time-consuming and difficult practice.[106] Women of all classes were more often frustrated if the obstetricians refused to be invasive. What has grown clear is that historical understanding of medicalization in general demands a combination of quantitative and qualitative research, and of the parallel consideration of gender and class and the changes in medical practice and the medical market from decade to decade.[107]

The same holds true for later developments. Through the first third of the twentieth century, patients' power in Germany and other German-speaking countries was expressed in a powerful movement of associations advocating "natural medicine," associations that sometimes boasted of tens of thousands of members.[108] Creating spaces for air and water cures in hundreds of locations and organizing conferences, participants expressed their own way to practice hygiene and self-treatment, ordinarily enlisting practices that were not controlled by physicians. The collective interests of these groups even had a certain impact on political decisions.[109] Homeopathic patients, while less powerful as a movement, organized self-help, medical care, and propaganda of this alternative medicine in Germany, Switzerland and the Austro-Hungarian empire.[110] The growing demand for alternative or complementary medicine in the present day may be sufficient to show that another presupposition of the earlier studies on medicalization—the monopolization of the medical market—has to be replaced by a vision of a pluralistic and continuously changing market that has to take patients' demand into serious account.[111]

Ideas, Ideologies, and the Praxeological Turn

I treat ideas at the end of this chapter because they are important to the social history of medicine only insofar as they have a social impact. This was well demonstrated in 1985 by Göckenjan in a study on the ideology of the bourgeois discourse on health.[112] He underlined the instrumental value of this ideology for the ascending physicians and the developing health care politics of the state, mainly during the nineteenth century. He also showed the importance of a changing discursive field as a precondition for an interested professional group—physicians and later the sanitation specialists—to achieve consensus for political action.

In his 1992 book on ideas about health, *Homo hygienicus*, Labisch wanted to show, that the central concept of a renewed history of medicine could only be health, not illness.[113] He therefore reconstructed the various notions of individual and public health from 1500 to the present by using the "great" texts and the Weberian concept of ideal types. What emerges are well-shaped notions that do not pretend to exist in "historical reality" but might correspond to what Michel Foucault would call different "epistemes." Labisch's ideal types of health are loosely linked to those who propagated them, and changes from one to the next ideal type are difficult to explain. As a normative discourse they define at the same time roles of physicians and medicine in society.

Several authors have sought to disclose the impact of normative ideas on everyday life. Geneviève Heller, for example, endeavored to do so by focusing on the implementation of property norms in the Swiss canton of Vaud.[114] In striking contrast to Labisch, her approach is highly empirical. Heller's study reveals a density of local details but completely lacks theoretical framing. At first sight, Michael Frey seems to strike a balance between theory and practice by conceptualizing the implementation of property norms as an element of cultural modernization.[115] After a supposed discovery of property by the bourgeoisie around 1800, property styles were, according to Frey, more and more unified during the nineteenth century. However, because he is presupposing neatly limited property styles harbored by such social groups as nobles, urban elites, and rural populations before 1800, he fails to take into account the ambivalences of the behavioral patterns of these groups. Frey uses sketchy empirical evidence from some cities to illustrate the modernization process. Normative texts or the sole introduction of a new institution like public conveniences, however, are too often presented as proof for the effective implementation of new behavioral patterns. Finally, Frey tries to illustrate the enduring cultural backwardness of most inhabitants of the countryside by building on sources like medical topographies of rural Bavaria. Although he is dealing with some of the missing links between the discourse and the practice of hygiene, he does not succeed in describing a process of reception. His approach can be said to be a doubtful return to cultural dichotomies, which never have been very instructive in conceptualizing cultural change.[116]

Some of the missing links may be better put into evidence using Foucault's concept of discourse-formation.[117] Philipp Sarasin, for example, in his pioneering 2001 study of the nineteenth-century discourse on the body, referred to this more specific notion of discourse, used material that went far beyond the "great" texts, and analyzed content, authorial strategies, and modes and media of diffusion.[118] Thus, the ambivalent and differentiated character of this discourse and its various shifts become evident. He put people's obsession with health and the fear of corporeal decline into context, and, more than in earlier studies, took into account effects of discourses as informing practice and emphasized the interests of the recipients of this discourse. Discourse—in the Foucauldian sense—empowers many more actors than earlier Marxist or Weberian theories of medicalization. This shift from elites and the state steering a top-to-bottom acculturation toward a discursive field serving many more interests is significant.

This discursive approach opens the field to a much larger history of health, liberated from the limited focus on medicine. Interest in the body had opened this horizon early on. In her groundbreaking 1987 study on women's body perception, Duden was looking for a specific female experience.[119] Inspired by the essentialist feminist body concepts of the time, she expected to discover women patients' own perceptions of their bodies in cases published by a physician. What she found instead was a form of popularized humoral pathology. This originally "scientific" medical idiom allowed the patients to construct their body through interactions with the physician. Such more or less unconscious appropriations of discourses can be observed in a variety of sources, including patients' letters, autobiographies, and diaries and, indirectly, even in literature.[120] Individual experience of pain—during plague, toothache, childbirth, and heroic therapies—have been studied.[121] Various authors recently have reconsidered the interaction between doctors and patients during their treatment by systematically studying patient letters.[122] What might be termed the praxeological turn toward discourse and practice in recent research will allow a more differentiated picture of health and its meanings over time. That many of these sources privilege the socioeconomically better off is a limitation of such studies.

The renewed interest in patients leads eventually back to medical practice. Beyond the issue of a simplistic power relation between physician and patient, the sociology of knowledge allows a renewed access to medical practice. Disputes between physicians, authorities, and the population about the meanings of the corpse during the early modern era is a particularly interesting case for the highly sensitive cultural issues—charged with high social, religious, and magical meanings—that were at stake.[123] Historically constructed knowledge instructs the physician's practice and creates power relations, but it is more complex than that alone. Using the example of the use of the stethoscope, Lachmund has analyzed the fascinating process by which a new medical knowledge about the interior of the patient's body was created entirely beyond the patient's control; with the stethoscope, physicians, exclusively decided about the "objective" meaning of the measures taken.[124] Practices situated in local contexts provide a more nuanced picture of the relations between doctors, instruments, and patients. Faure, in 1999, opened therapies as a field of research, situating it at the crossroads between medical theory, professional structures, and the doctor-patient relationship.[125] This praxeological turn seems, for example, to establish the important impact patients and drug producers had around 1800 on the acceptance of certain medications by physicians. This again invites revisions to traditional concepts of power and knowledge in the history of medicine.

Conclusion

During the last twenty years the social history of medicine has enormously enlarged its field of inquiry, mainly by developing new visions of long existing but little researched fields. The interaction between French and German research existed to a much lesser degree than influences from English language research. Despite that, as a general trend, the choice of subjects in both historiographies is very similar, which is even true for unresearched fields, as for example the economies of health, media (both popular media and that specific to the medical community),[126] nonmedical personnel and the diffusion of health standards, travel reports as a source for a comparative history of medical cultures,[127] and the anthropology of medication,[128] among others. Differences concern, for example, the way to conceptualize certain subfields. They result from the continuing influence of specific roots of the social history of medicine in the two countries, whereas general social—and now cultural—history and the history of science play influential roles in both countries.

Priorities changed in all fields from large visions and hypotheses about huge systems toward a more differentiated view of agency and to a larger interest in context. The median "subject" of the French history of mentalities or the statistical average physician or patient of historical demography or epidemiology is slowly passing to the background. Even in quantitative "hard-core" studies, the patient slowly emerges as a historical actor in his or her own right. Thanks to the earlier achievements, this actor is well embedded in a refined chronology and a deepened knowledge of social differentiation and the effect of discourses. This praxeological and sometimes cultural turn of a social history of medicine opens the larger field of a history of health, allowing fruitful interdisciplinary exchanges that have always been beneficial for the field.[129] Comparative research will further enrich the field, most especially when it transcends the pattern of comparison between two entities—two countries, for example—and advances to triangular comparisons.[130]

NOTES

1. Robert Jütte, "Sozialgeschichte der Medizin: Inhalte—Methoden—Ziele," *Medizin, Gesellschaft und Geschichte* 9 (1990): 149–164. Spree insists on a specific concept of "theory"; see Alfons Labisch and Reinhard Spree, "Neuere Entwicklungen und aktuelle Trends in der Sozialgeschichte der Medizin in Deutschland—Rückschau und Ausblick," *Vierteljahrschrift für Sozial- und Wirtschaftsgeschichte* 84 (1997): 171–210, 305–321, 190, 196f.

2. See *The Social Construction of Illness: Illness and Medical Knowledge in Past and Present*, Jens Lachmund and Gunnar Stollberg, eds. (Stuttgart: Steiner, 1992).

3. On epidemics, for example, see Martin Dinges, "Neue Wege in der Seuchengeschichte?" and "Pest und Staat: Von der Institutionengeschichte zur sozialen Konstruktion?" both in *Neue Wege in der Seuchengeschichte*, Martin Dinges and Thomas Schlich, eds. (Stuttgart: Steiner, 1995), 7–24 and 71–103.

4. See Jütte, "Sozialgeschichte"; Labisch and Spree, "Neuere Entwicklungen"; *Medizingeschichte: Aufgaben, Probleme, Perspektiven,* Norbert Paul and Thomas Schlich, eds. (Frankfurt: Campus, 1998); and Eberhard Wolff, "Volkskundliche Gesundheitsforschung. Medikalkultur-und 'Volksmedizin'-Forschung," in *Grundriß der Volkskunde*, Rolf W. Brednich, ed. (Berlin: Dietrich Reimer, 2001), 617–635. On France, see Olivier Faure, "Vingt ans d'histoire de la santé," *Revue historique Vaudoise* (1995): 315–327, and "The Social History of Health in France: A Survey of Recent Developments," in *Social History of Medicine* 3 (1990): 437–451.

5. On the years after World War I, see also *Medicine and Modernity: Public Health and Medical Care in Nineteenth- and Twentieth-Century Germany,* Manfred Berg and Geoffrey Cocks, eds. (Cambridge: Cambridge University Press, 1997), 17–33.

6. Dirk Blasius, *"Einfache Seelenstörung." Geschichte der Psychiatrie in Deutschland* (Frankfurt: Fischer, 1994).

7. *Maladies et société (XIIe–XVIIIe siècles)*, Neithard Bulst and Robert Delort, eds. (Paris: Editions du CNRS, 1989). Isabelle von Bueltzingsloewen, *Machines à instruire, machines à guérir. Les hôpitaux universitaires et la médicalisation de la société allemande 1730–1850* (Lyon: Presses universitaires, 1997); Calixte Hudemann-Simon, *L'État et la santé. La politique de santé publique ou "police médicale" dans les quatre départements rhénans, 1794–1814* (Sigmaringen: Thorbecke, 1995). While there are French studies on Germany, there are no German counterparts on France; see Francisca Loetz, "Histoire des mentalités und Medizingeschichte: Wege zu einer Sozialgeschichte der Medizin," *Medizinhistorisches Journal* 27 (1992): 272–291.

8. See now Franz-Werner Kersting, *Anstaltsärzte zwischen Kaiserreich und Bundesrepublik. Das Beispiel Westfalen* (Paderborn: Schöningh, 1996).

9. Alfons Labisch, *Geschichte, Sozialgeschichte, Historische Soziologie der Medizin* (Kassel: FB Sozialwesen der Gesamthochschule Kassel, 1990).

10. Reinhard Spree, *Health and Social Class in Imperial Germany: A Social History of Mortality, Morbidity, and Inequality* (1981; English trans., Oxford: Berg, 1988).

11. Ute Frevert, *Krankheit als politisches Problem 1770–1880. Soziale Unterschichten in Preußen zwischen medizinischer Polizei und staatlicher Sozialversicherung* (Göttingen: Vandenhoeck & Ruprecht, 1984); Claudia Huerkamp, *Der Aufstieg der Ärzte im 19. Jahrhundert. Vom gelehrten Stand zum professionellen Experten: Das Beispiel Preußens* (Göttingen: Vandenhoeck & Ruprecht, 1985); Annette Drees, *Die Ärzte auf dem Weg zu Prestige und Wohlstand* (Münster: Coppenrath, 1988).

12. Gert Göckenjan, *Kurieren und Staat machen. Gesundheit und Medizin in der bürgerlichen Welt* (Frankfurt: Suhrkamp, 1985).

13. Barbara Duden, *The Woman beneath the Skin: A Doctor's Patients in Eighteenth-Century Germany* (1984; English trans., Cambridge, Mass.: Harvard University Press, 1991); see now Maren Lorenz, *Leibhaftige Vergangenheit. Einführung in die Körpergeschichte* (Tübingen: Edition Diskord, 2000).

14. Arthur E. Imhof and Øivind Larsen, *Sozialgeschichte der Medizin. Probleme der quantifizierenden Quellenbearbeitung in der Sozial-und Medizingeschichte* (Oslo: Universitetsferlaget; Stuttgart: Fischer, 1976); Arthur E. Imhof, *Lebenserwartungen in Deutschland vom 17.–19. Jahrhundert* (Weinheim: VCH, 1990).

15. Ute Frevert, "Geteilte Geschichte der Gesundheit. Zum Stand der historischen Erforschung der Medizin in Deutschland, England und Frankreich," *Frankfurter Allgemeine Zeitung,* 28 January 1987, 31.

16. Gunter Mann, "Beschränktheit im Wissen. Eine Antwort auf Ute Freverts Thesen zur Medizingeschichte," *Frankfurter Allgemeine Zeitung,* 11 March 1987, 32.

17. *Anatomien medizinischen Wissens. Medizin, Macht, Moleküle,* Cornelius Borck, ed. (Frankfurt: Fischer, 1996). On the praxeological turn in laboratory studies, see Thomas Schlich, "Wissenschaft: Die Herstellung wissenschaftlicher Fakten als Thema der Geschichtsforschung," in Paul and Schlich, *Medizingeschichte,* 107–129; Thomas Schlich, *Die Erfindung der Organtransplantation. Erfolg und Scheitern des chirurgischen Organersatzes (1880–1930)* (Frankfurt: Campus, 1998); *Instrument—Experiment. Historische Studien,* Christoph Meinel, ed. (Berlin: Verlag für Geschichte der Naturwissenschaften und der Technik, 2000); *Normierung der Gesundheit. Messende Verfahren der Medizin als kulturelle Praktik um 1900,* Volker Hess, ed. (Husum: Matthiesen, 1997); and Volker Hess, *Der wohltemperierte Mensch. Wissenschaft und Alltag des Fiebermessens (1850–1900)* (Frankfurt: Campus, 2000). Andreas-Holger Maehle, *Kritik und Verteidigung des Tierversuchs: Die Anfänge im 17. und 18. Jahrhundert*

(Stuttgart: Steiner 1992) and *Drugs on Trial: Experimental Pharmacology and Therapeutic Innovation in Eighteenth Century* (Amsterdam: Rodopi, 1999). Christian Bonah, *Les sciences physiologiques en Europe, Analyses comparées du XIXe siècle* (Paris: Vrin, 1995); *Pathology in the 19th and 20th Century: The Relationship between Theory and Practice,* Cay-Rüdiger Prüll, ed. (Sheffield: EAHMH Publications, 1999); Johanna Geyer-Kordesch, *Pietismus. Medizin und Aufklärung in Preußen im 18. Jahrhundert. Das Leben und Werk Georg Ernst Stahls* (Tübingen: Niemeyer, 2000).

18. See, for example, Claudia Wiesemann, *Die heimliche Krankheit. Eine Geschichte des Suchtbegriffs* (Stuttgart: Frommann-Holzbog, 2000)

19. Jean-Pierre Peter, "Jacques Léonard, un historien face à l'opacité," in *Médecins, malades et société dans la France du XIXe siècle* (Paris: Sciences en situation, 1992), 9–19, 12ff.; *Pour l'histoire de la médecine. Autour de l'oeuvre de Jacques Léonard,* Michel Lagrée and François Lebrun, eds. (Rennes: Presses universitaires de Rennes, 1994).

20. On part of the background for these national differences, see Christiane Dienel, *Kinderzahl und Staatsraison: Empfängnisverhütung und Bevölkerungspolitik in Deutschland und Frankreich bis 1918* (Münster: Westfälisches Dampfboot, 1995).

21. Alain Corbin, *Le miasme et la jonquille: Odorat et l'imaginaire social XVIIIe–XIXe siècles* (Paris: Aubier Montaigne, 1982); Georges Vigarello, *Le propre et le sale: L'hygiène du corps depuis le Moyen-Âge* (Paris: Seuil, 1985); Jean Pierre Goubert, *La conquête de l'eau. L'avènement de la santé à l'âge de l'industriel* (Paris: Laffont, 1986).

22. Laurence Brockliss and Colin Jones, *The Medical World of Early Modern France* (Oxford: Clarendon Press, 1997).

23. Spree, *Health.*

24. See Patrice Bourdelais, ed., *Les hygiènistes: Enjeux, modèles et pratiques (XVIIIe–XXe siècles)* (Paris: Belin, 2001).

25. Sigrid Stöckel, *Säuglingsfürsorge zwischen sozialer Hygiene und Eugenik* (Berlin: De Gruyter, 1996).

26. Beate Witzler, *Großstadt und Hygiene. Kommunale Gesundheitspolitik in der Epoche der Urbanisierung* (Stuttgart: Steiner, 1995).

27. See François Delaporte, *Le savoir de la maladie. Essai sur le choléra de 1832 à Paris* (Paris: Presses universitaires de France, 1990), and Patrice Bourdelais and Jean Yves Raulot, *Une peur bleue: Histoire du choléra en France, 1832–1854* (Paris: Payot, 1987).

28. See Lutz Sauerteig, *Krankheit, Sexualität, Gesellschaft. Geschlechtskrankheiten und Gesundheitspolitik im 19. und frühen 20. Jahrhundert* (Stuttgart: Steiner, 1999).

29. Jörg Vögele, *Urban Mortality Change in England and Germany, 1870–1913* (Liverpool: University Press, 1998), 211ff.

30. See *Annales de démographie historique* (1990).

31. See Patrice Bourdelais, "Épidémies et population: Bilan et perspectives de recherches," *Annales de démographie historique* (1997): 9–26, 19.

32. See the complex model in Jörg Vögele, *Sozialgeschichte städtischer Gesundheitsverhältnisse während der Urbanisierung* (Berlin: Duncker & Humblot, 2001), 140.

33. Vögele, *Sozialgeschichte,* 237ff.; see Yankel Fijalkow, "Territorialisation du risque sanitaire et statistique démographique: Les 'immeubles tuberculeux' de l'îlôt insalubre Saint-Gervais (1894–1930)," *Annales de démographie historique* (1996): 45–60.

34. Sabine Barles, *La ville délétère. Médecins et ingénieurs dans l'espace urbain, XVIIIe–XIXe siècle* (Paris: Champ Vallon, 1999).

35. Michael Stolberg, *Ein Recht auf saubere Luft? Umweltkonflikte am Beginn des Indus-*

triezeitalters (Erlangen: Harald Fischer, 1994); Joachim Radkau, *Natur und Macht. Eine Weltgeschichte der Umwelt* (Munich: C. H. Beck, 2000), 274–283; *Le démon urbain,* Geneviève Massard-Guilbaud and Christoph Bernard, eds. (Clermont Ferrand: Presses universitaires, 2002).

36. Olivier Faure, "L'industrie et l'environnement à Lyon au XIXe siècle," *Cahier des annales de Normandie* 24 (1992): 299–311; Ulrike Gilhaus, *"Schmerzenskinder der Industrie." Umweltverschmutzung, Umweltpolitik und sozialer Protest im Industriezeitalter in Westfalen 1845–1914* (Paderborn: Schöningh, 1995).

37. See the special number of *Cahiers d'histoire* 42, no. 3–4 (1997), "L'animal domestique. XVIe–XXe siècle," Eric Baratay and Jean-Luc Mayaud, eds.

38. Olivier Faure, *Genèse de l'hôpital moderne. Les hospices civils de Lyon de 1802 à 1845* (Lyon: Presses Univeritaires, 1982).

39. Bueltzingsloewen, *Machines.*

40. On Foucault, see *Institutions of Confinement. Hospitals, Asylums, and Prisons in Western Europe and North America, 1500–1950,* Norbert Finzsch and Robert Jütte, eds. (Cambridge: Cambridge University Press, 1996).

41. *Kranke und Krankheiten im Juliusspital zu Würzburg 1819–1829. Zur frühen Geschichte des allgemeinen Krankenhauses in Deutschland,* Johanna Bleker, Eva Brinkschulte, and Pascal Grosse, eds. (Husum: Matthiesen, 1995).

42. See Barbara Leidinger, *Krankenhaus und Kranke. Die Allgemeine Krankenanstalt an der St. Jürgen-Straße in Bremen, 1851–1897* (Stuttgart: Steiner, 2000), and Ulrich Knefelkamp, *Das Heilig-Geist-Spital in Nürnberg vom 14.–17. Jahrhundert. Geschichte, Struktur; Alltag* (Nürnberg: Selbstverlag des Historischen Vereins der Stadt Nürnberg, 1989), 301–330.

43. Olivier Faure and Dominique Dessertine, *Populations hospitalisées dans la région lyonnaise aux XIXe et XXe siècles* (Oullins: Imp. Bosc Frères, 1991)

44. Isabelle von Bueltzingsloewen, "Pour une sociologie des populations hospitalisées: Le recours à l'hôpital dans l'Allemagne du premier XIXe siècle," *Annales de démographie historique* (1994): 303–316.

45. On measuring medical effectiveness, see Jörge Vögele, Wolfgang Woelk, and Barbara Schürmann, "Städtisches Armenwesen, Krankenkassen und Krankenhauspatienten während des späten 19. und frühen 20. Jahrhunderts in Düsseldorf," in *Krankenhaus-Report 19. Jahrhundert. Krankenhausträger, Krankenhausfinanzierung, Krankenhauspatienten,* Alfons Labisch and Reinhard Spree, eds., (Frankfurt: Campus, 2001), 405–426, 418ff.

46. *"Einem jeden Kranken in einem Hospitale sein eigenes Bett": Zur Sozialgeschichte des Allgemeinen Krankenhauses in Deutschland,* Alfons Labisch and Reinhard Spree eds. (Frankfurt: Campus 1996).

47. Gunnar Stollberg and Ingo Tamm, *Die Binnendifferenzierung in deutschen Krankenhäusern bis zum Ersten Weltkrieg* (Stuttgart: Steiner, 2001).

48. Barbara Elkeles, "Der Patient und das Krankenhaus," in Labisch and Spree, *Kranken,* 357–373.

49. Jens Lachmund and Gunnar Stollberg, *Patientenwelten. Krankheit und Medizin vom späten 18. bis zum frühen 20. Jahrhundert im Spiegel von Autobiographien* (Opladen: Leske & Budrich, 1995), 152–178, 20ff., 176.

50. See on morbidity in old age, see Christoph Conrad, *Vom Greis zum Rentner. Der Strukturwandel des Alters in Deutschland zwischen 1830 und 1930* (Göttingen: Vandenhoeck & Ruprecht, 1994), 80–94.

51. Marlene Ellerkamp, *Industriearbeit, Krankheit und Geschlecht* (Göttingen: Vandenhoeck & Ruprecht 1991).

52. Ingrid von Stumm, *Gesundheit, Arbeit und Geschlecht im Kaiserreich am Beispiel der Krankenstatistik der Leipziger Ortskrankenkasse 1887–1905* (Frankfurt: Peter Lang, 1995).

53. Bulst and Delort, *Maladies et société*.

54. Françoise Bériac, *Des lépreux aux cagots. Recherches sur les sociétés marginales en Aquitaine médiévale* (Bordeaux: Fédération Historique du Sud-Ouest, 1990); François-Olivier Touati, *Maladie et société au Moyen Age. La lèpre, les lépreux et les léproseries dans la province ecclésiastique de Sens jusqu'au milieu du XIVe siècle* (Paris: De Boeck & Larcier, 1998).

55. Kay Peter Jankrift, "'Up dat God sich aver uns verbarmen wolde.' Formen, Strukturen und Entwicklungen der Auseinandersetzung mit Seuchen in westfälischen und rheinischen Städten im Mittelalter" (Habil. Phil., Münster, 2001). Cf. Martin Dinges, "Süd-Nord-Gefälle in der Pestbekämpfung. Italien, Deutschland und England im Vergleich," in *Das europäische Gesundheitssystem. Gemeinsamkeiten und Unterschiede in historischer Perspektive,* Wolfgang U. Eckart and Robert Jütte, eds. (Stuttgart: Steiner, 1994), 19–51.

56. Jean Delumeau, *La peur en Occident XIV–XVIIIe siècles. Une cité assiégée* (Paris: Facard, 1978), 132–187.

57. Jean-Pierre Bardet, Patrice Bourdelais, et al., *Peurs et terreurs face à la contagion* (Paris: Fayard, 1988); Michael Stolberg, *Die Cholera im Großherzogtum Toskana. Ängste Deutungen und Reaktionen im Angesicht einer tödlichen Seuche* (Landsberg: ecomed, 1995).

58. Eberhard Wolff, *Einschneidende Maßnahmen. Pockenschutzimpfung und traditionale Gesellschaft im Württemberg des frühen 19. Jahrhunderts* (Stuttgart: Steiner, 1998).

59. Pierre Guillaume, *Du désespoir au salut: Les tuberculeux aux 19e et 20e siècles* (Paris: Aubier 1986). On patients more empirically, see Dominique Dessertine and Olivier Faure, *Combattre la tuberculose. 1900–1940* (Lyon: Presses universitaires, 1988), 155–174.

60. Sylvelyn Hähner-Rombach, *Sozialgeschichte der Tuberkulose. Vom Kaiserreich bis zum Ende des Zweiten Weltkriegs unter besonderer Berücksichtigung Württembergs* (Stuttgart: Steiner, 2000), 251–363, 324–328.

61. Flurin Condrau, *Lungenheilanstalt und Patientenschicksal. Sozialgeschichte der Tuberkulose in Deutschland und England im späten 19. und frühen 20. Jahrhundert* (Göttingen: Vandenhoeck & Ruprecht, 2000).

62. See Guillaume, *Désespoir,* 278ff., with interest in the "median patient."

63. Jacques Léonard, *Les médecins de l'Ouest au XIXe siècle* (Paris: Champion, 1978).

64. Chrisitan Barthel, *Medizinische Polizey und medizinische Aufklärung. Aspekte des öffentlichen Gesundheitsdiskurses im 18. Jahrhundert* (Frankfurt: Campus, 1989).

65. Frevert, *Krankheit;* Huerkamp, *Aufstieg;* Drees, *Ärzte*.

66. Calixte Hudemann-Simon, *Die Eroberung der Gesundheit 1750–1900* (Frankfurt: Fischer, 2000); Dominik Groß, *Die Aufhebung des Wundarztberufs. Ursachen, Begleitumstände und Auswirklungen am Beispiel des Königreichs Württemberg (1806–1918)* (Stuttgart: Steiner, 1999), 188ff.

67. Dominik Groß, *Die schwierige Professionalisierung der deutschen Zahnärzteschaft (1867–1919)* (Frankfurt: Lang, 1994). For critique on the dilution of the original concept, see Labisch and Spree, "Neuere Entwicklungen," 191.

68. Pierre Guillaume, *Le rôle social du médecin depuis deux siècles (1800–1945)* (Paris: Association pour l'étude de l'histoire de la sécurité sociale, 1996).

69. Olivier Faure, *Les Français et leur médecine* (Paris: Belin, 1993), 29, 40.

70. On their market share in 1913, see Spree, *Health,* 112, 215; Hedwig Herold-Schmidt, "Ärztliche Interessenvertretung im Kaiserreich 1871–1914," in *Geschichte der deutschen Ärzteschaft: Organisierte Berufs- und Gesundheitspolitik im 19. und 20. Jahrhundert,* Robert Jütte, ed., (Cologne: Deutscher Ärzteverlag, 1997), 43–95, 59ff.; Martin Dinges, "Patients in the History of Homoeopathy: Introduction," in *Patients in the History of Homoeopathy,* Martin Dinges, ed. (Sheffield: EAHMH Publications, 2002), 1–32, 8–9.

71. Sabine Sander, *Handwerkschirurgen-Sozialgeschichte einer verdrängten Berufsgruppe* (Göttingen: Vandenhoeck & Ruprecht, 1989); Sebastian Brändli, *"Die Retter der leidenden Menschheit." Sozialgeschichte der Chirurgen und Ärzte auf der Zürcher Landschaft (1700–1800),* (Zurich: Chronos, 1990).

72. Thomas Faltin, *Heil und Heilung. Geschichte der Laienheilkundigen und Struktur antimodernistischer Weltanschauungen in Kaiserreich und Weimarer Republik am Beispiel von Eugen Wenz (1856–1945)* (Stuttgart: Franz Steiner 2000), 242ff., 270 ff.

73. Jacques Léonard, "Femmes, religions, médecine. Les religieuses, qui soignent," *Annales E.S.C.* 32 (1977): 887–903; "Les guérisseurs," *Revue d'historie moderne et contemporaine* (hereafter, *RHMC*) 27 (1980): 501–516.

74. François Lebrun, *Se soigner autrefois. Médecins, saints et sorciers aux 17e et 18e siècles* (Paris: Temps actuels, 1983).

75. Olivier Faure, "Introduction: Les voies multiples de la médicalisation," *RHMC* 43 (1996): 571–577, 574ff., and *Français,* chap. 1. On nurses, see Joelle Droux, *L'attraction celeste. La construction de la profession d'infirmière en Suisse romande, XIXe–XXe siècles* (Geneva: Fac. de lettres, 2000); *Peu lire, beaucoup voir, beaucoup faire: Pour une histoire des soins infirmiers au 19e siècle,* François Walter, ed. (Geneva: Zoe, 1992); Jutta Helmerichs, "Krankenpflege im Wandel (1890–1933): Sozialwissenschaftliche Untersuchung zur Umgestaltung der Krankenpflege von einer christlichen Liebestätigkeit zum Beruf" (Ph.D. diss., Göttingen, 1992); Gabriele Dorffner, " . . . *ein edler und hoher Beruf." Zur Professionalisierung der österreichischen Krankenpflege* (Strasshof: Vier-Viertel, 2000); and Urs F. A. Heim, *Leben für andere. Die Krankenpflege der Diakonissen und Ordensschwestern in der Schweiz* (Basel: Schwabe 1998).

76. Isabelle von Bueltzingsloewen, "Confessionalisation et médicalisation des soins aux malades aux XIX s.," *RHMC* 43 (1996): 632–652, and "Femmes soignantes (XVIIIe–XXe siècle)," *Bulletin du Centre Pierre Leon* 2/3 (1995).

77. *La charité en pratique: Chrétiens français et allemands sur le terrain social: XIXe–XXe siècles,* Isabelle von Bueltzingsloewen and Denis Pelletier, eds. (Strasbourg: Presses universitaires, 1999).

78. Ronald Hubscher, *Les maîtres des bêtes: Les véterinaires dans la société française* (Paris: Jacob, 1999).

79. Francisca Loetz, " 'Medikalisierung' in Frankreich, Großbritannien und Deutschland, 1750–1850: Ansätze, Ergebnisse und Perspektiven der Forschung," in Eckart and Jütte, *Europäische Gesundheitssystem,* 123–161.

80. Göckenjan, *Kurieren;* Barthel, *Medizinische Polizey.*

81. Spree, *Health,* 178ff.; Frevert, *Krankheit,* 334.

82. See, for example, Huerkamp, *Aufstieg,* 41, 303.

83. Bettina Wischhöfer, *Krankheit, Gesundheit und Gesellschaft in der Aufklärung. Das Beispiel Lippe 1750–1830* (Frankfurt: Campus, 1991). On legislation and demography, see also Johannes Wimmer, *Gesundheit, Krankheit und Tod im Zeitalter der Aufklärung. Fallstudien*

aus den Habsburgischen Erbländern (Vienna: Böhlau, 1991), and Martin Dinges, "'Medicinische Policey' zwischen Heilkundigen und 'Patienten' (1750–1830)," in *Policey und frühneuzeitliche Gesellschaft,* Karl Härter, ed. (Frankfurt: Klostermann, 2000), 263–295.

84. Robert Jütte, *Ärzte, Heiler und Patienten. Medizinischer Alltag in der Frühen Neuzeit* (Munich: Artemis & Winkler, 1991); Annemarie Kinzelbach, *Gesundbleiben, Krankwerden, Armsein in der frühneuzeitlichen Gesellschaft. Gesunde und Kranke in den Reichsstädten Überlingen und Ulm 1500–1700* (Stuttgart: Steiner, 1995), 295ff.

85. Mary Lindemann, *Health and Healing in Eighteenth Century Germany* (Baltimore: Johns Hopkins University Press, 1996), 321; Wischhöfer, *Krankheit,* 90ff.

86. Francisca Loetz, *Vom Kranken zum Patienten. "Medikalisierung" und medizinische Vergesellschaftung am Beispiel Badens 1750–1850* (Stuttgart: Steiner, 1993).

87. Olivier Faure, "Demande sociale de la santé et volonté de guérir en France au XIXe siècle. Reflexions, problèmes, suggestions," *Cahiers du Centre de recherches historiques* 12 (1994): 5–11, based on a paper given at a Bielefeld conference in 1987; Faure, *Français.* Further demand induced by the offer stresses Michael Stolberg, "Heilkundige: Professionalisierung und Medikalisierung," in Paul and Schlich, *Medizingeschichte,* 69–86, 80.

88. *De la charité médiévale à la securité sociale,* André Gueslin and Pierre Guillaume, eds. (Paris: Editions ouvrières, 1992); *Démocratie, solidarité et mutualité, autour de la loi de 1898,* Michel Dreyfus, Bernard Gibaud, André Gueslin, eds. (Paris: Economica, 1999); Olivier Faure and Dominique Dessertine, *La maladie entre libéralisme et solidarités (1850–1940)* (Paris: Mutualité française, 1994).

89. Folker Förtsch, *Gesundheit, Krankheit, Selbstverwaltung. Geschichte der allgemeinen Ortskrankenkassen im Landkreis Schwäbisch Hall, 1884–1973* (Sigmaringen: Thorbecke, 1995), 110–121.

90. Josef Boyer, *Unfallversicherung und Unternehmer im Bergbau. Die Knappschafts-Berufsgenossenschaft 1885–1945* (Munich: C. H. Beck, 1995).

91. Catherine Duprat, *Usages et pratiques de la philanthropie. Pauvreté, action sociale et lien social à Paris au premier XIXe siècle,* 2 vols. (Paris: Comité d'historie de la sécurité sociale, 1996); Jean-Luc Marais, *Histoire du don en France de 1800 à 1939. Dons et legs charitables, pieux et philanthropiques* (Rennes: Presses universitaires, 1999).

92. Klaus-Dieter Thomann, *Das behinderte Kind. "Krüppelfürsorge" und Orthopädie in Deutschland 1886–1920* (Stuttgart: Gustav Fischer, 1995); Nicolas Postel-Vinay, "L'hypertension artérielle: Un chantier de travail pour l'historien?," *Cahiers d'histoire* 38 (1992): 231–246. Cf. Friedrich Dreves, ". . . Leider zum größten Theile Bettler geworden . . ." *Organisierte Blindenfürsorge in Preußen zwischen Aufklärung und Industrialisierung (1806–1860)* (Freiburg: Rombach, 1998).

93. Lindemann, *Health.*

94. Hudemann-Simon, *L'État,* 509–510.

95. Jacques Gélis, Mireille Laget, and Marie-France Morel, *Entrer dans la vie: Naissance et enfance dans la France traditionelle* (Paris: Gallimard, 1978); Marita Metz-Becker, *Der verwaltete Körper: Die Medikalisierung schwangerer Frauen in den Gebärhäusern des frühen 19. Jahrhunderts* (Frankfurt: Campus, 1997).

96. See Katrin Schmersahl, *Medizin und Geschlecht. Zur Konstruktion der Kategorie Geschlecht im medizinischen Diskurs des 19. Jahrhunderts* (Opladen: Leske & Budrich, 1998).

97. Jacques Gélis, *La sage-femme ou le médecin* (Paris: Fayard, 1988), 12.

98. Gunda Barth-Scalmani, "'Freundschaftlicher Zuruf eines Arztes an das Salzburgische

Landvolk': Staatliche Hebammenausbildung und medizinische Volksaufklärung am Ende des 18. Jahrhunderts," in *Rituale der Geburt. Eine Kulturgeschichte,* Jürgen Schlumbohm et al., eds., (Munich: C.H. Beck, 1998), 102–118, 113ff.

99. Eva Labouvie, *Andere Umstände. Eine Kulturgeschichte der Geburt* (Cologne: Böhlau, 1998).

100. See *Naissance, enfance et éducation dans la France méridionale du XVIe au XXe siècle,* Roland Andréani, Henri Michel, and Elie Pélaquier, eds. (Montpellier: Publications UPV, 2000).

101. Eva Labouvie, *Beistand in Kindsnöten. Hebammen und weibliche Kultur auf dem Land (1550–1910)* (Frankfurt: Campus 1999); *'Geschichte des Ungeborenen.' Zur Erfahrungs- und Wissensgeschichte der Schwangerschaft, 17.–20. Jahrhundert,* Barbara Duden, Jürgen Schlumbohm, et al., eds. (Göttingen: Vandenhoeck & Ruprecht, 2002).

102. Metz-Becker, *Verwaltete Körper.*

103. Jürgen Schlumbohm, "Der Blick des Arztes, oder: Wie Gebärende zu Patientinnen wurden. Das Entbindungshospital der Universität Göttingen um 1800," in Schlumbohm et al., *Rituale der Geburt,* 170–192, 184.

104. Verena Pawlowsky, *Mutter ledig—Vater Staat. Das Gebär- und Findelhaus in Wien, 1784–1910* (Innsbruck: Studienverlag, 2001); Schlumbohm, "Blick," 183, and " 'Verheiratete und Unverheiratete, Inländerin und Ausländerin, Christin und Jüdin, Weiße und Negerin': Die Patientinnen des Entbindungshospitals der Universität Göttingen um 1800," in *Struktur und Dimension (Fs. K.H. Kaufhold),* Hans Jürgen Gerhard, ed. (Stuttgart: Steiner, 1997), 324–343.

105. Hans-Christoph Seidel, *Eine neue "Kultur des Gebärens." Die Medikalisierung von Geburt im 18. und 19. Jahrhundert in Deutschland* (Stuttgart: Franz Steiner Verlag, 1998), 417ff.; cf. Scarlett Beauvalet-Boutouyrie, *Naître à l'hôpital au XIXe siècle* (Paris: Belin, 1999), 356.

106. On the earlier control of midwives, see Sibylla Flügge, *Hebammen und heilkundige Frauen. Recht und Rechtswirklichkeit im 15. und 16. Jahrhundert* (Frankfurt: Stroemfeld, 1998).

107. Seidel, *Kultur des Gebärens,* 425.

108. Cornelia Regin, *Selbsthilfe und Gesundheitspolitik. Die Naturheilbewegung im Kaiserreich (1889–1914)* (Stuttgart: Steiner, 1995).

109. *Medizinkritische Bewegungen im Deutschen Reich (ca. 1870–ca. 1933),* Martin Dinges, ed. (Stuttgart: Steiner, 1996); *Homöopathie. Patienten, Heilkundige und Institutionen. Von den Anfängen bis heute,* Martin Dinges, ed. (Heidelberg: Haug, 1996).

110. Eberhard Wolff, *Gesundheitsverein und Medikalisierungsprozess. Der homöopathische Verein Heidenheim/Brenz, 1886–1945* (Tübingen: Tübinger Verein für Volkskunde e.V., 1989); *Praticiens, patients et militants de l'homéopathie (1800–1940),* Olivier Faure, ed. (Lyon: Presses universitaires, 1992).

111. Martin Dinges, "Patients"; *Weltgeschichte der Homöopathie, Länder-Schulen—Heilkundige* (Munich: C. H. Beck, 1996); and "The Contribution of the Comparative Approach to the History of Homoeopathy," in *Historical Aspects of Unconventional Medicine: Approaches, Concepts, Case Studies,* Robert Jütte, Motzi Eklöf, and Mary Nelson, eds. (Sheffield: EAHMH Publications, 2001), 51–72.

112. Göckenjan, *Kurieren.* For the first part in medical knowledge, see Jacques Léonard, *La médecine entre les pouvoirs et les savoirs. Histoire intellectuelle et politique de la médecine*

française au XIXe siècle (Paris: Aubier, 1981), and *Archives du corps: La santé au XIXe siècle* (Rennes: Ouest France, 1986).

113. Alfons Labisch, *Homo hygienicus. Gesundheit und Medizin in der Neuzeit* (Frankfurt: Campus, 1992); cf. Georges Vigarello, *Le sain et le malsain. Santé et mieux-être depuis le Moyen-Âge* (Paris: Seuil, 1993).

114. Geneviève Heller, *"Propre en ordre." Habitation et vie domestique 1850–1930: l'exemple vaudois* (Lausanne: Editions d'en bas, 1979).

115. Manuel Frey, *Der reinliche Bürger. Entstehung und Verbreitung bürgerlicher Tugenden in Deutschland, 1760–1860* (Göttingen: Vandenhoeck & Ruprecht, 1997). The groundbreaking work on the topic is Vigarello, *Propre;* cf. Julia Csergo, *Liberté, égalité, propreté. La morale de l'hygiène au XIXe siècle* (Paris: Albin Michel 1988).

116. See Catherine Pellissier, "La médicalisation des élites lyonnaises au XIXe siècle," *RHMC* 43 (1996): 652–671.

117. See Maren Lorenz, *Kriminelle Körper—Gestörte Gemüter. Die Normierung des Individuums in Gerichtsmedizin und Psychiatrie der Aufklärung* (Hamburg: Hamburger Edition, 1999).

118. Philipp Sarasin, *Reizbare Maschinen. Eine Geschichte des Körpers, 1765–1914* (Frankfurt: Suhrkamp, 2001).

119. Barbara Duden, *Geschichte unter der Haut: Ein Eisenacher Arzt und seine Patientinnen um 1730* (Stuttgart: Klett Cotta, 1987).

120. On literary sources, see Udo Benzenhöfer and Wilhelm Kühlmann, *Heilkunde und Krankheitserfahrung in der frühen Neuzeit* (Tübingen: Niemeyer, 1992).

121. See the articles in *Medizin, Gesellschaft und Geschichte* 15 (1996); larger than the title suggests is *La douleur et le droit,* Bernard Durand, Jean Poirier, and Jean-Pierre Royer, eds. (Paris: Presses universitaires de France, 1997).

122. Michael Stolberg, "'Mein äskulapisches Orakel!' Patientenbriefe als Quelle einer Kulturgeschichte der Körper- und Krankheitserfahrung im 18. Jahrhundert," *Österreichische Zeitschrift für Geschichtswissenschaft* 7 (1996): 385–404. Philip Rieder and Vincent Barras, "Écrire sa maladie au siècle des Lumières," in *La médecine des Lumières: Tout autour de Tissot,* Vincent Barras and Micheline Louis-Courvoisier, eds. (Chêne-Bourg: Georg, 2001), 201–222; Martin Dinges, "Men's Bodies 'Explained' on a Daily Basis in Letters from Patients to Samuel Hahnemann (1830–1835)," in Dinges, *Patients,* 85–118.

123. Karin Stukenbrock, *"Der zerstückte Cörper." Zur Sozialgeschichte der anatomischen Sektion in der frühen Neuzeit (1650–1800)* (Stuttgart: Steiner, 2001).

124. Jens Lachmund, *Der abgehorchte Körper. Zur historische Soziologie der medizinischen Untersuchung* (Opladen: Westdeutscher Verlag, 1997); cf. Hess, *Wohltemperierte Mensch.*

125. Olivier Faure, *Les thérapeutiques: Savoirs et usages* (Oullins: Collection Fondation Marcel Mérieux, 1999).

126. First steps were taken in *Médecine et vulgarisation XVIIIe–XXe siècles,* Claude Langlois and Jacques Poirier, eds. (Créteil: Université de Paris XII, 1991); and *La médecine du peuple de Tissot à Raspail (1750–1850),* Daniel Teysseire, ed. (Créteil: Conseil général du Val-de-Marne, Archives départementales, 1995).

127. Compare the documentary approach of Christoph Mörgeli, *Europas Medizin im Biedermeier anhand der Reiseberichte des Züricher Arztes Conrad Meyer-Hofmeister 1827–1831* (Basel: Schwabe, 1997).

128. *La philosophie du remède,* Jean-Claude Beaune and Jacques Azéma, eds. (Seyssell: Champ Vallon, 1993); cf. Maehle, *Drugs on Trial.*

129. A well-conceptualized intercultural approach would lead to further innovations, for example, *Maladies, médecins et sociétés—approches historiques pour le présent,* François-Olivier Touati, ed. (Paris: L'Harmattan 1993).

130. See *Cultures of Neurasthenia: From Beard to the First World War,* Marijke Gijswijt-Hofstra and Roy Porter, eds. (Amsterdam: Rodopi, 2001).

CHAPTER ELEVEN

Trading Zones or Citadels?
Professionalization and Intellectual Change in the History of Medicine

Olga Amsterdamska and Anja Hiddinga

Ronald Numbers, a historian of American science and medicine, published an article in 1982 describing changes in the history of medicine during the previous two decades.[1] Numbers argued that in the 1960s, medical history in the United States already had begun its transformation from a subfield of medicine to a subfield of history. Linking the emergence of the social history of medicine to the replacement of physician-historians of medicine by general historians who took medicine as their subject, Numbers associated professional change with intellectual development. By the 1970s, "local, institutional, and biographical studies" were being replaced by works that situated "medical developments within their cultural context" and explored "connections between ideas and institutions."[2] The intellectual affinity between the history of medicine and other areas of cultural, intellectual, and social history meant that historians of medicine paid increased attention to the professional and economic interests of medical practitioners (not just the elite or the orthodox doctors); examined the broad social and cultural influences on the development of medical institutions such as hospitals and asylums, regarding their evolution as a process of change rather than necessarily progress; and began addressing issues such as gender and race. While applauding the professionalization of his discipline, Numbers also voiced a concern, shared by others he cited, that attention to various "socio-economic determinants" might obscure the role of medical science and intellectual change in the evolution of medicine.

Three years later, reviewing the relations between the history of medicine and the history of science, John Harley Warner also noted that the rise of the new social history of medicine diverted scholars' attention "from analysis of the content and internal logic of medicine," so that "close study of the development of medical science" came to be regarded as "antiquarian and more than likely positivistic."[3] According to Warner, however, this shift toward social history did not mean that the role of science in medicine was being ignored. Instead, the "historical study of the cognitive development of medical science" was being reconfigured in investigations of the cultural, social, and economic roles of science in medicine. Rather than following the development of scientific ideas, the new historians saw science in medicine as a source of authority and legitimacy or as a means of professional advancement.

In his article, Warner analyzed how historians of medicine came to question the progressive role of science in medicine by critically examining the social implications of technological and scientific change in medicine, criticizing the reductionist epistemology of the biomedical sciences, and relativizing the impact of developments in the biomedical sciences on improvements in the health of populations. Studies of the role of science as a means of social control and as a source of professional authority meant, according to Warner, that science in medicine came to be seen more as an ideological or—less polemically—more as "a cultural than as a technical resource."[4] Warner, in a sense, shared Numbers's concern that the new historiography of medicine tends to diminish the historian's attention to the "content of science." Warner, however, gave this problem a different twist. He argued that the problem is not so much due to the predisposition of the new historians, who are trained in history rather than medicine, to look at social rather than intellectual processes; instead, he attributed the relative inattention to the content of medical sciences to "historiographic reductionism." Such reductionism prevents historians from properly historicizing the very meaning of science and makes them hold onto the presentist identification of "medical science" with laboratory biomedical investigations. Warner called for more attention to the multiple meanings of science, which can be found among its practitioners at any given time and in the belief systems of the broader public, and to the historical changes in the notion of medical science espoused in different periods and settings.

Warner repeated virtually the same warning in a review paper written ten years later.⁵ Encouraged that history of medicine was more likely to be considered in the context of social, economic, or cultural history rather than as history of science, Warner worried that historians of medicine were still considering science in too narrow and restrictive a manner, reserving the term only for experimental laboratory sciences and not paying sufficient attention to the clinic as a place where scientific knowledge is not only "applied" but also established and developed. At the same time, Warner approved of other changes that had affected the history of medicine: a thorough rejection of Whiggism, the movement toward a moderate and "reasonable" version of constructivist epistemology regarding disease as both a social and a biological phenomenon, and an increasing attention to the medical views and experiences of groups other than the narrowly defined (elite) medical professions: women, different social classes and ethnic groups, nonorthodox physicians, and nurses, as well as nonelite doctors.⁶

Since the 1970s, history of medicine, history of science, and the social studies of science have undergone parallel changes by rejecting narratives that described the relentless and linear advance of science and embracing instead a thoroughgoing historicism that treats the developments of science and medicine in specific cultural and social contexts. Historians have come to regard scientific and technical changes less in terms of their contribution to the current state of science or medicine and more in terms of their significance for those actively involved in their introduction or implementation. Accordingly, both historians of science and historians of medicine tend to focus not as much on the structure and reconstruction of abstract knowledge or its implications for general epistemological problems as on the local activities of diverse groups of practitioners involved in the various aspects of their daily work. Scholars in the area of the social studies of science have actively debated a similar epistemic shift. Numerous case studies have emphasized the locally constructed character of knowledge, its social and cultural embeddedness, and the heterogeneity of the groups participating in its creation.⁷ Moreover, medicine and the biomedical sciences have become an important focus of research in the social studies of science. The link between history of medicine and the social studies of science could thus also be a substantive, rather than just an epistemological one. The growing interest of science and technology studies (STS) in medicine and the medical sciences is reflected in the number of articles dealing with medical issues in the social studies of science journals and in the number of contributions on medical topics to conferences of the professional STS societies (for example, papers on medical topics made up 11% of the presentations at conferences of the Society for the Social Studies of Science in 1988 and increased to 29% in 2001).⁸

Given these parallel developments in history of medicine, history of science, and the social studies of science, one would expect lively interchanges to be taking place between the three fields. History of medicine's focus on multiple meanings of science in medicine, on the perspectives of participants other than physicians, on the professional interests of doctors, and on local medical practices suggest a set of shared concerns with analyses of medicine stemming from the constructivist tradition in the history and social studies of science. Such a relationship has been suggested by Warner and actively advocated by Ludmilla Jordanova, who emphasized the need for greater theoretical awareness in the history of medicine.[9] According to Warner, the history of science and the history of medicine "have become integrated with the broader agendas of the humanities and social sciences," so that today "they are bound together by their shared exploration of how natural knowledge is produced, organized, and deployed in concrete historical settings."[10] Similarly, Jordanova has claimed that "if social historians of medicine attempt more than anecdotal or descriptive history, they frequently adopt social constructionism in one form or another."[11]

Numbers, Warner, and Jordanova base their conclusions about changes in the history of medicine on extensive familiarity with the field and the analysis of contributions they regard as the most indicative of the new directions. To a large extent their essays are programmatic in nature and structured accordingly. They focus on the social and intellectual features of the field that they consider to be the most problematic or the most promising and forward-looking. Accordingly, in their diagnosis of the history of medicine, they tend to emphasize the changed professional orientation of its practitioners and the adoption of new theoretical approaches that parallel broader developments in the history of science and science studies. They assume that the intellectual, social, and institutional conditions within history of medicine—its integration into general social and cultural history, for example—are propitious for promoting such developments.

What we attempt to do in the following pages is to examine some aspects of the development of American and British history of medicine using a scientometric approach.[12] Rather than focusing on theoretical innovations presented in outstanding texts of selected authors, we try to chart these changes more globally by looking at articles published in the four major history of medicine journals, examining whom they cite and how they are cited in turn. Following Numbers, Warner, and Jordanova, we ask how the professionalization of the history of medicine has been reflected in changes in its literature. How has it affected its relations to history of science and the social studies of medicine? How widespread and profound are the changes? To what extent and in what respects do professional historians today dominate the history of medicine? Has history of medicine really moved closer to general social and cultural history and away from practical medicine? To what extent can the intellectual changes identified by Numbers, Warner, and Jordanova be traced in publication and citation patterns?

We have selected two American and two British journals, which we expect to reflect the changes in the field that have taken place since the 1960s. Two of these journals, the *Bulletin of the History of Medicine* (*BHM*) and *Social History of Medicine* (*SHM*), are sponsored by professional organizations of historians of medicine, while the other two, *Journal of the History of Medicine and Allied Sciences* (*JHM*) and *Medical History* (*MH*), represent the mainstream of the field in both countries. In the history of medicine, just as in other fields in the humanities, books probably have a greater impact on the development of the field than journal articles do. However, looking at journal articles allowed us to survey these developments more widely, comparatively, and comprehensively. Moreover, articles can reflect intellectual changes in the field more rapidly because of their shorter production times.

In using these journals to examine the links between history of medicine and other fields, we are ignoring those contributions of historians of medicine published in journals in other disciplines, for example, general history, history of science, medical sociology, or health policy. While historians of medicine occasionally contribute to such journals and the work published there might give rise to interchange with scholars in these other fields, such contributions are too dispersed to be studied systematically.

Professionalization and Intellectual Change

According to the received view, history of medicine since the 1960s has become the province of professional historians rather than an avocation of retired physicians. This professionalization has been linked repeatedly to intellectual change. Numbers, for example, argues that the shift from heroic narratives of how great doctors conquered disease to the social history of medicine is a result of the domination of the field by professional historians. In studying this trend, Numbers looked at the authors of leading contributions to American medical history and their educational background. He concluded that by 1960–1980 historians significantly outnumbered physicians, so that "the time has passed when a physician untrained in history or the social sciences is likely to make a meaningful contribution to medical history."[13]

Unlike Numbers, we examined the professional backgrounds of *all* the authors of articles published in the *BHM, JHM,* and *MH,* comparing the proportions of M.D.s and Ph.D.s in two-year periods for each decade between 1960 and 2000. *SHM* was excluded because it only began publication in 1988, so a long-term comparison was not possible. Although the educational background of an author might not correspond fully to his or her professional identity, the proportions of M.D.s and Ph.D.s do give a rough indication of the extent to which participation in the field has come to be regarded as an activity distinct from medicine. An extra note of caution should be added here also because we were unable to separate those authors whose Ph.D.s are not in history but in some other field in the humanities or the social or natural sciences.

As expected, the proportion of authors whose highest degree is an M.D. has declined significantly in all three journals (see fig. 11.1). Whereas in 1960–1961 M.D.s accounted for 52% of the authors and Ph.D.s for 29%, in 2000–2001 the proportion of M.D.s declined to 13%, while Ph.D.s increased to 77%. As the figure shows, however, the shift from the predominance of M.D.s to the predominance of Ph.D.s occurred gradually and relatively slowly. As late as 1990, M.D.s still accounted for a quarter of the authors making presumably meaningful contributions to the literature in the field. The pattern appears roughly similar in all three journals, except that in the 1960s the dominance of doctors in the British journal was much more pronounced. While Numbers was correct in his observation that professionalization in the American history of medicine was well under way in the early 1960s, he seems to have overestimated the speed with which it was taking place.

In addition to examining the professional backgrounds of the contributors to the three journals, we were also interested in the relationship between professionalization and intellectual change. Do the professional historians address topics different from those addressed by the contributors to the history of medicine who were trained primarily in medicine? Has the professional transition resulted in changes in the focus of attention of articles published in the journals? Does this transition correspond to a shift in the way in which the subjects are treated?

A thorough investigation of the contents of all articles published in three history of medicine journals over a period of 40 years could not, of course, be carried out given the constraints of time and resources. Instead, we have chosen to focus our attention on the titles of articles, assuming that these give an indication of the kinds of subjects and issues that were addressed. Although titles of articles may not fully reflect their contents, we assume that titles do inform us about the manner in which authors want to present their work and engage their readers' attention. While titles do not tell us much about the intellectual frame-

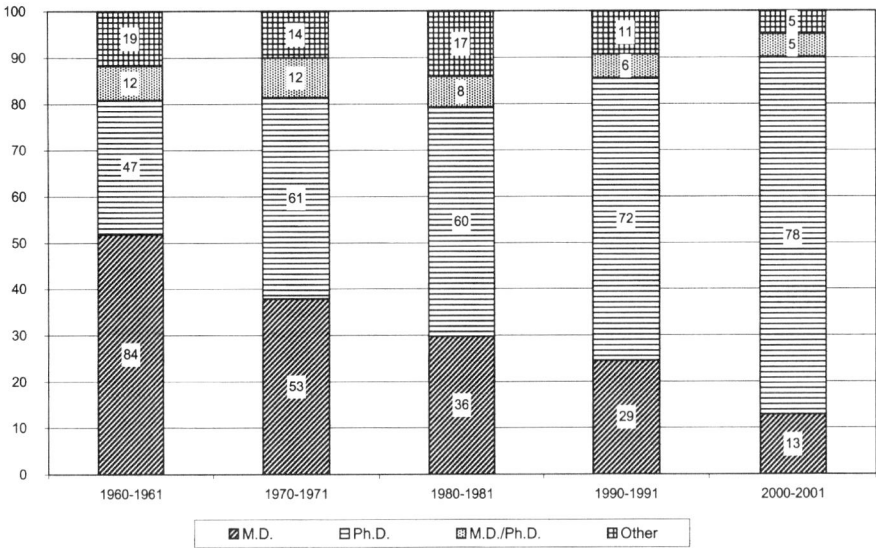

Fig. 11.1. Educational background of contributors to three history of medicine journals

work adopted by their authors, we assume that they can be treated as rough indicators of how the subject is being approached.

We have attempted to classify the titles of all the articles in terms of two orthogonal dimensions: we asked what subject was chosen as the article's main focus and whether or not the title also placed this subject in a social or intellectual context. We distinguished seven categories of subjects, classifying the articles as biographies, institutional histories, studies of specific diseases, studies of medical professions and practices, studies of patients, analyses of health policy or public health, and histories of scientific developments. So, for example, articles with titles such as "Empiricism in Nineteenth-Century French Surgery"[14] or "The Status of Physicians in Renaissance Venice"[15] or "The American Reception of Salvarsan"[16] would all be classified as dealing with professions and practices. Articles whose titles list the names of specific persons accompanied only by their identification as doctors or scientists were counted as biographies. When a title referred to the history of a single specific institution, such as "Reform at Harvard Medical School,"[17] or to the development to a particular class of institutions such as hospitals or medical schools, such as "Hospice to Hospital in the Near East: An Instance of Continuity and Change in Late Antiquity,"[18] we classified them as institutional histories. Articles on cultural or scientific conceptualizations of disease, "Measles in Fact and Fancy"[19] or " 'A Disease *Sui Generis*': The Origins of Sickle Cell Anaemia and the Emergence of Modern Clinical Research,"[20] as well as those dealing with diseases in specific settings, "Cholera in Newark, New Jersey,"[21] were categorized as studies of diseases. The category of studies on public health or health policy contains articles entitled "Civil Liberties and Public Good: Detention of Tuberculous Patients and the Public Health Act 1984,"[22] "Issues in the Anti-vaccination Movement in England,"[23] and "Medical Inspection of Prostitutes in America in the Nineteenth Century: The St. Louis Experiment and Its Sequel."[24]

Although most titles specified a clear subject, it was not always easy to decide on a single category for articles where titles contained a reference to more than one subject, or where the subject could be assigned to more than one category. A history of a public health institution, for example, could be classified as belonging either to the category of studies on public health or to that of institutional histories. We tried to resolve such ambiguities by looking at the rest of the title and comparing our respective views on the proper classification, but the procedure could occasionally result in a somewhat different assignment. Accordingly, the numbers can be seen as indicative only. The number of papers where the classification could be problematic was small. In addition, there were 42 articles that we had to leave out either because the titles were uninformative or because they dealt with subjects that were too idiosyncratic to fit into any category. Altogether we have classified 544 articles from the three journals that have been published continuously from the 1960s. These articles had 566 authors whose educational backgrounds we were able to determine. As before, we looked at two-year periods in each decade.

The categories were chosen so that we could trace whether and to what extent the changes in the history of medicine suggested by Warner, Numbers, Jordanova, and others can be confirmed by changes in the journal literature. For example, we were curious to see if we could track the transition from a focus on individuals to an interest in the medical professions in terms of a decline in the number of biographical articles and a rise in articles dealing with professional practice. Similarly, given the presumed changes in the field, one would expect to find a decline in attention to medical science and a rise in interest in patients and sufferers and their experiences. Because we expected a growth in interest in the social history of medicine, we anticipated an increase in the number of articles dealing with public health and health policy alongside an increase in the number of articles focused on the professions. Other categories, such as those grouping studies of specific diseases or institutional histories, reflect the traditional concerns of historians of medicine, and we did not expect to draw specific conclusions about the trends we would find in these titles.

The subject classification could not tell us much about the approaches that were adopted. One indication of an approach that writers take is given by whether they invoke a particular social, cultural, institutional, professional, economic, scientific, or philosophical context in presenting their findings. To some extent such a contextual treatment would be reflected in the titles. Accordingly, we tried to determine whether the subject addressed by a given author was also being placed in a specific context. Starting out with these categories, we soon found that many articles mentioned more than one such context and that it was sometimes impossible to distinguish between, for example, the social and the cultural contexts. Accordingly, we had to adopt a less fine-grained classification, distinguishing only among the titles that mentioned no context, those that appealed to social, cultural, professional, economic or political contexts, and those that placed their subject in a scientific or philosophical framework. For example, "The Medical Construction of Homosexuality and Its Relation to the Law in Nineteenth-Century England"[25] clearly places its discussion of homosexuality in the British legal context, whereas "The Odyssey of Smallpox Vaccination"[26] does not indicate any context. Given the rise of interest in the social and cultural history of medicine, we expected a large increase in the number of articles whose titles refer to a broadly defined "social context."

Because we were interested in the relationship between intellectual change and professionalization in the history of medicine, we cross-tabulated our classification of titles with the professional backgrounds of their authors. We expected contributions by physicians to be concentrated in the more traditional categories: biographical articles or institutional histories with no context mentioned and articles focusing on the development of the medical sciences, possibly in the scientific context. In contrast, historians of medicine with Ph.D.s, and thus usually with a background in general history, should be more likely to publish articles on professional practices, patients, or public health, and to place their subjects in a social context.

In the 1960s and 1970s, biographies of doctors or scientists were the most common category of articles published in the three journals (39% and 36%, respectively). By 1980–1981, however, this category was surpassed by articles with a focus on professional practices, which have remained the most popular to this day. In the 1990s, articles dealing with diseases also became more common than biographies, and by 2000–2001 the number of biographies had declined to 12%. This decline in interest in the lives of individuals corresponds to what observers of the history of medicine have repeatedly noticed. The steep decline in biographies was not, however, paralleled by an equally sharp increase in the proportion of articles dealing with professional practices or with public health. The proportion

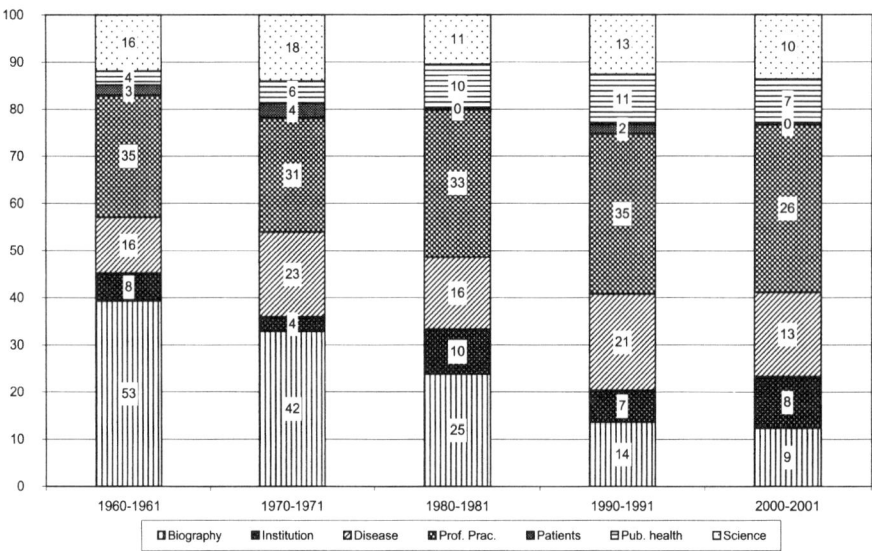

Fig. 11.2. Articles classified according to the subject specified in their title

of articles in these categories grew only from 26% in 1960–1961 to 36% in 2000–2001 in the case of professional practices and from 3% to 10% in the case of public health and health policy (given the small numbers of articles in the public health category, the tripling of percentages is misleading). There were small, but not striking, increases in the proportion of articles dealing with institutional histories or histories of disease (though, of course, how these articles address their subjects may have changed dramatically without being reflected in the titles). Contrary to our expectations, we found almost no articles focusing on patients and their experience, though of course it is possible that such subjects are addressed in articles whose titles refer to other subjects. Scientific developments remained a moderate but a relatively constant subject of interest. Contrary to Numbers's expectations, science does not appear to become increasingly marginalized as a focus in the history of medicine. Professional practices have indeed become the most important focus of attention in the history of medicine today, but other subjects—histories of science and disease, biographies, institutional histories, and histories of public health—remain significant foci of the historians' concern. The change we observe is more like a process of diversification than a radical shift of attention (see fig. 11.2).

As expected, there has been a significant decrease in the number of articles that do not mention any context in the title. In 1960–1961, 76% of all articles did not refer to either a social or a scientific context, but by 2000–2001 this had declined to 36%. At the same time, articles that referred to a social, professional, political, cultural, or economic context in their titles increased from 10% to 59% (see fig. 11.3). Placing the subject of an article in a scientific or philosophical context was never common; there seems to be a slight decrease in the titles that do so (from 13% in 1960–1961 to 5% in 2000–2001).

In order to examine more closely the relationship between intellectual shifts in the disciplines and changes in the professional backgrounds of contributors to the field, we have split up the classification of titles according to whether their authors had an M.D. or a Ph.D. degree. The 544 articles in the three journals we were able to classify had in total 566 authors whose professional backgrounds could be established. For the purposes of this comparison we have counted authors rather than articles, so an article co-authored by an M.D. and a Ph.D. would be counted twice. To simplify the number of categories, we have classified the small number of authors with double M.D.-Ph.D. degrees (less than 7% of the total) as Ph.D.s. The even smaller (and decreasing) category of "others" includes authors who reported no advanced degrees (beyond a B.A. or M.A.) in their biographical statements.

Somewhat surprisingly, there appear to be no dramatic differences in the main

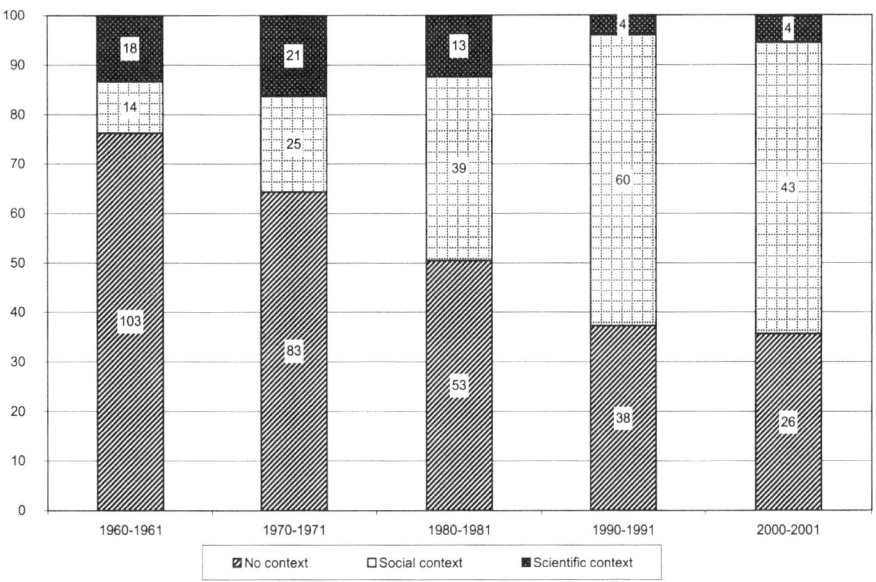

Fig. 11.3. Articles with titles placing the topic in a particular context

Table 11.1 Professional Backgrounds of Authors per Subject, Relative to the Total Number of Authors with the Same Type of Degree

	Biography			Institution			Disease			Professional Practice			Patients			Public Health			Science			Total		
	M.D.	Ph.D.	O	M.D.	Ph.D.	O	M.D.	Ph.D.	O	M.D.	Ph.D.	O	M.D.	Ph.D.	O	M.D.	Ph.D.	O	M.D.	Ph.D.	O	M.D.	Ph.D.	O
1960–61	33 43%	20 38%	2	3 4%	3 6%	2	12 16%	4 8%	1	17 22%	16 31%	3	2	0	1	1	2 4%	1	8 11%	7 13%	1	76	52	11
1970–71	17 35%	18 28%	8	2 4%	1 1%	1	16 33%	8 12%	0	7 14%	19 29%	5	2	0	0	0	6 9%	0	5 10%	13 20%	1	49	65	15
1980–81	8 23%	13 21%	8	4 11%	5 8%	1	6 17%	11 18%	3	10 29%	19 31%	4	0	0	0	1	8 12%	1	6 17%	5 8%	0	35	61	17
1990–91	4 15%	9 13%	1	2 7%	3 4%	3	8 30%	14 20%	1	9 33%	23 33%	4	0	0	0	2	8 13%	1	2 7%	11 16%	0	27	70	10
2000–01	2 14%	7 11%	0	3 21%	6 10%	0	3 21%	12 19%	1	3 21%	23 37%	0	0	0	0	2	5 8%	0	1 7%	9 14%	0	14	63	1
Total																						201	311	54

foci of interest as indicated by the titles of articles written by M.D.s and those authored by Ph.D.s. Over the years, both the physician-historians and those with Ph.D.s in history have become less likely to write biographical articles and more likely to devote their attention to subjects related to the medical professions and their practices, though as early as 1960, those with Ph.D. degrees were more likely than the M.D.s to write about the latter topic. The differences between the two groups, however, virtually disappeared by the 1980s, when 29% of all articles written by M.D.s and 31% of all articles written by the Ph.D.s fell in the category of professional practices. Articles dealing with the histories of institutions and histories of disease have been slightly more common among the M.D.s than the Ph.D.s, while histories of public health and health policy have been a more popular topic for those with a Ph.D. The overall number of articles in these categories, however, is too small to warrant firm conclusions. Contrary to the expectations of Numbers, who suggested that the professionalization of history of medicine would result in a lessening of attention to developments in medical science, the M.D.s were no more likely than the Ph.D.s to address this subject (see Table 11.1).

Somewhat more informative conclusions can be drawn when we compare the tendencies of the physicians and the Ph.D.s to indicate in their titles the contexts in which they are situating the subjects of their analyses. In the 1960s neither group of authors was likely to provide such indications, though Ph.D.s were more likely to do so than M.D.s (17% of the Ph.D.s and 5% of the M.D.s did so in 1960–1961). Such indications of contextual treatment became much more common for the Ph.D.s by 1980–1981, when more than half (52%) of the Ph.D.s placed their subjects in a specific "social" context in their titles. At the same time, M.D.s were still rather unlikely to spell out a contextual treatment of their subject in their titles (only 14% of M.D.s indicated such a context in 1980–1981, while 74% did not). This difference between the two groups of authors, however, narrowed by the 1990s, and by 2000–2001, fully half of the M.D.s and 59% of the Ph.D.s composed titles which indicated what we may call a social history perspective. In other words, it appears that over the years, the M.D.s lagged behind the Ph.D.s in adopting a contextual approach to the history of medicine, but over the last decade, this difference between the two groups has become insignificant. Without analyzing submissions to the journals rather than only the published articles, we cannot be sure, of course, whether the disappearance of this disparity in approach is a result of the diminishing role of differences in the professional and educational backgrounds of the historians of medicine or an outcome of changing editorial standards maintained by the journals (see Table 11.2).

Table 11.2 Professional Background of Authors per Type of Article, Relative to the Total Number of Authors with the Same Type of Degree

	No Context			Social Context			Scientific Context			Total		
	M.D.	Ph.D.	O	M.D.	Ph.D.	O	M.D.	Ph.D.	O	M.D.	Ph.D.	O	Total
1960–61	59 78%	37 71%	9 81%	4 5%	9 17%	2 18%	13 17%	6 12%		76	52	11	139
1970–71	37 76%	40 61%	7 47%	2 4%	16 25%	6 40%	10 20%	9 14%	2 13%	49	65	15	129
1980–81	26 74%	23 37%	6 35%	5 14%	32 52%	6 35%	4 11%	6 10%	5 29%	35	61	17	113
1990–91	11 41%	23 33%	5 50%	14 52%	45 64%	5 50%	2 7%	2 3%	0	27	70	10	107
2000–01	6 43%	22 35%	1 100%	7 50%	37 59%	0	1 7%	7 6%	0	14	63	1	78

Intellectual Configuration of the History of Medicine

The shift in the focus of history of medicine should be reflected not only in the manner in which authors frame their subject matter in article titles but also in the way they position their work relative to the work of other historians of medicine, historians of science, and other historians and social scientists. By referring to the work of others and to work published elsewhere, authors establish links between their own endeavors and those of their disciplinary peers and investigators in other related areas. While an analysis of the titles of articles can give us an indication of the structure of interests in the field at large, some contributions—whether in the form of articles or of books—are regarded as more significant, central, or exemplary than others. In the light of our findings on the intellectual changes and the professionalization of the field, we wanted to see whose work the historians of medicine referred to most often, and what group of authors formed the core of the new social history of medicine. How intellectually unified does the field appear from such a perspective? Has the change in the professional background of the contributors been reflected in this group of the most highly cited authors? Furthermore, the professionalization of the field was supposedly accompanied by a greater integration of the history of medicine into general history and by a growing theoretical similarity between the social history of medicine, the history of science, and the social studies of science. Can we discern these changes by examining how historians of medicine link their own work to that of others active in the different specialties?

We analyzed the manner in which historians of medicine position their work in this larger area by identifying the most frequently cited authors in the different journals in the history of medicine, and determining whether and to what extent articles in the medical history journals refer to the literature in history of science or the social studies of science. In our analysis we have included all the references in the articles published since 1988 in the four journals. The data for this analysis were drawn from the Science, Social Science, and Humanities Citation Indexes available on the Internet through the Web of Science (ISI).[27] Altogether we counted the citations in 582 articles, which appeared in these journals between 1988 and 1999 (the data for *SHM* were available only as of 1991).

From the references to all published works, books as well as articles, which were included in the articles in the four journals, we have constructed a list of names of the most highly cited historians of medicine, establishing the top-ten list separately for each journal. This resulted in a list of 31 names, indicating a very

Table 11.3 Most Highly Cited Historians of Medicine in Four Journals, 1988–1999 (times cited)

	BHM	JHM	MH	SHM	Total
Rosenberg, C.	87*	39*	20*	8	154*
Porter, R.	34*	13	46*	38*	131*
Warner, J. H.	53*	28*	7	14	102*
Bynum, W.	23*	21*	29*	19	92*
Ackerknecht, E.	40*	21*	18*	11	90*
Loudon, I.	13	9	36*	31*	89*
Risse, G.	17	21*	16	15	69*
Rosen, G.	28*	15	13	11	67*
Webster, C.	7	2	27*	31*	67*
Pelling, M.	13	5	25*	24	67*
McKeown, T.	18	11	7	31*	67*
Leavitt, J. W.	27*	16	6	4	53
Lawrence, C.	7	10	23*	12	52
Sigerist, H. E.	41*	3	4	3	51
Woods, R. I.	2	0	0	48*	50
King, L. S.	12	21*	12	4	49
Hardy, A.	5	2	10	30*	47
Szreter, S.	3	3	2	37*	45
Shryock, R.	22	17*	2	1	42
Arnold, D.	5	4	19*	14	42
Marland, H.	3	0	18*	21	42
Cassedy, J.	27*	9	1	3	40
Lewis, J.	6	1	6	26*	40
Stevenson, L. G.	6	25*	6	3	40
Foucault, M.	11	3	18*	7	39
Weindling, P.	8	2	4	25	39
Numbers, R.	23*	8	1	4	36
Riley, J. C.	5	2	1	28*	36
Ludmerer, K. M.	15	18*	1	2	36
Estes, J. W.	8	18*	7	2	35
Morris. J. N.	0	1	0	30*	31

Note: Top ten authors for each journal and for the total are marked with an asterisk.

small overlap between the four journals. Not a single author was among the ten most cited names for all the journals. Only four authors—Charles Rosenberg, Roy Porter, William Bynum, and Erwin Ackerknecht—were among the top ten authors in three journals. Another four authors—John Warner, Richard Shryock, Irvine Loudon, and Charles Webster—occur on two lists, the British authors on the lists from the two British journals and the Americans on those of the American journals. The very small overlap between countries and among the individual journals suggests a certain social, and perhaps also an intellectual, dispersion in the field. It is as if each of those journals had its own distinct culture (see Table 11.3).

It is perhaps not surprising that there are many historians of American medicine (Charles Rosenberg, John Harley Warner, Ronald Numbers, Judith Leavitt, James H. Cassedy, and Kenneth Ludmerer) among the highest cited authors in the *JHM* and the *BHM*, and many historians of British medicine and especially of British public health (Charles Webster, Christopher Lawrence, Margaret Pelling, Hilary Marland, Jane Lewis, Anne Hardy) among the highest cited authors in *MH* and the *SHM*. It is striking, however, that these national divisions are so manifest. Apart from the national biases, the group of highest cited authors in the *SHM* is dominated by participants in the debate about the effects of various medical and social interventions on long-term changes in mortality and morbidity. With the exception of Thomas McKeown, none of these authors are heavily cited in the other three journals. In this respect, the *SHM* might be said to have the clearest theoretical focus of the journals we examined.

We were rather surprised to find that the list of the highest cited authors includes so many classic authors whose main publications appeared many years ago. The founders of history of medicine in the United States such as Henry Sigerist, Erwin Ackerknecht, Richard Shryock, and George Rosen were particularly noticeable on the list of the highest cited authors in the *BHM*. Though the list includes many historians who would certainly regard themselves as social historians of medicine, it also includes authors who represent more traditional approaches to the field, and only a few authors dealing with such currently popular subjects as colonial medicine or gender studies. There is certainly very little trace of postmodernism or constructivism to be found on this list.

The educational background of the most highly cited authors does not diverge much from that of the authors publishing in the four journals in the relevant period, except for the overrepresentation of authors with double, M.D.-Ph.D. degrees. Altogether, some 42% (13 out of 31 authors) of the highest cited historians of medicine have a medical degree. In the early 1980s, Numbers claimed that doctors were only a small minority of the authors whose books were mentioned as the most influential works in the history of medicine, but some twenty years later, our evidence indicates that those trained in medicine still figure prominently among the most highly cited authors. The presence of M.D.s is particularly marked on the list of authors whose work has been cited most frequently in the *JHM*. Only three out of the ten authors on that list have no medical degree.

This image of the relatively traditional and conservative character of the field and of its dispersion is reinforced by an examination of the relations among the various journals in the history of medicine, as well as between history of medicine and related areas of scholarship. If history of medicine were a well-integrated area with a well-defined theoretical core and a body of canonical literature, we would expect to find a substantial number of references from one journal in the history of medicine to another. Moreover, given the presumed similarities in the theoretical approaches adopted in the history of medicine, history of science, and the social studies of science, we would expect that some reflection of the intellectual interchange among them could be found in citations to their respective literatures. Accordingly, we expected to find in our four journals a substantial number of citations to history of science and social studies of science journals. In order to estimate the extent of this interchange, we counted all citations to a number of journals in the history of science (*Isis, History of Science, Journal of the History of Biology, Studies in the History and Philosophy of Biology and the Life Sciences, Historical Studies in the Physical and Biological Sciences, Science in Context, British Journal of the History of Science,* and *Revue d'histoire de science*) and in the social studies of science (*Social Studies of Science; Science, Technology, and Human Values; Science and Culture;* and *Knowledge*). In order to establish the disciplinary unity of the history of medicine, we looked at journal-to-journal citations in each of our four journals, both of one another and of other journals in the history of medicine (for example, *Clio Medica* and *Janus*).

Contrary to suggestions in the historiographic literature, the presumed theoretical affinity between history of medicine, history of science, and the social studies of science was not to be found through an analysis of the journal-to-journal citation patterns. The 45,284 citations included only a miniscule number of citations to journals in these areas. (See Table 11.4). There were very few citations to the history of science journals (144), while those to the social studies of science journals were completely negligible (31 citations in 4 journals over 12 years). No names of historians of science or prominent scholars in the social studies of science community appeared on the most cited list, and when we extended our search for references to authors whose work we thought might be located at the interface between these various areas, we discovered that they were almost completely absent. There were also almost no references to theoretical work in the social studies of sciences, such as that of Bruno Latour, Barry Barnes, Harry Collins, or Karin Knorr. The intellectual interchange between these fields, if it exists at all, must take place so informally that it is not traceable in the literature.

Among the thousands of references in the articles in the four journals, even citations to journal articles in the history of medicine constitute only a small minority (about 3% of the 45,284 citations). Most citations seem to refer to source material rather than to the secondary (journal) literature. In the journal-to-journal analysis we did not, of course, include references to books in the field, but as the conclusions here reinforce the conclusions we have reached in our analysis

Table 11.4 References to Articles in Journals in History of Medicine and in Related Disciplines in Four Medical History Journals, 1988–1999

Journal	Number of Articles	Total References	References to Articles in Medical History Journals	Self-citations	References to Articles in History of Science Journals	References to Articles in STS Journals
BHM	202	14,345	499	223	79	6
JHM	183	10,391	430	101	35	4
MH	216	11,901	752	246	75	17
SHM	151	8,647	326	140	27	4
Total	582	45,284	1307	710	144	31

Note: Data for *SHM* from 1991 only.

of references to the highest cited authors, in which books were counted, it seems safe to conclude that the history of medicine is a rather dispersed and empirically driven area. There is no evidence of any body of core (theoretical) literature to which historians of medicine repeatedly orient themselves. It seems that individual articles stand each on their own and are not extensively or systematically linked to other literature in the field or engaged in an explicit discussion with other historical work.

This lack of strong connections between articles in journals in the history of medicine can be contrasted briefly with the situation in the social studies of science. We have examined references in the 89 articles dealing with medical issues published between 1988 and 2000 in the two core journals in this field, *Social Studies of Science* and *Science, Technology, and Human Values*. The five most frequently cited authors in these articles (Bruno Latour, Harry Collins, Michael Mulkay, Susan Leigh Star, and Adele Clarke) were cited in these 89 articles altogether 286 times; that is to say there were on average 32 references to these authors in every 10 articles. In contrast, the five most cited authors in our four history of medicine journals (Charles Rosenberg, Roy Porter, John Harley Warner, William Bynum and Erwin Ackerknecht) were cited 569 times in 778 articles; on average, only seven references were made to these authors in every ten articles. Social studies of science journals appear also to be more outwardly oriented than the history of medicine journals. We quickly compared how frequently articles in the social studies of science refer to work published in journals in history of science and history of medicine with the frequency of references to social studies of science and history of science in the medical history journals. The "medical" articles in the two social studies of science journals were twice as likely as articles in the history of medicine to cite work published in, respectively, history of science and history of medicine journals, and history of science and social studies of science journals.

Audiences of the History of Medicine

The finding that history of medicine articles seem to be so loosely linked to one another and so completely separated from articles in other cognate fields made us curious about the audiences addressed by the authors of articles in history of medicine journals. Where is the work published in history of medicine read and referred to? How often is it cited? In view of the often-claimed integration of history of medicine into general (social) history, is the work published in history of medicine journals cited in general history journals? How is the audience for history of medicine different from the audience for the social studies of science?

To address these questions, we selected the five most highly cited articles from each of our journals for the period 1988–2000. Because sometimes two or more articles were cited the same number of times, we ended up with a total of 24 articles. These 24 articles were cited altogether 339 times. We categorized the journals in which these citations to the five most cited articles appeared according to their disciplinary affiliation (see Table 11.5). Surprisingly, almost a third of the citations to these historical articles were made in medical journals (such as *JAMA* or *The Lancet*). The number of citations in such journals was slightly higher than the number of citations in history of medicine journals and much higher than the number of citations in general historical (or history of science) journals. It would

Table 11.5 Disciplinary Distribution of Citations to the Most Cited Articles in History of Medicine Journals

	SHM	JHM	BHM	MH	Totals
History of Medicine	30	17	23	30	100 29%
History of Science	9	5	12	11	37 11%
History (other than medicine and science)	7	5	21	8	41 12%
Medicine/Public Health/Epidemiology	24	29	21	34	108 32%
Social Sciences (including medical anthropology and sociology)	15	5	17	7	44 13%
Other	1	1	3	4	9 3%
Number of Articles	6	6	5	7	24
Total Number of Citations	86	62	97	94	339

Table 11.6 Disciplinary Distribution of Citations to the Most Cited Articles in Social Studies of Science Journals

Journals in the Field of	Social Studies of Science	Science, Technology, and Human Values	Total
Social Studies of Science	64	41	105 (37%)
History of Medicine	0	1	1
General History	1	0	1
Medicine and Public Health	3	10	13 (5%)
Social Sciences (including medical anthropology and sociology)	87	46	133 (47%)
History of Science	8	2	10 (4%)
Other	13	5	18 (6%)
Number of Articles	6	5	11
Total Number of Citations	176	105	281

appear that, contrary to the claims being made by historians of medicine, medical researchers remain a very significant part of the audience for history of medicine. In this context, it is perhaps worth mentioning that M.D.s account for 52% of the members of the American Association for the History of Medicine, even though, as we have seen, the proportion of doctors contributing to the literature in medical historical journals is much smaller. Moreover, despite the aspirations of prominent members of the history of medicine community, general and social history do not appear to be an important scholarly audience for journals in the history of medicine.

We have compared this audience for the history of medicine with the audience for articles on medical subjects that appear in the two major science studies journals by examining the citations to the five most highly cited medical articles in each journal. The 281 citations to 11 articles (in one journal, two articles were *ex equo* in the fifth place) were distributed mainly among journals in the social studies of science and in the social sciences more generally. Forty-seven percent of all these citations appeared in the literature in areas of the social science other than science studies (e.g., in general sociological journals or journals in medical anthropology). There was only one reference to any of these articles in the history of medicine journals and very few references (5% of the total) in medical journals (see Table 11.6).

Moreover, the rate of citation to articles in the social studies of science appears substantially higher than the rate of citation to history of medicine articles. The five most frequently cited articles in the history of medicine journals were cited on average 14 times, whereas the most cited articles in the science studies journals were cited on average 25.5 times. It is somewhat ironic that the medical articles in the social studies of science, which deal overwhelmingly with contemporary medicine, receive virtually no attention from medical researchers and lots of attention from other social scientists; whereas historians of medicine, who think they have moved away from medicine and toward history, continue to attract more attention in medicine than in history. The citation data confirm our earlier conclusion that the two fields interact very little and that the history of medicine journal literature seems relatively isolated.

Conclusions

In many respects, our scientometric analysis of developments in the history of medicine confirms commonly held beliefs about changes in the field. History of medicine today is indeed being written by historians rather than physicians, and the attention of its practitioners has shifted from heroic tales about individual physicians and their contributions to the progress of medical knowledge to more soberly told and contextualized stories of professionalization and the everyday practice of medicine. Topics such as the history of public health and health policies have also received more attention. Professional historians have spearheaded these changes, though by now the differences between authors with different kinds of educational backgrounds do not appear to be very pronounced.

Given the lack of a method for gauging the pace of change, it is impossible to say whether these changes in the history of medicine have been radical or not; however, a number of our findings indicate a relatively gradual transformation of the discipline. Professionalization has taken a long time, and the shifts in the selection of topics are better described as a process of diversification than as a dramatic refocusing of attention.

Some expectations, however, have not been confirmed. Contrary to what Numbers feared, there is no indication that the medical sciences have ceased to be a subject of concern. Although we cannot say whether historians have ceased to pay attention to "the content and internal logic of medicine," the development of the medical sciences, often examined in scientific or social contexts, continues to be of abiding interest to professional historians as well as physician-historians. At the same time, we could find no trace of the anticipated increase of attention to patient experiences and perspectives.

The biggest surprises, however, occurred when we looked at how the field was organized internally and how historians of medicine linked their work to the work of others. In many respects we found the field to be more traditional than we expected: the list of the most highly cited authors included a number of historians whose work was published many years ago, as well as a number of contemporary authors who adopt a rather orthodox approach to history. Very few of these highest cited authors seem to be working on subjects currently supposed to be fashionable. The proportion of M.D.s among them was also higher than among the contributors to journals in the history of science in general. There was no sign that this group represented a specific, theoretically coherent approach to the history of medicine.

Our data suggest that history of medicine is a dispersed and internally splintered field. Not only did we find little overlap among the highest cited authors in the various journals, but these journals also exhibited strong national orientations. Both the individual journals and the individual articles seem to stand on their own. Moreover, we discovered that historians of medicine both within the United Kingdom and the United States refer relatively little to each other's work, suggesting that there is no unified body of core literature and that historians of medicine are not discussing a shared set of theoretical concerns but rather orienting themselves overwhelmingly to specific empirical cases.

This internal dispersion appears to be accompanied by relative external isolation. Despite the presumed parallels in the theoretical orientations of history of medicine, history of science, and the social studies of science, we found no signs of linkage between the literatures in these fields. This lack of intellectual interchange is apparent in both citing and cited patterns. Jordanova's and Warner's hopes for a closer link with history of science and STS, and even more broadly with the social sciences and humanities, remain no more than a programmatic ambition.

One possible explanation for both the persistently traditional orientation of the field and for its apparent intellectual fragmentation might be found when we realize the extent to which medical professionals remain a vital part of the audience for the history of medicine. The domination of the history of medicine by physicians has usually been seen in terms of the characteristics of the contributors to the literature in the field. Once historians came to constitute a majority of the authors, there was an expectation—a hope or a fear—that the field would become a part of general social and cultural history. But the development of a field is shaped not only by its contributors but also by its audiences, and the audience for history of medicine seems still pervaded by medical authority. The extent to which historians of medicine are aware of this continuing presence of doctors as their audience, and how they relate to it in their own writings, remains an open question.[28]

NOTES

1. Ronald L. Numbers, "The History of American Medicine: A Field in Ferment," *Reviews in American History* 10 (1982): 245–263.

2. Ibid., 247, 250.

3. John Harley Warner, "Science in Medicine," *Osiris* 1 (1985): 37–58, 37.

4. Ibid., 42.

5. John Harley Warner, "The History of Science and the Sciences of Medicine," *Osiris* 10 (1995): 164–193.

6. Warner regarded these battles with the traditional history of medicine as already won and saw calls for the social history of medicine, such as that issued by Bryder and Smith in the introductory issue of the *Social History of Medicine* (hereafter *SHM*), as "rousing the troops to battle against an enemy long put to rout, magnifying the danger of stragglers in order to sustain momentum of a crusade no longer novel" (Warner, "The History of Science," 174). See also, Lynda Bryder and David Smith, "Editorial Introduction," *SHM* 1 (1988): v–vii.

7. For a recent review of parallel changes in the history and social studies of science, see for example, Jan Golinski, *Making Natural Knowledge* (Cambridge: Cambridge University Press, 1998), John Pickstone, *Ways of Knowing: A New History of Science, Technology, and Medicine* (Chicago: University of Chicago Press, 2001), and the essays in *Constructing Knowledge in the History of Science,* edited by Arnold Thackray, *Osiris* 10 (1995).

8. A similar increase in the share of articles on medical subjects can be seen in the two main STS journals, *Social Studies of Science* and *Science, Technology, and Human Values*. In the 1988–1990 period there were nine articles on medical subjects out of a total of 115 articles published in these two journals (8%); in 1997–1999, there were 31 such articles out of a total of 127 (24%).

9. Ludmilla Jordanova, "The Social Construction of Medical Knowledge," *SHM* 8 (1995): 361–381 (reprinted as chap. 15 in this volume); see also "Has the History of Medicine Come of Age?" *Historical Journal* 36 (1993): 437–449. A call for the introduction of social constructivism into the sociology of medicine was made by Malcolm Nicolson and Cathleen McLaughlin, "Social Constructionism and Medical Sociology: A Study of the Vascular Theory of Multiple Sclerosis," *Sociology of Health and Illness* 10 (1988): 234–261.

10. Warner "The History of Science," 165

11. Jordanova, "The Social Construction," 361.

12. For work using scientometric data as a means of investigating disciplinary development rather than evaluating performance or rate of growth, see Steven Cole, Jonathan Cole, and Lorraine Detrich, "Measuring the Cognitive State of Scientific Disciplines," in *Towards a Metric of Science: The Advent of Science Indicators,* Yehuda Elkana, Robert K. Merton, Arnold Thackray, and Harriet Zuckerman, eds. (New York: John Wiley, 1978), 209–252; Jona-

than Cole and Harriet Zuckerman, "The Emergence of a Scientific Specialty: The Self-Exemplifying Case of the Sociology of Knowledge," in *The Idea of Social Structure: Papers in Honor of Robert K. Merton,* Lewis A. Coser, ed. (New York: Harcourt, Brace, Jovanovich, 1975), 139–174; Loet Leydesdorff and Peter van den Besselaar, "Scientometrics and Communication Theory: Towards Theoretically Informed Indicators," *Scientometrics* 38, no. 1 (1997): 155–174. See also Paul Wouters, "The Citation Culture" (Ph.D. diss. University of Amsterdam, 1999).

13. Numbers, "The History of American Medicine," 251.

14. T. Gelfand, *Bulletin of the History of Medicine* (hereafter *BHM*) 44 (1970): 40–49.

15. Guido Ruggiero, *Journal of the History of Medicine and Applied Sciences* (hereafter *JHM*) 36 (1981): 168–84.

16. Patricia Spain Ward, *JHM* 36 (1981): 44–62.

17. Kenneth Ludmerer, *BHM* 65 (1981): 343–370.

18. Nigel Allen, *BHM* 64 (1990): 446–462.

19. Samuel X. Radbill and Gloria R. Hamilton, *BHM* 34 (1960): 430–486.

20. Keith Wailoo, *BHM* 65 (1991): 185–208.

21. Stuart Gallishoff, *JHM* 25 (1970): 438–448.

22. Richard Coker, *Medical History* (hereafter *MH*) 45 (2001): 341–358

23. Ann Beck, *MH* 4 (1960): 310–321.

24. John C. Burnham, *BHM* 45 (1971): 203–218.

25. Ivan Dalley Crozier, *MH* 45 (2001): 61–82.

26. John Z. Bowers, *BHM* 65 (1981): 17–34.

27. Data available electronically do not go further back than 1988.

28. For two contrasting answers to this question, see Charles E. Rosenberg, "Why Care about the History of Medicine?" in *Explaining Epidemics and Other Studies in the History of Medicine* (Cambridge: Cambridge University Press, 1992), 1–6, and Roger Cooter, " 'Framing' the End of the Social History of Medicine," chap. 14 in this volume.

CHAPTER TWELVE

The Power of Norms
Georges Canguilhem, Michel Foucault, and the History of Medicine

Christiane Sinding

What is the origin of norms, whether vital or social, and from where do they derive their power? The same question runs throughout the work of George Canguilhem (1904–1995) and Michel Foucault (1926–1984), on the boundary between knowledge and ethics. Both took as a starting point, not norms as such, but the violation of norms—what is considered to be "abnormal" by societies: disease to illuminate health, madness to illuminate reason. By addressing the issue of norms, especially medical norms, they ipso facto raised ethical and political questions as well. Canguilhem contrasted *normality* with *normativity,* which for him characterizes the capacity of the living to invent new norms. He also developed a critical analysis of social norms. Foucault was impressed by his mentor's reflections and further developed Canguilhem's idea of bivalence of norms. For Foucault, norms were to be seen not only as constraints or a source of power but as productive forces. Both authors drew on history: Foucault used a wide variety of material from social and political history; Canguilhem, who limited himself to the narrower field of medicine and the life sciences, was less a historian than a philosopher trying to provide philosophy with concrete questions. To address the ethical or political issues of interest to them, these two authors practiced a critical history that was an outright attack on the traditional history of medicine. From

1943 in the case of Canguilhem and 1963 in the case of Foucault, this controversial history ran counter to the positivist, continuist, and retrospective history written by doctors and other hagiographers of medicine. To understand the creation of a new impulse for medical history by both authors, I argue that we must grasp the way a particular sensibility to the power of norms informed and animated their work.

Despite an initially awkward meeting,[1] Canguilhem agreed to be the official supervisor of Foucault's Ph.D. thesis on the history of madness, the importance of which he immediately grasped.[2] Thereafter, he constantly supported Foucault. In 1966, he wrote one of his best articles to defend Foucault against violent criticism triggered by the publication that year of *Les mots et les choses* (a title not convincingly translated in English as *The Order of Things*).[3] Foucault wrote the preface to the American edition of *Le normal et le pathologique* (1978) to highlight the often unrecognized importance of Canguilhem for an entire generation of French philosophers. He also pointed out that in France, the philosophy of knowledge, of rationality, and of concepts (i.e., that which was most theoretical and apparently distant from immediate political concerns) produced intellectuals who participated directly in World War II as *résistants*. He explained this paradox by demonstrating that *Aufklärung* assumed a particular form in France, focusing on the question of the power of reason through positivism and its critiques. He added the following phrase, which illuminated his own work as much that of Canguilhem: "Reason as both despotism and enlightenment."[4] This bivalence of reason was the main concern of both Canguilhem and Foucault, which explains why their history of medicine had ethical and political more than historiographical aims.

The Normativity of the Living

Georges Canguilhem, born at Castelnaudary (near Toulouse) in 1904, attended the Henri IV secondary school in Paris, where he prepared for the entrance examination to the École Normale Supérieure. His fellow students included Raymond Aron and Jean-Paul Sartre. In 1943 he obtained his M.D. degree, choosing to become a doctor in order to further his understanding of medicine and to "acquire a medical culture first-hand."[5] In 1955, he succeeded Gaston Bachelard at the Sorbonne and at the Institut d'histoire des sciences et des techniques de l'université de Paris, where he remained until his retirement in 1971. Canguilhem died in 1995. His work consists of short essays, almost exclusively on the life sciences and medicine. Their conciseness was deliberate, for Canguilhem's aim was more to problematize and define the main features of a method than to describe. His historical work as such is limited to a few articles and to the history of the concept of reflex, the subject of what, considering the requirements of French universities, was an exceptionally short Ph.D. thesis in philosophy.[6] The rest of his articles were devoted to the philosophy of medicine and the life sciences and drew on history essentially to illuminate philosophical questions.

The unity of Canguilhem's work stems from his insistence on the originality of the living, which cannot be reduced to a sum of physicochemical constants and, above all, produces its own inherent norms.[7] It was in *Le normal et le pathologique* (1966), translated in 1988 as *The Normal and the Pathological,* that for the first time he presented this conception. In this major book, Canguilhem "expected medicine to provide an introduction to concrete human problems"[8] and defined medicine as an art or technique at the meeting point of several sciences. The first part of the book, dated 1943 and headed "Essay on Some Problems Concerning the Normal and the Pathological," is in fact his M.D. thesis. In it, Canguilhem asserted that health is defined by the capacity of the living to invent new norms and to overcome obstacles, a capacity that he qualifies as *normativity* of the living. The "Essay" of 1943 led to a critique of the quantitative conception of pathology defended by positivists, especially Claude Bernard, François Broussais, and Auguste Comte. Broussais was the first to propose a quantitative definition of disease by likening it to an excess or lack of excitability of tissues. Bernard used the example of diabetes to defend a similar conception. Having observed increased rates of sugar in the blood and urine of diabetics, he established the principle of the identity of the normal and the pathological, which differ only quantitatively. This doctrine was used to identify health and normality (later assimilated with statistical averages) and to turn disease into an object of science that in principle could be remedied by means of appropriate techniques. The foundations of modern medicine thus were laid.

According to Canguilhem, this positivist conception raised a number of questions. If there is continuity between the normal and pathological states, at what point can one talk of a pathology? In the example of glycemia in diabetes, one can probably easily talk of a pathology when the glycemia level is 3 g/liter, but are glycemia levels considered as pathological at 1.1 g/l, 1.2 g/l, 1.3 g/l, or more? Since normally there is no sugar in the urine of healthy subjects, of what is glycosuria an excess? Does the latter example not indicate the radical discontinuity between normal and pathological states? To these empirical questions, Canguilhem added a theoretical problem: Bernard and Comte had to use qualitative terms to define disease. They spoke of disharmony, imbalance, and alteration, and characterized the state of health by the term "harmony," and Canguilhem emphasized the aesthetic and moral rather than the scientific senses of that term.

Canguilhem's insistence on qualitative changes that affect an organism—as the result of disease but also of aggression or a change of environment and living conditions—was intended to refute reductionism and to show that what is living is characterized above all by its *normativity*. Disease, stress, and changes in the living environment are an opportunity to reveal normativity. "Being healthy means being not only normal in a given situation but also normative in this and other eventual situations," Canguilhem asserted. "What characterizes health is the possibility of transcending the norm."[9] Disease reduced this bodily capacity to innovate, causing sick persons to live in a reduced world, inventing for their lives new norms that vary from one patient to the next. Biological evolutionism also supports the thesis of the normativity of the living, as the only organisms that survive certain major environmental changes are those capable of innovating.

Vital norms also have social components that Canguilhem illustrated with examples drawn from epidemiology. Each society afforded its own living conditions—altitude, temperature, habitat, diet, and so forth—and the biological norms of the individuals comprising that society largely correspond to those conditions. Hence, Canguilhem later used the concept of normativity, which first referred to biological properties of the living, to challenge social norms. Canguilhem discussed this "extension" of the concept in "New Reflections Concerning the Normal and the Pathological (1963–1966)," the second part of *Le normal et le pathologique*.

For Canguilhem, the elaboration of the concept of normativity had a number of consequences. First, physiology is based on the study of the pathological, not its opposite. Disease attracts attention to the physiological functions of the injured organ, and experimental physiology is in fact an experimental pathology. Second, quantitative explanations disregard the fact that disease affects and transforms the entire organism. Third, and perhaps most important, the pathological is never entirely objective; it always retains a relation to subjectivity. Before being an object of science, disease is above all an experience reported by the patient to her or his doctor. If, in practice, diagnoses can be made solely on objective examinations, historically this is because one day a patient drew a doctor's attention to certain symptoms and because the doctor methodically examined those symptoms and drew preliminary conclusions. If the patient died, the comparison of clinical observations with the autopsy report led to new conclusions said to be "objective," yet still related to the patient's initial subjective descriptions. As Canguilhem put it, "We think *that there is nothing in science that did not first appear in consciousness,* and . . . [that] it is the patient's point of view that is basically correct."[10]

The philosopher overturned a number of prevailing beliefs. He affirmed the primacy of the pathological over the physiological, the subjective over the objective, the qualitative over the quantitative, vitalism over mechanism. It is hardly surprising that, in his "New Reflections," he focused a great deal of attention on the concept of errors of metabolism that Archibald Garrod elaborated around 1909. In 1966, Canguilhem reverted to that focus when molecular biology informed interpretations of disease in terms of errors in the reading of genetic messages.[11] Foucault characterized Canguilhem's philosophy as *a philosophy of error:* "In the extreme life is what is capable of error," he wrote, adding that "it must be that error is at the root of what makes human thought and its history."[12] This valorization of the "negative," or of what is considered as such, is probably what best characterizes Canguilhem's thinking.[13]

In the introduction to his 1943 essay, Canguilhem noted that he wished to explore two questions: norms and the relation between technique and science. For him, the question of technique was closely bound to that of ethics. When he drew attention to Bernard's statements on man becoming "a veritable supervisor of creation," on whom no limits can be set regarding "the power he can acquire over nature through future advances in the experimental sciences," it was to highlight that this demiurgic conception is of an ethical order.[14] Canguilhem took a stand against this philosophy and these ethics. For him, it is not up to science to govern individuals' behaviors: "One does not scientifically dictate norms to life."[15]

His philosophy views technique as being rooted in life, of which it is an extension. Therapeutics is "a technique for establishing or restoring the normal which cannot be reduced entirely and simply to a single form of knowledge."[16] Therapeutics has almost always drawn on a large common pool of empirical recipes to develop its remedies and often precedes and informs theoretical knowledge.

Canguilhem's positions on normality and pathology derive from his conception of the living, which he himself characterized—not without a touch of provocation—as vitalist. To him, vitalism was simply the recognition of the originality of the living or "an imperative rather than a method and more of an ethical system, perhaps, than a theory."[17] Moreover, in the history of the knowledge of the living, vitalism has often been a driving force rather than the hindrance it has been made out to be. Canguilhem undertook to demonstrate this with particular vigor in the domain that seems related least to vitalism and most to mechanism: reflexology.

In his history of the concept of reflex, published in 1955,[18] Canguilhem openly criticized historians of biology who want to find the present in the past. In the nineteenth century, physiologists and historians presented reflex as a mechanist notion because social and cultural interests prompted them to do so. They therefore sought its roots in Descartes. But nothing in Cartesian philosophy allows us to elaborate a notion such as reflex. The essential idea in the formation of the concept of reflex, "the homogeneity between incident movement and the reflected movement," is found in a contemporary of Descartes, the Englishman Thomas Willis.[19] This vitalist physician and philosopher likened life to light, and he applied optical laws of reflection to describe reflexive movement. In the nineteenth century, social and political interests erased the vitalist Czech Georg Proschaska from the history of the concept of reflex and turned that history into an entirely mechanist one. The twentieth century, with its focus on speed, efficiency, and automation, consolidated that falsified history. At issue in this historical reconstruction by Canguilhem was a certain conception of human beings, impossible to reduce to a set of mechanisms governed by coupled stimuli and responses. For Canguilhem, human biology and medicine were the elements needed for an "anthropology," and in 1951 he wrote that there was "no anthropology which does not imply a moral code, so that the concept of 'normal' in human order always remains a normative concept with a strictly philosophical meaning."[20]

Political issues became clearer as Canguilhem's work progressed. This was probably the result of personal maturity, and possibly of the acquisition of institutional weight, but also due to his reading of Foucault.

Birth of Clinical Norms

Born in Poitiers in 1926 into a well-to-do family of doctors, Michel Foucault was accepted into the École Normale Supérieure in 1946, where he met Pierre Bourdieu, Jean-Claude Passeron, and Paul Veyne. His professor was Jean Hippolyte, a specialist in Hegel, whose dazzling lectures fascinated Foucault and prompted him to do his Ph.D. thesis on this great German philosopher. He was subsequently also taught by Maurice Merleau-Ponty, and later Louis Althusser. Foucault also studied psychology, moved in psychiatric circles, and considered studying medicine.[21] After 1955 he lived for more than ten years in foreign countries, then lectured at the university of Clermont-Ferrand and the nonconformist University of Paris-VIII (Vincennes). In 1970 he was appointed to the chair of the history of systems of thought at the Collège de France.

Foucault's first short book, *Maladie mentale et personnalité,* published in 1954 in a collection edited by Althusser, was an attempt to illuminate the historical dimension of madness by showing how it had been the result of a practice of social exclusion and incarceration before being a medical object.[22] Written in a Marxist context, the work did not question the existence of madness; it accused capitalist societies of being responsible for it. It was in a second book, *Folie et déraison* (*Madness and Civilization*), that he developed these early theses to the full.[23] In this first major work, Foucault addressed the distinction between normal and pathological, and the relation between the two, in order to identify common points between reason and insanity. How, he wondered, did society go from the medieval humanist experience of madness to what is now our experience of it, limiting madness to mental disease and excluding and alienating it? Foucault undertook to describe the changes of "perception" of madness from the Middle Ages to the advent of positive psychology. In this development, the main "invention" was incarceration, peculiar to our civilization, which confined the insane with vagrants, idlers, and beggars. What really interested Foucault, rather than the idea of incarceration that so struck readers at the time, was the intervention of the administration and the underlying ethical intention. Canguilhem wrote in his report on the dissertation, "This administrative practice and surveillance is also an ethical practice. Incarceration mixes idlers, rogues and libertines together in the same spatial limits of reprobation."[24]

Legal problems of banning and incarceration required doctors to formulate criteria defining madness, which, however, remained ruled by prevailing laws until the nineteenth century. The "release" of the insane during the French Revolution was simply a transfer to new institutions of incarceration that retained the "reform" aspect of general hospitals. Cures were also punishments, ranging from cold showers to deprivation of food or corporal punishments. According to Foucault, it was only after the insane were placed together in institutions that "positive" knowledge of madness could develop, a knowledge that nevertheless retained the trace of these operations of relegation, discipline, and exclusion: "Is it not important in our culture that insanity was able to become an object of knowledge only in so far as it was first an object of excommunication?" Foucault asked.[25] It is suggestive to compare this sentence with a passage found in the conclusion to *Naissance de la clinique* (1963), translated in 1973 as *The Birth of the Clinic: An Archaeology of Medical Perception*: "It will no doubt remain a decisive factor about our culture that its first scientific discourse concerning the individual had to pass through this stage of death. Western man could constitute himself in his own eyes as an object of science . . . only in the opening created by his own elimination."[26] Taken together, these passages show just how convinced Foucault was of the need to question history in order to identify and understand the models of thinking, representation, and action that we live with today.

Birth of the Clinic is a difficult book, often highly technical and written in a language that is sometimes as obscure as it is beautiful. It poses almost insoluble problems of translation, which worsen the opacity of the language. In it, we follow the laborious path that leads from a totally incomprehensible (for us) citation by Dr. P. Pomme (1769), to a detailed "objective" description—so much more familiar to us than the first one—of A. L. J. Bayle in 1925. The juxtaposition of these two citations was intended to make clear that an important break took place between those two dates. Foucault started with the medicine of nosology implemented by Sydenham but of which the most complete expression is found in *Nosographie philosophique* by Pinel.[27] For this medicine of species, disease is an entity that exists outside its inscription in a sick body and that can and must be classified in a system based on the botanical model. During the Revolution, the political, institutional, and social conditions were met for the reorganization of medical knowledge. First, a new type of hospital was instituted as a place for administering care and training; there, the priority was the constitution of a system of knowledge based on individual cases. The patient was no longer the subject of her or his disease; the patient was an example of the universal. Next, a tacit agreement between rich and poor allowed the latter to be used as examples in exchange for free care. By gathering together large numbers of patients in the same place, it was possible to observe individuals continuously and comparatively and thus to establish new *codes of knowledge,* which were to allow the creation of a "protoclinic." Two of these codes were essential: first, the transformation of the symptom into a significant element, making it possible to express the disease and its truth fully. For Foucault, Condillac's analysis was the linguistic model that was then applied to deciphering disease.[28] Symptoms had the value of a sign for the initiated, whose gaze was "sensitive to difference, simultaneity or succession and frequency."[29] Second, the importation of probability thinking into medicine in the time of Laplace, "either under his influence or within a similar movement of thought,"[30] allowed every recorded event to take place in a random and indefinitely open sequence. The idea of imperfection constituting medical knowledge and its uncertainty was then rejected.

The myth of the alphabetic structure of disease made it appear as a combination of simple elements that had only a nominal reality. The doctor's aim was therefore to free these elements, like the chemist frees elements by combustion. Death carried this release through to its conclusion. Corpses were dissected, and it was with the birth of Bichat's anatomoclinical medicine that the invisible was finally made visible. The inventor of the concept of *tissue* rediscovered the pathological anatomy of Morgagni, but through the clinical model that detached pathological analysis from the Italian doctor's localism. He showed general pathological forms in the human body that affected tissues before organs and thus remained faithful to the lessons of clinical medicine because in pathological analysis, tissue was the element used for deciphering.

The question remained of the *being* of the disease and its relation to the lesion. Broussais struck the final blow at the ontological conception of disease by stating that local disease is not the point of *insertion* of disease. He explained diseases in terms of an excess or reduction in the irritability of tissues, and he was violently criticized for this return to Haller and to medical monism. Yet it was these "errors" that enabled him to discount any difference between the organic and functional phenomenon and, above all, to establish the necessity of physiological medicine. Broussais, moreover, introduced to medicine the question of *causality*, with the idea that an external irritating factor is the cause of disease.

We thus reach the end of an inquiry into conceptual, institutional, and political configurations that made it possible to constitute anatomoclinical medicine as a new experience of disease. "For clinical experience to become possible as a form of knowledge," Foucault asserted, "a reorganization of the hospital field, a new definition of the status of the patient in society, and the establishment of a certain relationship between public assistance and experience, between help and knowledge, became necessary; the patient has to be enveloped in a collective, homogeneous space. It was also necessary to open up language to a whole new domain: that of perpetual and objectively based correlation of the visible and the expressible."[31]

In many respects Foucault's first two major works are very similar: the modern subject is constituted by distinguishing her- or himself from negative figures: the insane, the diseased, and, later, in *Surveiller et punir. Naissance de la prison* (1975)— translated in 1977 as *Discipline and Punish: The Birth of the Prison*—the delinquent.[32] The acceptance of "myths" of the constitution of positive knowledge owing to the advances of reason and humanity is replaced by the search for the ambiguous conditions that made such knowledge possible. Just as the release of the insane was simply a new form of exclusion, so *Birth of the Clinic* shows that the clinical experience was not the result of a straightforward confrontation, without any concept, of a gaze and a face, a glance and a mute body, that is, of the singular mythical medical encounter.[33] Nor was it the Enlightenment thinkers, who played a decisive part in the birth of clinical medicine by lifting the ban on the dissection of corpses, who made a new form of pathological anatomy necessary; it was epistemic requirements. Finally, both books emphasized the complex play of institutional regulations and political and ethical choices that governed the constitution of new knowledge.

Birth of the Clinic reveals obvious traces of Canguilhem and has strands of continuity with *The Normal and the Pathological*. These strands are thematic, since in both cases the idea was to establish a new approach to the history of medicine, as well as chronological, for the last chapter in *Birth of the Clinic* treats of the birth of a physiological medicine with Broussais, with which *The Normal and the Pathological* starts.[34] Above all, the two books have related ideas. Canguilhem showed that ill health illuminated the physiological, that is to say, life. Foucault extended the scope and meaning of this principle by showing how death, and not only disease, illuminated life. Finally, with Canguilhem the question of the statistical norm as the result of observation contrasted with the norm resulting from a choice. This central thesis involved issues of freedom, which Foucault clearly perceived. In *Birth of the Clinic,* the appearance of a collective and normative medical conscience announced the developments found in subsequent work in the concept of biopower.

Important divergences separated the two authors, too. Although the book is Foucault's most epistemological work—giving the impression of witnessing the lifting, with time, of certain obscurities, illusions, and confusions—scientific concepts were not his central preoccupation. He wanted to analyze the social and political process that conditioned the birth of clinical medicine. Medical norms are produced by doctors as a social body belonging to an institution (the hospital). Patients are put into the background in this history, and with them the subject of normativity, a concept which does not appear in Foucault's writings. Experience, which for Canguilhem referred to the patient's lived experience, became with Foucault anonymous, collective, and controlled by the institution. In this context, the patient becomes the object of the medical corps' gaze—constituting a norm because they are recognized as being competent—in an institution that socially legitimizes the relationship between watched and watcher.

Birth of the Clinic is one of the first works on the history of medicine to entangle scientific, social, and political factors so closely. Foucault noted that when he was a student, from 1950 to 1955, "One of the great problems that arose was that of the political status of science and the ideological function it could serve." Medicine had the advantage of having "a much stronger scientific structure than psychiatry but also of being very deeply embedded in social structures."[35] Foucault's subsequent analyses of medicine (in interviews, talks, lectures, and articles) constantly maintained a link with this first work and should not be dissociated from it, as they so often are. One of Foucault's major theses is set out in the book: the decisive role of medicine and of the distinction between normal and pathological in the constitution of the humanities.[36]

The Genealogy of Social Norms

It was between 1963 and 1966, that is, after reading Foucault, that Canguilhem outlined a genealogy of social norms. "The normalization of the technical means of education, health, transportation for people and goods," he asserted, "expresses collective demands which, taken as a whole, . . . in a given historical society, defines its way of referring its structure, or perhaps its structures, to what it considers its own good."[37] He maintained that there existed bodies or systems of norms, ranging from the technical to the legal.

It was primarily Foucault, however, who developed the question of the normalization of societies and the concepts of "biopolitics" and "biopower" that have been among those in Foucault's work most extensively used and commented on. As indicated above, their roots and legitimacy lie in *Birth of the Clinic*, where we see the appearance of the notion of medical police, legitimized by the new scientific authority claimed by the clinic.[38] In the nineteenth century, medicine assumed a *public health* function, which centralized and coordinated medical care and normalized knowledge. It took into account death, reproduction, and birth and, between individuals and society, formed a new intermediate object of knowledge, *population*. Hence, it proposed a collective solution to the problem of randomness. Above all, unlike medicine in the seventeenth century, it was regulated in accordance with normality more than health.

Initially motivated by the concern to limit epidemics, this policy aimed at observing and registering epidemics and death rates and at developing a real population technology. In the eighteenth century, it already had initiated the medicalization of society and especially of the family, the first point of application of health measures. The maintenance of health turned medicine into an authority for social control, intended to maintain the national labor force—an objective that was overturned after the publication in 1942 of the Beveridge report in England. With this report, which outlined the first model of the welfare state, the state was put at the service of the individual's health and not vice versa.[39] But this reversal did nothing to undermine the normalizing function of medicine; on the contrary. The medicine that was then established was a medicine of the indefinite medicalization of societies.

It is important to note that while Foucault himself always stressed the constraining power of norms, thus favoring an interpretation that he sometimes regretted, we should not forget that he always highlighted the fact that medicine could not be considered only in terms of its disciplinary effects. "What enables medicine to function with such force," he stated, "is the fact that, unlike religion, it is inscribed in the scientific institution. We cannot be content simply to point out the disciplinary effects of medicine. Medicine can function as a mechanism of social control, but it also has other, technical and scientific, functions."[40] By emphasizing the fecundity and positive effects of medicine, he was eager to distinguish himself from Ivan Illich, in a lecture delivered in Brazil in 1974.[41] Norms are powerful in both a positive and negative way, in the sense that they create individuals subjected to power but also able to resist constraints.

The question remains of "self-care" (*le souci de soi*), in which some have tried to see a return to the subject of former work and a questioning of it. In *Le souci de soi* (1984), translated as *The Care of the Self* (1986), Foucault is very clear about the close correlation between self-care and thinking, and medical practice: "Medicine," he argued, "was not conceived simply as a technique of intervention, relying, in case of illness, on remedies and operations. It was also supposed to define, in the form of a corpus of knowledge and rules, a way of living, a reflective mode of one self."[42]

Most "positive" interpreters of Foucault's work have seen no break between the last two volumes of the history of sexuality and his earlier work. After Foucault's death, Georges Canguilhem concluded his analysis of the reception of *Les mots et les choses* and *Naissance de la clinique* and their "explosive charge" by noting that "It was normal, in the strictly axiological sense, for Foucault to undertake the elaboration of an ethics. Faced with normalization and against it: *the care of the self.*"[43] Care of the self can be understood as contributing toward "practices of freedom," a term that Foucault preferred to that of liberation because the latter referred to "repressive locks" that simply had to be broken for humans to find their real nature.

Canguilhem and Foucault as Medical Historians

How can our two authors' practice of medical history be characterized? Theirs is, first and foremost, a controversial practice that hunts down the false continuities and fabricated chronologies concealing the real conditions of the production of new modes of knowledge in order to replace them with a history that focuses on discontinuities and rejects positivism.[44]

Canguilhem focused on the history of concepts and not of theories, as Bachelard did. With the history of concepts one can retain the possibility of some *continuism* in history while criticizing a history that judges the past in terms of the present.[45] Focusing on concepts allows one to "reveal the permanence of a question and illuminate its current meaning."[46] In addition, a single concept can alternately bring into play opposing theories or philosophies, like vitalism and mechanism or atomistic and holistic conceptions of the living. This dialectical movement corresponds to the very movement of life, so that the history of concepts is particularly well suited to its object, the living. Thus, the "nonscientific" is involved in the sciences; Canguilhem called this *scientific ideology*. Science must not be separated from that which is not yet scientific, like myths and images, which play a key part in the elaboration of a concept. The concept of scientific ideology also makes it possible to show that the relation between true and false constantly changes throughout history.

We know that the only designation accepted by Foucault for himself until about 1975 was *archaeologist of knowledge*, devoted to the reconstruction of that which accounts for a culture *in depth*, and which he initially called the *épistème*. He wanted to find the tacit, unconscious rules that in each era govern the constitution of the objects of knowledge and of the different types of knowing subjects. For Canguilhem, the history of the sciences remained controlled by the current norms of the sciences, whereas Foucault, even if he did not reject the idea of cores of truth in the sciences, tried above all to account for that which makes the constitution of knowledge possible, and to that end he developed his own working methods, constantly going back to amend, correct, or reshape them. These methods were to be criticized by many historians, particularly the scarcity of secondary references in Foucault's writings and his use of archival material such as the regulations of institutions, police notes, anonymous testimonies, architectural plans, and utopian projects.

These methods were crafted to suit an ambitious project consisting not in studying the constitution of a scientific object, as Canguilhem did, but in wondering why the human subject had taken itself as an object of possible knowledge, through what forms of rationality and in what historical conditions.

Reception in Medical History

Canguilhem remains relatively little known outside France, especially in the English-speaking countries. *Le normal et la pathologique* was translated only in 1978. Among the Americans who helped to diffuse his work were Stuart F. Spicker, Gary Gutting, and Paul Rabinow.[47] In France, Canguilhem is read mainly by a circle of specialists, philosophers or historians of the sciences, and is almost unknown in the medical profession. The history of medicine in France is hardly institutionalized at all, and its practice is dispersed. Despite this situation, Canguilhem did train some students, mostly philosophers, some of whom had also studied medicine. Among the latter, François Dagognet published a seminal work on the philosophy of remedy,[48] a reference book on Pasteur, and a large number of books on the philosophy and history of medicine. Claire Salomon-Bayet applied what she calls an "institutional epistemology," that is, a complex play between an institution, concepts, and methods, to study the constitution of a strictly biological rationality under the auspices of the Académie Royale des Sciences.[49] She also edited an important collective work on Pasteur.[50] The other "disciples" of Canguilhem in France were oriented more toward philosophy and the history of biology, and thus abandoned what, in my view, made this philosopher so original, that is, a close alliance between a philosophy of knowledge and a philosophy of values and action.[51]

The reception of Foucault's work on medicine cannot be separated from that of the rest of his work, about which I can simply mention a few salient points. This contrasting reception was tumultuous from the early 1960s until the publication of *Les mots et les choses* (1966), which really unleashed passions. From eulogies to violent attacks, through more moderate criticism, the critics in France and English-speaking countries accused this philosopher-historian of limiting to a minimum his critical apparatus and of giving only very few secondary references. More seriously, his work was said to be strewn with "empirical" mistakes. Finally, what probably bothered readers most was the constant revision of his analyses, that is, "the absence of system established for once and for all" and of a "limited field of objects."[52]

The reception of his work on medicine displays the same difficulties. At first, *Madness and Civilization* was fairly well received in France, but after 1968 it was "taken over by social movements," which were to give it a political impact that it did not have when it first appeared.[53] From then on, French psychiatrists attacked the book. In the United States, critical reviews of this work and books criticizing its methods appeared fairly early on, especially in the *New York Review of Books*.[54] British historians expressed similar criticisms, although less controversial. Michael MacDonald has pointed out that "Ever since the publication of *l'Histoire de la folie,* in 1961, rival groups of historians have retold their version of the history of madness and shouted imprecations at each other."[55] Colin Jones and Roy Porter have pointed out that a number of scholars (including Porter himself) have shown the book to be "beset by fundamental empirical flaws, or at least by hasty generalizations from the exceptional case of France."[56] However, they have also acknowledged that "Foucault has had often highly beneficial impact on the study of the past without necessarily authorizing his exact interpretation or exhibiting willful blindness to his empirical errors."[57]

By comparison with the two-step reception of *Madness and Civilization*, *Birth of the Clinic* had little echo among historians of medicine and the public at large. In France, Dagognet noted, with some reservations, "the extreme importance of these analyses which caused an upheaval in the history of medicine."[58] Eribon pointed out Jacques Lacan's interest in *Birth of the Clinic*,[59] but the vast majority of doctors remained unaware of the book's existence, as did most historians. Foucault was surprised that the political dimension of his book was not perceived from the outset. In 1963, the highly technical content of the book overshadowed the political aspect. A new reading of the book after the "protest" years, especially as regards medical power, helped to highlight its political dimension. However, *Birth of the Clinic* did not really have the "second life" that *Madness and Civilization* had. In the English-speaking world, despite a few reviews, the publication of *Birth of the Clinic* went virtually unnoticed.[60] When Foucault started to be known for his views and Brazilian lectures on the medical institution and "medicalization" of our societies, it was these scattered works that were to resound among historians of medicine, together with *Discipline and Punish*, his most successful book, which deals with the hospital as a disciplinary institution.

Recently however, Foucault's thesis has been challenged by Caroline Hannaway, Ann La Berge, and some of their co-workers, in *Constructing Paris Medicine*.[61] This enriching book deconstructs in part the "myth" of the "Paris school" of medicine. For matter of space, it is impossible to discuss its general aim here, but one can regret how little the contributors make explicit use of Foucault. His theses are summarized in only two pages of an otherwise illuminating review of the literature on the Paris school. Characterized as the "intellectual guru for English-speaking academe," Foucault is always associated in this book with Erwin Ackerknecht, as if the theses, aims, and methods of the two men were comparable.[62] To be sure, the two had some points in common: they both thought that a major breakthrough occurred at the beginning of the nineteenth century, linked to the creation of a new sort of hospital during the French Revolution. The new hospitals brought together a great number of patients who could be thoroughly examined and compared. If they died, it was easy to perform an autopsy and to correlate clinical symptoms with anatomical lesions. But here the comparison ends. Whereas Ackerknecht was interested mainly in medicine and its technical and theoretical progress, Foucault wanted to grasp the entanglement of the effects of power and knowledge with regard to medicine. More generally, he conceived his book as part of a more ambitious project centered on the genealogy of the modern subject on which he spent all his life. In *Constructing Paris Medicine,* only scattered historiographical points made in *Birth of the Clinic* are discussed, never the more global thesis.[63]

Foucault never pretended to be a historian of medicine or even a historian, and he had "emphasized the anti-historical features of his orientation" (his lack of interest in studying historical "periods" and even capturing historical "reality").[64] As François Delaporte pointed out, he thought that "documents do not enable us to reconstitute the past. The mass of documents must itself be searched for unifying structures, series and relations."[65] Two British historians have emphasized that his aim was to defamiliarize, "to expose seemingly natural categories as constructs, articulated by words and discourse, and thus to underline the radical contingency of what superficially seems normal. Nothing in history could be taken for granted; all history was culturally fabricated; everything had therefore to be questioned."[66] Perhaps this interpretation gives more than its due to the vocabulary of cultural studies and the social sciences, of which Foucault was sometimes considered the precursor, no doubt mistakenly. But it clearly highlights what a French reader would call "the fantastic power of agitation in the revealing of games of truth between subjects, objects, knowledge and powers."[67]

In France, a recent symposium reviewed Foucault and medicine.[68] Its aim was to take a new look at Foucault's thinking from the point of view of medicine, to move away from the prevailing idea that this philosopher above all promoted a critical reading of the medical institution. The goal was thus to draw attention to the fact that Foucault had a plurality of thoughts on medicine that were considered from multiple angles, ranging from an analysis of the medical institution and its practices to what he called the medicalization of our societies. The organizers of the symposium also asserted that Foucault's works had contributed substantially to the "new conquest" of the history of medicine by historians, but they gave very few examples to ground this affirmation.[69] Indeed Foucault's influence on history of medicine might be considered pervasive, and it is not always easy to document. For reasons of space, I will limit myself to a few examples.

Among historians of medicine who made extensive use of Foucault, one might cite John Pickstone who claimed that "Foucault is a mainstay for medical and cultural historians" and developed his "ways of knowing" explicitly in reference to Foucault's épistèmes.[70] Roy Porter also frequently quoted Foucault, both in a critical and positive manner. Historical projects concerned with sexuality and reproduction are often deeply informed by Foucault.[71] Feminist studies, feminist historical projects among them, have also drawn on Foucault.[72] Historians of psychiatry have been for a long time divided between the one obsessed with the "great confinement" and the one who defended the "march of progress."[73] More recently, the history of postcolonial medicine seems to be influenced by Foucault.[74] Apart from these specific domains, however, I believe that Foucault is not often used by historians of medicine.[75] Robert Markley expressed a similar opinion regarding Foucault's significance to the history of science, that he is either treated as an icon, "more invoked than studied seriously," or completely ignored.[76]

Sociologists, anthropologists of health and medicine, and interdisciplinary workers have shown much more interest in Foucault's work on medicine, which is understandable if we bear in mind that the philosopher's aim was above all to understand the present. David Armstrong, a sociologist and a historian, is one of those who explicitly has situated himself in the tradition of Foucault's work, where he focuses above all on medicalization and medical power.[77] In the United States Paul Rabinow helped to make Foucault known very early on, especially with the book he published with Dreyfus and his publication of a collection of writings translated into English.[78] David Horn, an anthropologist like Rabinow, also used Foucault in his analyses of medicine, especially on the history of the medicalization of criminality.[79] Adele Clarke believes that "there are multiple incarnations of historical research and, today, deeply historically informed research by scholars from related disciplines such as medical sociology and anthropology that draw upon Foucault." She goes as far as to argue that "medical sociology and medical anthropology have been transformed by Foucault and have become *much* more historical because of him. The phrase 'history of the present' resounds."[80]

On the whole, as far as history of medicine is concerned, Foucault has certainly been more commented on, criticized, and utilized in Anglo-American countries than in France. The dispersion of French studies was highlighted fifteen years ago by Geof Bowker and Bruno Latour in an article in which they set out to explain French researchers' lack of interest in science studies.[81] Subsequently, science studies have developed considerably in France, but Bowker and Latour's underlying analysis is still relevant. They maintained that the longstanding French tradition of critical, ethical, and political discourse disappears as soon as science is concerned and that love for rationality prevented French historians, or a sociologist such as Pierre Bourdieu, from including science in a sociologizing or historicizing analysis.

Yet when it came to Canguilhem or Foucault, that analysis is less convincing. Bowker and Latour presented Canguilhem as "the philosopher of radical discontinuity" whose "task in all his books has been to separate ideologies from science."[82] I have tried to show in this chapter just how much more nuanced Canguilhem's positions were.[83] As for Foucault, the authors accused him of refusing on principle to criticize the hard sciences, and of having removed the social from the construction of scientific disciplines, insofar as Foucault's "social" would never be anything but theory. *Discipline and Punish* was nevertheless spared this criticism because it "could easily be read as social history." Finally, according to the authors, Foucault is considered by the "Anglo-Saxons" as far more "radical" than he really was. On this point, the authors were probably right. But in my view there is a major absence in this article's analyses, and that is the history of medicine.[84] It seems that we have here a classic case in which biology is considered as a science that purely and simply spawned medicine, which consequently is in itself treated as being of very little interest as a subject of research.[85]

Canguilhem and Foucault studied medicine itself and the reasons for which it claimed to be a science at one stage by trying to eliminate its moral and political components. That is probably why their analysis still has strong critical significance, especially when the works of these two authors are examined together. Despite the many questions they left open, they deserve serious rereading, and, in the case of Foucault, corrections of misreading.

NOTES

I wish to thank Luce Giard, Ilana Löwy, Isabelle Baszanger, and Luc Berlivet for their careful reading of this text and their criticisms. I am especially indebted to Adele Clarke, who provided helpful suggestions for the section on the reception of the work of Canguilhem and Foucault among historians of medicine. For reasons of space, I was not able to develop fully her suggestions. Finally, I wish to thank Elizabeth Libbrecht for her excellent assistance with the translation of this text.

1. Didier Eribon, *Michel Foucault,* Betsy Wing trans. (Cambridge, Mass.: Harvard University Press, 1991), 102.

2. The thesis was published under the title *Folie et déraison: L'histoire de la folie à l'âge classique* (Paris: Plon, 1961) and later under the title *L'histoire de la folie à l'âge classique* (Paris: Gallimard, 1972); it was translated as *Madness and Civilization: A History of Insanity in the Age of Reason,* R. Howard, trans. (New York: Vintage, 1973). Contrary to what some Anglo-American scholars believe, Foucault never studied under Canguilhem and submitted his manuscript after its completion.

3. Georges Canguilhem, "Mort de l'homme ou épuisement du cogito," *Critique* 24 (1967) 599–618, translated by Catherine Porter as "The Death of Man, or Exhaustion of the Cogito?" in *The Cambridge Companion to Foucault,* Gary Gutting, ed. (Cambridge: Cambridge University Press, 1994), 71–91.

4. Michel Foucault, "La vie: L'expérience et la science," *Revue de métaphysique et de morale* 90 (1985) 3–14, 7. This article is the French version of Foucault's introduction to the American translation of Canguilhem's *Le normal et le pathologique*. It was translated by Robert Hurley as "Life: Experience and Science" in Michel Foucault, *Essential Works of Foucault, 1954–1984*, vol. 2: *Aesthetics*, James D. Faubion, ed. (London: Penguin Books, 200), 465–478, 470. The *Essential Works of Foucault, 1954–1984* are selections from *Dits et écrits, 1954–1988*, Daniel Defert and François Ewald, eds., with the collaboration of J. Lagrange (Paris: Editions Gallimard, 1994).

5. Georges Canguilhem, *The Normal and the Pathological* (New York: Zone Books, 1989), 34.

6. Georges Canguilhem, *La formation du concept de réflexe aux XVII et XVIIIe siècles*, (Paris: Presses Universitaires de France, 1955). His most important historical articles include "La théorie cellulaire," in *La connaissance de la vie* (1952; reprint, Paris: Vrin, 1980), 43–80, and "Pathologie et physiologie de la thyroïde au XIXe siècle" in *Études d'histoire et de philosophie des sciences* (Paris: Vrin, 1979), 274–294.

7. Pierre Macherey, "De Canguilhem à Canguilhem en passant par Foucault," in *Georges Canguilhem, philosophe, historien des sciences* (Paris: Albin Michel, 1993), 286–294, 288.

8. Canguilhem, *The Normal*, 34.

9. Ibid., 115.

10. Ibid., 92–93. Author's emphasis.

11. Georges Canguilhem, "Le concept et la vie," *Revue philosophique de Louvain* 64 (1996): 193–223, reproduced in *Études d'histoire*, 335–364.

12. Foucault, "Introduction," in Canguilhem, *The Normal*, 22.

13. However, Canguilhem did not really clarify the relationship between this "error" of life and the old concept of lesion.

14. "According to Claude Bernard, experimental method is more than a code for a laboratory technique, it is an idea for a moral code"; Georges Canguilhem, "Théorie et technique de l'expérimentation chez Claude Bernard," in *Études*, 143–155, 154.

15. Canguilhem, *The Normal*, 226.

16. Ibid., 34.

17. Canguilhem, in *A Vital Rationalist: Selected Writings from Georges Canguilhem*, François Delaporte, ed., Arthur Goldhammer, trans. (New York: Zone Books, 1994), 288. A complete critical bibliography compiled by Camille Limoges is at the end of the volume, 385–454.

18. Canguilhem, *La formation*.

19. Canguilhem, "The Concept of Reflex" in *A Vital Rationalist*, 179–202, 184.

20. Georges Canguilhem, "Le normal et le pathologique," in *La connaissance*, 169.

21. Eribon, *Foucault*, 42.

22. Michel Foucault, *Maladie, mentale, et personnalité* (Paris: Presses Universitaires de France, 1954).

23. Foucault, *Madness*.

24. "Rapport de M. Canguilhem sur le manuscrit déposé par M. Michel Foucault, directeur de l'Institut français de Hambourg, en vue de l'obtention du permis d'imprimer comme thèse principale de doctorat ès lettres." Canguilhem's report (19 April 1960) on Foucault's Ph.D. thesis.

25. Foucault, *Madness*.

26. Michel Foucault, *Naissance de la clinique,* (1963; 6th ed., Paris: Presses Universitaires de France, 2000), 200–201, translated by A. M. Sheridan as *The Birth of the Clinic: An Archaeology of Medical Perception* (London: Tavistock, 1973). The subtitle "Une archéologie du regard médical" was deleted by Foucault in 1972.

27. For a more detailed analysis of *The Birth of the Clinic,* the reader is referred to C. Sinding, "La méthode de la clinique," in *Michel Foucault. Lire l'oeuvre,* Luce Giard, ed. (Grenoble: Jerôme Millon, 1992), 59–82.

28. For Condillac (1714–1780), thoughts stem from elementary feelings or from initial thoughts, as in Locke. His contribution was to affirm that language serves not only to express thoughts but also to shape them and that it is invented by humans living in groups. Furthermore, the self is not an independent thinking and self-conscious substance, the result of a combination of feelings with different origins. It is thus an atomistic conception of the individual.

29. Foucault, *Birth,* 94.

30. Ibid., 97.

31. Ibid., 196.

32. Michel Foucault, *Surveiller et punir* (Paris: Gallimard, 1975), translated by A. Sheridan as *Discipline and Punish: The Birth of the Prison* (New York: Pantheon, 1977).

33. Foucault, *Birth,* xiv–xv.

34. Note that *Naissance de la clinique* was published by Presses universitaires de France in the Gallien collection edited by Canguilhem.

35. Michel Foucault, "Truth and Power," translation of an interview with Alessandro Fontana and Pasquale Pasquino, *Microfisica del potere: Interventi politici,* (Turin: Einaudi, 1977), 3–28; reproduced in *Essential Works,* vol. 3: *Power,* 111–133.

36. Foucault, *Birth,* 36.

37. Georges Canguilhem, "From the Social to the Vital," in *The Normal,* 237–256, 238.

38. Here the word "police" does not have the negative connotation it now has but was used for all mechanisms used to ensure order and the maintenance of health in general.

39. Michel Foucault, "Crise de la médecine ou crise de l'antimédecine?" *Dits et écrits,* Vol. 3:40–58.

40. Foucault, "L'extension," 76.

41. Foucault, "Crise." In particular, Foucault criticized "the radical and bucolic rejection of medicine in favor of a nontechnical reconciliation with nature," 48.

42. Michel Foucault, *Histoire de la sexualité. III. Le souci de soi* (Paris: Gallimard, 1984), 136, translated by Robert Hurley as *History of Sexuality,* vol. 3: *The Care of the Self* (New York: Pantheon, 1986), 99.

43. Georges Canguilhem, "Sur l'histoire de la folie en tant qu'événement," *Le débat* 41 (1986): 37–40.

44. Dominique Lecourt, *Pour une critique de l'épistémologie. (Bachelard, Canguilhem, Foucault)* (Paris: Maspero, 1973) 7, translated by Ben Brewster as *Marxism and Epistemology: Bachelard, Canguilhem, Foucault,* (London: NLB, 1975).

45. As some authors have noted, Foucault shared with Canguilhem a certain continuism and the rejection of the uniquely negative conception of epistemological obstacles developed by Bachelard. See, for example, Gary Gutting, *Michel Foucault's Archaeology of Scientific Reason* (Cambridge: Cambridge University Press, 1989), 53, and *Reassessing Foucault: Power, Medicine, and the Body,* Colin Jones and Roy Porter, eds. (London: Routledge, 1994), 7.

46. Pierre Macherey, "La philosophie de la science de Georges Canguilhem," *La pensée* 113 (1964): 50–74.

47. Stuart F. Spicker, "An Introduction to the Medical Epistemology of Georges Canguilhem: Moving beyond Michel Foucault," *Journal of Medicine and Philosophy* 12 (1987): 397–411; Gutting, "Bachelard and Canguilhem," in *Michel Foucault's Archaeology*, 9–54; Paul Rabinow, "Introduction," in Canguilhem, *A Vital Rationalist*, 11–22.

48. François Dagognet, *La raison et les remèdes* (1964; reprint, Paris: Presses universitaires de France, 1984).

49. Claire Salomon-Bayet, *L'institution de la science et l'expérience du vivant* (Paris: Flammarion, 1978).

50. *Pasteur et la révolution pastorienne,* Claire Salomon-Bayet, ed. (Paris: Payot, 1986).

51. In France, only François Delaporte works in the continuation of both Canguilhem's and Foucault's work. See, for example, François Delaporte, *Disease and Civilization: The Cholera in Paris, 1832,* Arthur Goldhammer, trans. (Cambridge, Mass.: MIT Press, 1986).

52. Luce Giard, "Foucault, lecteur de ses critiques," in *Au risque de Foucault,* Dominique Franche, Sabine Prokhoris, Yves Roussel, eds. (Paris: Centre Georges Pompidou, Centre Michel Foucault, 1997), 193–201. Luce Giard analyzes the reasons for difficulties in the reception of the work of this strange historian by American readers.

53. Eribon, *Foucault,* 124.

54. The *New York Review of Books,* in Eribon's opinion, celebrated any book attacking French intellectuals. See Eribon, *Michel Foucault et ses contemporains* (Paris: Fayard, 1994), 70.

55. Michael MacDonald, "Madness, Suicide and the Computer," in *Problems and Methods in the History of Medicine,* Roy Porter and Andrew Wear, eds. (London: Croom Helm, 1987), 207–229, 208.

56. Jones and Porter, *Reassessing,* 4.

57. Ibid., 5.

58. François Dagognet, "Archéologie ou histoire de la médecine? Michel Foucault, Naissance de la clinique," *Critique* 216 (1965): 436–447, 443.

59. Eribon, *Foucault,* 154.

60. F. N. L. Poynter, review of *Naissance de la clinique. Une archéologie du regard médical, History of Science* 3 (1964): 140–143, 142; Karl Figlio, "Review of *The Birth of the Clinic. An Archaeology of Medical Perception,*" *British Journal for the History of Science* 10 (1977): 164–167.

61. *Constructing Paris Medicine,* Caroline Hannaway and Ann La Berge, eds. (Amsterdam: Rodopi, 1998).

62. Hannaway and La Berge, "Paris Medicine: Perspectives Past and Present" in ibid., 1–69, 32–33.

63. For a thoughtful review of the volume, the reader is referred to George Weisz, "Reconstructing Paris Medicine," *Bulletin of the History of Medicine* 75 (2001): 105–109. Having reviewed the different contributions and acknowledged their quality and the challenge that some of them brought to the "standard view" of the Paris school, Weisz nevertheless states that the "revisionist" authors had to recognize "that Paris was a remarkable and unique center of medical research in the early decades of the nineteenth century."

64. Jan Goldstein, "Foucault among the Sociologists: The 'Disciplines' and the History of the Professions," *History and Theory* 23 (1984): 170–192.

65. François Delaporte, "The History of Medicine According to Foucault," in *Foucault and the Writing of History,* Jan Goldstein, ed. (Oxford: Blackwell, 1994), 137–149, 141.

66. Jones and Porter, *Reassessing,* 4.

67. Giard, "Foucault, lecteur," 197.

68. *Foucault et la médecine: Lectures et usages,* Philippe Artières and Emmanuel da Silva, eds. (Paris: Kimé, 2000).

69. Philippe Artières and Emmanuel da Silva, "Introduction," in ibid., 14.

70. John Pickstone, *Ways of Knowing: A New History of Science, Technology, and Medicine* (Manchester: Manchester University Press, 2000), 22.

71. See, among others, Adele E. Clarke, *Disciplining Reproduction: Modernity, the American Life Sciences, and the Problems of Sex*, (Berkeley: University of California Press, 1998); Alice Domurat Dreger, *Hermaphrodites and the Medical Invention of Sex* (Cambridge: Harvard University Press, 1998); and Thomas Laqueur, *Making Sex: Body and Gender from the Greeks to Freud* (Cambridge: Harvard University Press, 1990).

72. Nelly Oudshoorn, *Beyond the Natural Body: An Archeology of Sex Hormones* (New York: Routledge, 1994); Anne Fausto-Sterling, *Sexing the Body: Gender Politics and the Construction of Sexuality* (New York: Basic Books, 2000).

73. See John C. Burnham, "Jack Pressman and the Future of the History of Psychiatry," *Bulletin of the History of Medicine* 74 (2000): 778–785, 779. Burnham quotes Elizabeth Lunbeck, *The Psychiatric Persuasion: Knowledge, Gender, and Power in Modern America* (Princeton: Princeton University Press, 1994), as a sophisticated work in the Foucauldian tradition. Here, again, it should be pointed out that Foucault had insisted on the production of positive knowledge by the confinement of the mad, a point generally understated by his commentators.

74. Alexander Butchart, *The Anatomy of Power: European Constructions of the African Body* (New York: Zed Books, 1998). For a review, see Warwick Anderson, "Where Is the Postcolonial?" *Bulletin of the History of Medicine* 72 (1998): 522–530. I wish to thank Adele Clarke for having drawn my attention to this emergent body of scholarship in which, she thinks, "questions of biopower are vivid."

75. This impression is confirmed by Olga Amsterdamska and Anja Hiddinga, "Trading Zones or Citadels? Professionalization and Intellectual Change in the History of Medicine," chap. 11 in this volume. In the four analyzed journals, Foucault ranks twenty-fifth of the thirty-one most highly cited historians. He was cited thirty-nine times between 1988 and 1999, whereas Canguilhem was cited thirteen times in two of the journals (personal communication by Olga Amsterdamska).

76. Robert Markley, "Foucault, Modernity, and the Cultural Study of Science," *Configurations* 7 (1999): 153–176, 153.

77. David Armstrong, *Political Anatomy of the Body: Medical Knowledge in Britain in the Twentieth Century* (Cambridge: Cambridge University Press, 1983) and "Bodies of Knowledge/Knowledge of Bodies," in Jones and Porter, *Reassessing,* 17–27.

78. Hubert Dreyfus and Paul Rabinow, *Michel Foucault: Beyond Structuralism and Hermeneutics,* 2d ed. (Chicago: Chicago University Press, 1983); Paul Rabinow, *A Foucault Reader,* (Harmondsworth: Penguin, 1986).

79. David G. Horn, *Social Bodies: Science, Reproduction, and Italian Modernity* (Princeton: Princeton University Press, 1994).

80. Adele Clarke, personal communication, 29 May 2002. For a more thorough analysis of Foucault's reception among sociologists of medicine, see *Foucault, Health, and Medicine,* Alan Petersen and Robin Burton, eds. (London: Routledge, 1997).

81. Geof Bowker and Bruno Latour, "A Booming Discipline Short of Discipline: (Social) Studies of Science in France," *Social Studies of Science* 17 (1987): 715–748.

82. Ibid., 725.

83. To support my analyses, an article by Malcolm Nicolson shows that the researchers in the social studies of science working on the history of medicine could use Canguilhem's contributions. See Malcolm Nicolson, "The Social and the Cognitive: Resources for the Sociology of Scientific Knowledge," *Studies in the History and Philosophy of Science* 22 (1991): 347–369.

84. Medicine is missing from Latour's work. See Bruno Latour and Steve Woolgar, *Laboratory Life: The Construction of Scientific Facts,* (Princeton: Princeton University Press, 1986), and Bruno Latour, *The Pasteurization of France* (Cambridge: Harvard University Press, 1988). Both books are based on the observation or the history of biological work and almost totally overlooked medical doctrines and practices. Contrary to Latour, Bowker produced work on medical practices and, notably, medical classifications, together with Leigh Star. See Geoffrey C. Bowker and Susan Leigh Star, *Sorting Things Out: Classification and Its Consequences* (Cambridge: MIT Press, 1999). This book refers explicitly to the work of Foucault.

85. "The doctor is called by the patient. It is the echo of this pathetic call which qualifies as pathological all the sciences which medical technology uses to aid life"; Canguilhem, *The Normal,* 226.

CHAPTER THIRTEEN

Postcolonial Histories of Medicine

Warwick Anderson

In the first issue of the *Bulletin of the History of Medicine,* Henry E. Sigerist urged his readers to consider carefully the spatial distribution of disease, not just its history. Since the late nineteenth century, pathological and physiological studies in medicine had nudged aside investigations of geographical influences on the character and distribution of disease, except perhaps in the tropics. Similarly, while one place or another was presupposed in most histories of medicine, it had become unfashionable to accentuate the territorial limits, the situation, of historical subjects. Yet, as Sigerist insisted in 1933, "whenever we trace the history of a disease, we do it in a definite country. We cannot study the history of the Plague at large." The history and geography of disease were always inseparable, and the nexus should be made clear. Sigerist hoped that the new journal would bring together the "good work" on these linked subjects from "all over the world," but in this matter at least it was to prove sorely disappointing.[1]

Fielding H. Garrison, the glum representative of an older generation of American medical historians, may have differed in historical sensibility from Sigerist, yet he too shared the Swiss émigré's interest in promoting a spatially informed, and dispersed, history of medicine. In 1932, the year before the founding of the *Bulletin,* Garrison argued that the medical historian of the future "will be concerned, not only with the achievement of a few advanced civilizations, but with the medicine and sanitation of the whole world." Displaying his overpacked mind, Garrison then discussed the great nineteenth-century geographies of disease: Laveran on the distribution of paludism and trypanosomiasis; Rockefeller hookworm surveys; the pattern of illness in Russia after the revolution; effects of disease on settlement in the new world; Berber medicine in North Africa; and Tupi healing practices in Brazil. He reported at length on disease in the "Kenya colony" and speculated on the adaptability of the "Negro" to Western ways. He recalled the proceedings of the International Congress of Tropical Medicine, held in Singapore in 1923, which had indicated that the history of disease in the Torrid Zone was "by no means reducible to the pattern implicit in textbooks on tropical medicine. Each of these areas has, in fact, its own peculiar type of tropical medicine."[2] After this patchy international comparison of disease distributions and medical interventions, Garrison went on to recommend the journals of anthropology and ethnology to aspiring medical historians.

The global vision of Sigerist and Garrison seems sadly to have faded by the 1970s, when North American and European scholars wrote their manifestos for the social history of medicine. Admittedly, the interest of Sigerist and Garrison in other parts of the world—derived from an admiration of the great nineteenth-century works of historical and geographical pathology, and expressed mostly in programmatic statements—always remained in tension with their own research, which concentrated still on the history of medical ideas and on scientific biography. For Sigerist, the history of Western medicine as an intellectual enterprise and the life stories of the great doctors certainly required social and geographical context, but the settings that fascinated him were found generally in Europe or, less frequently, in the neo-Europe of North America. With the emergence of the social history of medicine after World War II, and its proliferation during the 1970s, fine-grained local or national studies, usually located in the West, became ever more common. Sigerist and Richard Shryock, his successor at Johns Hopkins, were often invoked as founders of the new social history of disease and health care. According to Susan Reverby and David Rosner, these precursors had prompted a "growing understanding of both the social causation of health and disease and the way in which science is embedded in society's social relations."[3] Through pioneering social histories, we would learn more and more about specific medical sites—laboratories, hospitals, community clinics, public health departments, to name a few—and gain a richer understanding of the social basis of health and healing. Even so, for many years the geographical ambit of conventional medical history would not extend much beyond a few sites in western Europe and the United States east of the Mississippi.

It is perhaps easy now to criticize Sigerist and Garrison's 1930s prospectus for a global historical geography of disease and medicine. They disregarded deepening disciplinary fissures; they clung to a diffusionist schema; and, in Garrison's case at least, human difference was figured in crude racial typologies. Moreover, their own historical practice repeatedly belied their global ambitions. Yet, unlike the later generation of social historians of medicine, they continued to value, if only in theory, a comparative imagination, demanding multisited and more dispersed histories, narratives that would focus on the global transmission of medical ideas and practices. The social history of medicine, even in the geographically contracted form that emerged in the 1970s, has given us much, but it was not the only path to which the historians of the 1930s were pointing.

It seems that nature and politics, regardless of historiographic trends, are continually plotting to reassert the importance of geography, of spatial patterning, in our understanding of disease and health care. In the nineteenth century, a combination of European expansion and repeated cholera epidemics inevitably drew more attention to historical and geographical aspects of pathology. In the twentieth century, Garrison readily admitted that the influenza pandemic and global depression, helped by his stint in Manila in the early 1920s, had made him think more seriously about the distributions of disease and medical science in the world. Since the 1980s, emergent diseases such as AIDS, along with growing corporate and cultural globalization and the intellectual appeal of insights from anthropology, have reminded historians of medicine that no clinic and no laboratory, however comfortably enclosed in one national edifice or another, is quarantined, socially or intellectually. Historians of colonialism have for a long time studied complex exchanges between the West and the rest, charting a spatial distribution of power and influence, but unfortunately few of them until recently were interested in disease or medicine. During the past two decades, however, numerous historical studies of the conjunction of medicine and colonialism have tried to address the perceived need for a more encompassing, diverse, and comparative comprehension of disease and health care, and to do so without abandoning the earlier achievements of densely situated social history.

In this chapter I outline the development of new histories of colonial medicine and discuss their significance for medical history more generally.[4] If they do nothing else, many of these studies of medicine and colonialism have belatedly applied to the rest of the world—to the majority of the world's population—the analytic resources of the social history of medicine, ensuring that local and national histories of health and healing can crop up anywhere. But I argue that such studies are in fact doing more than that, or rather, they might do still more. David Arnold, among others, has claimed that, in a sense, all medicine is "colonial" in its relation to the body.[5] Thus, the various colonial histories of medicine should have the potential to reshape our understanding of medicine and public health and to redirect attention to the structuring of race and place in biomedical science and health care wherever they are studied, whether in the tropics or the chilly North. That is, the history of medicine and colonialism might helpfully insert the "colonial" as an analytic category even into the hitherto rather insular, irrelative medical history of Europe and North America. One can only imagine how surprised Sigerist and Garrison would be to find their calls for a greater geographic sensibility in medical history harnessed in this way to the politicized spatiality of postcolonial studies.

Tropical Sites, Tropical Medicine, and Tropical Diseases

Before it became the history of medicine and colonialism, some knew it, in a different, more narrowly instrumentalist form, as the history of tropical medicine and disease, or the history of military or naval medicine abroad. When the first textbooks of tropical medicine were issued in the 1890s, the specialty seemed too young to warrant a brief introductory history, even though Europeans had been studying the diseases of warm climates for centuries. In 1920, Aldo Castellani and Alfred J. Chalmers prefaced their *Manual of Tropical Medicine* with a perfunctory yet nonetheless novel history of the discipline, asserting that it had emerged only in the late nineteenth century in response to discoveries of the parasitic causes of the diseases prevalent in equatorial regions. Scientific research thus seemed to distinguish modern tropical medicine from the clinical routinism of the old medicine of warm climates. Accordingly, the history of tropical medicine entailed a listing of recent discoveries in helminthology, protozoology, mycology, entomology, and, somewhat more rarely, bacteriology.[6] But when H. Harold Scott compiled his immense *History of Tropical Medicine* (1939–1942) a few decades later, he was more liberal and eclectic than Castellani and Chalmers, and he focused instead on the long histories of the various diseases that had come finally to constitute the subject of tropical medicine. Of course, the history of each disease culminated in its scientific elucidation, but Scott was also prepared to describe its geographical distribution and the earlier, apparently superseded, beliefs about its causation. "It is a most fascinating occupation," Scott reflected, "to study the early vague, empirical ideas as to the causation of a disease, to trace the beginnings of rational thought there-anent, the interpenetration of scientific notions and empiricism, the progressive clarification of the haze of doubt to the final solution of the problem."[7] His work, still useful today, was an example of what happens when the history of ideas attempts to revive an almost moribund tradition of geographical and historical pathology.

Scott concluded with a series of biographies of the founders of the scientific study of the diseases of warm climates—that is, the founders of "modern tropical medicine." Biographies of great doctors and great institutions would dominate the history of tropical medicine for another forty years or more. Thus, Charles Morrow Wilson organized *Ambassadors in White* (1942), an aptly titled history of American tropical medicine, around the life stories of its leading scientists.[8] Significantly, both Scott and Wilson were writing at the end of the 1930s, looking back on twenty years of memoirs and reminiscences of the first generation of specialists in tropical medicine. Tales of swashbuckling disease fighters in distant jungles, these stories often managed to combine evangelical zeal, gory particulars, and pious self-satisfaction. Books such as *An American Doctor's Odyssey* (and later, *Siam Doctor, Burma Surgeon, A Yankee Doctor in Paradise*) proved immensely popular, mostly as a form of inspirational literature.[9] Other aging tropical experts, perhaps less self-regarding, turned instead to proxy biographies of the institutions they had built.[10] In all of these genres, colonialism, that essential predicate of tropical medicine, was scarcely mentioned, and on the rare occasions it intruded, it was still a wholly benign presence.

With the spate of decolonization after World War II came a number of institutional histories of medicine and public health, delimited by the borders of emergent nation-states. For a time it seemed that every nation must have its official history of public health, just as later every nation might look forward to its own social history of medicine. Again, these early nationalist historians tended to have been participants in the events they described, and, not surprisingly, they emphasized local agency as part of a more general shedding of the imperial legacy. Moreover, they were inclined to review the responses to a broad range of local diseases, not just the typically "tropical" ailments, such as malaria, which so excited and preoccupied elite researchers and teachers in European centers. Many of these accounts repay reading today; others are little more than tedious administrative catalogues and have not aged gracefully.[11] Recently, area specialists have taken to writing more critical "national" histories of public health in Africa, and less commonly Asia, in order to demonstrate the character and reach of the colonial state.[12] Since the 1990s, scholars with some disciplinary allegiance to the social history of medicine, such as Mark Harrison, Lenore Manderson, and Heather Bell, also have contributed significantly to our understanding of public health and colonialism in emergent nation-states.[13] More than their precursors, these area specialists and medical historians have engaged with political economies of health or else provided a revised calculus of the social costs and benefits of what had usually been regarded as the most rewarding aspect of the imperial enterprise. I shall return to a few salient features of these works later.

In a classic essay, published in 1976, Michael Worboys described the emergence of tropical medicine in Britain in the late nineteenth century. The first historian of medicine to study the intellectual development of tropical expertise, Worboys claimed that his subject "lucidly demonstrates the part played by 'external' social and economic factors in the establishment of scientific specialities."[14] The conceptual framework, derived from sociology of science, may look a little worn now, but for the following decade Worboys was virtually alone in considering critically the emergence of the British specialty of tropical medicine. In a series of influential papers, he linked tropical medicine tightly with "constructive imperialism," examining the politics of the different approaches and recommendations of Patrick Manson and Ronald Ross. By the 1990s, Worboys's studies were stimulating other British medical historians, many from a younger generation, to examine more critically the histories of the British schools of tropical medicine, as well as the international health services more generally.[15] Of course, much of this important work revealed more about the structuring of British—and sometimes broadly European and, on rare occasions, even North American—medical research and policy-making than it illuminated the actual entanglements of clinical practice and tropical hygiene in specific imperial settings. Above all, these studies indicated clearly, and repeatedly, the impact of the imperial imagination and mundane colonial needs on "metropolitan" ideas and institutions. It now seems remarkable the degree to which the history of tropical medicine became a British enthusiasm during this period, perhaps in part a reflection of lingering imperial nostalgia or discomfort.

A few specific diseases have long been associated with the tropics, or with warmer climates. August Hirsch and Scott, and later Erwin Ackerknecht, traced the history of many of these, and disease increasingly has come to join colony and discipline as a category of historical analysis.[16] Individual diseases may be considered as biological constants across space and time, thus offering the historian a social, intellectual, or political "sampling device" and conferring on the narrative a certain coherence, neatly bridging issues of causation, experience, and response.[17] Additionally, the historian's interest in the circumstances of a specific disease often matches, or overlaps, the ecological approach to understanding and managing disease that was such a distinctive feature of much tropical research.

Malaria, the most typical of "tropical" ailments, has prompted the most historical investigation, ranging from Gordon Harrison's *Mosquitoes, Malaria and Man* (1978) to the recent collective endeavors of the network for the history of malaria.[18] Using insights from disease ecology and medical anthropology, Maryinez Lyons has charted the history of trypanosomiasis, or sleeping sickness, in the Belgian Congo; outbreaks were the result of social dislocation, and control strategies for this complex disease could shape colonial economic and political structures.[19] Other scholars have found that controversies over the identity of a disease such as yellow fever can, in effect, give a sagittal section through the discipline, revealing previously hidden aspects.[20] Similarly, John Farley, in his ambitious, multisited history of schistosomiasis (or bilharzia) research, provides a richly textured study of the development of tropical medicine in "center" and "periphery," and in both the British and American empires.[21] It seems too that scrutiny of a chronic, disfiguring disease like leprosy, not necessarily limited to the tropics, can indicate, in various locations, the colonial framing of embodiment, race, and citizenship.[22] The history of specific diseases thus can retrace colonial boundaries, such as the "Belgian Congo," or it can track their causes and effects across a number of different imperial sites. Either way, disease provides a powerful analytic tool for the historian, but there may be times when a focus on one of the more excitingly "tropical" of ailments will obscure ordinary, unexotic patterns of disease prevalence. After all, diarrhea was generally more of a problem in the tropics than yellow fever, for example, but it has rarely captured the historian's imagination.

In attending to the historiography of medicine and colonialism, one is struck by how little the repertoire has changed since the 1940s. We still hark back to biography, disciplinary and institutional history, nationalist teleology, and the history of disease. But if genre is so sustained, content is not. The histories of colonial medicine written since the 1980s place a heavier emphasis on social and political setting; they are less likely to assume that tropical medicine was simply a variation on the theme of scientific progress, played in a minor key; and they tend to challenge the codependence of medicine and the colonial state. Another striking difference is simply the greater quantity of studies of colonial and postcolonial medicine emerging in the past fifteen years or so, most of it situated not in the European "center" but in the formerly colonial "periphery." The depth of interest is especially impressive when one includes other disciplines that have contributed to knowledge of health and healing in the colonial, or temporally postcolonial, world. Medical anthropologists, for example, explain health beliefs and non-Western medical systems in Asia and Africa and increasingly give their accounts historical grounding.[23] Still others, from sociology or economics, have studied the political economy of health, especially the entanglement of disease and development, in what came to be known, after the early 1950s, as the "Third World."[24] It is worth asking why so many scholars, at the end of the twentieth century and the beginning of the twenty-first, have been attracted to the study of medicine and colonialism.

Bitten by the Social History Bug

The impetus for writing social histories of colonial medicine came not from medical historians but from regional historians, in particular those associated with "area studies" in North America or "imperial and commonwealth history" in Britain. Regional historians eventually recognized that the study of disease, and the medical responses to it, would give them access to previously undisclosed aspects of personal experience, social life, and cultural influence. Moreover, the study of disease laid bare the ecologies of imperialism, and the operation of public health authorities could provide a particularly telling demonstration of the functioning of the colonial state. In Europe and North America, social history might supervene on an established history of medicine, but elsewhere, disease and medicine, if subjected to historical analysis at all, often were studied from the beginning as domains of general social history.

A recurring theme of the study of colonial disease, especially in Africa, is the linking of epidemics to imperial processes. Building on extensive scientific documentation of the impact of colonialism on local ecology, historians have shown repeatedly in various settings how disease can spread through increased contact and communication, war, agricultural change, and urbanization.[25] "Ecological" historians, such as Alfred W. Crosby and William McNeill, by the early 1970s had argued, on a global scale, that European invaders disrupted the environmental stability of the New World, spreading disease to vulnerable populations.[26] Following them, regional social historians produced ever more nuanced local studies of the impact of colonialism on disease patterns. In 1989, Randall M. Packard, for example, described how the South African reserve system by the 1930s was leading to increasing impoverishment and disease. "Changing patterns of capitalist production, and shifts in the demand and nature of labor supplies within colonial economies" caused much African disease and gave rise to disparaging representations of Africans.[27] Similarly, in an ambitious study of disease in colonial Malaya,

Lenore Manderson argued that "illness and deaths occurred which for many were shaped by the inequities, powerlessness and poverty produced by the structures of colonialism."[28] In an unusually intricate demographic reconstruction of the effects of colonial warfare, Ken de Bevoise has demonstrated that widespread disease in the Philippines, which Americans attributed to Filipino customs and habits, actually derived largely from social disruption, the result of a brutal American military campaign.[29] Ironically, Philip Curtin, in the other major reconstruction of colonial epidemiology, has argued that by the end of the nineteenth century public health efforts were capable of preventing some of the damaging effects of imperial rule, though local populations would rarely receive these benefits.[30] Medical intervention was still generally designed to protect the welfare of colonizers, whether settlers or sojourners; not until the twentieth century was it directed extensively, and often intrusively, to aid colonial laborers and the mothers of future workers.[31]

Colonial public health measures most commonly, it seems, produced further social and cultural disruption in local communities, and reinscribed older racial stigma, this time with even greater pathological depth, on the bodies of the colonized. Initially, colonizers had imagined themselves vulnerable to foreign environments, susceptible to a perceived incongruity of their race and new surroundings, but with broad acceptance of microbial pathology at the end of the nineteenth century, colonizers began to focus instead on the dangers of contact with local fauna, which included native races, refashioned as natural "reservoirs" of germs. "Natives" had supposedly adapted to the local disease environment, and if they became ill despite this evolutionary concord, then they must, it was supposed, either have behaved very badly or else become maladapted, a consequence perhaps of underpowered efforts to acquire "civilization." Medical theory thus exonerated socioeconomic conditions and the effects of imperialism and, instead, blamed the victims. Packard has described the myth of the "dressed native," which "placed responsibility for the apparent physical and mental failings of urban Africans, reflected in high mortality rates, alcoholism, family separations and crime, on African inexperience with the conditions of urban industrial life."[32]

It seemed that even apparently "healthy natives" might secretly carry germs that threatened white health.[33] Therefore, the colonial state, especially in southern Africa, increasingly would seek to limit contact between vulnerable whites and feckless or meretriciously healthy "natives." M. W. Swanson and others have documented systematic racial segregation, rationalized on sanitary grounds, which became common in many colonial societies and remained in place so long as it was consistent with local labor arrangements.[34] Colonial medical theory might lend support to a variety of population policies, depending on the local racial economy. Where whites were few and transient, they would often find sanitary excuses to isolate themselves in cantonments or compounds; where they dominated and planned to stay, as in Australia, they might justify the repatriation and exclusion of other races for microbial reasons.[35] In another setting, "progressive" and optimistic imperial powers—and those more concerned with maintaining a cheap, and relatively healthy, labor force—might emphasize the education and retraining of supposedly disease-dealing "natives." In the Philippines, colonial public health, long modeled on the military campaign, came to assume the character of an evangelical movement, converting Filipinos to the gospel of hygiene and thereafter relentlessly monitoring and disciplining their adherence to the new faith.[36] Although colonial medicine was deployed in different ways, depending on local economic and demographic resources, the ideology of empire was becoming, by the early twentieth century, increasingly medicalized, and colonial interventions were more likely than before to find expression in public health activity.

Recently, however, some historians have come to question assertions of the hegemonic character of colonial medicine and public health. When historians of colonial medicine, often using scanty archival sources, tried to reconstruct "history from below," they found that local health practices were often surprisingly resistant to Western medical intervention, and local identities and relationships could be relatively impervious to the colonial state's attempts to reform or reorganize them. It is not hard to detect here the influence of E. P. Thompson, Eric Hobsbawm, and other British social historians, but a growing awareness of anthropological studies of the continuity and adaptation of indigenous cultures also clearly played a part. In South Asia, though, the interest in decentering self-serving elite testimony derived perhaps most directly from the subaltern studies group, which claimed allegiance to the theories of Antonio Gramsci. Thus, David Arnold describes the reluctance of the Indian Medical Service to reshape social life and intimate behavior in response to cholera and smallpox epidemics. As Gramsci would have put it, British medical practices may have become dominant in nineteenth-century India, but they were not hegemonic or consensual.[37] Yet, even in India, the state eventually began to intervene more viscerally, attempting by 1914—the point at which Arnold ends his narrative—to discipline and "reform" subordinate populations. Elsewhere, the colonial state often intervened much earlier and more effectively in social arrangements. Medical management became more intense and intimate than it ever did in India, and although admittedly it was nowhere fully hegemonic, medicine in other colonies by the early twentieth century did generally come to exert more leverage on identity and behavior than on the subcontinent. Still, we know that, even in this later period, the disciplines of colonial medicine were frequently contested, or appropriated and adapted by local elites, or simply ignored.[38] Even in the Philippines—the government of which was among the most iatrocratic of colonial regimes—it was possible to maintain indifference to the health service or else channel its energies into projects that the colonized elite most desired.[39]

As disease and medicine were inserted into established genres of regional social history, the social history of colonial medicine, not surprisingly, assumed a form significantly different from the social history of medicine in Europe and North America. Recent histories of colonial medicine are more likely to bear the imprint of social or cultural anthropology and political economy; they focus more often on race and state formation; and they more commonly engage with a long colonial tradition of ecological reasoning. In explaining the social and environmental structuring of disease patterns, the impact of sanitary intervention on social boundaries and subject positions, and the varying cultural value and influence of Western medicine, historians of colonialism are thus offering medical historians, whatever their location, a rich vocabulary to express longstanding, if hitherto muted, concerns.

Medicine and the Making of Colonial Identities

Toward the end of the 1990s, many historians of colonial medicine chose to focus on biomedical contributions to the framing of the identities of colonizer and colonized. Cultural historians took up an older strand in the history of imperialism, weaving it together with disease and medicine into studies of the patterning of social boundaries and human difference, emphasizing the medical production of "subject positions," including racial and gender identities. Although many of the terms, and a little of the conceptual framework, were borrowed from Michel Foucault and Edward Said, the idea that images of disease and associated health practices might shape colonial subjectivity was hardly new.[40] In 1964, Philip Curtin had discussed the influence of disease theory on colonial European self-fashioning and on the representation of Africa and Africans.[41] Frantz Fanon, a psychiatrist regarded as one of the founders of "postcolonial studies," had often described, in work rarely cited by historians, how Western medicine in Algeria reshaped and disabled the colonized.[42] At the end of the twentieth century, many historians of colonial medicine would recognize and amplify the concerns of these earlier scholars, addressing more explicitly the biomedical framing of race, gender, embodiment, mentality, and citizenship.

In an early, and unusually intense, engagement with Foucault's theories, Megan Vaughan argued that "medicine and its associated disciplines played an important part in constructing 'the African' as an object of knowledge, and elaborated classification systems and practices which have to be seen as intrinsic to the operation of colonial power."[43] But in her studies of madness, leprosy, and health education, Vaughan questioned the extent to which colonial power in Africa created the individual subjectivities that Foucault had described for Europe. Biomedicine in its African colonial form seemed more commonly to produce not individuals but group identities, each with a distinctive collective psychology and body.[44] Even so, other historians of colonial medicine have found practices of hygiene, at different times and places, shaping individualized identities among the colonized. Arnold, for example, has argued that colonial South Asia demonstrates in an "exceptionally raw and accentuated" manner, the importance of medicine in the cultural and political construction of individual subjectivity, even though its operations on the body were always contested.[45] Unlike Foucault, Arnold suggests the ways in which state power directly promoted biomedical subject formation in the twentieth century, and he takes more seriously the impact of local resistance on the articulation of discourses of hygiene. Arnold's critical revision of Foucault, in which state power and subaltern agency are incorporated, has become more or less the standard version in the history of colonial medicine.

The recent "cultural turn" in histories of medicine and colonialism has directed attention to the importance of hygiene in disciplining, and even remaking, bodies, races, and citizens. Thus, for the Philippines, I have argued that a racializing of germ theories in the early twentieth century allowed the construction of a binary typology, in which a clean, ascetic American body represented one pole, and an open, polluting Filipino body the other. But having acknowledged and confessed their "putrescence," Filipinos were then expected to announce their desire for "civilization" or modernity and make themselves available for reformation. Progressive colonial health authorities would then monitor the perilous passage from "native" to proletarian, or citizen, and discipline those who strayed off track. In a sense, the "copy"—in this case, the development of the modern subject—was, in the twentieth century, becoming as interesting and functional as any typological construction of bodily difference. Of course, during the colonial period the successful completion of this "benevolent assimilation" into an ideal of American adulthood was always deferred, and its incompleteness would seem to justify continuing colonial surveillance and control. But in the 1930s, the local elite took up the development project with nationalist optimism so that increasingly public health would be defined in terms of "human rights" and not just the management of dangerous bodies.[46] This argument, even in such a simplified form, indicates how a conventional understanding of race, subjectivity, and citizenship may be refigured in studies of colonial medicine. Moreover, it should be clear that it describes a process that is not uniquely colonial, though it may be that in the colonial setting, as Arnold puts it, one finds its most "raw and accentuated" form.

The colonial asylum has been a major site for the study of the successes and failures of biomedical "subject formation"—especially the failures. Vaughan describes the inmates of the Nyasaland asylum as those who no longer conformed to the notion of the African subject and were beyond rehabilitation.[47] Similarly, Jock McCulloch argues that British asylums in Rhodesia and East Africa were little more than places for the disposal of dangerous African employees; practicing in such settings, colonial psychiatrists developed theories that disparaged the normal "African mind."[48] In a detailed social history of madness in colonial Nigeria, Jonathan Sadowsky documents the appalling condition of the local asylums, which ironically were taken to represent the "civilizing mission." Sadowsky attempts to read the patients' madness, "inchoate articulations of the stresses of colonial society," as a gauge of "social pressures and contradictions."[49] Through understanding colonial "pathology," it seems, one might understand the politics of colonial "normality" too.

In some colonies, especially after 1920, the "talking therapy" of psychoanalysis might on occasion be used to salvage elite neurotics, whether colonizers or colonized, and rehabilitate them as suitably modern subjects. India was perhaps the most "psychologized" of colonies, yet even there relatively few Indians or British became analysands. Fifteen medical doctors, two of them British, established the Indian Psychoanalytical Association in 1922; among its leading figures were Girindrasekhar Bose, a Bengali physician, and Owen Berkeley Hill, a socially defiant member of the Indian Medical Service and later the director of Ranchi Mental Hospital in Bihar. Ashis Nandy has observed that because psychoanalysis came to India along with colonialism, "the new psychological man had to be by definition, a colonial subject." It was "as if psychology had to be, by definition again, the latest in a series of techniques of retooling Indians into a prescribed version of the nineteenth-century European."[50] According to Sudhir Kakar, colonial psychoanalysis produced "images of people who talk the European language with a quaint engagingness but whose inner life is bland and certainly far less complex than that of the writers' middle-class English, French or American friends."[51] Nonwestern subjects of psychotherapy thus became, in a sense, second-class citizens in psychological modernity, poorly understood and crudely enculturated.

Even if the degree of biomedical influence on the identity of the colonized is still in dispute, there is little doubt that it played a crucial role in the self-fashioning of those who attempted to colonize them. Historians have eagerly reported on how colonial medicine shaped the corporeal and mental self-perceptions of European and North American settlers and sojourners. Colonizers generally were more prepared than the colonized to accept medical advice, which must have appeared to them an important resource for making sense of their bodies and their mentality in trying circumstances. They learned what to fear in a foreign milieu and how to behave with propriety in their colonial careers, and should they ever break down or succumb to tropical neurasthenia, psychiatrists would carefully work to restore their sense of white mastery. Medicine also constructed and reinforced the boundaries of gender in the colonies, delimiting, and if necessary recuperating, the qualities of white masculinity and femininity. Until recently, much of the historical analysis of the contributions of Western medicine to framing the subjectivities of colonizers has concentrated on theories of acclimatization, the study of racial adaptation to foreign climates and places.[52] Some historians, though, are now examining the more general permeation of medical ideas and practices through the cultures of the colonizers, describing the emergence of a mobile and happily alienated type of white body and mentality in the early twentieth century.[53] Much of this historical scholarship has helped to reveal potentially destabilizing anxieties and uncertainties in the identities of colonizers, showing the tensions and flaws in a supposedly hegemonic project. But if this approach were to prove more attractive, there is a danger that the study of colonial medicine would be reduced again to the study of colonial white men, thus reflecting the narcissistic demands of a part of the European and North American academy.

Postcolonial Histories of Medicine?

In 1988, Roy MacLeod observed that "in ways yet to be fully established, European medicine, playing upon a distant stage, formed a lasting part of the colonial legacy, the discourse of conquest, and the quixotic politics of deference and resistance."[54] The collection that MacLeod and Milton Lewis edited, which was quickly followed by Arnold's equally influential compilation, signaled the beginning of a new academic industry and the proliferation of critical histories of medicine and colonialism.[55] Many of the monographs that I discuss in this essay are found in embryonic form as contributions to these collections or can be traced to articles referred to in their extensive bibliographies. But the mature figures that emerged in the 1990s were often quite different from earlier expectations, and by the end of the decade, as we have seen, they were joined by a new generation of scholars that has further transformed the study of colonial medicine. Although the historical relations of medicine and colonialism are still not "fully established," as MacLeod discerned, our understanding of them is much more deeply rooted than it was fifteen years ago.

Do these new studies of the relations of medicine and colonialism constitute a postcolonial history of medicine? In the most elemental sense, most of these works are postcolonial: they are written after formal decolonization, and they often imply a trajectory from colony to new nation. In this sense, even the critical history of colonial medicine can be, as much as medicine itself, a means through which a new nation is imagined. Eventually every member of the United Nations will have its own history of germ theories and public health, its own social history of medicine. And why not? Surely we cannot have too much of such a good thing. Every nation-state has the right to view, and even to construe, its own history of medicine. But many of these multiplying national histories seem disappointingly enclosed and self-sufficient. We may observe how some ideas and practices come in with colonialism but rarely appear to circulate thereafter. There is frequently a predictable tone and teleology in these proliferating, meticulously situated, social histories of medicine. Writing of national histories more generally, Dipesh Chakrabarty has argued that "there is a peculiar way in which all these other histories tend to become variations on a master narrative that could be called 'the history of Europe.'"[56] Often it seems that histories of colonial medicine, predicated on the emergence of a national medical system, delimited by the borders of the future nation, also tend to become local variations of a master narrative called "the development of modern medicine."

There is, however, another sense in which the study of the relations of medicine and colonialism might become postcolonial. Among the myriad definitions of postcolonialism, I think that Stuart Hall's analytic framework may be the most pertinent here. Hall has suggested that the "post" in postcolonial might usefully be considered "under erasure," thus making the function of colonialism available as an analytic resource in any location. He argues that postcolonial studies should enable "decentered, diasporic, or 'global' rewriting of earlier nation-centered imperial grand narratives"—a "re-phrasing of Modernity within the framework of 'globalisation.'"[57] That is, medical historians, better perhaps than regional historians, might try to hold the histories of medicine in Europe, North America, and "the colonies" within the same analytic frame, thus dispersing and "colonializing" the standard history of Western medicine.[58] This would require historians of medicine to become more nomadic themselves, investigating disease, biomedical science and health care at a number of sites, tracing the passage of metaphor, practice, money, and career between them.[59] Our histories of medicine generally might become more realistic were they to incorporate into their narratives the dispersive economics and logic of colonialism, accounting for the mobility of ideas, models, and practices, recognizing the circulation of careers between imperial outposts and European and North American urban centers. Medicine was never, and nowhere, as insulated or bounded as it appears in conventional social histories of medicine. A postcolonial history of medicine in this sense would entail what Chakrabarty has called "the project of provincializing 'Europe,' the 'Europe' that modern imperialism and (third world) nationalism have by their collaborative ventures and violence made universal."[60] It might also mean that we are finally getting to the point where we can attempt, in our more critical style, the dynamic, multisited histories of medicine that Sigerist and Garrison urged us to do back in the 1930s.

NOTES

I would like to thank John Harley Warner and Frank Huisman for their comments and encouragement. James Vernon, Sarah Hodge, Ilana Löwy, and Rosa Medina also gave me helpful advice. An earlier version of this chapter was presented in the Sociology Department of National Taiwan University, thanks to Angela Leung and Shang-jen Li.

1. Henry E. Sigerist, "Problems of Historical-Geographical Pathology," *Bulletin of the History of Medicine* (hereafter *BHM*) 1 (1933): 10–18, 17–18. In particular, Sigerist expressed his admiration for August Hirsch, *Handbuch der Historisch-Geographischen Pathologie*, 2 vols., 2d ed. (Stuttgart, 1881–1886).

2. Fielding H. Garrison, "Medical Geography and Geographic Medicine," *Bulletin of the New York Academy of Medicine* 8 (1932): 593–612, 612, 607. See also the abbreviated version: "Geomedicine: A Science in Gestation," *BHM* 1 (1933): 2–9.

3. Susan Reverby and David Rosner, "Beyond 'the Great Doctors,'" in *Health Care in America: Essays in Social History,* David Rosner and Susan Reverby, eds. (Philadelphia: Temple University Press, 1979), 3–16, 4. On reading this sentinel work again, it is remarkable to find how little critical attention is given to the "situatedness" of social history—despite the broad sweep of the title, all of the studies collected here deal with sites in the United States east of the Mississippi and north of the Mason-Dixon line.

4. In this essay I will emphasize work on Africa and the Asia-Pacific region, though on occasion I refer to studies of Latin American medicine. Nonwestern medical systems, such

as Ayurveda and Chinese medicine, have separate historical literatures, which are not considered here.

5. David Arnold, *Colonizing the Body: State Medicine and Epidemic Disease in Nineteenth-Century India* (Berkeley: University of California Press, 1993), 9.

6. Aldo Castellani and Albert J. Chalmers, "History of Tropical Medicine," in *Manual of Tropical Medicine,* 3d ed. (New York: William Wood & Co., 1920), 3–38.

7. H. Harold Scott, *A History of Tropical Medicine,* 2 vols. (Baltimore: Williams & Wilkins, 1939–1942), vol. 1:v. For the French empire, see Paul Brau, *Trois siècles de médecine coloniale française* (Paris: Vigot Frères, 1931). See also Erwin Ackerknecht, *History and Geography of the Most Important Diseases* (New York: Hafner, 1965).

8. Charles Morrow Wilson, *Ambassadors in White: The Story of American Tropical Medicine* (New York: Henry Holt, 1942).

9. Victor Heiser, *An American Doctor's Odyssey: Adventures in Forty-Five Countries* (New York: W. W. Norton, 1936); S. M. Lambert, *A Yankee Doctor in Paradise* (Boston: Little, Brown, 1941); Gordon S. Seagrave, *Burma Surgeon* (New York:W. W. Norton, 1943); Jacques M. May, *Siam Doctor* (Garden City, N.Y.: Doubleday, 1949). The biographies included Howard A. Kelly, *Walter Reed and Yellow Fever* (New York: McClure, Phillips & Co., 1906); Ronald Ross, *Memoirs, with a Full Account of the Great Malaria Problem and its Solution* (London: J. Murray, 1923); M. D. Gorgas and B. J. Hendrick, *William Crawford Gorgas: His Life and Work* (New York: Doubleday, Page, 1924); Philip Manson-Bahr and A. Alcock, *The Life and Work of Sir Patrick Manson* (London: Cassell, 1927); and Hermann Hagedorn, *Leonard Wood: A Biography* (New York, Harper Bros., 1931). Among the more nuanced recent biographies is Douglas M. Haynes, *Imperial Medicine: Patrick Manson and the Conquest of Tropical Disease* (Philadelphia: University of Pennsylvania Press, 2001).

10. See, for example, Philip Manson-Bahr, *The History of the School of Tropical Medicine in London, 1899–1940* (London: School of Hygiene and Tropical Medicine, 1956); Edmond Sergent and L. Parrot, *Contribution de l'Institut Pasteur d'Algérie à la connaisance humaine du Sahara, 1900–1960* (Algiers: Institut Pasteur, 1961); and Willard H. Wright, *40 Years of Tropical Medicine Research: A History of the Gorgas Memorial Institute of Tropical and Preventive Medicine, Inc., and the Gorgas Memorial Laboratory* (Baltimore: Reese Press, 1970).

11. Among examples of the former are Ralph Schram, *A History of the Nigerian Health Services* (Ibadan: Ibadan University Press, 1971); P. W. Laidler and Michael Gelfand, *South Africa: Its Medical History, 1652–1898* (Cape Town: Struik, 1971); and Michael Gelfand, *A Service to the Sick: A History of the Health Services for Africans in Southern Rhodesia, 1890–1953* (Gwelo: Mambo Press, 1976). A precursor in this genre was D. G. Crawford, *A History of the Indian Medical Service, 1600–1913,* 2 vols. (London: W. Thacker, 1914).

12. For example, Ann Beck, *A History of the British Medical Administration of East Africa* (Cambridge, Mass.: Harvard University Press, 1970); K. David Patterson, *Health in Colonial Ghana: Disease, Medicine and Socio-Economic Change, 1900–55* (Waltham, Mass.: Crossroads Press, 1981); Nancy E. Gallagher, *Medicine and Power in Tunisia, 1780–1900* (Cambridge: Cambridge University Press, 1983); Donald Denoon, *Public Health in Papua New Guinea: Medical Possibility and Social Constraint, 1884–1984* (Cambridge: Cambridge University Press, 1989); and Arnold, *Colonizing the Body.*

13. Mark Harrison, *Public Health in British India: Anglo-Indian Preventive Medicine, 1859–1914* (Cambridge: Cambridge University Press, 1994); Lenore Manderson, *Sickness and the*

State: Health and Illness in Colonial Malaya, 1870–1940 (Cambridge: Cambridge University Press, 1996); Heather Bell, *Frontiers of Medicine in the Anglo-Indian Sudan, 1899–1940* (Oxford: Clarendon Press, 1999). See also Anne Marcovich, *French Colonial Medicine and Colonial Rule: Algeria and Indochina* (London: Routledge, 1988); Wolfgang Eckart, *Medizin und Kolonialimperialismus: Deutschland, 1884–1945* (Paderborn: Schoningh, 1997); and Julyan G. Peard, *Race, Place and Medicine: The Idea of the Tropics in Nineteenth-Century Brazilian Medicine* (Durham, N.C.: Duke University Press, 1999).

14. Michael Worboys, "The Emergence of Tropical Medicine: A Study in the Establishment of a Scientific Specialty," in *Perspectives on the Emergence of Scientific Disciplines,* Gérard Lemaine, Roy MacLeod, Michael Mulkay, Peter Weingart, eds. (The Hague: Mouton, 1976), 76–98, 76. See also "The Origins and Early History of Parasitology," in *Parasitology: A Global Perspective,* K. S. Warren and J. Z. Bowers, eds. (New York: Springer Verlag, 1983), 1–18; "Manson, Ross and Colonial Medical Policy: Tropical Medicine in London and Liverpool, 1899–1914," in *Disease, Medicine and Empire,* Roy MacLeod and Milton Lewis, eds. (London: Routledge, 1988), 21–37; and "British Colonial Medicine and Tropical Imperialism: A Comparative Perspective," in *Dutch Medicine in the Malay Archipelago, 1816–1942,* G. M. van Heteren, A. de Knecht-van Eekelen, and M. J. D. Poulissen, eds. (Amsterdam: Rodopi, 1989), 153–167. The antecedents of modern tropical medicine are described in *Warm Climates and Western Medicine: The Emergence of Tropical Medicine, 1500–1900,* David Arnold, ed. (Amsterdam: Rodopi, 1996). For an earlier effort to distinguish "external" and "internal" influences, see R. H. Shryock, "The Interplay of Social and Internal Factors in Modern Medicine: An Historical Analysis," in his *Medicine in America: Historical Essays* (Baltimore: Johns Hopkins University Press, 1966), 307–332.

15. Helen Power, *Tropical Medicine in the Twentieth Century: A History of the Liverpool School of Tropical Medicine, 1898–1990* (London: Kegan Paul International, 1999); *International Health Organisations and Movements, 1918–1939,* Paul Weindling, ed. (Cambridge: Cambridge University Press, 1995).

16. Ackerknecht, *History and Geography.* See also *Disease in African History: An Introductory Survey and Case Studies,* G. W. Hartwig and K. D. Patterson, eds. (Durham, N.C.: Duke University Press, 1978).

17. Charles E. Rosenberg, "Cholera in Nineteenth-Century Europe: A Tool for Social and Economic Analysis," *Comparative Studies in Society and History* 8 (1966): 135–162, and *The Cholera Years: The United States in 1832, 1849, and 1866* (Chicago: University of Chicago Press, 1962).

18. Gordon Harrison, *Mosquitoes, Malaria, and Man: A History of the Hostilities since 1880* (New York: E. P. Dutton, 1978); Mary Dobson, Maureen Malowany, and Darwin Stapleton, eds., "Dealing with Malaria in the Last Sixty Years," *Parasitologia* 42 (2000): 3–182. See also Erwin H. Ackerknecht, *Malaria in the Upper Mississippi Valley, 1760–1900* (Baltimore: Johns Hopkins Press, 1945), and P. F. Russell, *Man's Mastery of Malaria* (London: Oxford University Press, 1963).

19. Maryinez Lyons, *The Colonial Disease: A Social History of Sleeping Sickness in Northern Zaire* (Cambridge: Cambridge University Press, 1992). See also J. Ford, *The Role of Trypanosomiasis in African Ecology* (Oxford: Clarendon Press, 1971); Michael Worboys, "The Comparative History of Sleeping Sickness in East and Central Africa, 1900–1914," *History of Science* 32 (1994): 89–102; and Luise White, "Tsetse Visions: Narratives of Blood and Bugs in Colonial Northern Rhodesia, 1931–9," *Journal of African History* 36 (1995): 219–245.

20. For example, François Delaporte, *The History of Yellow Fever: An Essay on the Birth of Tropical Medicine,* trans. Arthur Goldhammer (Cambridge, Mass.: MIT Press, 1991). See also Nancy Stepan, "The Interplay between Socio-Economic Factors and Medical Science: Yellow Fever Research, Cuba and the United States," *Social Studies of Science* 8 (1978): 397–423; Rosa Medina-Doménech, "Paludismo, explotación y racismo científico en Guinea ecuatorial," in *Terratenientes y parásitos: La lucha contra el paludismo en la España del siglo XX,* E. Rodríguz-Ocaña, ed. (Granada: University of Granada Press, 2001); and Ilana Löwy, *Virus, moustiques et modernité: La fièvre jaune au Bresil entre science et politique* (Paris: Archives d'histoire contemporaine, 2001).

21. John Farley, *Bilharzia: A History of Imperial Tropical Medicine* (Cambridge: Cambridge University Press, 1991). See also G. W. Hartwig and K. D. Patterson, *Schistosomiasis in Twentieth-Century Africa: Historical Studies in West Africa and Sudan* (Los Angeles: Crossroads Press, 1984).

22. Warwick Anderson, "Leprosy and Citizenship" *Positions* 6 (1998): 707–730; Suzanne Saunders, "Isolation: The Development of Leprosy Prophylaxis in Australia," *Aboriginal History* 14 (1990): 168–191; Megan Vaughan, "Without the Camp: Institutions and Identities in the Colonial History of Leprosy," in *Curing Their Ills: Colonial Power and African Illness* (Stanford: Stanford University Press, 1991), 77–99; Rita Smith Kipp, "The Evangelical Uses of Leprosy," *Social Science and Medicine* 39 (1994): 165–178, 176; Harriet Jane Deacon, "A History of the Medical Institutions on Robben Island, Cape Colony, 1846–1910" (Ph.D. thesis, University of Cambridge, 1994), chap. 6; Sanjiv Kakar, "Leprosy in British India, 1860–1940: Colonial Politics and Missionary Medicine," *Medical History* 40 (1996): 215–230.

23. Among influential anthropological accounts are J. M. Janzen, *The Quest for Therapy in Lower Zaire* (Berkeley: University of California Press, 1978); Steven Feierman, "Struggles for Control: The Social Roots of Health and Healing in Modern Africa," *African Studies Review* 28 (1985): 73–147; and John and Jean Comaroff, *Ethnography and the Historical Imagination* (Boulder: Westview, 1992). See also the essays in *The Social Basis of Health and Healing in Africa,* Steven Feierman and John M. Janzen, eds. (Berkeley: University of California Press, 1992).

24. For example, Vicente Navarro, *Medicine under Capitalism* (London: Croom Helm, 1976); E. Richard Brown, "Public Health and Imperialism in Early Rockefeller Programs at Home and Abroad," *American Journal of Public Health* 66 (1976): 897–903; Lesley Doyal with Imogen Pennell, *The Political Economy of Health* (London: Pluto Press, 1979); and Meredith Turshen, *The Political Ecology of Disease in Tanzania* (New Brunswick, N.J.: Rutgers University Press, 1984).

25. See, for example, Ford, *Role of Trypanosomiasis,* and C. C. Hughes and J. M. Hunter, "Disease and Development in Africa," *Social Science and Medicine* 3 (1970): 443–493. Helen Tilley has described the ecological orientation of much tropical research in Africa in a "living laboratory": "The African Research Survey and the British Colonial Empire: Consolidating Environmental, Medical, and Anthropological Debates, 1920–1940," (D.Phil. thesis, Oxford University, 2001), chap. 5. More generally, see Warwick Anderson, "Ecological Vision in Biomedicine: Natural Histories of Disease in the Twentieth Century," *Osiris* (forthcoming, 2003).

26. Alfred W. Crosby Jr., *The Columbian Exchange: Biological and Cultural Consequences of 1492* (Westport Conn.: Greenwood Press, 1972); William H. McNeill, *Plagues and People*s

(New York: Anchor Books, 1976). For an earlier historical account of these processes, see P. M. Ashburn, *The Ranks of Death: A Medical History of the Conquest of America* (New York: Coward-McCann, 1947).

27. Randall M. Packard, "The 'Healthy Reserve' and the 'Dressed Native': Discourses on Black Health and the Language of Legitimation in South Africa," *American Ethnologist* 16 (1989): 686–703, 688, and *White Plague, Black Labor: Tuberculosis and the Political Economy of Health and Disease in South Africa* (Berkeley: University of California Press, 1989).

28. Manderson, *Sickness and the State*.

29. Ken de Bevoise, *Agents of Apocalypse: Epidemic Disease in the Colonial Philippines* (Princeton: Princeton University Press, 1995).

30. Philip D. Curtin, *Death by Migration: Europe's Encounter with the Tropical World in the Nineteenth Century* (Cambridge: Cambridge University Press, 1989).

31. See Manderson, *Sickness and the State*.

32. Packard, "The 'Healthy Reserve' and the 'Dressed Native,'" 687.

33. Warwick Anderson, "Immunities of Empire: Race, Disease, and the New Tropical Medicine, 1900–1940," *BHM* 70 (1996): 94–118.

34. M. W. Swanson, "The Sanitation Syndrome: Bubonic Plague and Urban Native Policy in the Cape Colony, 1900–1909," *Journal of African History* 18 (1977): 387–410. See also Philip D. Curtin, "Medical Knowledge and Urban Planning in Tropical Africa," *American Historical Review* 90 (1985): 594–613, and the response by John W. Cell, "Anglo-Indian Medical Theory and the Origins of Segregation in West Africa," *American Historical Review* 91 (1986): 307–333.

35. On the Australian physiocratic state and its quarantine mentality, see Alison Bashford, "Quarantine and the Imagining of the Australian Nation," *Health* 2 (1998): 387–402, and Warwick Anderson, *The Cultivation of Whiteness: Science, Health, and Racial Destiny in Australia* (New York: Basic Books, 2003).

36. Warwick Anderson, "Excremental Colonialism: Public Health and the Poetics of Pollution," *Critical Inquiry* 21 (1995): 640–669, and "Leprosy and Citizenship." For an earlier segregationist discourse in American medicine in Philippines, see Reynaldo C. Ileto, "Cholera and the Origins of the American Sanitary Order in the Philippines," in *Imperial Medicine and Indigenous Society,* David Arnold, ed. (Manchester: Manchester University Press, 1989).

37. Arnold, *Colonizing the Body* and "Gramsci and Peasant Subalternity in India" [1984], in *Mapping Subaltern Studies and the Postcolonial,* Vinayak Chaturvedi, ed. (London: Verso, 2000), 24–49. See also Roger Jeffrey, "Recognizing India's Doctors: The Institutionalization of Medical Dependency, 1918–39," *Modern Asian Studies* 13 (1979): 301–326, and Gyan Prakash, *Another Reason: Science and the Imagination of Modern India* (Princeton: Princeton University Press, 1999).

38. *Western Medicine as Contested Knowledge,* Andrew Cunningham and Bridie Andrews, eds. (Manchester: Manchester University Press, 1997).

39. Warwick Anderson, "Going through the Motions: American Public Health and Colonial 'Mimicry,'" *American Literary History* 14 (2002): 686–719.

40. Many historians found Foucault's work, in particular *Discipline and Punish: The Birth of the Prison,* trans. Alan Sheridan (Harmondsworth: Penguin, 1991), heuristically useful and stimulating, but the impact of Said's *Orientalism* (New York: Random House, 1978) did

not spread much beyond South Asia, where in any case it was frequently dismissed as facile and narrowly textual.

41. Philip Curtin, *The Image of Africa: British Ideas and Action, 1780–1850* (Madison: University of Wisconsin Press, 1964).

42. See for example, Frantz Fanon, "Medicine and Colonialism," in *The Cultural Crisis of Modern Medicine,* John Ehrenreich, ed. (New York: Monthly Review Press, 1978), 229–245.

43. Vaughan, *Curing Their Ills,* 8.

44. Vaughan's thesis contrasts with Comaroff and Comaroff, *Ethnography and the Historical Imagination.*

45. Arnold, *Colonizing the Body,* 9. Unfortunately, Arnold's analysis stops just when notions of hegemony and discipline might be employed more fruitfully. See also Nicholas Thomas, "Sanitation and Seeing: The Creation of State Power in Early Colonial Fiji," *Comparative Studies in Society and History* 32 (1990): 149–170.

46. Anderson, "Going Through the Motions." For a more extensive survey of work on race and embodiment in colonial medicine, see Warwick Anderson, "The 'Third World' Body," in *Medicine in the Twentieth Century,* Roger Cooter and John Pickstone, eds. (Amsterdam: Harwood Academic, 2000), 235–246. See also P. M. E. Lorcin, "Imperialism, Colonial Identity and Race in Algeria, 1830–1870: The Role of the French Medical Corps," *Isis* 90 (1999): 653–679; Alexander Butchart, *The Anatomy of Power: European Constructions of the African Body* (London: Zed Books, 1998); and Nancy Scheper-Hughes, "AIDS and the Social Body," *Social Science and Medicine* 19 (1994): 991–1003.

47. Vaughan, *Curing Their Ills,* chap. 5. See also Sander L. Gilman, *Difference and Pathology: Stereotypes of Sexuality, Race, and Madness* (Ithaca: Cornell University Press, 1985).

48. Jock McCulloch, *Colonial Psychiatry and "the African Mind"* (New York: Cambridge University Press, 1995).

49. Jonathan Sadowsky, *Imperial Bedlam: Institutions of Madness in Colonial Southwest Nigeria* (Berkeley: University of California Press, 1999), 75, 77. See also James H. Mills, *Madness, Cannabis, and Colonialism: The "Native Only" Lunatic Asylums of British India, 1857–1900* (New York: St. Martin's, 2000). For a perceptive review essay, see Richard Keller, "Madness and Colonization: Psychiatry in the British and French Empire, 1800–1962," *Journal of Social History* 35 (2001): 295–326.

50. Ashis Nandy, "The Savage Freud: The First Non-Western Psychoanalyst and the Politics of Secret Selves in Colonial India," in *The Savage Freud, and Other Essays on Possible and Retrievable Selves* (Princeton: Princeton University Press, 1995), 81–144, 139. See also Christiane Hartnack, "British Psychoanalysts in Colonial India," in *Psychology in 20th-Century Thought and Society,* Mitchell G. Ash and William R. Woodward, eds. (Cambridge: Cambridge University Press, 1987), 233–251, and "Vishnu on Freud's Desk: Psychoanalysis in Colonial India," *Social Research* 57 (1990): 921–949.

51. Sudhir Karkar, "Culture in Psychoanalysis," in *Culture and Psyche: Selected Essays* (Delhi: Oxford University Press, 1997), 1–19, 15. See also Françoise Vergès, "Chains of Madness, Chains of Colonialism: Fanon and Freedom," in *The Fact of Blackness: Frantz Fanon and Visual Representation,* Alan Read, ed. (Seattle: Bay Press, 1996), 47–75.

52. David N. Livingstone, "Human Acclimatisation: Perspectives on a Contested Field of Inquiry in Science, Medicine, and Geography," *History of Science* 25 (1987): 359–394, "Climate's Moral Economy: Science, Race, and Place in Post-Darwinian British and American

Geography," in *Geography and Empire,* Anna Godlewska and Neil Smith, eds. (Oxford: Blackwell, 1994), 132–154, and "Tropical Climate and Moral Hygiene: The Anatomy of a Victorian Debate," *British Journal of the History of Science* 32 (1999): 93–110; Dane Kennedy, "The Perils of the Midday Sun: Climatic Anxieties in the Colonial Tropics," in *Imperialism and the Natural World,* John M. McKenzie, ed. (Manchester: Manchester University Press, 1990), 118–140; Mark Harrison, *Climates and Constitutions: Health, Race Environment, and British Imperialism in India, 1600–1850* (New Delhi: Oxford University Press, 1999).

53. Warwick Anderson, "'The Trespass Speaks': White Masculinity and Colonial Breakdown," *American Historical Review* 102 (1997): 1343–1370, and *Cultivation of Whiteness.* For earlier understandings of European "breakdown," and the means to preserve white prestige, see Waltraud Ernst, *Mad Tales from the Raj: The European Insane in British India, 1800–1858* (New York: Routledge, 1991).

54. Roy MacLeod, "Introduction," in MacLeod and Lewis, *Disease, Medicine and Empire,* 1–18, 4.

55. Arnold, *Imperial Medicine and Indigenous Society.*

56. Dipesh Chakrabarty, "Postcoloniality and the Artifice of History: Who Speaks for 'Indian' Pasts?" *Representations* 37 (1992): 1–26, 1.

57. Stuart Hall, "When was 'the Post-Colonial'? Thinking at the Limit," in *The Post-Colonial Question: Common Skies, Divided Horizons,* Iain Chambers and Lidia Curti, eds. (London: Routledge, 1996), 242–260, 247, 250.

58. Shula Marks has made a similar point in "What is Colonial about Colonial Medicine? And What Has Happened to Imperialism and Health?" *Social History of Medicine* 10 (1997): 205–219.

59. Some of these intercolonial and transnational passages are prefigured in the work of Farley, *Bilharzia;* Philippa Levine, "Modernity, Medicine and Colonialism: The Contagious Diseases Ordinances in Hong Kong and the Straits Settlements," *Positions* 6 (1998): 675–705; and Mary P. Sutphen, "Not What, but Where: Bubonic Plague and the Reception of Germ Theories in Hong Kong and Calcutta," *Journal of the History of Medicine* 52 (1997): 81–113, 112. A coordinated study of distribution of hookworm expertise and the associated circulation of hookworm experts (mostly from the Rockefeller Foundation) in the Americas and Asia is long overdue. This would build on *Missionaries of Science: The Rockefeller Foundation and Latin America,* Marcos Cueto, ed. (Bloomington: Indiana University Press, 1994); Stephen Palmer, "Central American Encounters with Rockefeller Public Health, 1914–21," in *Close Encounters of Empire: Writing the Cultural History of U.S.–Latin American Relations,* Gilbert Joseph, Catherine C. LeGrand, and Ricardo Salvatore eds. (Durham: Duke University Press, 1998), 311–332; Anne-Emanuelle Birn, "A Revolution in Rural Health? The Struggle over Local Health Units in Mexico, 1928–1940," *Journal of the History of Medicine and Allied Sciences* 53 (1998): 43–76; Anne-Emanuelle Birn and Armando Solorzano, "Public Health Policy Paradoxes: Science and Politics in the Rockefeller Foundation's Hookworm Campaign in Mexico in the 1920s," *Social Science and Medicine* 49 (1999): 1197–1213; and Anderson, "Going through the Motions" and *The Cultivation of Whiteness,* chap. 6.

60. Chakrabarty, "Postcoloniality," 20.

Part III / After the Cultural Turn

CHAPTER FOURTEEN

"Framing" the End of the Social History of Medicine

Roger Cooter

Apart from involving me in something close to writing my own obituary, framing the end of the social history of medicine is difficult for the simple reason that the subdiscipline's roll into the grave was far from obvious or straightforward. The end was more in the manner, literally, of a passing—a sluggish, uneventful meltdown, nowhere much noticed or commented on. Indeed, the walking dead are still many among us: explicitly "social" histories of medicine continue to be written, and undergraduate courses in the social history of medicine (not to mention a journal by the name) continue to be subscribed to. This is odd because, in somewhat more than literal fashion, the end was heralded over a decade ago when Charles Rosenberg—the acknowledged doyen of the field—proposed the banishment of "the social" in the social history of medicine and its replacement with "the frame."[1] The substitution was widely endorsed.[2] However, the expressed motives for it (as well as the unexpressed ones I will come to) were different from the historiographical ones that *might* have been posed. Thus, the actual ending of the social history of medicine was obscured at the same time as the literal ending was (in that other sense of the word) "framed." Such is my contention. This chapter, therefore, is mainly concerned with the wider context for the demise of "the social" in the history of medicine—with what might be styled (though I wish to imply no conspiracy) the framing of the framing.

That our concern has to do with more than merely the substitution of words first became apparent to me in the mid-1990s when I was approached to write a social history of medicine. The prospect was enticing. Here would be a chance to pull together some of the many different themes that had come to comprise the field: the politics of professionalization, alternative healing, the study of patient narratives, welfare strategies, constructions of sexuality and gender, madness, deviance, diseases and disability, public health and private practice, ethics, epidemiology, experimentation and education, tropical medicine and imperialism, along with the various actual and rhetorical relations between medicine and war, medicine and technology, medicine and art, medicine and literature, and so on. More than that, the book would afford an opportunity under such headings, chapter by chapter, to review how, over the past two or three decades, historical and historiographical understandings had broadened, sharpened, and deepened. There was need for such a book, and there probably still is. But when pen came to paper, paralysis set in. The problem wasn't the survey of the literatures involved, formidable though they had become. Rather, it was more the nature of, and need for, the overall packaging. A *social history of medicine?* In 1992, Andrew Wear proudly declared that it had "come of age," to which Ludmilla Jordanova roundly responded that it was "still in its infancy."[3]

What caused my pen to falter, though, was the realization that somehow, somewhere along the way in the 1980s and 1990s, all the key words had lost their certainty of meaning, and some (*pace* Rosenberg) had even been threatened with excision. No longer could they be taken for granted: "history" and "medicine," no less than "the social," had become deeply problematic. The hinges of the triptych were rusting up and coming unstuck; historiographically, the whole was no longer the sum of its elemental parts. Such a realization was more than a little worrying politically, inasmuch as the implied loss of disciplinary coherence could only disable the critical, if not socialist, thrust of an enterprise that had been honed over the previous quarter-century in alliance with social medicine, medical sociology, and social history—an unusual mix of radical politics and policy, critique and theory, and empirical practice, represented by Thomas McKeown, Ivan Illich, and E. P. Thompson, respectively.

I have no desire to defend the use of "the social" in the social history of medicine. It might be as well to admit that the subdiscipline has had its day, done its bit, much in the manner of older forms of political and economic history. Given that much social history of medicine has been intellectually flatfooted and theoretically unreflective—at best revisionist, at worst dominated by empiricism and even scientism (if a vast number of demographic contributions are taken into account)—moving on may be no bad thing. The intention here is more modest: merely to capture something of the wider intellectual context that has in fact necessitated this moving on and ought, I believe, further to inform it. Thus, this chapter has been conceived mainly as a record of what happened, a chronicle of an "ending" insofar as history (the business of history writing) ever has closure.[4] It seems worth doing, for while the history wars have been conducted over the past few decades as vigorously as the culture wars and the science wars of which they are a part, it is not at all clear what exactly has changed, or how and why. Least of all is this clear to new entrants to the history of medicine.

Admittedly, exactitude is easier called for than accomplished, given the almost ineffable (for being so recent, radical, and multifaceted), inevitably partial, and certainly very messy nature of the context within which the meaning and practice of history writing has been challenged. For these reasons, among others, it is not possible to provide anything like a social constructivist account of the historiography of the end of the social history of medicine—an analysis, that is, in terms of a situated body of knowledge mediating socioeconomic and political interests. (The subdiscipline was never "schooled" enough for that; indeed, it was always in tension with the aspiration to de-ghettoize itself by merging into the historical mainstream.) Nor, for essentially the same reasons, is it possible to provide either an analysis of the discourse of the historiographical body or a deconstructivist, semiotic account of "the text," since neither the infrastructural body nor the text as such exist. We can, however, lay out some of the features of the subject's problematization with reference to its wider intellectual and, to some extent, sociopolitical context. This is easiest done by focusing first on the key words, starting with "medicine" itself. We can then turn to an analysis of Rosenberg's article on framing, locating it within the politics of historiographical change. Finally, by way of a postscript on the postmortem, as it were, some observations can be tendered on the possible place of "the social," "the political," and "the medical" in history now.

"Medicine," "History," and "the Social"

"Medicine" has always been troublesome, if one stopped to think of it. Recently described by John Pickstone as a convenient omnibus term, resembling in this respect "agriculture," or maybe "engineering" or "electronics,"[5] it invariably consists of more than merely the professional practice of licensed healers in all their economic, political, and social settings. It is more, too, than just the knowledge of diseases and processes affecting the body in sickness and health and the prevailing technologies for corporeal intervention.[6] Worldviews and ideologies like humoralism, evolutionism, and environmentalism or Christianity, communism, and fascism have necessarily been as much a part of it as more specific sets of shifting discourses, rhetorics, and representations. "Cultures of healing" and "healing in culture" as objects of study can be as conceptually broad or narrow as the analyst chooses to make them. As knowledge and as practice, medicine (like religion) can be experienced at the most intimate level of being and politics, as well as, simultaneously, through precipitates from the highest reaches of global ideology and economics. And not just "experienced" in any simple sense; as Foucault appreciated, medical sites and personnel historically have been bound up with mutations of political thought into their modern governmental form. In the course of such processes, individuals have come to describe themselves in the languages of health and illness, and to accept the norms of "the normal" and "the pathological" as the basis for circumscribing their mortal and moral existence.[7] Obviously, then, as Jordanova has insisted, we cannot treat medicine simply as knowledge, or merely as "another form of science."[8]

At root, medicine is about power: "the power of doctors and of patients, of institutions such as churches, charities, insurance companies, or pharmaceutical manufacturers, and especially governments, in peacetime or in war."[9] Recently, however, there has been a seismic shift in the nature and exercise of the powers that constitute "medicine." Whether our gaze is on the disarray of health services in post–cold war Eastern Europe, China, and the poorer countries of postcolonial Africa, or on the fast-changing reorganization of medical systems in the West, things look far different from what they did ten or twenty years ago. Crucially, welfare medicine has been felled. National health services now embrace the logic of private, multinational corporations.[10] In Britain's National Health Service this thinking is evident in the adoption of private finance initiatives for the building of new hospitals and the turning of general practitioners into NHS "fund holders" operating according to the rules of internal markets. The shift is further made apparent and exercised through the emphasis since the 1980s on "evidence-based medicine," the mechanism enabling the state to pay only for treatments for which there is statistical evidence of benefit—a concept and practice of particular appeal to managers and accountants. Many of these changes have been pioneered in the United States, where doctors now queue to obtain degrees in business administration in order not to be irrelevant in the so-called medical reform process.[11] As this suggests, fears by the medical profession (especially in Britain and the United States) of losses of autonomy no longer stem directly from the state, as they were thought to in the past, but from the rationalizing forces of business management, albeit encouraged by governments. In the United States, more than one-half of all doctors are now salaried employees of medical corporations, and the consequent rhetoric of an "embattled profession" has been taken as compelling evidence of the decline of professional authority in general.[12]

Professional authority—power—in medicine has also been seriously challenged by outside interference in the hitherto professionally sacrosanct area of clinical decision-making. Both cost calculating managers and evidence-weighing governments now act as third parties in such decisions, even though doctors remain the principal targets of complaint in what they perceive as a "culture of blame." Through sensational exposure of medical malpractice and incompetence, further reason is found for withdrawal of the state's former compliance in the profession's self-regulation (existing in Britain since the Medical Registration Act of 1858). Now, the profession's ethical governance is increasingly given over to the courts through legal regulation.

Such redistributions in the power around medicine are reflected in, and inseparable from, equally profound changes in doctor-patient relations—to the extent that the word "patient," too, has lost much of its certainty of meaning and has become open to contestation.[13] In the face of fragmentation, depersonalization, and multiprofessionalization in the delivery of medical care, calls have been made for "narrative-based medicine," the ethics of which allege "the primacy of the patient's voice."[14] Meanwhile, increasing numbers of "health consumers" have turned to untested therapies by unregulated practitioners while at the same time demanding more evidence-based regulated medicine. Many of the same health consumers (wealthy, white, and Western for the most part) partake in what Roy Porter once dubbed the "MacDonaldization of medicine."[15] They demand as a "right" the freedom to shop unfettered in the "supermarkets of life," picking and choosing reproductive technologies as readily as vaccines, kidneys, hearts, and assisted suicides. Much of the laity has also been medically reskilled and empowered through web sites and illness support-groups, and it now comes to act as a jury for a medicine on trial.[16] Clearly, in the pluralized medical marketplace there is not one but multiple sources of authority, and the idea of the passive patient is noticeably passé.

This rearrangement of power is further reflected in the refutation by some medical ethicists of the very existence and operation of a universalistic morality, such as that purported for the late twentieth century by the philosopher John Rawls. The idea of a moral consensus (Rawls's "reflective equilibrium") presupposed social homogeneity. But this has become increasingly difficult to imagine and decreasingly desirable to sustain in liberal democracies where multiculturalism and multifaith are hyped as politically correct. Hence, in theory at least, different, equally "rational," interpretations of what is medically ethical have come to coexist based on religion, culture, and ethnicity.[17] Alongside, and not entirely separate from, this development has been the undermining of the instrumental rationality of modern medicine. Models of linear progress, and belief in rational control over the processes of medicine, have been seriously eroded by the manifestation of previously unforeseen risks and the negative side-effects of biomedical "progress"—erosions that mark, in Ulrich Beck's terms, the shift from "simple" to "reflexive modernity."[18] Once these risks began to be registered in the minds of insurance companies, medical professionals, medical management teams, the state, and potential individual "clients," the would-be rationality of biomedical research began to be demystified and demonopolized. (Contributing to this view was a heightened awareness of how multinational and monopolistic-tending pharmaceutical companies controlled much of the research, as well as the allocation of products.) Increasingly, the public began to decide between different plausible or probable scientific claims. In this situation, as has been pointed out in reference to the pressures for redefining brain death, political groups came to make use of scientific expertise and counter-expertise in order to push forward their own favorite practical and legal solutions.[19]

Unsurprisingly, such fundamental change in the relations, organization, consumption, and overall conception of modern medicine has had an impact on how medicine is thought about historically. Specifically, the notion of the social control of docile bodies, which was basic to the social history and historical sociology of medicine as it developed in the 1970s and 1980s, has come to seem dated as an analytical imperative. No longer is it quite so obvious to regard medicine simply as a powerful means of imposing social order though "disciplinary normalization." Perhaps in the past as in the present, the relationship between medicine and the laity entailed wider interactions between self, society, and knowledge, all according to competing priorities and the different material constraints of everyday life.[20] To arrive at this conclusion in no way necessitates discarding Foucault's insights on the crucial role of modern medical language and practice as a medium for the policing and self-policing of bodies and desires. Required, rather, is discarding crude or vulgar histories of medical surveillance, social control, and the deskilling of patients—1970s and 1980s contributions to the historical narratives of professionalization and medicalization. In part, these narratives were hoist on their own petard in the 1980s and 1990s when the self-serving antiprofessionalism of radical feminists and critics came to support dialectically the interests of free market ideologues and antiabortionists, along with eco-activists, neofascists, ravers, "hacktivists," and the others who now make up "do-it-yourself" culture.[21]

Although medicine as an epistemological and discursive concern should not be conflated with medicine in the service of professional power, it is fair to say that the very idea of "medicine" or "the medical" has been destabilized. As Nikolas Rose has pointed out, "What we have come to call medicine is constituted by a series of associations between events distributed along a number of different dimensions, with different histories, different conditions of possibility, different surfaces of emergence."[22] Medicine is no longer the self-contained entity that it once seemed; as it is technically more complex, so it is correspondingly more multifocal and multivocal in its material relations. Its boundaries less clear and more porous than formerly thought, it consequently has become less sharp a category for analysis. In extending everywhere, it might be seen as everything and nothing—like Foucauldian "power," everywhere and nowhere. What, then, is the thing to which the analytical tool of history is to be applied? And how can medicine be an analytical tool for the history of society? Not only were all the assumptions that historians made about medicine in the 1960s and 1970s with respect to the state, professional power, and science called into question by the realities of the 1990s,[23] so too was the very object or category of study.

And "history"? While it may be true that "from the time of Herodotus and Thucydides, historians have vehemently disagreed about the purposes, methods, and epistemological foundations of the study of the past,"[24] before the 1980s and 1990s these debates were never so extensive or intensive. Echoing Marx, Robert Putman in 1993 insisted that "history matters" because "individuals may 'choose' their institutions, but they do not choose them under circumstances of their own making, and their choices in turn influence the rules within which their successors choose."[25] Of course. What was new in the 1980s and 1990s was the urgency to defend this commonplace—not against an old, reactionary right, or against a perceived-to-be overdeterministic ("choice"-denying) Marxist left, but against a radically iconoclastic postmodern avant-garde of philosophers, literary theorists, and cultural critics. According to these, history was nothing more than the "invention" of historians dealing in images and representations of the past to which they could not possibly have any direct access. Further, the much-loved periodization of historians was nothing more than a strategy for narrative closure. Thus, historical "truth" and historical practice were to be regarded as no less contingent and subjective ("shaped not found") than the scientific "truth" discerned by sociologists of scientific knowledge. In new and more extreme ways than in the past, the old hoary question of objectivity was back on the agenda.[26]

Professional historians were stunned and deeply threatened by such attention to their methods and assumptions. A torrent of defensive publications were issued by (to name but a few) Arthur Marwick; Bryan Palmer; Joyce Appleby, Lynn Hunt and Margaret Jacob; Richard Evans; Geoff Eley; Eric Hobsbawm; Raphael Samuel; and Gareth Stedman Jones.[27] Despite the very different political orientations of these historians, they all shared some of the same professional and political suspicion of "postmodern postures,"[28] while to differing degrees confessing to their own complicity in the old rationalist and increasingly demoralized Enlightenment search for objectivity. As the political culture veered to the right, confusion, apathy, and uncertainty set in.

Nor was the "postmodern challenge"[29] the only cause for concern. In new and far more extensive ways than in the past, "history" (like medicine) was being "managed" in both crude and subtle ways by contending communities of opinion—a feature of the present to which, in fact, postmodern writers drew attention. In the museum in the bowels of the Statute of Liberty, as in countless other public sites, political battles for control over representations of history were fiercely fought.[30] Allied to this (again not unlike in medicine), history was increasingly subject to naked market forces, both within the academy (for example, though assessment exercises linked to research funding), and outside it in the expanding commercial heritage and leisure industries.[31] Ideologically hand-in-hand with the "rationalization" of history departments went the global expansion of capital-intensive "Disney history," the latter providing the new discipline of museology with an unlimited supply of case studies in the manipulation of historical representation. Forces of a slightly different nature, more sinister for being less public, were also at work in the history panels of grant-giving bodies, not least in the history of medicine. Increasingly, the tendency of such bodies was away from humanities-style appraisals to more science or social science models, with emphases on practical applicability, "relevance" (to short-term political interests), directed goals, and publicly accessible outputs. In Britain especially, the idea of the historian as a devoted, critically minded intellectual was increasingly derided as a relic of ivory-towered times. The fetid breath of managerialism hung heavy.

As for "the social," by the mid-1980s it was deemed by Francophone semiologists as absorbed into "the cultural."[32] A decade later, the prospect of an "end of social history" had not only been raised, it had been realized.[33] In what superficially appears in retrospect as an intellectual parody of the thinking of the then dominant political parties, Francophone poststructuralists thoroughly underpinned Margaret Thatcher's would-be class consensual claim that "there is no such thing as society." Although Thatcher and her free market cronies were scarcely like Jean-François Lyotard and his disciples in their quest to deprivilege the economic and the political, the effect was much the same: "a loss of political appetite for the old frameworks of social analysis"[34] and, in particular, for the validity and relevance of Marxism.[35] While the new political elite effected their ideological cleansing in the name of a new "end to ideology,"[36] poststructuralists put to flight all notion of structure, agency, and social determinism. Operating from the aesthetic critique of modernity first elaborated by Nietzsche in the late nineteenth century, the French "new philosophers," as they were often called in the English-speaking world (notably, Jacques Derrida, Giles Deleuze, Lyotard, and Jean Baudrillard)—or the "young conservatives" as Jürgen Habermas referred to them because of their abandonment of all hopes of social change[37]—demanded freedom from political forms of life and the rejection of the "tyranny of reason," technocratic rationality, and the old emphasis on the economic.[38]

Although the pre-"postmodern" Foucault—the Foucault concerned with liberating the revolutionary process from ritualized and dogmatized Marxism[39]—had only invited *consideration* of whether power was "always in a subordinate position relative to the economy,"[40] Derrida and other linguistic deconstructivists argued forcefully for the impossibility of reducing technologies of violence to instrumental political power, economic interests, and social control.[41] The Foucault of *Surveiller et punir: Naissance de la prison* who appeared to offer a critique of capitalism ("it is largely as a force of production that the body is invested with relations of power and domination")[42] was increasingly refashioned as an *avant-garde* literary theorist.[43] This other Foucault, identified largely with the *Histoire de la sexualité*, doubted that power was always in the service of, and ultimately answerable to, the economy. Instead, he insisted on the grandiloquence and rules of discourse as *constructing* bodies.[44] Here, as everywhere in what became the flight into cultural studies, sociological categories were dust-binned in favor of semiotic ones, which were often inflected psychoanalytically, as in the bogey of "narrative *fetishism*." The master narratives of modernity provided by Marx, Durkheim, and Weber were shed through the literary turn to the discursive—what Bryan Palmer dubbed the "descent into discourse."[45] Although sophisticated Marxist theoreticians, such as Ernesto Laclau, were as involved as anyone in deconstructing "society" as an intelligible and essentialist totality,[46] it was above all literary and linguistic theorists who compelled Western intellectuals to problematize the relationship between discourse, structure, and knowledge and to question whether "structure" and "knowledge" had status at all within the analysis of discourse.[47] It was they who compelled historians to attend to the historicity of concepts such "class," "the people," and "the social"—that is, to see them not as inevitable, natural, or god-given but as historical products that had become "naturalized" over time.

Thus, while capitalism restructured itself monopolistically and globally in the wake of the collapse of socialist modes of production and social systems, while Islamic fundamentalism arose and ethnic cleansings swept the planet, and while (closer to home) the academic job market shrank from its 1960s heyday, Western intellectuals were increasingly inclined to reflexivity, the play of signifiers without referents, free actors, representations, "irreducible ambiguities," and evocative explorations into the interrelations, reverberations, and tensions within and between disparate political and fictional genres.[48] Within the professional practice of history, in the face of despair at the collapse of the old totalizing social discourses, there emerged compensatory fascination with constructions of identity, interiority, and other "imaginative geographies." As the language of historians became more subtle and self-referential, the more the cord to political action was severed. Having come to doubt the modernist interest in politics and rationality, intellectuals were impaled on the spikes of their own profound sense of skepticism and powerlessness. In cultural studies, at the same time as an emancipatory politics was raised to new heights through the pursuit of *différance* or "the Other" (abetted much by the work on gender and "orientalism" of Donna Haraway and Edward Said, among others),[49] the difference between murder and murder fiction was not only increasingly blurred (as Robert Darnton was overheard to remark),[50] but, thanks in part to Hayden White, was ceasing much to matter. By some reckonings, the abattoir and Auschwitz were not a great deal different historically, and "the body"—resurrected not merely as a textual site of contestation and struggle but *the* locus on which power was to be seen as inscribed (that is, outwith a framework of social class and race, or even social context)[51]— became so discursive, flexible, and fragmented as to be immaterial.

All "reality"—not just messianic historical materialism—was in a hopeless mess,[52] whether or not (à la Baudrillard) one saw it as having given way to virtual reality or to a "hyperreality" of endless copies without originals (with the video as the icon). Power was no longer something found at the end of a gun or under the heel of a boot, or even something negotiated through the exercise of social relations in a material world. It was now a monolithic object dispersed through discursive fields of knowledge production. All the world had become a text, the author of which was unclear and of no great concern. Along with the concepts of "class" and "society" and other such scientific pretensions of modernity, "ideology" (however devulgarized) virtually disappeared from the historical vocabulary. By Francis Fukuyama's *The End of History and the Last Man* (1992), which in fact contested the notion of a new consensual "end of ideology" in politics, social history as the forum for the study of such things was an embarrassed bygone—a residue on the outer edges of a vastly expanding and hyperinflating, pastiche-reducing, postmodern marketplace of infinite diversity, eclecticism, and discontinuities (in some eyes, anarchic, nihilistic, and narcissistic, if not entirely weightless, decorative, and commercial). Certainly social history's political anchor was weighed by it. Already by 1980 the political thrust of the social history seeded by E. P. Thompson was perceived as "marooned on a sea of increasingly diffuse cultural analysis."[53] By 1990, the tide had overwhelmed it; like the "social," it had sunk into salient silence.

Not only in history generally but specifically in relation to the history of medicine, there was no call for "the social." On the contrary; reviewing twenty-five years of the Society for the Social History of Medicine in 1995, Dorothy Porter dealt in understatement when she concluded that "New historiographical trends in cultural history may make a society dedicated to the social history of medicine redundant."[54] Jordanova in "The Social Construction of Medical Knowledge," published in the same issue of *Social History of Medicine* as Porter's paper, suggested a disciplinary relabeling to "what might best be called a cultural history of medicine."[55] Lynn Hunt's *New Cultural History* (1989) may have been more in mind than the likes of Jonathan Sawday's *The Body Emblazoned* (1995), Judith Butler's *Bodies that Matter* (1993), Elizabeth Grosz and Elspeth Probyn's *Sexy Bodies* (1995), Athena Vrettos' *Somatic Fictions* (1995), and the psychomedical studies of visual and literary representations of disease and illness pioneered and pursued above all by Sander Gilman.[56] But by then, Foucault was under everyone's skin, whether you liked it or not;[57] beyond sex and madness and far into the intricacies of gender, literary-turned-cultural excavations of the body were extending rapidly. Grumbling there was, and criticism too, such as at the disconnection of the critical analysis of modernist mentalities from the historical and sociological examination of modernity itself.[58] But, without doubt, the pundits of what was nebulously referred to as the New Historicism in literary and cultural studies (main domicile Berkeley, main journal *Representations*) had taken the sunshine from social historians of medicine while borrowing their wares. The "somatic turning" mopped up, providing "a new organizing principle within Anglo-American intellectual activity."[59]

Body studies were in many ways the epitome of the linguistic turn in cultural studies,[60] and they were fundamentally at odds with social histories of medicine, however constructivist, Foucauldian, and contingent-emphasizing the latter might aspire to be. This was not simply because they engaged with synecdoche, metaphor, analogy, "close reading," and other techniques honed in English departments. Nor was it because they privileged texts over contexts and preferred to deal with inscriptions of modernity on the body and embodied cultural practices over narratives of modernity in medicine and health. (To a degree, the pioneering social historians of ideas in science and medicine, Owsei Temkin, George Rosen, and Erwin Ackerknecht, might be accused, or praised, for having done similar.) No, somatic studies challenged social histories of medicine because they undermined the reductive or determinist assumption at their heart—the notion that everything is ultimately socially constitutive. Following Derrida, somatic studies asserted, on the contrary, that nothing is reducible to anything (while proceeding to reduce most everything to discourse). The thinking was not only nonpositivist and nonteleological (as in the best social history of science and medicine) but nonontological: fundamental or absolute structures, inalienable human natures, and essentialist categories like "society" were denied; everything was to be seen as emergent or immanent. "Instantiation" became a favored word.

Social historians of medicine could thus be accused (as by Patrick Joyce) of intellectual naïveté for treating the body merely as a corporeal entity and for failing entirely to recognize "the social" as itself a product of modernity.[61] Even in their most novel pursuits (such as the "understudied" experiences of patients and the role of the laity in medical practice),[62] social historians of medicine were left to look like mindless empiricists—a look particularly fixed on the faces of those unschooled in Foucault and the epistemology and sociology of scientific and medical knowledge, whether à la Frankfurt, Edinburgh, or Paris. Retreat was on the cards. At best, market-wise social historians of medicine sought to put old wine into new bottles through books and courses relabeled "the body."[63]

By the 1990s, moreover, there appeared no longer to be any *need* for the "social" in the social history of medicine. According to Charles Rosenberg in 1989, enough had been written to convince all but the most moronic that "every aspect of medicine's history is necessarily 'social' whether acted out in laboratory, library, or at the bedside."[64] Since this "all-is-social" theme had been the historiographical mission of the context-celebrating subdiscipline since the mid-1970s, there was no need to go on about it. Mission accomplished. As Rosenberg submitted, the "social" in the social history of medicine and science had become tautological—"as tautological as the 'social construction of disease.'" Harking back to Erving Goffman's *Frame Analysis* (1975), Rosenberg proposed that we speak instead of "framing disease." This then became the main title of the edited volume in the *cultural* history of medicine in which his 1989 essay was reprinted. Thus, social historians of medicine, far from being forced into historiographical worry over the retheorizing of "the social," were provided with a means to ignore it while carrying on business as usual.

As I have said, it is not my intention here to defend "the social" in the social history of medicine; I merely wish to register the synchronicity of the postmodern turn outlined above with Rosenberg's insistence on exchanging "the social" for "the frame." Arguably, the contextualization of his essay on framing disease is as important as the historiographical claims made within it, and as important as the possible impact of those claims on the writing of the history and historical sociology of medicine (so that now, ironically, we find the writing of social constructivist accounts of disease left to members of the medical profession).[65] Rosenberg's essay deserves close attention, then, despite the fact that it was neither intended nor received as an important theoretical paper.

Rosenberg's Frame

Written in the late 1980s for the policy outreach journal the *Milbank Quarterly*, the immediate context of Rosenberg's essay was the intellectual burden imposed by AIDS. Indeed, only a few years before, Rosenberg had written specifically on this subject for the same journal.[66] To many observers, the biological realities of AIDS confounded the assertion of diseases as "mere" social constructs. As such, Rosenberg's essay was also an intervention in the much wider and more heated "science wars" between the defenders of scientific realism, rationalism, and Truth, on the one hand, and the philosophers and sociologists of science promoting varieties of relativism, on the other.

Although Rosenberg's essay did not touch on contemporary controversy over the extreme relativism of the literary deconstructivists that challenged the sacred objectivity claims and assumptions of historians, to some extent that backdrop was understood. In the late 1980s, discussion of relativism was extensive, especially in America, and especially among historians—in part because of a debate unleashed in 1988 by Peter Novick's magisterial history of the "objectivity question" in the practice of American history.[67] The fourth and final part of Novick's book, "Objectivity in Crisis," recorded an experience that was only too familiar to those like Rosenberg in departments of the history and sociology of science where post-positivist distinctions between fact and value; science, scientism, and ideology; nature and culture; and biology and society had been much discussed, especially after the publication in 1962 of Thomas Kuhn's *Structure of Scientific Revolu-*

tions. In such places, Lyotard was received not simply as another postmodern philosopher but as one claiming to pronounce specifically on the nature of science. (Dependent in part on the insights of Kuhn and philosopher of science, Paul Feyerabend, Lyotard's claim in *The Postmodern Condition: A Report on Knowledge* [1979; translated, 1984] was that science was no longer in the business of truth-seeking but rather in the manufacture of incommensurable theories.)[68] As everywhere, these debates were emotional, increasingly public, and politically charged.[69] Indeed, there was more than a whiff of McCarthyism to them. Setting the pace, Hilary Putnam in *Reason, Truth, and History* (1981) saw the cutting edge of relativism as deriving explicitly from Marx, Freud, and Nietzsche, who taught that "below what we are pleased to regard as our most profound spiritual and moral insights lies a seething cauldron of power drives, economic interests, and selfish fantasies."[70] By the 1990s, relativism of any type or strength was tending to be equated with radical relativism and to have many of the same "damning associations as Communism, whether you're a party member or not."[71]

Rosenberg's essay was by no means simply a reactionary response to the alleged naked biological realities of the disease that smote Foucault. A moderate relativist himself, Rosenberg was fully aware of the provisional nature of knowledge. His essay can be read, rather, as an effort to deprivilege "radical" or "hyper"-relativism and hence spare the social history of medicine, and social constructivism in particular, from the arrows of political outrage then surrounding it. Necessarily, therefore, in this context, the renunciation of "the social" and its replacement with "the frame" was a calculated political act; the verb "to construct" was in fact being substituted by "to frame" as "a less programmatically [i.e., politically] charged metaphor."[72] While shifting attention away from the political implications of social constructivism, Rosenberg's "frame" emphasized the relations between biological events and their individual and collective experience and perception. In other words, it was as a strategy both for a compromised relativism and for a pluralistic approach to history. Through "framing," the fragments of historical evidence might be disciplined but not overnarrativized; likewise, agency could still be ascribed to social, economic, intellectual, and political forces without granting overdeterminacy to any of them, and the role of individuals could be preserved. (Rosenberg's harking back to Goffman's use of the frame should not therefore be regarded as incidental, since Goffman had held to a social constructivist view of the self.)[73] As a descriptive category, "the frame" avoids responsibility being attached to any particular interest group or set of historical actors (including historians).[74] A versatile metaphor, then, "the frame" has the look of structure but commits one to no particular theoretical architecture or politics.

Yet, by these very means, and in common with a great deal of American neo-left liberalism, Rosenberg's formulation can be seen as a part of an effort to return to a "commonsense" pragmatism around which consensus can be built. In that sense, it can be regarded as part of an act to *reframe* the writing of the history of medicine within a political philosophy that, while it may not be as ideologically distinctive as Marxism, is no less ideological. One of the characteristics of this political philosophy is to make any retreat from it impossible by burning all the bridges to the old conceptual machinery while purporting (to a degree) to incorporate them. It is noteworthy how Rosenberg in his essay both praises the pioneering social history of medicine of the socialist Henry Sigerist (to illustrate how social history can be written without regard to relativist social constructivism), even as he draws attention to the antiquated positivist nature of Sigerist's regard of medical knowledge. Links between medicine (and the production of medical history) and politicoeconomic purpose are thereby, in effect, dismissed as vulgar (positivistic) Marxism, at the same time as (to cite Simon Schaffer dealing with the same "facetious blackguarding" in the history of science), over-robust "connections between natural knowledge and social interests are damned as sociological relativism."[75] Note, too, that Rosenberg's highly qualified endorsement of the provisionality of knowledge ("Knowledge *may be* provisional") is made in the course of staking a claim for his own historical revision (" . . . *but* its successive revisions are no less important for that").[76] This maneuver conforms to what Stuart Hall has called the "discursive struggle" over the delegitimation of opposing ideologies (or discourses), where the "older" machinery is presented as, at best, "optional."[77] In common with the postmodernists critiqued by Fredric Jameson, it's all within the spirit of pluralism: "As with so much else, it is an old 1950s acquaintance, 'the end of ideology,' which has in the postmodern returned with a new and unexpected kind of plausibility."[78]

Thus, Rosenberg's pseudostructural notion of "the frame" might appropriately be described as "within the frame" of the depoliticizing thrust of Francophone postmodernity. Given that "framing" was also a concept deployed by the antirealist Derrida for deconstructivist purposes,[79] it is tempting to suggest some closer affinity. The temptation must be resisted, however, for no two projects were less alike philosophically and practically. Nevertheless, Rosenberg's framing does go with the loosely Foucauldian (skeptical-relativist) culturalist flow. Without entirely dismissing either the social constructivist approach to medical knowledge, or the left-wing political tradition behind both social medicine (from Ryle to McKeown) and the social history of medicine (from Sigerist to Charles Webster and beyond), the notion of framing occludes both while appearing merely to encourage the historical analysis of yet more "full and appropriate contexts."[80] If the production of historical and historiographical knowledge is but one form of the production of knowledge that accompanies the restructuring of capital (along with the restructuring and rearticulation of practices),[81] then the substitution of "the frame" for "the social" in the history of medicine may be regarded as constitutive. Perhaps this is only to state the obvious: that, to mimic Marx, Gramsci, and Putman, historians don't choose their history under circumstances of their own making. That the act of writing history creates texts and constructs knowledge seems obvious, but there is never an intellectual free market (any more than there is a "free" economic one). Like Rawls's assertion of an ethical consensus for the late twentieth century, or Thatcher's "end of society," the would-be historiographically hegemonic—the invented "consensual"—must inevitably do political work, if only through the process of occluding. The displacement of "the social" by "the frame" was one such act.

This is not the only way to think about Rosenberg's "frame." As germane might be a view of it as seeking to reembody a history whose analytical categories had become destabilized—as destabilized (or generally messy) as the medicine and society of "late capitalism." Thus, the illusion of "the frame" may be less that of structure than that of stability, and less that of a totalizing contextual view than that of putative coherence. The warm embrace of the idea of the frame by historians of medicine in the 1990s might be explicable in terms of this illusion; at the very least it enables writing the history of medicine without having seriously to question the terms of analysis. It places a comforting blanket over the conceptual diaspora of modern history. This partly explains why Jordanova's call to render the social history of medicine a territory for critical engagement and debate has gone largely unheeded.[82] Jordanova, in effect, sought to explore the categorical messiness of the historiography of medicine as it was emerging in the 1990s. Hers was an invitation to reexamine the repertoire of categories supplied through the social history of medicine's response to the old positivist and doctor-centered, or tribal legitimating history of medicine—categories such as medicalization, professionalization, culture, representations, health, disease, hygiene, sexuality, and the family. Rosenberg's "frames," on the other hand, can be seen as the effort to put some kind of lid on that messiness and its investigation—in part to contain, in part to avoid the ever more apparent instability of the categories. If Jordanova's mission was one of disciplining the social history of medicine and bringing it to "maturity," a part of Rosenberg's strategy was to avoid the subdiscipline having any crisis of identity.

Post-Frame, or the End of the Beginning of the End

History, it might almost be said, has rendered the frame as superfluous as "the social" it sought to replace. The conditions for its possibility have surrendered to others, on the whole less politically despairing and historiographically fraught. This is not to suggest that with the change of millennium the slate has been wiped clean, the epistemological agonies of the last two or three decades magically erased. On the contrary, notched-up but still grating against the old modernist certainties and categories of "social" analysis are such postmodern insights as that on the impossibility of sustaining universal categories and truths (like "the social"); the fallacy of essentialism or fundamental causes; the idea that power resides in the making of discourse and that language has the capacity to shape what it represents (be it the "orient," "sex," or "disablement"); and the inability to distinguish in principle between scientific rationality and the stuff of religious belief. We have not heard the last of the theory wars, and the search for "the soul of history" carries on.

But just as there are signals from Francophone intelligence and elsewhere of a return to the political,[83] so there are signs in the practice of history that we have passed beyond the notion of a postmodern declension and the attendant fear that "a radical skepticism could yet defeat us all" (as Mary Douglas put it in the mid-1980s and the philosopher Peter Sloterdijk struggled against in his *Critique of Cynical Reason*).[84] There are indications that we have superceded the would-be hegemonic neodeterministic view of language as the root of "what it is possible for people to think and do."[85] To a considerable extent "postmodernism" has been desanctified, and there are clear signs that the "civil war" between discursivity and historicity, or more broadly between "the cultural" and "the social," has entered a more accommodating phase, less privileging of the cultural.[86] For some historians, this largely means the recognition and reconciliation of these different categories, as in *Beyond the Cultural Turn* (1999) and *Reconstructing History* (1999),[87] although these still attempt to rescue history from the perceived tight clutches of postmodern cultural studies. For others, it opens the way to de-polemicizing the debate between "discursivity" and "reality" in order to repoliticize it in terms of the political and ideological context in which it occurred.[88] For still others, such as Antoinette Burton, campaigns like these serve only to validate the old disciplinary demarcations and to foster anew static, stolid, and unitary understandings of their natures. Burton demands, rather, that we inquire into the naturalizing use to which the categories "the social" and "the cultural" continue to be put in contemporary historiography.⁸⁹ This idea cannot be developed here, but as it suggests, historiographical discourse has already moved beyond the postapartheid politics of the social *versus* the cultural. Indeed, as early as 1995, Catherine Casey, in *Work, Self, and Society after Industrialism,* provided a compelling illustration of how one might "return the social to critical theory" without necessarily discarding discourse analysis or returning to a conventional social analysis that merely privileges material social relations.⁹⁰ In many ways, this is the drift of some of the most engaging recent work in the history of medicine, though to call it "history of medicine" is to force it into increasingly anachronistic boxes.⁹¹

A review of this literature cannot be undertaken here, but three general points are worth making. The first is that it owes large debts to somatic cultural studies and, in particular, to Mary Poovey's *Making a Social Body* (1995)—a work that specifically challenged the idea that "the textual" and "the social" are antithetical or mutually exclusive domains of inquiry.⁹² Going beyond the earlier insights of Foucault and Jacques Donzelot on the formative role of medicine in the invention of "the social,"⁹³ Poovey has demonstrated how the modernist abstract of the "social body" was itself generated in the early Victorian state in response to cultural and political anxieties about anatomy and contagion, poverty and disease. Subsequent studies, such as Erin O'Connor's *Raw Material* (2000), take this further, revealing the operation of constraining somatized metaphors not only in the Victorians' own social critiques (as in the writings of Thomas Carlyle) but also in such contemporary cultural practices as postcolonial discourse and *its* critique and, indeed, in the practices of cultural studies as a whole.⁹⁴

Second, while none of this literature is intended to contribute to the social history of medicine or even to the history of medicine (despite publisher's classifications), it might be said to fulfill the erstwhile ambition of the subdiscipline to join the mainstream of history. As Jim Epstein remarked in a review of Poovey and similar studies: in the face of worry over postmodernism's threat to social history, such work powerfully testifies to "the very real openings available for writing new kinds of social and cultural history."⁹⁵ Seen from this vantage, "the end" of the social history of medicine might almost be considered as that running from nemesis to omniscience.

Third, and finally, this literature brings us back to politics. For one thing, it marks a turning away from the over self-referential, hall-of-mirrors-type cultural studies of recent years, much of which was written for the sake of its own disciplinary ends or for the sake of illustrating interesting tensions in the literary history of modernity. It is not just that the new work firmly registers Foucault's point about the body as the place where (as O'Connor puts it) "power has historically assumed its most monstrous and its most liberatory incarnation."[96] It is also that the literature amplifying this point now often passionately embraces a belief in the transformative potential of cultural theory to force thinking beyond categorical constraints. Thus, for O'Connor (drawing strength from Haraway), cultural studies are a "radical and necessary form of activism," which is all the more driven by the realization that "genuinely searching academic work is fast becoming a vestigial structure, a useless and hence expendable appendage to a culture that neither values nor understands it."[97] In a more straightforward way, Lawrence Driscoll, for instance, maps the Victorian discourse on drugs to expose how its cultural construction continues (devastatingly) to constrain social thought and political action. The point of Driscoll's discourse analysis is directly to effect political reform.[98]

Just as this literature helps us understand how "the social" was very largely invented and problematized within and through the vocabularies of medicine, so at a deeper level (if only by extension) it compels us to think about "the political" by encouraging the problematization of "the medical" as constitutive of "the social."[99] To be blunt, we can no longer speak of "the political" or "the medical" or "the social" because we no longer know what holds these categories together. Such categories must now be seen as hypothetical at best; like "nature," they are labels that do not explain so much as beg explanation. Hence, the complex and diverse phenomenon called medicine cannot be said to exist inside "the political" or "the social," anymore than the political or the social can be said to exist within (or to structure) "the medical." It is only through the material organization of the objects and resources having to do with medicine that "the medical" and "the political" can be seen as held together or given agency, that is, through technologies, expertise, texts, architectures, and the material (actual) social relations that go with them. The latter would include not just the relations between doctors and patients but between doctors and doctors, patients and patients, doctors and families, researchers and the state, pharmaceutical companies and the law, and so on, and so on. Admittedly, these material relations of medicine are not the only connections between people, and they can be distinguished from those around, say, education, consumerism, religion, the military, diplomacy, and the law, even if they may often overlap or be in tension with them. Whether they are *primary* relations in a world as "medicalized" as ours may be open to debate; what is more important to stress here is the lack of any need to privilege them over discursive and epistemological considerations. The need, rather, is to understand that it is these material social relations that actually produce the political of which the emergent discourses and epistemologies of medicine (themselves capable of acting as material forces) are a part. In other words, like "the social," "the political" cannot be regarded as a transcendent category with assumed inherent force.

Happily, this leaves the territory and the practice of the history of medicine wide open, even if it collapses the old disciplinary boundaries. There is no basis for privileging "the cultural" over "the social" any more than there is reason to lord medical ethics over medical economics. And equally, there is the opportunity to revisit such old sites as professionalization, the idea of medical elites, the technical content of medical knowledge, and so on, whose recent loss to the history of medicine has been lamented.[100] The prospect, then, at the end of the social history of medicine is not a return to what was lost when "the social" got retheorized and "framed" but rather to a different kind of poststructural "political" understanding of the phenomenon that is medicine within a historiographical frame that is more critically aware of its own values, perspectives, and aims.

NOTES

For helpful comments on an earlier version of this chapter, I'm grateful to John Arnold, Rhodri Hayward, Elsbeth Heaman, Rickie Kuklick, Bill Luckin, Eve Seguin, Sally Sheard, Steve Sturdy, and the participants in the conference at Maastricht where it was first aired. As ever, I am indebted to the Wellcome Trust for their generous support.

1. Charles Rosenberg, "Framing Disease: Illness, Society, and History," introduction to *Framing Disease: Studies in Cultural History,* Charles Rosenberg and Janet Golden, eds. (New Brunswick, N.J.: Rutgers University Press, 1992), xiii–xxvi. The introduction was a revised version of Rosenberg's "Disease in History: Frames and Framers," *Milbank Quarterly* 67 (1989): 1–15.

2. A sure sign was the commonplace appearance in the 1990s of "frames" and "framing" in applications for history of medicine funding. Further reflecting the term's vogue, as well as some of the significance of its use or nonuse, is Michael Worboys's *Spreading Germs: Disease Theories and Medical Practice in Britain, 1865–1900* (Cambridge: Cambridge University Press, 2000), 12–13, where he submits that, to describe his approach, he has "chosen to use 'construction' rather than the currently more popular terms of *social construction* and *framing*."

3. Andrew Wear, "Introduction," in *Medicine in Society: Historical Essays,* Andrew Wear, ed. (Cambridge: Cambridge University Press, 1992), 1–13, 1; Ludmilla Jordanova, "Has the Social History of Medicine Come of Age?" *Historical Journal* 36 (1993): 437–449. Of Jordanova's four criteria for determining a field's "maturity," her most substantial one—the conducting of sophisticated debates to encourage, refine, and, if necessary, radically alter interpretations—has remained largely unfulfilled within the social history of medicine (as

Godelieve van Heteren also noted in "Pourquoi Pas? The Absence of Radical Constructionism in the Social History of Medicine & SHM's Critical Potentials," paper presented at "Medical History: The Story and Its Meaning," Maastricht, Netherlands, 16–18 June 1999.

4. See Carolyn Steedman, *Dust* (Manchester: Manchester University Press, 2002), esp. chap. 7, "About Ends: On How the End Is Different from an Ending."

5. John Pickstone, *Ways of Knowing: A New History of Science, Technology, and Medicine* (Manchester: Manchester University Press, 2000), 6.

6. Ludmilla Jordanova has repeatedly drawn attention to the definitional problem with the word, most recently in *Nature Displayed: Gender, Science, and Medicine, 1760–1820* (London: Longman, 1999), esp. part 2, "Body Management," 101. In much of the historiography of medicine from Sprengel to Sigerist "medicine" was debated in terms of whether it was a "science" or an "art." See Charles Webster, "The Historiography of Medicine," in *Information Sources in the History of Science and Medicine,* Pietro Corsi and Paul Weindling, eds. (London: Butterworth Scientific, 1983), 29–43.

7. Nikolas Rose, "Medicine, History, and the Present," in *Reassessing Foucault: Power, Medicine and the Body,* Colin Jones and Roy Porter, eds. (London: Routledge, 1994), 48–72, 49.

8. Ludmilla Jordanova, "The Social Construction of Medical Knowledge," *Social History of Medicine* 8 (1995): 361–381, 362. See also John Harley Warner, "The History of Science and the Sciences of Medicine," *Osiris* 10 (1995): 164–193.

9. John Pickstone, "Medicine, Society, and the State" in *Cambridge Illustrated History of Medicine,* Roy Porter, ed. (Cambridge: Cambridge University Press, 1996), 304.

10. Rudolf Klein, "The Crises of the Welfare States," in *Medicine in the Twentieth Century,* Roger Cooter and John Pickstone, eds. (Amsterdam: Harwood Academic, 2000), 155–170; see also in this volume, John Pickstone, "Production, Community, and Consumption: The Political Economy of Twentieth-Century Medicine," 1–19.

11. James S. Kuo, "Swimming with the Sharks—the MD, MBA," *The Lancet* 350 (20 September 1997), 828.

12. Henrika Kuklick, "Professionalization and the Moral Order," in *Disciplinarity at the Fin de Siècle,* Amanda Anderson and Joseph Valente, eds. (Princeton: Princeton University Press, 2002), 126–152.

13. R. J. Lilford et al., "Medical Practice: Where Next?" *Journal of the Royal Society of Medicine* 94 (November 2001): 559–562.

14. Anne Hudson Jones, "Narrative Based Medicine: Narrative in Medical Ethics," *British Medical Journal* 318 (23 January 1999), 255. The use of the phrase in relation to ethics is symptomatic of the dominance in contemporary medical discourse of evidence-based medicine. See, for example, "Evidence-Based Art," *Journal of the Royal Society of Medicine* 94 (June 2001): 306–307.

15. Roy Porter, review of *Foucault: Health and Medicine, Social History of Medicine* 12 (1999): 178.

16. Simon Williams and Michael Calnan, "The 'Limits' of Medicalization?: Modern Medicine and the Lay Populace in 'Late' Modernity," *Social Science and Medicine* 42 (1996): 1609–1620, 1616.

17. Leigh Turner, "Medical Ethics in a Multicultural Society," *Journal of the Royal Society of Medicine* 94 (November 2001): 592–594. The political correctness of multiculturalism is of

course another form of would-be universalistic morality. Rawls's notion of "moral consensus" compares with Richard Rorty's notion of "solidarity" or "community" among Western philosophers, a notion denounced as "an exemplar of the nationalist philosophy of a new world"; Michael Billig, *Banal Nationalism* (London: Sage, 1995), 11–12, 162ff.

18. Ulrich Beck, *Risk Society: Towards a New Modernity,* M. Ritter, trans. (London: Sage, 1992), and "The Reinvention of Politics: Towards a Theory of Reflexive Modernization" in *Reflexive Modernization: Politics, Tradition, and Aesthetics in the Modern Social Order,* U. Beck, A. Giddens, and S. Lash, eds. (Cambridge: Polity Press, 1994), 1–55.

19. Claudia Wiesemann, "Defining Brain Death: The German Debate in Historical Perspective," in *Coping with Sickness,* John Woodward and Robert Jütte, eds. (Sheffield: European Association for the History of Medicine and Health Publications, 2000), 149–169, 151.

20. Williams and Calnan, "The 'Limits' of Medicalization."

21. See Tim Jordan, *Activism! Direct Action, Hacktivism, and the Future of Society* (London: Reaktion Books, 2002).

22. Rose, "Medicine, History, and the Present," 50.

23. John Pickstone, "The Development and Present State of History of Medicine in Britain," *Dynamis* 19 (1999): 457–486, 484.

24. Lynn Hunt, "Does History Need Defending?" *History Workshop Journal* 46 (1998): 241.

25. Robert D. Putman, *Making Democracy Work* (Princeton: Princeton University Press, 1993), 8; Karl Marx, "The Eighteenth Brumaire of Louis Bonaparte [1858]," *Pelican Marx Library: Political Writings,* vol. 2: *Surveys from Exile* (Harmondsworth: Penguin, 1973), 146.

26. Peter Novick, *That Noble Dream: The "Objectivity Question" and the American Historical Profession* (Cambridge: Cambridge University Press, 1988).

27. Arthur Marwick, *The Nature of History,* 3d ed. (Basingstoke: Macmillan, 1989), and the sequel, *The New Nature of History: Knowledge, Evidence, Language* (Basingstoke: Palgrave, 2001); Bryan Palmer, *Descent into Discourse. The Reification of Language and the Writing of Social History* (Philadelphia: Temple University Press, 1990); Joyce Appleby, Lynn Hunt, and Margaret Jacob, *Telling the Truth about History* (New York: Norton, 1994); Richard Evans, *In Defence of History* (London: Granta Books, 1997); Geoff Eley, "Is All the World a Text? From Social History to the History of Society Two Decades Later," in *The Historical Turn in the Human Sciences,* Terrence J. McDonald, ed. (Ann Arbor: University of Michigan Press, 1996), 193–243; Eric Hobsbawm, "On History," in *On History* (London: Abacus, 1998), 351–366; Raphael Samuel, "Reading the Signs: Fact Grubbers and Mind Readers," *History Workshop Journal* 32 (1991): 88–109 and 33 (1992), 220–251; Gareth Stedman Jones, "The Determinist Fix," *History Workshop Journal* 42 (1996): 19–35; David Mayfield and Susan Thorne, "Social History and Its Discontents: Gareth Stedman Jones and the Politics of Language," *Social History* 17 (1992): 165–188.

28. The phrase is Daniel Cordle's; *Postmodern Postures: Literature, Science, and the Two Cultures Debate* (Aldershot: Ashgate, 1999).

29. Georg G. Iggers, *Historiography in the Twentieth Century: From Scientific Objectivity to Postmodern Challenge* (Hanover, N.H.: Wesleyan University Press, 1997). See also *Poststructuralism and the Question of History,* Derek Attridge, Geoff Bennington, and Robert Young, eds. (Cambridge: Cambridge University Press, 1987).

30. Mike Wallace, "Hijacking History: Ronald Reagan and the Statue of Liberty," *Radical*

History Review 37 (1987): 119–130. See also in this journal in the 1980s the regular feature on the "abusable past."

31. See Robert Hewison, *The Heritage Industry: Britain in a Climate of Decline* (London: Methuen, 1987); David Lowenthal, *The Heritage Crusade and the Spoils of History* (New York: Free Press, 1996); Tony Bennett, *The Birth of the Museum: History, Theory, Politics* (London: Routledge, 1995); Ludmilla Jordanova, "The Sense of a Past in Eighteenth-Century Medicine," The Stenton Lecture 1997 (University of Reading, 1999); idem, *History in Practice* (London, Arnold, 2000). See also *Reconstructing History: The Emergence of a New Historical Society*, Elizabeth Fox-Genovese and Elizabeth Lasch-Quinn, eds. (New York: Routledge, 1999), xvii.

32. Jean Baudrillard, *In the Shadow of the Silent Majorities . . . or the End of the Social, and Other Essays* (New York: Semiotext(e), 1983). The social has been absorbed into the cultural, he argued, because "there is no longer any social signifier to give force to a political signifier," 19. Quoted in Catherine Casey, *Work, Self, and Society* (London: Routledge, 1995), 10.

33. Patrick Joyce, "The End of Social History," *Social History* 20 (1995): 73–91. For a lucid account of "the ending" in terms of its epistemological bankrupting, see Christopher Kent, "Victorian Social History: Post-Thompson, Post-Foucault, Postmodern," *Victorian Studies* 40 (1996): 97–133. With reference more to the epistemological wasteland of socialist historians in the wake of the end of the Marxist epic, see Raphael Samuel, "On the Methods of *History Workshop:* A Reply," *History Workshop Journal* 9 (1980): 162–176, and discussion in Steedman, *Dust,* 79ff. For a more mundane account, see Evans, *Defence of History,* 168ff.

34. Ulrich Beck, "How Modern is Modern Society," *Theory, Culture & Society* 9 (1992): 163.

35. This is not to suggest that the Francophone philosophers and linguists were anti-Marxists. Raphael Samuel ("Reading the Signs") hinted that many were, and the exposure in 1987 of one of the leading postmodern literary theorists, Paul de Man, as having had fascist connections during the occupation of Belgium during World War II (see Evans, *Defence of History,* 233ff) was seized upon by many as part of the hidden reactionary agenda of literary deconstructionists in their insistence on the irrelevance of authorial intentions in textual interpretations. But the attempt to lump postmodernists together on the right is no more convincing that the effort to tar them with leftist associations, as historian Arthur Marwick wants to suggest by his blanket dismissal of "Postmodernist/Marxist junk"; "All Quiet on the Postmodern Front: The 'Return to Events' in Historical Study," *Times Literary Supplement,* 23 February 2001, 13–14. Many of the francophone founders of discourse analysis *were* Marxists (for example Régine Robin, Michel Pêcheux, Denise Maldidier, Jean-Baptiste Marcellesi, Jacques Guilhaumou, and Jean Pierre Faye), though after May 1968 they became uncomfortable with the workers' movement and the preconceived historical explanations offered by dogmatic Marxists. There may be more truth to Marwick's assertion that "much of postmodernism appealed profoundly to those who were by no means politically radical." Marxist literary theorist Terry Eagleton had said as much in 1983, accusing deconstructionists of being intellectual elitists who savored signifiers over "what ever might be going on in the Elysee Palace or the Renault factories"; cited in Novick, *That Noble Dream,* 567, who also concedes that postmodern thought "was on the whole quite apolitical," 566. On the anti-Marxist tendency of poststructuralists, see Tony Bennett, "Texts in History: The Determina-

tions of Readings and Their Texts" in Attridge, Bennington, and Young, *Poststructuralism*, 66–67. Nevertheless, the assault on modernity under the banner of différance was a part of a critique of the reifying rationality culture of late capitalism. Moreover, as the reaction to the "linguistic turn" suggests, there was nothing nonpolitical about adopting avant garde literary theory in the wake of the perceived exhaustion of the "political" 1970s.

36. Daniel Bell's classic *The End of Ideology: On the Exhaustion of Political Ideas in the Fifties* (New York, Free Press, 1960), appeared laughably anachronistic after 1968, but in 1988 it could be credibly reissued. For the context of Bell's work, see Job Dittberner, *The End of Ideology and American Social Thought, 1930–1960* (Ann Arbor: UMI Research Press, 1979). Albert Camus apparently first used the phrase "end of ideologies" in 1946, referring to absolute utopias such as Marxist ones that destroy themselves. On the history of the utopian politics of an "end to politics," see Jacques Rancière, *On the Shores of Politics*, Liz Heron, trans. (London: Verso, 1995).

37. "The *young conservatives* embrace the fundamental experience of aesthetic modernity—the disclosure of a decentered subjectivity, freed from all constraints of rational cognition and purposiveness, from all imperatives of labor and utility—and in this way break out of the modern world. They thereby ground in intransigent antimodernism through a modernist attitude. They transpose the spontaneous power of the imagination, the experience of self and affectivity, into the remote and the archaic; and in manichean fashion, they counterpose to instrumental reason a principle only accessible via 'evocation': be it the will to power or sovereignty, Being or the Dionysian power of the poetic. In France this trend leads from Georges Bataille to Foucault and Derrida. The spirit [*Geist*] of Nietzsche that was reawakened in the 1970s of course hovers over them all"; Jürgen Habermas, "Modernity versus Postmodernity," *New German Critique* 22 (1981): 3–14, cited in Richard Wolin, "Introduction," Habermas, *The New Conservatism: Cultural Criticism and the Historians' Debate*, Sherry W. Nicholsen trans. and ed. (Cambridge: Polity Press, 1989), xxi–xxii.

In France, those called the "new philosophers" were André Glucksman, Alain de Benoist, Bernard-Henri Lévy, and Pascal Bruckner. All were opposed to totalitarianism in Europe, the seeds of which they saw in Marx and Hegel's philosophy; and some were deeply involved in human rights in Bosnia and elsewhere. My thanks to Eve Seguin for this information.

38. Much of this is recognizably pre-Thatcher 1960s thinking (indeed, in some ways it echoes the postrationalist thought of the 1930s), but in general, in Anglo-American circles, the ideas of the "new philosophers" were taken up in the late 1970s and 1980s.

39. See Michel Foucault, *Power/Knowledge: Selected Interviews and Other Writings, 1972–1977*, Colin Gordon, ed. (Brighton: Harvester Press, 1980), 51, 57–58, 76. Foucault eschewed the label "postmodernist," and legitimately so, for as Rose points out, his "rejection of unities was not done in the name of a post-modern metaphysics that celebrates diversity, [but rather] in the light of a more sober and, dare one say, more historical conviction that that which 'is' is much less determined, much more contingent, than we think"; Rose, "Medicine, History and the Present," 70; see also Steven Best, *The Politics of Historical Vision: Marx, Foucault, Habermas* (New York: Guilford Press, 1995).

40. Foucault, *Power/Knowledge*, 89.

41. See Jonathan Joseph, "Derrida's Spectres of Ideology," *Journal of Political Ideologies* 6 (2001): 95–115.

42. Michel Foucault, *Discipline and Punish*, Alan Sheridan, trans. (London: Penguin, 1979), 25–26.

43. Simon During, *Foucault and Literature* (London: Routledge, 1992), 3. On the place of Foucault within the intellectual transformation of the 1980s, see Kent, "Victorian Social History," and Mitchell Dean, *Critical and Effective Histories: Foucault's Methods and Historical Sociology* (London: Routledge, 1994).

44. Michel Foucault, *The History of Sexuality,* Vol. 1: *An Introduction,* Robert Hurley, trans. (London: Penguin, 1990), 7–12.

45. Palmer, *Descent into Discourse.*

46. Ernesto Laclau, *New Reflections on the Revolution of Our Time* (London: Verso, 1990), see, in particular, chap. 2, "The Impossibility of Society."

47. John Law, "Editor's Introduction: Power/Knowledge and the Dissolution of the Sociology of Knowledge" in *Power, Action, and Belief: A New Sociology of Knowledge?* John Law, ed. (London: Routledge & Kegan Paul, 1986), 3. See also Stuart Sim, *Derrida and the End of History* (Trumpington: Icon, 1999).

48. Two brilliant examples of the application of some of these postmodern concerns to historical material are Judith Walkowitz, *City of Dreadful Delight: Narratives of Sexual Danger in Late-Victorian London* (London: Virago, 1992), and Daniel Pick, *War Machine: The Rationalisation of Slaughter in the Modern Age* (New Haven: Yale University Press, 1993).

49. Donna Haraway, *Simians, Cyborgs, and Women: The Reinvention of Nature* (London: Free Association, 1991); Edward Said, *Orientalism* (New York: Pantheon, 1978) and *Culture and Imperialism* (New York: Knopf, 1993).

50. At a conference on "Dissolving the Boundaries: Historical Writing Towards the Third Millennium," University of Warwick, July 1997.

51. Mark Jenner, "Body, Image, Text in Early Modern Europe," *Social History of Medicine* 12 (1999): 143–154.

52. Williams and Calnan, "Limits of Medicalization," who conclude: "Without wishing to sound too post-modern, reality, in truth is a mess, and we would do well to remember this as we edge ever closer towards the twenty-first century!" Some explication of this "mess" is provided by Paul Barry Clarke, "Deconstruction," in the *Dictionary of Ethics, Theology, and Society,* P. B. Clarke and A. Linzey, eds. (London: Routledge, 1996), 216–223.

53. Geoff Eley and Keith Nield, "Why Does Social History Ignore Politics?" *Social History* 5 (1980): 267.

54. Dorothy Porter, "The Mission of Social History of Medicine: An Historical View," *Social History of Medicine* 8 (1995): 345–359, 359. Cf. Charles Webster's call in 1983 for a social history of medicine "that would place its primary emphasis on the changing pattern of health of the population as a whole"; "The Historiography of Medicine," 40.

55. Ludmilla Jordanova, "The Social Construction of Medical Knowledge," *Social History of Medicine* 8 (1995): 363, reprinted as chap. 15 in this volume.

56. Among them are Sander Gilman, *Difference and Pathology: Stereotypes of Sexuality, Race, and Madness* (Ithaca, N.Y.: Cornell University Press, 1985); *Disease and Representation: Images of Illness from Madness to AIDS* (Ithaca, N.Y.: Cornell University Press, 1988); *The Jew's Body* (London: Routledge, 1991); *Inscribing the Other* (Lincoln: University of Nebraska Press, 1991); *Health and Illness: Images of Difference* (London: Reaktion Books, 1995); and *Making the Body Beautiful: A Cultural History of Aesthetic Surgery* (Princeton: Princeton University Press, 1999).

57. Responding to one of the American critics of Adrian Desmond and Jim Moore's *Darwin,* Jim Moore remarked, "The 'F' word today is flung about by critics like navvies

flourish theirs.... It would be a solecism to suppose that either of us owes anything to Foucauldian 'archaeology' or 'epistemic shifts'. Desmond himself has never cracked a book by Foucault; Moore has repeatedly faulted Foucault-like accounts.... We didn't need his advice on sucking epistemologcal eggs, nor indeed did the scholars whose work made *Darwin* possible." *Journal of Victorian Culture* 3 (1998): 152. The latter point is also made by Jordanova in "The Social Construction of Medical Knowledge," 368–369. See also Jones and Porter, *Reassessing Foucault.*

58. Roger Cooter and Steve Sturdy, "Of War, Medicine, and Modernity: Introduction" in *War, Medicine, and Modernity,* Roger Cooter, Mark Harrison, and Steve Sturdy, eds. (Stroud: Sutton, 1998), esp. 5.

59. Jenner, "Body, Image, Text," 143–154; see also Mark Jenner and Bertrand Taithe, "The Historiographical Body," in Cooter and Pickstone, *Medicine in the Twentieth Century,* 187–200.

60. Evidence of which lies with the nature of most of the books sympathetically reviewed by Martin Wiener in terms of their contribution to modern British history: "Treating 'Historical' Sources as Literary Texts: Literary Historicism and Modern British History," *Journal of Modern History* 70 (1998): 619–638.

61. Patrick Joyce, "The Return of History: Postmodernism and the Politics of Academic History in Britain," *Past and Present* 158 (1998): 212 n.18, cited (in more user-friendly format) in "The Challenge of Poststructuralism/Postmodernism" in *The Houses of History: A Critical Reader in Twentieth-Century History and Theory,* Anna Green and Kathleen Troup, eds. (Manchester: Manchester University Press, 1999), 297.

62. See, for example, the case for the study of the laity in Cornelie Usborne and Willem de Blécourt, "Pains of the Past: Recent Research in the Social History of Medicine in Germany," *Bulletin of the German Historical Institute London* 21 (1999): 5–21.

63. See Jenner, "Body, Image, Text" 143. A more recent example is Roy Porter's *Bodies Politic: Disease, Death, and Doctors in Britain, 1650–1900* (London: Reaktion Books, 2001), which, while nodding in the direction of "the body as text," is preoccupied with "contextualizing [visual material] within the wider cultural pool," 12. In this respect, as in others, *Bodies Politic* is, as Porter submits, a sequel to his previous social histories of the sick and the sick trade, 35.

The most explicitly Foucauldian studies of the body were conducted not by social historians of medicine but by historically minded sociologists of medicine such as David Armstrong, Nikolas Rose, Deborah Lupton, and Bryan Turner: Armstrong, *Political Anatomy of the Body: Medical Knowledge in Britain in the Twentieth Century* (Cambridge: Cambridge University Press, 1983); Lupton, *Medicine as Culture: Illness, Disease, and the Body in Western Societies* (London: Sage, 1994) and *The Imperative of Health: Public Health and the Regulated Body* (London: Sage, 1995); Rose, *Governing the Soul: The Shaping of the Private Self* (London: Routledge, 1989); Turner, *Medical Power and Social Knowledge,* 2d ed. (London: Sage, 1995). See also, *The Body: Social Process and Cultural Theory,* Mike Featherstone, Mike Hepworth, and Bryan Turner, eds. (London: Sage, 1991). A notable exception among social historians writing on the body (and drawing heavily on sociology, anthropology, and political theory) was Barbara Duden, *The Woman beneath the Skin: A Doctor's Patients in Eighteenth-Century Germany,* T. Dunlop, trans. (Cambridge, Mass.: Harvard University Press, 1991), see esp. chap. 1, "Toward a History of the Body."

64. Rosenberg, "Framing Disease," xiv.

65. For example, Robert A. Aronowitz, *Making Sense of Illness: Science, Society, and Illness* (Cambridge: Cambridge University Press, 1998).

66. Charles Rosenberg, "Disease and Social Order in America: Perceptions and Expectations," *Milbank Quarterly* 64, suppl. 1 (1986): 34–55, reprinted in *AIDS: The Burdens of History,* Elizabeth Fee and Daniel Fox, eds. (Berkeley: University of California Press, 1988), 12–32.

67. Novick, *That Noble Dream.* Reference to many of the reviews of the book are noted in its discussion by Allan Megill: "Fragmentation and the Future of Historiography," *American Historical Review* 96 (1991): 693–698.

68. See Christopher Norris, *Against Relativism: Philosophy of Science, Deconstruction, and Critical Theory* (Oxford: Blackwell, 1997), 102. On Lyotard, see also the special issue of *Parallax* 6 (2001): 1–145.

69. For many historians, Novick observes, "what has been at issue is nothing less than the meaning of the ventures to which they have devoted their lives, and thus, to a very considerable extent, the meaning of their own lives. 'Objectivity' has been one of the central sacred terms of professional historians, like 'health' for physicians, or 'valor' for the profession of arms"; *That Noble Dream,* 11, see also 564ff.

70. Cited in Mary Douglas, "The Social Preconditions of Radical Skepticism," in *Law, Power, Action, and Belief,* 68–87, 81.

71. Liz McMillen, "The Science Wars Flare at the Institute for Advanced Study," *Chronicle of Higher Education,* 16 May 1997. The spark for this particular conflagration was Gerald Geison's demythologizing, contextualizing, and social-constructivist-tending biography of Pasteur, *The Private Science of Louis Pasteur* (Princeton: Princeton University Press, 1995).

72. Rosenberg, "Framing Disease," xv. As it emerged, substituting "frame" for "social" was all the more politically expedient for those applying for research funding in this charged environment.

73. See, in particular, Erving Goffman, *The Presentation of Self in Everyday Life* (New York: Doubleday, 1959). "Framing," of course, had and has other lives in medicine and elsewhere; see, for example, Harry Collins and Trevor Pinch, *Frames of Meaning: The Social Construction of Extraordinary Science* (London: Routledge, 1982), and Adrian Edwards et al., "Presenting Risk Information: A Review of the Effects of 'Framing' and Other Manipulations on Patient Outcomes," *Journal of Health Communication* 6 (2001): 61–82.

74. For an illustration of this use of framing as a descriptive category within an alleged historical case of "framing" for the avoidance of attributing blame or attaching responsibility to any particular interest group, see Chris Feudtner, " 'Minds the Dead Have Ravished': Shell Shock, History, and the Ecology of Disease Systems," *History of Science* 31 (1993): 377–420. Feudtner draws explicitly on Rosenberg's concept of framing to avoid what "seems irresponsible[,] to dismiss shell shock as 'myth' or 'social construct,' " 380.

75. Simon Schaffer, "A Social History of Plausibility: Country, City, and Calculation in Augustan Britain," in *Rethinking Social History: English Society 1570–1920 and Its Interpretation,* Adrian Wilson, ed. (Manchester: Manchester University Press, 1993), 133.

76. Rosenberg, "Disease and Social Order," 29 (my italics).

77. Cited in Frederic Jameson, *Postmodernism, or the Cultural Logic of Late Capitalism* (London: Verso, 1991), 397.

78. Ibid., 398.

79. Nick J. Fox, "Derrida, Meaning, and the Frame," in *Beyond Health: Postmodernism and*

Embodiment (London: Free Association, 1999), 134–135. See also John Frow, who has deployed "the frame" to mark off a literary space which establishes "the particular historical distribution of the 'real' and the 'symbolic' within which the text operates." For Frow the frame organizes the "inside" and the "outside" of a text and the relations between them; the function of the frame is culturally dependent. Frow, "The Literary Frame," *Journal of Aesthetic Education* 16 (1982): 25–30.

80. For a critique of the social history of medicine as merely elaborating "full and appropriate contexts," see Roger Cooter, "Anticontagionism and History's Medical Record," in *The Problem of Medical Knowledge: Examining the Social Construction of Medicine,* Peter Wright and A. Treacher, eds. (Edinburgh: Edinburgh University Press, 1982), 87–108.

81. See Karl Figlio, "Second Thoughts on 'Sinister Medicine,'" *Radical Science Journal* 10 (1980): 159–166, 165.

82. A partial and peculiar recent exception is David Harley, "Rhetoric and the Social Construction of Sickness and Healing," *Social History of Medicine* 12 (1999): 407–435, to which Paolo Palladino has replied "And the answer is . . . 42," *Social History of Medicine* 13 (2000): 142–151.

83. See *Democracy and Nature* 7 (March 2001), especially the contributions by Simon Tormey, "Post-Marxism," and Takis Fotopoulos, "The Myth of Postmodernity," 27–76.

84. Douglas, "Social Preconditions," 86. "The intellectual position of the relativists," she added, "can be shown to be contingent on their sense of futility or immorality of exercising power and authority, and this contingency rests in turn on their place in a social structure." Peter Sloterdijk, *Critique of Cynical Reason* (Minneapolis: University of Minnesota Press, 1987).

85. Bryan Palmer, "Is There Now, or Has There Ever Been, a Working Class?" in *After the End of History,* Alan Ryan, ed. (London: Collins and Brown, 1992), 100. See also Appleby et al., *Telling the Truth about History,* 230.

86. See, for example, *A Cultural Revolution? England and France, 1750–1820,* Colin Jones and D. Wahrman, eds. (Berkeley: University of California Press, 2002).

87. *Beyond the Cultural Turn,* Victoria E. Bonnell and Lynn Hunt, eds. (Berkeley: University of California Press, 1999); Fox-Genovese and Lasch-Quinn, *Reconstructing History.* See also, Gill Valentine, "Whatever Happened to the Social? Reflections on the 'Cultural Turn' in British Human Geography," *Norwegian Journal of Geography* 55 (2001): 166–172.

88. Geoff Eley and Keith Nield, "Farewell to the Working Class?" *International Labour and Working-Class History* 57 (2000): 1–30, and "Reply: Class and the Politics of History," ibid. 57 (2000): 76–87.

89. Antoinette Burton, "Thinking Beyond the Boundaries: Empire, Feminism, and the Domains of History," *Social History* 26 (2001): 60–71. See also Carolyn Porter, "History and Literature: 'After the New Historicism,'" *New Literary History* 21 (1990): 253–272.

90. Casey, *Work, Self, and Society,* 11. Casey's study is also one of the most lucid on the relations between postmodern theory and the politics of late capitalism.

91. Among examples of this work, not cited below, are Andrew Aisenberg, *Contagion: Disease, Government, and the "Social Question" in Nineteenth-Century France* (Stanford: Stanford University Press, 1999), and Ian Burney, *Bodies of Evidence: Medicine and the Politics of the English Inquest, 1830–1926* (Baltimore: Johns Hopkins University Press, 2000).

92. Mary Poovey, *Making a Social Body: British Cultural Formation, 1830–1864* (Chicago: University of Chicago Press, 1995). Ian Burney's debts to Poovey are made explicit in his

contribution to the roundtable discussion "The Making of a Social Body," *Journal of Victorian Culture* 4 (1999): 104–116.

93. Michel Foucault, "An Ethics of Pleasure," in *Foucault Live: Interviews, 1966–84,* S. Lotringer, ed. (New York: Semiotext(e), 1989), 261, cited in Gavin Kendall and Gary Wicham, "Health and the Social Body" in *Private Risks and Public Dangers,* Sue Scott et al., eds. (Aldershot: Avebury, 1992), 8–18, 9–10; Jacques Donzelot, *L'invention du social* (Paris: Fayard, 1984).

94. Erin O'Connor, *Raw Material: Producing Pathology in Victorian Culture* (Durham: Duke University Press, 2000). Cholera, breast cancer, amputations, and monsters are among the tropes she critically explores.

95. "Signs of the Social," *Journal of British Studies* 36 (1997): 473–484, 483.

96. O'Connor, *Raw Material,* 215.

97. Ibid., 214.

98. Lawrence Driscoll, *Reconsidering Drugs: Mapping Victorian and Modern Drug Discourses* (New York: Palgrave, 2000).

99. The best discussion of these matters is still Rose, "Medicine, History, and the Present."

100. By John Harley Warner, who perceives their study as having been delegitimized, not by the force of postmodernism, but by too great an emphasis on the history of science. Warner, "History of Science," 173.

CHAPTER FIFTEEN

The Social Construction of Medical Knowledge

Ludmilla Jordanova

In the 1970s, social constructionist approaches were the subject of fierce debate, especially in English-speaking communities that were struggling to absorb "theory." Now, although debates continue, social constructionism is regarded with less suspicion; indeed, it has become institutionalized in much that goes under the name of sociology of science, and perhaps by that token it is less subversive. It is not my purpose here to analyze how that change came about or to summarize the philosophical issues social constructionism raises. I suggest that further debates among historians about whether such approaches are right or acceptable are unlikely to be productive. My position is simple: if social historians of medicine attempt more than anecdotal or descriptive history, they frequently adopt social constructionism in one form or another, even if they have been less explicit than historians of science about their conceptual maneuvers. However, it is vital to be clear about the historical perspectives that social constructionism opens up. In order to do so, we can assume that it is useful and then probe the nature of its fruitfulness for historical practice.

The chapter is divided into five parts. The first deals with the enabling potential of social constructionism, and with the intellectual currents that made it possible. The second deals with some common misapprehensions about it. The next part sets out some interpretative issues to be addressed. In part four, I compare the historiographies of science and medicine. The final part attempts to draw the threads together.

It has not proved possible to formulate a neat definition of social history of medicine.[1] As the contributions to *Social History of Medicine* have shown, there are many social histories of medicine. Social constructionism has particular relevance for those interested both in medical thinking broadly conceived and in conceptualizing the relationships between such thinking and the settings in which it occurs. The phrase "medical knowledge" in my title reflects the centrality of knowledge claims in social constructionist approaches.[2] The term "knowledge" is hardly neutral, since it implies claims that have been validated in some way and foregrounds the cognitive dimensions of medical and scientific practice. It is a mistake to separate the knowledge claims of medicine from its practices, institutions, and so on. All are socially fashioned, and so it may ultimately be more helpful to think in terms of mentalities, modes of thought, and medical culture than in terms of "knowledge," which implies the exclusion of what is inadmissible, while the former are looser, more capacious categories.

The currency of the phrase "medical knowledge" reveals the debt of historians of medicine to writings on science. Social constructionists who study the natural sciences have been concerned to show how even the most dense theoretical claims of the so-called hard sciences are amenable to social explanation—that is, they have needed a rather strict interpretation of "knowledge." The agenda of those who work on medicine has been somewhat different, partly because commentators have become more scrupulous in distinguishing between medicine and science and partly because those who study science have been more preoccupied with the philosophical claims implicit in their work, not least because of the existence of history and philosophy of science as a field.[3] Some scholars have come to feel that "medicine" is a problematic term precisely because it has been unthinkingly treated as another form of science. They are concerned that "medicine" suggests orthodox knowledge systems thereby implicitly marginalizing not only healing practices but the whole range of behavior and representations associated with health. While this was a fair criticism of some earlier approaches to the field, we can now feel confident that social perspectives, in making the importance of medical practice their cornerstone, permit us to speak of medicine and to have far more than theory in mind.

I have used the word "knowledge" in the title to signal that thinking about ideas remains an important goal in the social history of medicine. In fact, together social constructionism and an attention to medical ideas constitute what might best be called a cultural history of medicine. The recent growth of interest in cultural history has been stimulated by many of the currents of thought discussed below and a significant proportion of its exponents adopt a constructionist position.[4]

Enabling Potential for Historical Practice

It may be fruitful to think of social constructionism as delineating a space that the social history of medicine can occupy. By stressing the ways in which scientific and medical ideas and practices are shaped in a given context, it enjoins historians to conceptualize, explain, and interpret the processes through which this happens. The old Whiggish history permitted no such spaces to exist. In a progressivist narrative, the search for truth was told in terms of blind alleys and right answers; the model was a journey, and the main emphasis was on content. Since ideally there was a tight fit between explanans and explanandum, few questions were asked about the mediating processes between them or about how problems requiring explanation were defined. By stressing that knowledge is produced in and through social processes, social constructionism encouraged historians to conceptualize the constituent processes and to come up with imaginative ways of recreating them, including through the use of a wide range of primary sources. More than this, it became important to talk about these processes using the ideas and frameworks that other historians employed, such as state, class, imperialism, patronage, and so on. Put like this, it all sounds rather simple, but we are still novices and many of the potentials remain under- or unexplored.

Many intellectual strands have contributed to social constructionism over the last two decades or so. I would like to mention briefly eight styles of thinking, some of which overlap. Nonetheless, it is analytically useful to distinguish between them.

First, philosophy of science played a central role in providing tools for questioning the epistemological status of scientific and, by extension, medical theories and for reconstructing the intellectual maneuvers scientists employed. It was not that settled answers were arrived at but that permission was given to explore a number of different models of how natural knowledge was acquired. Attempts to uncouple "nature" and theories about it were particularly helpful: between the material world and our representations of it there now appeared to be a space, which it was the job of historians (and sociologists and philosophers) to examine. Social constructionist approaches conceptualize that space and thereby generate ideas about what is in it. Many figures played an important part in the process initiated by philosophers of science, especially after the first edition of Thomas

Kuhn's *The Structure of Scientific Revolutions* in 1962.[5] Some scholars interpreted Mary Hesse's writings as especially useful, although she did not herself adopt social constructionism.[6] In particular, her *Models and Analogies in Science* suggested some ways in which language acted as a mediator between nature and science.[7] Indeed, analyzing the languages—both verbal and visual—of science and medicine has been a major tool in the social constructionist project, which accordingly is similar to and has been influenced by new critical methods in art history and literature.[8] Into that space was also inserted a concern with the practices of scientists and medical practitioners, encouraged initially by an interest in how discoveries took place, and specifically in the distinction between logics of discovery and those of justification.

Second, and closely related to this, was the general revolt against Whiggish history, which had been especially prevalent in writings on science and medicine and indeed had been sustained by a realist philosophy of science. It was also sustained by the dominance of scientists and medical doctors over the history of their fields. The urge to applaud the "right" and castigate the "wrong" is still all too common as is clear in the continued use of the use/abuse model mentioned below. It follows from the critique of Whig history that historians should be critical of the use and perpetuation of canons and of the unthinking forms of heroization they imply. While it has long been fashionable to decry these forms of triumphalism, they remain entrenched. The task of sympathetically understanding actors' perspectives has proved more appealing than has the challenge to rethink the notions of "genius" and "great thinker" or the very concepts of "canon" and "hero"—all are equally central to an anti-Whig perspective. Interestingly enough, general historical writings on this issue have made less impact on the history of science and medicine than those emanating from more theoretical domains. Herbert Butterfield was better known for his book on the scientific revolution than for his *Whig Interpretation of History*.[9]

Third, the sociology of knowledge was especially important, and was a major impulse behind what became known as the Edinburgh school. There, sociologists, philosophers, and historians together examined the ways in which the sociology of knowledge could be applied to science and to medicine—Barry Barnes, David Bloor, Steven Shapin, and Donald MacKenzie are perhaps the best-known figures.[10] In the more general literature, Peter Berger and Thomas Luckman's *The Social Construction of Reality* is a convenient example of this perspective.[11] Other sociological approaches, such as studies of professionalization and power, have also been influential.[12] Bruno Latour has been unusually successful in bringing his orientation to wide audiences by developing sociological models others could put to use and by demonstrating their value through his own detailed case studies. His stress on the practice of science, on the need to observe the details of *Science in Action*, has captured the interest of those sympathetic to social constructionism not only through forceful and vivid writing but by foregrounding the processes of knowledge-production.[13] Perhaps the best-known single sociological contribution for the history of medicine remains Nicholas Jewson's pair of articles, which have been followed up by many scholars, especially those who work on the eighteenth century.[14] My contention is that the sociology of knowledge inspired, especially among historians of science, scholarship that is so outstanding it enjoins us to explore ways in which those working on medicine can make better use of it.

Fourth, and in keeping with the trends already outlined, was the impact of social and cultural anthropology.[15] Robin Horton and Mary Douglas can stand as examples of this, although they were certainly not alone among anthropologists in claiming the relevance of what they were doing for the social nature of science and medicine. As a result of the influence of anthropology, statements about belief systems became commonplace, even if few took the trouble to spell out what the pay-off might be. In general, it seemed more plausible to treat magic and medicine not as successive stages in a progressivist trajectory but as somehow equivalent, if different *Modes of Thought*.[16] The interpretation of magic has become a major historiographical issue, not just for historians of science and medicine, but for all those who study what had been deemed the "irrational."[17] Indeed it was historians of witchcraft who drew the attention of the historical community as a whole to the value of anthropology.[18] The writings of Frances Yates along with the work of scholars such as Piyo Rattansi and Charles Webster, for whom magic, medicine, natural philosophy, and utopianism were intertwined, also suggested the direct relevance of taking magic seriously.[19] Mary Douglas's writings were especially influential in rekindling an interest in cosmology and in encouraging fresh reflections about the nature of fundamental social boundaries, as well as the symbolic density of all the realms that mediate nature and human nature.[20] Anthropology was inspirational because, as a reflexive discipline, it was pledged to understand sympathetically ways of being that were other. By that token, it offered far more than intellectual frameworks; it embodied relationships between knower and known that appeared exceptionally relevant to historians of science and medicine.

Fifth, a politicized reaction to scientific and medical power did much to nurture the sense that these were major arenas of conflict and struggle, in which concepts were contested precisely because they were forged and deployed as "social relations."[21] Particularly influential were feminist critiques of medicine, which examined the complex relationships between practitioners and patients before there was a substantial feminist literature on the natural sciences.[22] It was precisely because there was a variety of social settings in which medical practitioners interacted with women as colleagues, subordinates, and competitors, and as "customers," that a social analysis of the resulting ideas about femininity and women's bodies was possible. Here was an instance where ideas and practices were analyzed together.[23] For just this reason, radical critiques of technology have also been important. They had an immediate relevance to contemporary political issues, such as nuclear power, pollution, imperialism, and scientific management. As a result, those who work on the twentieth century have developed forms of social constructionism for which political critiques of technology are central.[24]

Sixth, a number of these approaches encouraged scholars to look at how interests shaped the theory and practice of medicine. Clearly, this was not new, since in a sense it was precisely what Robert Merton had done for puritanism and science in the 1930s.[25] The question of how "interests" are studied remains a central issue for historians. What was laid down in the 1970s remains true today of the history of science and medicine. Interests were interpreted in terms of professional advancement; religious affiliation; political allegiance the quest for power, money, and authority; and patronage and networks. In practice, many historians of medicine try to consider a range of interests, but there is still a tendency to give weight to some interests and play down others, according to one's orientation. On the whole, an emphasis on professional advancement appeals to those most open to sociological theory, while those who see their work as closer to the historical mainstream prefer religious and political interests. The authors of political critiques lay particular stress on expertise—the interests that emerge with specialized, authoritative knowledge—but it has largely been left to those who study institutions to explore patronage and networks. The most obvious concept to invoke in this context is "class," yet how this is to be defined and to which periods it is best applied remain contentious. Thus, an anxiety about using notions such as class endures, and this is linked with a tension between understanding interests in individual and in collective terms. We still pay far too little attention to the complex manner in which interests are experienced, shaped, and expressed.

Seventh, critiques were developed, often in a self-consciously Marxian mode, of the use of nature and its cognates.[26] The argument went something like this: when scientists or the medical profession talked about nature they constructed that term, not just in their own interests, but in such a way as to veil its thoroughly social provenance. This enabled them to use their expertise in an authoritative way by insisting on their exclusive, privileged access to a domain that was above and beyond society. "Nature" was thus a prescriptive category masquerading as a descriptive one and an instrument of class struggle.[27] This point was well exemplified in the power of orthodox medicine to designate a person fit or unfit, well or ill, and if unfit or ill to intervene as was thought appropriate. The formidable complexities of how this worked in practice can be seen in the now vast secondary literature on eugenics and in the resulting historiographical debates.[28] There are economic variants of this position for which medicine becomes a tool for controlling the size and composition of the work force.[29]

All of these strands opened up spaces for historians to explore; they encouraged an emphasis on the constituent processes of medicine and science; and they required the elaboration of infinitely more complex frameworks and models to help imagine the social placement of that amalgam we call "medicine."

Finally, we have a distinctive approach, which I shall call localism. It arose when historians worked on health and medicine who had neither been medically trained nor had their ideas shaped by history and philosophy of science. They, by contrast, had been trained to see particularities of all kinds, to detect the forces that shaped institutions and medical provision outside institutions. Never having believed in the tight fit between nature on the one hand and science or medicine on the other, they tended to take it for granted that social processes filled the gap. In a sense, they were social constructionists without knowing it. However, they often stopped short of analyzing how societies shape medical theories and beliefs, preferring to look at "popular" practices, charities, and so on, leaving theoretical or technical matters to one side.[30]

Social constructionist approaches can usefully be applied to all aspects of medicine. One reason why they have not is the continuing prevalence of the use/abuse model. It is vital that we consider this approach and at the same time dispose of some of the main misunderstandings concerning social constructionism. Inevitably, philosophical debates about social constructionism continue, and it is worth distinguishing between social constructionism as a heuristic device for practicing historians and its theoretical status. While attempts to clarify its status can have a bearing on historical practice, they do not necessarily do so, largely because discussion is pitched at such an abstract level. My interest is in its heuristic capacities.

Misapprehensions

Making a distinction between the use of scientific and medical ideas and their abuse serves both to affirm their value-neutrality and to effect a separation between the production of knowledge and its deployment. Social constructionists want to show, through theoretical argument and empirical claims, that ideas necessarily carry or mediate values, that making and using knowledge cannot be so neatly separated, and that understanding the social meanings of natural knowledge is preferable to making moral judgments about the propriety of practitioners. A recent example of the use/abuse approach is Cynthia Russett's *Sexual Science* in which she pokes fun at the excesses of nineteenth-century science and medicine in relation to women and to the understanding of gender differences, using examples largely from the United States and Britain.[31] When "sexist" pronouncements were made or practices undertaken, practitioners were, for Russett, not employing very good science. Liberal feminism has become a reason for condemning scientific "abuse." The book simply assumes that earlier approaches to the biology of gender are at once risible and depressing. I read the use of this model as highly defensive. It leaves science and medicine intact epistemologically and politically. Two kinds of historical laziness result. First, the implicitly moral framework appears to make the work of understanding how people could think so differently unnecessary because it offers up simple answers—either they were not good scientists or their work was used by others for bad ends. Such moralizing is antithetical to the social constructionist project. Second, the use/abuse model does not challenge historians to unravel the mediating processes involved in the creation of knowledge, leaving the "best" science and medicine as unhistoricized, because true and acceptable, and capable of being used for worthy purposes.

This is the point to dispose of three misapprehensions related to social constructionism. The first concerns "medicalization." This refers to the process whereby domains of life that were not previously so came under the aegis of medical practitioners and/or medical theories. The implication is that since the eighteenth or nineteenth centuries, the periods most often taken to mark the onset of medicalization, medical power has slowly but inexorably grown. French historians especially have taken the number of trained practitioners for a given number of people as an index of medicalization.[32] This in turn implicitly privileges professionalization as a key historical process. Most English-speaking historians who use the term are less concerned with the number of practitioners than with the qualitative growth of medical power. In fact, there is no inherent link between "medicalization" and a social constructionist approach. Indeed, in some ways they are historiographically at odds in that the former has a teleology—that of modernization—built into it, while the latter opposes this and is generally quite hostile to present-centered approaches.[33] Exponents of a medicalization approach tend to see medical power as gained with relative ease or even simply appropriated. A social constructionist will more likely look for the points of tension, for negotiations and conflicts through which particular kinds of authority may or may not be gained, and specify rather precisely the social groupings involved.

The second misapprehension concerns a charge that is still sometimes made and was mentioned frequently when social constructionist approaches were first debated, namely that they ignore the material dimensions of life. This is based on a misunderstanding of the philosophical claims that underlie social constructionism. It is often caricatured by critics, who impute to it the claim that diseases are not real and who associate it with a denial that science and medicine really work. The implication is that social constructionism deals with what is evanescent, epiphenomenal, precisely with what is not "real." There is no logical basis for these assertions. On the contrary, the material world is constantly shaped and interpreted through human actions and consciousness. Social constructionism takes this as one of its main tenets and without the dynamic relationship just described, it would have no meaning. It is not a form of idealism. But it does insist that there is room for a variety of interpretations and meanings, that behind consensus or "knowledge" lie social processes and that such processes involve negotiations and conflict, both overt and implicit. That is, to use the metaphor I employed earlier, it presents spaces that can be filled in diverse ways. It follows that forms of knowledge and the social processes whereby they are created are given intellectual priority. It does not follow that materiality and physical embodiment are denied.

The third misapprehension concerns the distinction between so-called internalist and externalist approaches.[34] Contained in this distinction is the assumption that what is "social" is what is also "external" to the heart of medicine and science, since their core is taken to be knowledge claims and these are assigned to the category non-social. Thus, as in the use/abuse model, something special is saved from the perceived taint of sociological analysis. While it is now evident that separating content and context in this way was profoundly unhelpful, integrating them is less easy than it looks, hence the importance of highly specific case studies.[35] Social constructionism is not allied with externalism, although this has sometimes been assumed; on the contrary, exponents of this approach have generally been skeptical about facile dualisms such as internal/external.

Before leaving the question of misapprehensions, I want to say something about Michel Foucault, who is, somewhat misleadingly, credited with a founding role in social constructionism. The trends that encouraged social constructionism were broader and more general than the influence of any one figure, and they were well-established among English-speaking historians long before Foucault had a major impact. *Madness and Civilization* was one of a clutch of works that sought to dismantle the pretensions and deceits of psychiatry in the 1960s.[36] Far more important for the topic under discussion were *The Birth of the Clinic, The Order of Things,* and *The Archaeology of Knowledge.*[37] It was the assertion contained in all three that ways of seeing/knowing (epistemes) contained their own logics and limitations and that radical shifts in epistemes occurred over a relatively short period of time, which was both influential and contentious. A revulsion toward scientific and medical power was certainly present in these works, as it had been in *Madness and Civilization* and in his less well known earlier work *Mental Illness and Psychology,* but the claims about epistemes were and remain more intellectually challenging.[38] Described in this way we can see why the impact of Foucault and that of Kuhn via the notion of paradigm pointed in similar directions. Yet for practicing historians there is a very important difference between them. Kuhn tried to show through a series of case studies how one paradigm gave way to another, and he conceptualized the processes involved, if not as inherently social, at least as having a significant social dimension. Foucault, by contrast, tended to create a very powerful but somewhat static sense of what each episteme consisted of, with the transformations from one to another appearing more like a gestalt switch than complex processes of transformation. Indeed, he does not seem to have been interested in the processes that permitted the emergence of epistemes or in precisely how shifts between them occurred.

Some would say that Foucault nurtured an interest in discourse, and especially in medical discourse.[39] It is undoubtedly true that he drew the attention of, above all, literary critics to these topics and that his work in effect brought the history of medicine to a wider audience, but those interested in the cultural history of medicine had and have a wide range of intellectual models that may serve as inspiration. In practice, history inspired by Foucault tends to be somewhat "flat," lacking in historical texture, and it could be argued that considerable effort has had to be expended on correcting the more common "truths" he put into the public domain, especially about the history of hospitals.[40] The best work inspired by a Foucauldian orientation succeeds precisely because it does more than discourse analysis, because it is not flat. The growing interest in scientific and medical discourse has been nurtured as much by scholars such as Raymond Williams as by Foucault.[41] The realization that texts are dense, and that they should be interpreted accordingly, has been productive. While just what counts as a text and what larger historical insights texts afford remains a subject for debate, a more vivid appreciation of the ways in which languages carry, shape, and change ideas is extremely fruitful. The problems arise when discourse is detached from other historical processes, and when a small number of texts are taken out of their context and made to carry a heavy explanatory load.

Interpretive Issues

I have presented the potential of social constructionism in such a way as to suggest that it raises a number of issues that require further debate and elaboration. These discussions are particularly important if we wish to practice the social history of medicine as part of the general discipline of history and to explore its kinship with cultural history. The four issues that have been selected for brief discussion here reflect this agenda.

First, interests: these are often invoked to explain changes in beliefs and to account for situations where there appears to be a good fit between the content of beliefs and social attributes. An insistence on the shaping role of interests has been one way of breaking down the rather globalized approach of Foucault, for whom an episteme had a special meta-status; it somehow had the capacity to structure consciousness in general. Interests can do this too, but the concept suggests above all particularities and differences—to invoke the term is to imply conflict and competition between groups or ideologies with definable characteristics. Conventionally used, interests can also easily be construed as "external" to medicine, so that they become social forces outside medicine proper, which is equated with medical theories. Hence, there was a danger that interests were deemed, above all by those hostile to social constructionism, simply social trappings, irrelevant to the real business in hand. Even those sympathetic to social constructionist approaches tend to treat interests as independent variables or as simply economic at their base. For instance, Malcolm Nicolson's interesting argument about the social construction of pathological ideas in the eighteenth century, presents a particular theory—metastasis—as serving, in his words, "business interests."[42] Admittedly, he does widen the scope by referring to the metastatic theory as "a tool of professional interest and social control," but even here a sense of instrumentalism is conveyed.[43] All too easily the result of giving priority to interests is mechanistic explanations. This is not to deny the importance of interests, only to suggest that we develop more complex models of how they work, especially given that individuals and groups have innumerable actual or potential interests, some of which may be at odds with each other.

The growing concern with religion among historians of science and medicine has proved especially effective in opening up this issue.[44] We can identify a spate of work—for example by Margaret Jacob, Jim Jacob, and Charles Webster—which insisted that religious-cum-political concerns (for these two could not be separated) shaped the content of medical and scientific beliefs and practices.[45] Thus, the precise variety of Christianity adopted and the manner in which this was integrated with assumptions about political power and legitimacy was virtually the same as the form of natural philosophy that was favored. I say Christianity because the most compelling writings in this genre have concerned seventeenth- and eighteenth-century England. This approach is still evident in recent work.[46] For instance, in his article in *The Medical Enlightenment of the Eighteenth Century*, Robert Kilpatrick pays particular attention to the Quaker John Coakley Lettsom and his medical-cum-philanthropic activities. Speaking of dissenting physicians in late-eighteenth-century London, he states, "I will show that it is possible to account for their medicine—and the institutions and societies they founded to advance it—in terms of their religious and political beliefs."[47] He mounts a compelling argument, which shows the complex role interests, interpreted not just in professional terms but as religious-cum-political commitments, could play in medical practice and organization. Yet Kilpatrick also says, "Lettson was a Quaker first, a physician second and a philanthropist third."[48] What kind of claim is this? Biographical perhaps? About depth of commitment? And what kind of evidence sustains it (no footnote is given)? I feel uneasy about this disaggregating of variables. Does the reified concept of interests, as it is generally employed, serve well the social history of medicine?

Interests are generally used to link the content of beliefs with biographical attributes. One problem is that historians of medicine have relatively few detailed modern biographies to draw upon, in marked contrast to historians of science, who have become eager biographers especially over the last ten years or so.[49] A larger and more sophisticated biographical literature would enable scholars to provide more precise reconstructions of the relations between the constituent elements of the lives of medical practitioners.[50] Although a traditional form of scholarship, biography can help with other interpretative challenges thrown down by social constructionism. A specific example may help.

Take William Hunter (1718–1783). His ideas about the anatomy and physiology of the human body and about how these should be taught and represented were shaped by his institutional situations (teaching anatomy in his own school and in the Royal Academy) and by his collaborations with other medical men, students, artists, printers, and engravers. They were also molded by the multiple patronage relationships he experienced, both as client and as patron. It is by no means irrelevant to his medical work that he was an influential art collector, that he sat for the court painter Allan Ramsay, or that his artistic connoisseurship enabled him both to work with highly skilled artists, engravers, and printers in producing his obstetric atlas, and to articulate the aesthetic vision that lay behind it. Thus, a full biography would provide significant insights into the relationship between medical practice and court life, the patronage networks of the period in medicine, politics, and the visual arts, medicine as a business, the relations between different professional groupings and their institutional bases, collaborations between artists, anatomists, and printers, and so on. Some of this is known, but the larger picture has never been drawn together in a systematic way.[51] It could be said that Hunter is atypical, but in fact he was by no means unique in the wide number of ways in which his medicine articulated with institutions, professions, and elites, indeed with many aspects of polite society. How can we judge him and his works until we have a range of studies of medical contemporaries to use as comparisons and to generate models of how knowledge was shaped? It would be possible to mount a similar argument for Sir Charles Bell (1774–1842).

The second issue that arises, then, is context. Context is not precisely the concept we need because it has dualistic connotations. In pairing context with content, the implication is that although related, they are distinct. As a result, affinities with the couple internal/external can be evoked, which is unfortunate and misleading. Nonetheless, context is the best available term. I have suggested that biographical studies, because of the sharpness of focus they can achieve, are capable of providing a more elaborate sense of context, of how things were seen at a given time and place—the equivalent of Michael Baxandall's "period eye."[52] Of course there are many other ways of generating context; the most successful studies of institutions can have the same effect, as can studies of specific conditions, illnesses and diseases, specialisms, and so on. By tracing as many of the threads as possible that lead from and to any given medical focus, by being conscious of doing so, and by conceptualizing the relationships discovered in this way, social constructionist approaches to medicine will be enhanced. There is a difficult balance to be achieved here between drawing in a context so specific that the findings do not readily bear on any other case study and using a schema that is not detailed or particular enough to have explanatory power.

The third issue is chronological specificity. Those who work on the twentieth century tend to think about context quite differently from those doing research on, say, the seventeenth century. A notion akin to the period eye can be particularly helpful because it is designed to elicit what is special about a given historical setting both as a whole and in nonteleological terms. Teleology still dominates the social history of medicine—even if we reject triumphalism, what is known about changes in morbidity and mortality, together with assumptions about professionalization and state intervention, makes the assumption that medicine moves toward a modernist goal almost irresistible. One positive outcome of paying more attention to conceptualizing chronologies specifically in relation to medicine would be that, at last, we could have survey histories of medicine written in thoroughly social terms. Furthermore, the frameworks such writing would produce could constitute a first step toward generating a serious comparative literature in the field. It is fair to assume that the resulting histories would have a quite different shape from the more fragmentary accounts we have now. This will only be possible, however, if research on all historical periods is actively encouraged. The general drift of historical fields seems to be toward the study of the recent past; while this is understandable, it can lead to the loss of a bigger picture. Detailed historical work in the social constructionist mode has tended to take the form of case studies, which enhance our sense of period specificity, but there is no logical reason why it should not also deploy other frameworks that are more ambitious in their chronological, and geographical, scope.

The final interpretative issue is the use of models. Because of its closer links with philosophy of science, the history of science has had a range of models of scientific processes and scientific thinking to draw upon, and has on the whole been enthusiastic about "testing" such models in detailed historical research. This then predisposed the field to consider models from anthropology and sociology. The journal *Social Studies of Science* shows how especially those looking at contemporary natural science and technology have adopted this strategy. But it has appealed less to social historians of medicine, with good reason. The mismatch between model and case study can be a problem; often, convincing "translation" between the two is difficult. Hence, it seems attractive to use a more eclectic approach, to proceed more in terms of grounded theory than in terms of models, strictly speaking. Yet there is a case to be made for the use of models, not so much for the purposes of prediction, but more loosely as heuristic devices. Rather than turning to other disciplines for models, historians should be more active in generating their own. The "Me Tarzan, you Jane" approach that was once common, whereby historians provided data for a macho theoretical domain to crunch on, has long gone. Indeed, a few historians of medicine have written in such a way that their schema could be applied to other cases.

Historiographies of Science and Medicine

In turning to further examples, both of work done and of areas for future research, I want to come back to a topic that has run through this paper—a comparison between the historiographies of science and medicine.[53] I indicated above that the differences between the history of science and the history of medicine were partly explainable in terms of differences between the two domains that have been identified with increasing clarity by some recent scholars, who would argue that they hold true across a long time span. The key to this lies in two related aspects of medicine. First, a very significant proportion of most populations have experienced medicine at first hand through social interactions with practitioners of one kind or another. Second, these interactions had an immediate significance for them, which was understood in the most intimate terms of bodily well-being, as well as through the languages and images that pervade societies.[54] Thus, in most societies, matters of health and sickness were far more dispersed, commonplace, and immediate than any aspect of science. In this respect, the history of medicine is more akin to the history of technology than to the history of science, hence the approaches found in the journal *Technology and Culture* and in radical writings on technology might be of particular value to social historians of medicine.[55]

One result of the dispersed, socially integrated nature of medicine is that certain kinds of historiographical approaches, especially those that are localist in character, are easier to develop. This partly derives from the nature of the primary sources—aside from official and semiofficial sources, they consist of numerous scattered and fragmentary references. Work is only just beginning on the almost infinite number of relevant materials in local archives. These generally relate to three aspects of medicine: the lives of practitioners, the experiences and behavior of patients, and the interaction of medicine with agencies that operated at a local level, such as the law, charities, the church, and relief of the poor. Other kinds of archives can be equally revealing. For example, the eighteenth-century State Papers Domestic in the London Public Record Office contain references to medicine when it impinges on government business—plague on mainland Europe or the serious illnesses of politicians, for example. The riches of the Public Record Office at Kew in relation to medicine are slowly coming to light. It is now becoming clear how valuable a source newspapers can be. One difficulty, however, is that unless a wide range of these materials is thoroughly analyzed, a small number of sources are given a paradigmatic status, as has happened to Ralph Josselin's diary.[56] The need to avoid anecdotalism is paramount. Superficially, there is no necessity for research on such sources to adopt a particular theoretical orientation or to work through the philosophical basis of its claims. In other words, unmodified "localism" might be thought to work perfectly well. But now local studies in the social history of medicine are emerging—Bristol has been particularly well served in this respect—that reveal the value of a more analytical approach in addressing the issues outlined in the section on interpretive issues.[57]

Historians of science, by contrast, tend to proceed quite differently. Their sources are likely to be more systematic by their very nature; on the whole, they feel the need to spell out their logical maneuvers rather explicitly, and to explain the implications of these for "science." The result, I believe, is that in fact more rigorous work along social constructionist lines has been done in relation to science than to health and medicine; furthermore, the best work on the latter has come from those with an interest in history, philosophy, and sociology of science. Historians of science have been able to endow their writings with a special kind of conceptual coherence that others could emulate. So, despite the differences between science and medicine, it may be that the benefits of research with a strong conceptual focus, a clear sense of the theoretical issues at stake, and a systematic set of models to pursue can be adapted to the medical case.

Much classic work in the history of science has centered on fierce yet apparently clear-cut disputes, such as those between biometricians and Mendelians, Cuvierians and Lamarckians, natural theologians and materialists.[58] This has been effective because a definable set of claims, however complex these turned out to be, could be mapped against other variables: generation, politics, religion, and class. Although medicine is no stranger to controversy, its more fragmentary and dispersed quality makes it less amenable to study through polarized debates, or through processes of discovery, unless the thought processes of practitioners and researchers are to occupy center stage. While laboratory-based medical research can be examined using the scientific model, most medicine is not like this. It is therefore not surprising that social constructionism has worked particularly well in the history of medicine where specific conditions have been studied.[59] These are akin to debates, controversies, or theories in their capacity to act as a conceptual focus, but they are at the same time "dispersed" in medical practice, in cultures around illness, in literature, art, film, and in the social arrangements and reactions the condition elicits. This permits clarity of focus and historical richness to complement each other.

Orthodox Western medicine shares with the natural sciences the centrality of a self-consciously systematic and specialized group of theories. The organization of medical care and medical education embodies, if in highly complex ways, the ideas that lie at the heart of the medical endeavor. Some alternative systems also proceed in this way. This is to reiterate the point that the social history of medicine cannot avoid ideas, whether these are theories, common assumptions, part of a "collective unconscious," or representations. The social history of ideas, medical or otherwise, remains an underdeveloped field, especially in Britain, and has perhaps been superseded by cultural history, which has never enjoyed more interest than it does now.[60] Yet distinctions between the two fields may be made. A social history of ideas would examine how ideas move about in societies and do justice to esoteric and highly technical ideas that enjoy a more restricted existence. Culture, however difficult it is to define, is more than ideas. The so-called new cultural history is not engaging with difficult ideas in the same philosophical depth as did scholars in the Lovejoy mould.[61] Nonetheless, if ideas are given due prominence, some form of cultural history will probably generate the most productive approaches to science and medicine, but it will need to be a form capable of analyzing ideas and social processes together. There is some evidence that this is beginning to happen, partly because cultural history is an umbrella under which literary critics with an interest in the social history of medicine feel comfortable.[62]

Processes

It has been implicit here that we need a historiography capable of explaining the imaginative reach of ideas of health, healing, and sickness. Medical ideas inform how people experience those states, react to them, act upon them, and construct their significance. These ideas have a primal quality in two senses: they underpin experiences and actions, and they are deeply embedded in our consciousness and hence only partly available for conscious manipulation. The challenge is to explain how such ideas change and to resist the temptation to attribute uniform beliefs to large groups of people. If we take this to be an area for cultural history, it may become easier to see how the processes through which disease construction takes place can be disentangled. Cultural history is a broad church that can encompass a wide range of methods and approaches, including attempts to do justice to the "collective unconscious" around health and disease. Models can accordingly be developed, their exact form depending on the period and the condition in question, whereby the interactions between a number of domains can be conceptualized. The range of interactions is likely to be formidable, and they will only make sense if the verbal and visual languages that mediate them are deconstructed. This is likely to work well where there are shared languages of health and sickness, and conditions that can be or have been fairly clearly defined.

There can be no doubt that social constructionism has represented a significant historiographical advance. In the absence of any coherent alternatives, we have no choice but to test its potential to the full. Its great contribution, I suggest, has been to make historians think much harder about processes and interactions that were previously invisible, denied, or thought unproblematic. Three types of processes in particular have come to the fore. The first is often summed up in the phrase "the patient's perspective," but it is really more about understanding what goes on between patients and practitioners.[63] The supposition that patients were absolutely or even relatively passive has lost credibility, and it is precisely in contexts where patients had a certain economic bargaining power that it has been easiest to study these dynamics. Expanding the range of sources and interpretations used to explore these relationships must be a high priority.

The second concerns the ways in which divisions of labor are shaped and reshaped. I am using "division of labor" loosely here to include distinctions between mental and manual work; definitions of discrete skills; the divisions found in organizations, specialties, and disciplines; and the social and cultural forms they give rise to. Work on earlier periods has shown how very specialized many practitioners were.[64] Over a lifetime, even a person of quite modest means would be treated by many different hands, especially if we include those from whom they bought remedies, where, as Pelling has reminded us, the very idea of a remedy was extremely loose, often inseparable from cosmetics.[65] Overt competition between practitioners, in the eighteenth century for example, was fought out through definitions of quackery and over the issue of specialization. The situation reveals how central the division of labor was to negotiations about the nature and causes of illness and about therapy, as *Health for Sale* demonstrates.[66] The amount of time, money and effort expended on occupying the medical high ground indicates how fragile and labile everything to do with medicine was and how elaborate the processes were through which "proper," orthodox medicine was contested and constructed.[67]

There is indeed a kind of division of labor here, although very different from the one that is found in, say, late nineteenth-century Europe. Since the division of medical labor followed strikingly different patterns in North America, there is a significant opening here for comparative history. Indeed, social constructionism implies a comparative method, since a single instance or case study is a rather weak way, logically speaking, of demonstrating the claims upon which social constructionism rests. Knowledge and skills are molded by shifts in the division of labor—most obviously, the changing relationships between nurses and doctors have, since the late nineteenth century, demanded constant shifts in definitions of care, in the apportionment of skills, in the understanding of danger and risk, in "patient management," and in the social relations of medicine more generally.[68]

Along with paying greater attention to the division of medical labor would go more careful analyses of the full range of medical occupations and of the processes of specialization.[69]

The third type of process concerns the cognitive dimensions of medicine, especially the processes whereby health and disease are conceptualized. Many parts of a society contribute to such processes, which are often worked through in popular culture, advertising, art, and literature, as well as in the more obvious areas such as politics and policy-making around health issues. Social constructionists will want to look more at the cultural resources available for fashioning medicine. It is an approach that works especially well for the visual dimensions of clinical work. Medical practice is visual at its heart; hence, it would be to ignore a central historical issue not to study how looking shapes virtually all aspects of medicine.[70] The centrality of visual experience in medicine demands an interdisciplinary approach, learning from art history, film studies, and cultural studies. While social constructionism nurtures interdisciplinary history, it does not necessarily involve imposing frameworks from other disciplines on historical materials.

Of particular importance for the future will be the integration of the "top-down" approach that is so often implicit in social constructionist methods with "bottom-up," localist ones. I have suggested that focusing on specific conditions, although not necessarily or only on diseases as defined by official medicine, is one way of bringing these two rather different styles of analysis together. In effect, this achieves another important goal: a historical analysis that both explains even the most technical, arcane aspects of health and medicine and does justice to the pervasiveness of medical languages and ideas, by conveying how, when, and why they elicit reactions.

Social constructionism is a valuable perspective for historians of health, medicine, and healing. Far from neglecting material life, it is the only approach that integrates this with ideologies, images, ideas. It is effective partly because it eschews the rigid polarities that weaken other approaches: here, theories and archives are totally compatible; here, ideas are not separated from practices; here, an emphasis on process undercuts unproductive distinctions between internal and external factors, content and context, good and bad science. There is an important difference between a social history of these questions, which implies few theoretical claims, and social constructionism, which does. Because practices surrounding healing and sickness are so dispersed, there is a danger that untempered localism will lead to anecdotal history. Another possibility—subsuming such matters under a nonmedical rubric—religion, social policy, charity (the most favored alternatives)—may work in particular cases but perhaps loses us the chance of seeing how health and illness work across a whole society. This is surely important since medicine generates some of the most powerful vehicles for disapproval and approbation, for expressing moral values in naturalized forms. When images of disease and of health are potent (when are they not?) not just in relation to individuals and to collectivities, but to nations, even continents, a method for analyzing ideas is essential. Certainly, social constructionists emphasize ideas, not to deny the importance of everything else or to exaggerate the influence of elite groups, but because they have a vivid appreciation of their primal power. Ideas are primal by virtue of their capacity to act as mediators, to shape both conscious and unconscious experience, and to play a dynamic role in organizations and social life. Uncovering the fully social and cultural nature of medicine in all its facets requires a historical approach that takes ideas, be they about health, illness, or healing, very seriously indeed. So far, only social constructionism has applied for that job: the future will consist of seeing how well it performs in it.

Afterword

In rereading what I wrote in 1995 about the social construction of medical knowledge, I am struck by how many themes were later developed in *History in Practice*.[71] Although in the book I sought to speak of historical practice more generally, the ideas developed earlier, specifically in relation to science and medicine, became central. One conclusion I draw from this is that, for reasons mentioned in the chapter, we had no choice but to think through a wide range of conceptual issues in relation to the history of science and of medicine that turned out to have wider applicability.

My book is considerably more explicit than this essay could be about how we understand historical fields. In it, I suggest that we always need to be historians of our own practices, to understand their moments of origination, their life histories. I also discuss the professional investments in journals, the names of fields and subfields and institutional arrangements. Thus, we should see that notions such as "social history" and "cultural history" are more or less convenient labels and that we use them to effect other kinds of business, such as affirming professional identities. What they refer to is underdetermined. When I wrote this essay I held a position in cultural history, a field that was then receiving a lot of attention. While I do not think that its allure has gone away, to me the idea of cultural history has shifted in its significance. I still think of myself as a cultural historian, but that is no more than a convenient, sheltering umbrella. A huge shift in my professional practice occurred in 1996 when I took up a post in art history. Since then, I have given careful consideration to the relationships between these various forms of history. That is a topic in its own right, but here I want to signal that this change of disciplinary location has made me acutely aware of how important what we call "social history" is.

Art, like science and medicine, has some chunky concepts at its heart, as well as a distinctive type of aura and its own elaborate gatekeeping mechanisms. All these phenomena are social—although that may not sound like a particularly strong claim. In fact, it is shorthand for the careful attention to practice and process that this chapter advocated. Since writing it, I have become a lot more interested in networks, including in medicine. The work I have done on portraits of medical men has demonstrated their value for me most vividly. Their specifically *cultural* density can only be fully released through carefully building up a *social* account. That is, in order to understand what and how they mean, it is necessary to know and to piece together intelligently a great deal of social information about patronage, commissioning, sitting, owning, and displaying. Medical institutions paid a great deal of attention to these activities, and they need to be charted painstakingly. Then it becomes possible to appreciate the cultural richness of the items of visual and material culture that they owned, their capacities to carry associations and lineages through which individual and collective medical identities are forged.

This example is designed to illustrate the point that social and cultural history are natural allies and that nothing is to be gained by drawing too strong a distinction between them. I can perhaps affirm the broader point by mentioning the status of economic history, which has been declining over the past twenty years. The fact is that this field, a bit like social history, is not seen as exciting as it used to be. I realize this is a generalization, but markers such as student demand bear me out. There are quarters where more "empirical" fields, as well as fields that are perceived to be old-fashioned and even abstruse (as is sometimes claimed about economic history), have become dowdy. It is commonly assumed too that interdisciplinarity and theory are somehow "good." I want to refuse those polarities. The social constructionist approach still has much to offer precisely because it does engage with the complexities of historical detail; because it can find, if you like, the cultural in the social; because it is flexible enough to see that, without warm engagement with human practices, theory is no good at all. All I might add is that for historians of medicine, material and visual culture are likely to be become increasingly important, not out of some abstract commitment to being interdisciplinary, but because (like medicine itself) they lie right at the heart of social-cum-cultural worlds.

NOTES

Earlier versions of this chapter were given at Clare College, Cambridge, the Science Museum, London, the University of Essex, and the University of Alberta. My thanks to all concerned, to the Wellcome Trust for financial support, and to Cathy Crawford, Luke Davidson, Karl Figlio and Richard Smith for their help and advice. It is to be read as a position paper rather than as a survey article. In citing works I have emphasized those that have contributed to my own ideas; the original authors do not necessarily figure themselves as "social constructionists."

1. *Companion Encyclopedia of the History of Medicine,* 2 vols., W. F. Bynum and Roy Porter, eds. (London: Routledge, 1993); Ludmilla Jordanova, "Has the Social History of Medicine Come of Age?" *Historical Journal* 36 (1993): 437–449; *Medicine in Society: Historical Essays,* Andrew Wear, ed. (Cambridge: Cambridge University Press, 1992); *The Problem of Medical Knowledge: Examining the Social Construction of Medicine,* Peter Wright and Andrew Treacher, eds. (Edinburgh: Edinburgh University Press, 1982).

2. David Armstrong, *Political Anatomy of the Body: Medical Knowledge in Britain in the Twentieth Century* (Cambridge: Cambridge University Press, 1983); Wright and Treacher, *The Problem of Medical Knowledge.*

3. *Information Sources in the History of Science and Medicine,* Pietro Corsi and Paul Weindling, eds. (London: Butterworth Scientific, 1983); *Companion to the History of Modern Science,* R. Olby, G. N. Cantor, J.R.R. Christie, and M.J.S. Hodge, eds. (London: Routledge, 1990).

4. Roger Chartier, *Cultural History: Between Practices and Representations* (Cambridge: Polity Press, 1988); *The New Cultural History,* Lynn Hunt, ed. (Berkeley: University of California Press, 1989).

5. Thomas S. Kuhn, *The Structure of Scientific Revolutions* (1962; reprint, Chicago: University of Chicago Press, 1970).

6. Michael A. Arbib and Mary B. Hesse, *The Construction of Reality* (Cambridge: Cambridge University Press, 1986).

7. Mary Hesse, *Models and Analogies in Science* (Notre Dame: University of Notre Dame Press, 1966).

8. Gillian Beer, *Darwin's Plots: Evolutionary Narrative in Darwin, George Eliot, and Nineteenth-Century Fiction* (London: Routledge & Kegan Paul, 1983); Evelyn Fox Keller, "From Secrets of Life to Secrets of Death," in *Body/Politics: Women and the Discourses of Science,* Mary Jacobus, Evelyn Fox Keller, and Sally Shuttleworth, eds. (London: Routledge, 1990), 177–191; Bynum and Porter, *Companion Encyclopedia,* chaps. 9 and 51; Evelyn Fox Keller, *Secrets of Life, Secrets of Death: Essays on Language, Gender, and Science* (London: Routledge, 1992); Olby et al., *Companion to the History of Modern Science,* chap. 65; Mary Poovey, *Uneven Developments: The Ideological Work of Gender in Mid-Victorian England* (London: Virago, 1988); Steven Shapin and Simon Schaffer, *Leviathan and the Air Pump: Hobbes, Boyle, and the Experimental Life* (Princeton: Princeton University Press, 1985); Raymond Williams, *Keywords: A Vocabulary of Culture and Society,* 2d ed. (London: Fontana, 1983).

9. H. Butterfield, *The Whig Interpretation of History* (1931; reprint, Harmondsworth: Penguin, 1973) and *The Origins of Modern Science* (London: G. Bell, 1949).

10. Barry Barnes, *Scientific Knowledge and Sociological Theory* (London: Routledge & Kegan Paul, 1974) and *Interests and the Growth of Knowledge* (London: Routledge & Kegan Paul, 1977); *Natural Order: Historical Studies of Scientific Culture,* Barry Barnes and Steven Shapin, eds. (Beverly Hills: Sage, 1979); David Bloor, *Knowledge and Social Imagery* (London: Routledge & Kegan Paul, 1976); David MacKenzie, *Statistics in Britain 1865–1930: The Social Construction of Scientific Knowledge* (Edinburgh: Edinburgh University Press, 1981); *The Social Shaping of Technology,* David MacKenzie and Judy Wacjman ed. (Milton Keynes: Open University Press, 1985); Shapin and Schaffer, *Leviathan and the Air Pump.*

11. Peter L. Berger and Thomas Luckmann, *The Social Construction of Reality: A Treatise in the Sociology of Knowledge* (Harmondsworth: Penguin, 1967).

12. Bynum and Porter, *Companion Encyclopedia,* chap. 47; Terence J. Johnson, *Professions and Power* (London: Macmillan, 1972); Olby et al., *Companion to the History of Modern Science,* chap. 64; Noel Parry and José Parry, *The Rise of the Medical Profession: A Study of Collective Social Mobility* (London: Croom Helm, 1976); M. Jeanne Peterson, *The Medical Profession in Mid-Victorian London* (Berkeley: University of California Press, 1978).

13. Bruno Latour, *Science in Action: How to Follow Science and Engineers through Society* (Milton Keynes: Open University Press, 1987); Bruno Latour and Steve Woolgar, *Laboratory Life: The Construction of Scientific Facts* (Princeton: Princeton University Press, 1986); Bruno Latour, *The Pasteurization of France* (Cambridge, Mass.: Harvard University Press, 1988).

14. Nicholas Jewson, "Medical Knowledge and the Patronage System in Eighteenth Century England," *Sociology* 8 (1974): 369–385, and "The Disappearance of the Sick Man from Medical Cosmology," *Sociology* 10 (1976): 225–244; Malcolm Nicolson, "The Metastatic Theory of Pathogenesis and the Professional Interests of the Eighteenth-Century Physician," *Medical History* 32 (1988): 277–300.

15. Bynum and Porter, *Companion Encyclopedia,* chap. 60; Corsi and Weindling, *Information Sources,* chap. 4.

16. *Modes of Thought: Essays on Thinking in Western and Non-Western Societies,* Robin Horton and Ruth Finnegan, eds. (London: Faber, 1973).

17. Olby et al., *Companion to the History of Modern Science,* chap. 37; *Occult and Scientific Mentalities in the Renaissance,* Brian Vickers, ed. (Cambridge: Cambridge University Press, 1984).

18. Alan MacFarlane, *Witchcraft in Tudor and Stuart England: A Regional and Comparative Study* (London: Routledge & Kegan Paul, 1970); Keith Thomas, *Religion and the Decline of Magic: Studies in Popular Beliefs in Sixteenth- and Seventeenth-Century England* (London: Penguin, 1971).

19. P. Rattansi, "The Social Interpretation of Science in the Seventeenth Century," in *Science and Society 1600–1900,* Peter Mathias, ed. (Cambridge: Cambridge University Press, 1972), 1–32; Charles Webster, *The Great Instauration: Science, Medicine, and Reform, 1626–1660* (London: Duckworth, 1975) and *From Paracelsus to Newton: Magic and the Making of Modern Science* (Cambridge: Cambridge University Press, 1982); Frances Yates, *Giordano Bruno and the Hermetic Tradition* (London: Routledge & Kegan Paul, 1964).

20. Mary Douglas, *Purity and Danger: An Analysis of Concepts of Pollution and Taboo* (Harmondsworth: Penguin, 1970) and *Natural Symbols: Explorations in Cosmology* (Harmondsworth: Penguin, 1973).

21. Robert Young, "Science *Is* Social Relations," *Radical Science Journal* 5 (1977): 681–705.

22. *Clio's Consciousness Raised: New Perspectives on the History of Women,* Mary S. Hartman

and Lois Banner, eds. (New York: Harper & Row, 1974); Carroll Smith-Rosenberg, *The Male Midwife and the Female Doctor: The Gynecology Controversy in Nineteenth Century America* (New York: Ayer Company, 1974).

23. Olby et al., *Companion to the History of Modern Science*, chap. 8.

24. MacKenzie and Wacjman, *The Social Shaping of Technology; Gender and Expertise*, Maureen McNeil, ed. (London: Free Association Books, 1987); David F. Noble, *America by Design: Science, Technology, and the Rise of Corporate Capitalism* (New York: Knopf, 1977), Judy Wacjman, *Feminism Confronts Technology* (Sydney: Polity Press, 1991).

25. Robert K. Merton, *Science, Technology, and Society in Seventeenth-Century England* (1938; 2d ed., New York: Harper and Row, 1970).

26. Olby et al., *Companion to the History of Modern Science*, chap. 6; *Changing Perspectives in the History of Science: Essays in Honour of Joseph Needham*, Mikulas Teich and Robert K. Young eds. (London: Heinemann, 1973); Young, "Science *Is* Social Relations"; Robert K. Young, *Darwin's Metaphor: Nature's Place in Victorian Culture* (Cambridge: Cambridge University Press, 1985).

27. Olby et al., *Companion to the History of Modern Science*, chap. 57.

28. Bynum and Porter, *Companion Encyclopedia*, chaps. 20 and 51; Olby et al., eds., *Companion to the History of Modern Science*, chaps. 33 and 67; Paul Weindling, *Health, Race, and German Politics between National Unification and Nazism, 1870–1945* (Cambridge: Cambridge University Press, 1989); *Health, Medicine, and Mortality in the Sixteenth Century*, Charles Webster, ed. (Cambridge: Cambridge University Press, 1979).

29. Karl Figlio, "Sinister Medicine? A Critique of Left Approaches to Medicine," *Radical Science Journal* 9 (1979): 14–68.

30. Anne Digby, *Madness, Morality, and Medicine: A Study of the York Retreat, 1796–1914* (Cambridge: Cambridge University Press, 1985); Hillary Marland, *Medicine and Society in Wakefield and Huddersfield, 1780–1870* (Cambridge: Cambridge University Press, 1987).

31. Cynthia Eagle Russett, *Sexual Science: The Victorian Construction of Womanhood* (Cambridge, Mass.: Harvard University Press, 1989).

32. *La médicalisation de la société Française, 1770–1830,* Jean-Pierre Goubert, ed. (Waterloo: Historical Reflections Press, 1982).

33. Antony D. Smith, *The Concept of Social Change: A Critique of the Functionalist Theory of Social Change* (London: Routledge & Kegan Paul, 1973).

34. Ludmilla Jordanova, "The Social Sciences and History of Science and Medicine," in Corsi and Weindling, *Information Sources,* 81–96.

35. Adrian Desmond, *The Politics of Evolution: Morphology, Medicine, and Reform in Radical London* (Chicago: University of Chicago Press, 1989); Karl Figlio, "Chlorosis and Chronic Disease in Nineteenth-Century Britain: The Social Constitution of Somatic Illness in a Capitalist Society," *Social History* 3 (1978): 167–197, and "How Does Illness Mediate Social Relations? Workmen's Compensation and Medico-Legal Practices, 1890–1940," in Wright and Treacher, *The Problem of Medical Knowledge,* 174–224; Latour, *The Pasteurization of France;* McKenzie, *Statistics in Britain;* Simon Schaffer, "States of Mind: Enlightenment and Natural Philosophy," in *The Languages of Psyche, Mind, and Body in Enlightenment Thought,* G. S. Rousseau, ed. (Berkeley: University of California Press, 1990), 233–290.

36. Michel Foucault, *Madness and Civilization: A History of Insanity in the Age of Reason* (New York: Pantheon Books, 1965).

37. Michel Foucault, *The Order of Things: An Archaeology of the Human Sciences* (London:

Tavistock Publications, 1970), *The Archaeology of Knowledge* (London: Tavistock Publications, 1972), and *The Birth of the Clinic: An Archaeology of Medical Perception* (London: Tavistock Publications, 1973).

38. Michel Foucault, *Mental Illness and Psychology* (New York: Harper & Row, 1976).

39. Jacobus, Fox Keller, and Shuttleworth, *Body/Politics;* Olby et al., *Companion to the History of Modern Science,* chap. 9.

40. *William Hunter and the Eighteenth-Century Medical World,* William F. Bynum and Roy Porter, eds. (Cambridge: Cambridge University Press, 1985), chap. 8; Bynum and Porter, *Companion Encyclopedia,* chap. 49.

41. Raymond Williams, *Culture and Society, 1780–1950* (London: Chatto & Windus, 1958) and *Keywords.*

42. Nicolson, "The Metastatic Theory," 298.

43. Ibid., 300.

44. Bynum and Porter, *Companion Encyclopedia,* chap. 61; Olby et al., *Companion to the History of Modern Science,* chap. 50.

45. J. R. Jacob, *Robert Boyle and the English Revolution* (New York: B. Franklin, 1977); Margaret C. Jacob, *The Newtonians and the English Revolution, 1689–1720* (Hassocks: Harvester, 1976), *The Cultural Meaning of the Scientific Revolution* (New York: Knopf, 1988), and *Newton and the Culture of Newtonianism* (Atlantic Highlands, N.J.: Humanities Press, 1995); Webster, *The Great Instauration.*

46. *The Medical Enlightenment of the Eighteenth Century,* Andrew Cunningham and Roger French, eds. (Cambridge: Cambridge University Press, 1990); *The Medical Revolution of the Seventeenth Century,* R. French and A. Wear, eds. (Cambridge: Cambridge University Press, 1989).

47. Robert Kilpatrick, "'Living in the Light': Dispensaries, Philanthropy, and Medical Reform in Late-Eighteenth Century London," in Cunningham and French, *The Medical Enlightenment of the Eighteenth Century,* 254–280, 257.

48. Ibid., 259.

49. Janet Browne, *Charles Darwin: A Biography* (London: Jonathan Cape, 1995); A. Desmond, *Huxley, the Devil's Disciple* (London: Michael Joseph, 1994); Adrian Desmond and James Moore, *Darwin* (London: Michael Joseph, 1991).

50. Roy Porter, *Doctor of Society: Thomas Beddoes and the Sick Trade in Late-Enlightenment England* (London: Routledge, 1992); K. Wellman, *La Mettrie: Medicine, Philosophy, and Enlightenment* (Durham, N.C.: Duke University Press, 1992).

51. Bynum and Porter, *William Hunter; Dr. William Hunter at the Royal Academy of Arts,* Martin Kemp, ed. (Glasgow: University of Glasgow Press, 1975).

52. Michael Baxandall, *Painting and Experience in Fifteenth-Century Italy* (1972; reprint, Oxford: Clarendon Press, 1988).

53. Bynum and Porter, *Companion Encyclopedia,* chap. 1; Corsi and Weindling, *Information Sources,* chap. 2; Olby et al., *Companion to the History of Modern Science,* chap. 3.

54. Bynum and Porter, *Companion Encyclopedia,* chaps. 30 and 53; François Loux, *Le Jeune Enfant et son corps dans la médecine traditionelle* (Paris: Flammarion, 1978).

55. McKenzie and Wacjman, *The Social Shaping of Technology;* Noble, *America by Design.*

56. Lucinda M. Beier, "In Sickness and in Health: A Seventeenth Century Family's Experience," in *Patients and Practitioners,* Roy Porter, ed. (Cambridge: Cambridge University Press, 1985), 101–128; Lucinda M. Beier, *Sufferers and Healers: The Experience of Illness in*

Seventeenth-Century England (London: Routledge & Kegan Paul, 1987); Alan MacFarlane, ed., *The Diary of Ralph Josselin, 1616–1683* (London: Oxford University Press for the British Academy, 1976).

57. Jonathan Barry, "Piety and the Patient: Medicine and Religion in Eighteenth-Century Bristol," in Porter, *Patients and Practitioners,* 145–175; Mary E. Fissell, *Patients, Power, and the Poor in Eighteenth-Century Bristol* (Cambridge: Cambridge University Press, 1992).

58. Toby Appel, *The Cuvier-Geoffroy Debate: French Biology in the Decades before Darwin* (New York: Oxford University Press, 1987); Desmond, *The Politics of Evolution; Biology, Medicine, and Society,* Charles Webster, ed. (Cambridge, 1981).

59. Figlio, "Chlorosis"; C. Rosenberg, *Explaining Epidemics and Other Studies in the History of Medicine* (Cambridge: Cambridge University Press, 1992).

60. Roger Chartier, *Cultural History: Between Practices and Representations* (Cambridge: Polity Press, 1988); Hunt, *The New Cultural History.*

61. Arthur O. Lovejoy, *Essays in the History of Ideas* (New York: Capricorn Books, 1960).

62. Poovey, *Uneven Developments.*

63. Bynum and Porter, *Companion Encyclopedia,* chap. 34; Fissell, *Patients, Power, and the Poor;* Porter, *Patients and Practitioners; Medicine in Society. Historical Essays,* A. Wear, ed. (Cambridge: Cambridge University Press, 1992).

64. Margaret Pelling and Charles Webster, "Medical Practitioners," in Charles Webster, ed., *Health, Medicine, and Mortality in the Sixteenth Century* (Cambridge: Cambridge University Press, 1979), 165–235; Margaret Pelling, "Appearance and Reality: Barber-Surgeons, the Body, and Disease," in *London 1500–1700: The Making of the Metropolis,* Lucinda Beier and Roger Finlay, eds. (London: Longman, 1986), 82–112.

65. Pelling, "Appearance and Reality."

66. R. Porter, *Health for Sale: Quackery in England, 1660–1850* (Manchester: Manchester University Press, 1989).

67. *Medical Fringe and Medical Orthodoxy, 1750–1850,* W. F. Bynum and Roy Porter, eds. (London: Croom Helm, 1986); Bynum and Porter, *Companion Encyclopedia,* ch. 28.

68. Bynum and Porter, *Companion Encyclopedia,* chaps. 54 and 55.

69. Roger Cooter, *Surgery and Society in Peace and War* (London: Macmillan Press, 1993); George Rosen, *The Specialization of Medicine with Particular Reference to Ophthalmology* (New York: Froben Press, 1944).

70. *Medicine and the Five Senses,* W. F. Bynum and Roy Porter, eds. (Cambridge: Cambridge University Press, 1993).

71. Ludmilla Jordanova, *History in Practice* (London: Arnold; New York: Oxford University Press, 2000).

CHAPTER SIXTEEN

Making Meaning from the Margins
The New Cultural History of Medicine

Mary E. Fissell

Once I'd hit on the phrase "making meaning from the margins" as a way of encapsulating the project of cultural history, it annoyed me even though it captured my thoughts quite economically. My title irritates me because it points to a tendency in cultural history toward fetishizing the marginal that invites caricature as the history of the weird or bizarre. A friend and I were working in the British Library and met up for tea, and he told me he'd seen a book and immediately thought of me. The book was an early history of embalming. What could I say? Cultural history has attracted a fair amount of hostility, some of which comes from its predilection for the marginal or the transgressive (just as social historians were often attracted to the category of deviance), and I don't want to provide ammunition for further attack. In this chapter I want to explore how cultural history and the history of medicine have intersected, and I want to proselytize about the possibilities for new work that cultural history affords.[1] As will become clear, this is preaching from the converted. I began my historical career as a social historian committed to "history from below" and have, along with many others, become as interested in cultural processes as social ones. In part, this chapter gives me the chance to reflect on how we got here and what we may have lost as well as gained in the process.

What Is This Thing Called Cultural History?

By cultural history, I refer to the boom in historical scholarship influenced variously by anthropology, cultural materialism, the history of *mentalités,* and so forth that took off in the late 1970s and 1980s. That was the moment when what Donald Kelley has called the "old cultural history"—the history of high culture—became "old."[2] However, what makes "old" cultural history old is not just its attention to the production of canonical aesthetic or artistic works but also its methods. In what follows, I will highlight some of the methodological characteristics of new work in the cultural history of medicine, emphasizing the shift from social to cultural histories.

When I'm in an irreverent mood, I sometimes define cultural history as "the intellectual history of the semi-literate and their friends." Those who tease me about writing the history of the seventeenth-century equivalent of the sensationalist tabloid *Weekly World News* endorse this definition. In more serious moments, I argue that the core of cultural history is its attention to the making of meaning—to how people in the past made sense of their lives, of the natural world, of social relations, of their bodies. This definition suggests that meaning is not inherent, that it does not reside within a text or a practice, waiting to be called on. Meaning is not uniform or transhistorical or even apparent. It must be made, and "making" is not an easy or simple process; it admits of struggle, perhaps even of contest. Meanings that are made can be unmade and remade. On another level, "the making of meaning" can, in postmodern fashion, point to our own making of history itself. All of these aspects of "making meaning" imply that we must think about methods as well as topics.

My jokey definition about the intellectual history of the unintellectual, however, contains a grain of truth. Cultural history can be understood as an attempt to take some of the methods and questions of intellectual history (Why did he think that? Where did she learn this?) and apply them to members of social groups whose thoughts had not previously been considered of historical interest. Carlo Ginzburg's *The Cheese and the Worms* (first English edition, 1980) is paradigmatic of this attempt, as it is of so much else.[3] A heretical miller's thought is analyzed with great care—What books did he read? How did he frame his cosmology? With whom did he talk? Why did he think what he did?

This intellectual history of developments and prospects in our field is also a more personal story of generational change. I am at a kind of midpoint both in my career as a historian and in understanding where cultural history fits into the larger field of history. I was trained in the heyday of the social history of medicine by one of its leading practitioners, Charles Rosenberg. My first book was a study of health care for the poor in eighteenth-century Bristol, a tale influenced by Rosenberg's compelling analyses of social dimensions of nineteenth-century hospitals, as well as by a "history from below" commitment derived from social history more generally.[4]

But now I'm working in a different mode and in a different time. My current project is a study of women's bodies in cheap print in early modern England, an exploration of how women's bodies became a cultural site for the articulation and discussion of historical changes such as the Protestant Reformation and the English Civil War. Some of the same concerns that motivated my first book drive this one: How does power work in minute and quotidian ways? How does the human body (or ideas about it) work within these microstructures of power? How does the inside of the body get imagined, diagnosed, and described when patients and practitioners have to rely on the outside of the body for information? But some concerns are new: What is being female and how do its meanings change? How do gender relations structure our imaginings of our own bodies? Are gender relations a pivot between political change and ideas about the body? In part, like many a midlife practitioner of any art, I'm experiencing the never-graceful transition from young Turk to establishment figure. Also, however, I am bothered that the commitments that exercised me and many of my contemporaries are quite alien to our students. I worry that my bright and talented students do not seem to have much of an idea about what the category "the social" might entail and explain.

Let me clarify what I mean by this generational change with a reminiscence. I recall quite clearly when I experienced an "a-ha" moment about the potential power of cultural analysis, but that insight was grounded in the minutiae of social history. In my first book, I had made extensive use of accounts generated by overseers of the poor under the Old Poor Law, England's pre-1834 welfare system. Each parish collected a tax or a "rate" and then disbursed those monies to the "deserving" poor. Social historians were acutely aware of the many ways in which the category "deserving" was constructed and reconstructed over time. My focus was on the health care aspects of this system and the ways in which the Old Poor Law made fairly generous provision for medical care for the poor. In the days before laptops, I reconstructed all of the payments made by one parish on a series of index cards, one card per recipient. I argued that medical care was an important component of social welfare and that welfare often provided for life-cycle exigencies. Welfare propped up families in trouble because they had too many young children and a sick head-of-household or assisted an elderly widow whose children could not provide for her. My analysis was primarily social, grounded in the demographic facts I gleaned from parish registers of baptisms, marriages, and burials.

Sometime after finishing the book, however, I happened to read an essay by Leonore Davidoff in which she remarked in passing that the ideology of separate spheres was not just adumbrated in prescriptive literature but was also embodied in legal codes such as the New Poor Law of 1834.[5] All of a sudden, I saw the social facts in my account books as cultural constructions. On one level, the overseers were dealing with social problems, often created by demographic mishap: children suddenly orphaned, unwed mothers deserted by men who had promised marriage, and elderly people unable to work were supported by their neighbors through the redistribution of very local resources. At another level, however, the overseers were performing and enacting ideas about what "the family" was. By providing resources to those whose families were dead or gone, the overseers propped up the idea that the family, headed by a male breadwinner, was the basic economic unit of society. The overseers, in effect, created a temporary fictive family when they hired an older woman to cook for, clean, and nurse a sick man who lacked a spouse or paid the indenture for an orphan girl to be apprenticed to learn housewifery. They created more permanent families when they urged a man to marry the mother of his child or got the midwife to question an unwed mother in labor about the father of her baby. Their efforts thus underwrote intact, "real" families by creating substitutes for those that were lacking. At the same time, they reproduced a particular idea of "the family" through their disbursements of shillings and pence.

In what follows, I will emphasize this transition from social to cultural history as a means of clarifying what cultural history is, and as a reflection on the gains and losses attendant upon this shift. Of course, the division between social and cultural histories is not as clear-cut as my analysis will suggest. Carlo Ginzburg's germinal book was written as social history, explicitly oriented toward debates about the existence of a genuine popular culture that could be distinguished from an elite culture's attempts to squash it or reform it. Two essays by Margaret Pelling illustrate how cultural analyses have begun to occupy the space formerly inhabited by social histories of medicine.[6] In the first essay, which appeared in 1987, Pelling questioned the "rise of the profession" narrative that characterized early modern medicine as a few forward-thinking College of Physicians members lost in a kind of netherworld of ineffectual and superstitious practitioners. Practitioners, she showed, moved in and out of medical work, apothecaries for instance doubling as grocers, spicers, and cosmetic salespersons. Hers is an exemplary social history that looks at ordinary practitioners and questions the sociological category of "profession."

In her subsequent (1996) essay, Pelling explored the rich set of meanings associated with medical work, and came to some unexpected insights. She argued that issues of status for medical practitioners were intimately related to questions about gender. In her discussion of the elite of practitioners, members of the College of Physicians, she illuminated many status and gender ambiguities, which may help us to understand some of the actions and behaviors of this group. The College of Physicians was resolutely all male, but its image was one of a certain gender ambiguity and effeminacy. The college distanced itself from the masculine world of guilds and political participation in the City of London, linking its fortune with the court through the membership of royal physicians. But court physicians occupied a curious niche; they were part of the royal household, and throughout the long reign of Elizabeth such body-servants tended to be female; on the accession of James I, court culture as a whole was characterized by gender ambiguities, as male favorites of the king vied for power. So, too, service was increasingly feminized. Pelling shows how cookery, long linked with physic, became associated with foreign cooks, luxurious and decadent food, and thence a certain effeminacy. Physicians' presumed familiarity with poisoning linked them with a method of killing seen as sneaky, underhanded, and quintessentially female. A medical practitioner, in other words, even one socially demarcated as a member of the College of Physicians, did not occupy a fixed social identity. Rather, he negotiated among an array of possible representations, from cook to spy, confounded by gender and status ambiguities that would only begin to be resolved over a century or more. Pelling's later essay is a good example of a cultural history that examines meaning as it was constructed by historical actors. Those meanings are not obvious—I for one had never thought about the womanly connotations of a physician's work, let alone the ominous court doctor/poison subtext.

With the clarity of hindsight, cultural history's ascendance seems overdetermined. Often, the publication of Clifford Geertz's *The Interpretation of Cultures* in 1973 is taken as a kind of germinal moment for the creation of cultural history, and it was certainly the anthropology book most likely to be seen on a historian's bookshelf when I entered graduate school in the early 1980s.[7] Historians took from Geertz the idea that an event could be read like a text, that a riot or a parade or a massacre of cats or other nonliterary production had symbolic meanings that a historian could recover and analyze. At the same time, however, English Marxist historians were also coming to an interest in the symbolic aspects of social interactions. The noted collection *Albion's Fatal Tree* (1975) included a discussion of the ways in which the English ruling classes used the criminal courts as a kind of theater of power, reproducing social roles through symbolic means.[8] French Annalistes had been developing sophisticated analyses of mentalités, arguing about the nature of the category "popular culture" while doing much of the work that made it a category worth arguing about. Italian historians, such as Ginzburg, created the form of "microhistory," a close study of one particular person or event, often interpreted in relation to a large body of similar examples, such as Inquisition cases.[9]

These various strands converged in Anglo-American historiography, and along with a new commitment to thinking about the role of language as productive rather than reflective (the so-called linguistic turn), created the conditions for the inauguration of the "new" cultural history.[10] The 1984 publication of the collection of essays entitled *The New Cultural History* can serve as a marker of the subdiscipline's arrival.[11] When I told a friend that I was writing an essay on cultural history, he muttered something like "well, aren't we all?" In a sense, he is correct. The kinds of social history or intellectual history that dominated history-writing twenty years ago are rarely encountered in their pure forms anymore. It is as if aspects of social and intellectual histories have collapsed into each other.

However, when I say that I am proselytizing for cultural history, I mean something more specific than just this middle ground created by the intersection of social and intellectual histories. In what follows, I sketch out three aspects of a "new" cultural history that seem most pertinent to historians of medicine: the making of meaning, the shift from pattern to process, and an attention to forms of storytelling, including lapses, errors, and omissions.

Cultural History and Medical History

Before turning to these themes, however, we need to look more closely at the history of medicine's recent trajectory, which is somewhat peculiar in relation to cultural history. Cultural history has been especially significant to two areas of the history of medicine: the sociology of disease and the history of the body. In the 1970s, sociology was often the paradigm other discipline to which historians turned, be they students of urban riots or of behaviors in a scientific laboratory. The sociology of knowledge provided historians of science with a sword to cut their ties with internalist accounts of scientific developments. Their work was influential in histories of alternative medicine and ideas about disease.[12] I spent the entire first year of graduate school completely under the spell of Steven Shapin's phrenology articles, in which he showed how the structure of the brain, as understood by phrenologists, was a correlate of their understanding of the social world they inhabited.[13] My first few graduate school papers tried to find other examples where I could make bodies and social structures so tidily parallel.

Not surprisingly, historians of medicine began to apply the kinds of deconstruction of scientific knowledge practiced by Shapin and others to disease categories. Some of this work was "soft" in the sense that it looked at "easy" diseases, such as masturbation and neurasthenia.[14] I do not mean "easy" or "soft" pejoratively; these essays were important to me and provided me with many happy hours of teaching. Instead, these adjectives point toward a kind of Comtean hierarchy, in which it is assumed (rightly or wrongly) that it is easier to show social forces shaping the knowledge of botany or phrenology than that of particle physics or radio astronomy. Essays in Wright and Treacher's 1982 collection, *The Problem of Medical Knowledge*, however, moved away from "easy" diseases to ones like miners' nystagmus, asthma, and genetic diseases.[15]

Charles Rosenberg's 1992 collection *Framing Disease* can be read as a reply to the social construction of disease that proposes a cultural model in its stead.[16] Rosenberg defanged the social construction of disease by claiming that it is a tautology, "a specialized restatement of the truism that men and women construct themselves culturally." He then did the same with the phrase "social history of medicine," saying that "every aspect of medicine's history is necessarily 'social.'"[17] He went on to fault the social construction of disease for focusing on a handful of culturally resonant diagnoses whose biological basis is unknown or contested (those "soft" diagnoses). Moreover, he dismissed a social-construction-of-disease model because he claimed it partakes of an outdated understanding of the ways in which knowledge circulates socially, casting purveyors of knowledge as agents of oppression.[18]

Instead of "constructing" diseases, Rosenberg chose what he considered to be a less-charged metaphor of "framing" diseases. He elegantly outlined the different ways in which disease entities are social players, for an individual sufferer, for a healer, and for social critics. In an interesting move, he talked about Owsei Temkin's germinal 1963 essay in which Temkin explained ontological vs. physiological models of disease and then put Temkin into dialogue with Arthur Kleinman, particularly on the topic of Kleinman's emphasis on the difference between disease (that entity understood by medicine) and illness (what the patient experienced).[19] Every year in the survey course, I write "ontological" and "physiological" on the chalkboard and, in a different lecture, "illness" and "disease," but until I reread Rosenberg, I had not thought of them together. Rosenberg is interested in the messy parts in between: the spaces between a practitioner and her patient, between the patient's family and a practitioner, between an institution and practitioners—those places where disease concepts "mediate and structure relationships." In many ways, what Rosenberg proposed is a fuller social understanding of disease that takes the practice of medicine seriously, rather than a "new cultural" history, but the richness of his model reminds us that "cultural" and "social" are heuristic structures, not exclusive commitments.[20]

The cultural analysis of disease has taken on a life of its own, distinct from the kinds of analysis pursued in *Framing Disease*.[21] In a 1996 essay, Colin Jones argued that we have lots of social histories of the plague but we lack an understanding of the meanings of plague. Interestingly, both Jones and Rosenberg point to a tension between historians' emphasis on bursts of infectious diseases compared with more quotidian miseries, such as chronic illness (Rosenberg) or infant mortality (Jones). Jones's mission was to "chart the metaphors and symbols" in the chief site where plague was discussed and debated, namely plague tracts. He worked with 264 of these books, many of which, as he acknowledged, are massively repetitive and even "arid." In other words, Jones explored the symbolic dimensions of the plague, rather than focusing on the patterns of transmission or the ways in which the disruptions of epidemic disease reveal social structures. Three languages characterize these texts: the medical, the religious, and the political. Jones argued that in the mid-seventeenth century, religious and medical discourse about the plague begins to converge toward a political script written in the language of high absolutism. However, Jones did not want to argue that this phenomenon was merely the triumph of the absolutist state *tout court*. Instead, traffic was "messy, complex, and two-way."[22] The state, for example, adopted the medical method of *cordon sanitaire* (quarantine enforcement to stop waves of desertion from the army). Both absolutism and the plague shaped each other.

The second aspect of the history of medicine that has a particular relationship with cultural history in its larger sense is the history of the body. Here, the relationship is largely one of absence. Most historians of medicine have shied away from considerations of the body as a cultural site, for reasons that I do not fully understand, while most of the history of the body has been written by literary scholars such as Jonathan Sawday or by historians who do not think of themselves as "medical," such as Caroline Bynum.[23] Thomas Laqueur provides a kind of crossover example. His first book was a social history of Sunday schools in nineteenth-century Britain. Even in his early days, the body beckoned; he wrote an essay on bodies, death, and pauper funerals published in the first number of

Representations (1987). When he wrote his blockbuster on the body, however, he situated himself within medicine—explaining how he attended the first two years of medical school—rather than in its history.[24] Ludmilla Jordanova's *Sexual Visions* or Katharine Park's essays on bodies, death, anatomy, and sexuality show us how the history of medicine and that of the body can each inform the other.[25] In each case, the author's deep knowledge of medicine makes a history of the body that helps us break down interpretive barriers between analyses of "medical" and "nonmedical" bodies. Barbara Duden's book is another of the rare ones that tries to integrate an account of "medical" bodies (those in a doctor's case book) with classic themes in the history of the body, such as the development of the closed, polite, bourgeois body.[26] Too often, sick or suffering bodies seem to be structured completely differently from healthy ones—sick bodies being the preserve of medical historians, while well ones are up for grabs by literary critics, art historians, and cultural studies professors.

Making Meaning: Microhistory and Ethnography

Harold Cook's award-winning book about Johannes Groenevelt tells the story of a practitioner who fell afoul of the College of Physicians.[27] It is a rare example of the explicit adoption of the methods of microhistory by a historian of medicine. Cook cited the classic example of this genre, Ginzburg's *The Cheese and the Worms*, and some readers may feel a tinge of disappointment that Groenevelt does not lend himself to some of the characteristics of the miller Mennocchio. As Cook acknowledged, we have no diaries or intimate letters, no subjective recording or construction of experience, that might illuminate Groenevelt's inner world. We lack the enticing, and sometimes distracting, exoticism of Ginzburg's miller, whose Inquisition testimony revealed a highly individual, not to say bizarre, syncretism.

However, Cook compensates for the lack of these customary microhistorical elements by writing a very personal book that invites the reader to speculate along with the author. It is personal in an old-fashioned way—no postmodern author peeking out of the text at every turn—that is highly satisfying. For example, we cannot know how Groenevelt understood or interpreted the fate that befell him, but Cook reminds us of some of the frameworks available to the Dutch physician. As the author thinks about how Groenevelt's Calvinist notions of the elect or his profoundly classical education might have given shape to his suffering, we too are drawn to wonder about the links between ideologies and individuals. As Cook tells us: "Time takes all things, and no modern magic allows us to conjure up the real voices or shapes of people from the seventeenth century. All we have left are the words and objects, largely in the form of ink on paper: the tangible signs of vanished spirits. Yet with a bit of sympathy, and a collection of words, we can begin to sense something of the world in which Groenevelt and his acquaintances walked."[28] This interpretive modesty makes the book very appealing. While Cook's narrative talents may remind us of an older generation of scholars, his willingness to write a story with holes and gaps and questions into which we must enter is sophisticated and subtle.

Groenevelt, a learned man and member of the Royal Society, did not fall afoul of the college by accident. Cook shows how Groenevelt, along with Richard Browne and Christopher Crell, sought to reform both therapeutics and medical hierarchies. While experience and experiment were essential to the reformation of therapeutics, Groenevelt and his companions were careful to distinguish themselves from mere empirics when they opened a sliding-scale-fee, group-practice clinic in 1687. This group published *The Oracle for the Sick,* the first multiple-choice popular medical book in English. A prospective patient read questions usually asked by a physician and underlined which of the multiple answers most closely matched his or her affliction; patients with wounds or other external manifestations of disease drew them on the human figures provided in the book. Then the patient brought the book to the clinic, where the physician quickly read it and prescribed accordingly. Not only did Groenevelt openly dismiss the hierarchical divisions among physicians, surgeons, and apothecaries, he also streamlined the clinic's practice in a way that might implicitly criticize the customary detailed and highly individualized personal consultations that were the physician's habit. In lesser hands, this episode would have been read merely as a random moment when the College of Physicians cracked down on one John Greenfield (as Groenevelt was known in England), an unauthorized practitioner, but Cook shows how a web of personal connections, intellectual beliefs, and the political meanings ascribed to them produced this incident.

Although this story is billed as a microhistory, Cook borrows from a range of methods and positions, each of which derives from a different national context. In some ways, its careful explication of an individual life is more like Paul Seaver's reconstruction of the life and beliefs of London wood-turner Nehemiah Wallington than it is a microhistory in the Italian context.[29] Both Seaver and Cook employ a wealth of knowledge about their subjects' contexts, creating a rich miniature of religious or medical ideas and practices, and both are methodologically diverse, drawing on different strands of cultural history to make sense of their subjects. Guido Ruggiero's recent essay on the death of Margarita Marcellini represents a more canonical medical microhistory in terms of method.[30] Ruggiero has read thousands of inquisition records and uses the breadth of his knowledge to contextualize this one woman's story. When Marcellini died in 1617, witchcraft was suspected. As Ruggiero shows, just when we think we understand what may have happened, there is another layer of meaning, another set of possible interpretations. Was it Margarita's sister Grazimana who had bewitched her, greedy for the return of her dowry to her natal family? Or did she die from the *mal francese*, what we might call syphilis, contracted from her husband? Ruggiero shows us how competing stories can be found in the inquisitorial records, forcing us, like the inquisitors themselves, to make sense of confusing and contradictory signs. We are all engaged in the making of meaning here.

In the American context, microhistory is often blended with or put alongside a home-grown tradition that derives from very different commitments, such as ethnography. Robert Darnton, who taught at Princeton with the anthropologist Clifford Geertz for a number of years, sought to translate some aspects of anthropological method into historical practice. Geertz advocated reading an event like a text, by which he meant pursuing a kind of formal symbolic analysis of something as seemingly evanescent as a cock-fight. Anthropologists, of course, have events to observe, while historians rarely participate in the events they study. The "text-analogy" is thus applied to events recorded in texts, such as Darnton's famous analysis of the cat massacre.[31]

Historians of medicine have not adopted this approach, although I do not know why it has not found more favor. A close reading of a surgical operation, or an anatomy riot, or the ritual of Grand Rounds might be undertaken on these lines. Given the ways in which the practice of ethnography has come under fire in anthropology itself, it may be that the moment for such a methodological borrowing has come and gone.[32] One of the few analyses of the rituals of medicine is Terri Kapsalis's study of the ways in which medical students are taught to perform gynecological examinations. This thought-provoking work is grounded in its author's training in performance theory, however, not anthropology.[33]

From Pattern to Process: Feminism, Poststructuralism, and Appropriation

Another theme in the cultural history of medicine has to do with how categories and concepts are made and remade constantly. This is a shift from pattern to process, from understanding social categories as fairly static entities to analyzing how cultural categories work as ongoing sets of negotiations. At least three quite different strands of scholarship have promoted an interest in process. First, Judith Butler's work, among others, on gender roles as performance has made male and female into moving targets rather than fixed entities. What used to be fairly static—say, masculinity in early modern London—is now a role or category that is enacted, built, and rebuilt through the actions of thousands and thousands of historical actors.[34] Second, in part due to the so-called linguistic turn, class and race, those other crucial categories of social-historical analysis, have been reunderstood as categories that were constantly being produced and reproduced.[35] A third strand derives from Roger Chartier's germinal 1984 essay on appropriation, in which he argues that consuming a cultural artifact is a form of production, that different groups can and do read the same texts in radically different ways.[36] In this sense, the printing of a ballad is just one moment among many that need to be understood by the historian. The pinning-up of the ballad in an ale-house, the singing of it in the street, the borrowing of its title by a playwright for a "serious" piece of theater, the ballad composed in reply or in parody to the original—again, we have a sense of movement, even instability, compared to an earlier more static focus. These three inspirations for a shift from pattern to process each derive from a variety of historiographic traditions: feminism, deconstruction, poststructuralism, English cultural Marxism, and French studies of popular culture. I feel a certain sense of interpretive violence in wrenching them from their contexts, but the emergence of each has helped to push us toward a much more fluid sense of social process.

Carolyn Steedman's work provides an example of a poststructuralist attention to process and the remaking of identities and categories. In her book *Strange Dislocations,* Steedman explored the ways in which selfhood and interiority became connected to ideas about childhood in the nineteenth century by using the figure of Mignon, the strange child acrobat in Goethe's *Wilhelm Meister,* as a kind of shifting signifier.[37] Steedman showed how a mid-nineteenth century physiological fascination with the topic of growth and development and a subsequent evolutionary turn were the preconditions for Freud's version of interiority, in which childhood lives within us in repressed and forgotten ways. Thus far, we might mistake this for an intellectual history. But then Steedman shifts to an analysis of how these meanings of childhood were enacted on the stage and in the streets of London. For it was in the ceaseless social reporting about poor children and, in the 1870s, the campaign to "rescue" child acrobats that we can see these ideas about childhood being employed and recognized. The fascination with child acrobats, which produced tearjerkers such as *Pantomime Waif* (1884) or *The Little Acrobat and His Mother* (1872), was part of a deeper cultural inquiry into the meanings of childhood and the connotations of loss ascribed to it. One of the accomplishments of Steedman's book is to focus on the places and processes through which these meanings of childhood were made and remade.

In a very different way, Nancy Tomes's article on the private side of public health shows us this making and remaking of categories that look the same and yet are different—or look different and yet are the same.[38] While both Steedman and Tomes cut their intellectual teeth as social historians, the intellectual roots of their cultural histories are quite different. Steedman's poststructuralist analysis grows from feminist inquiry and cultural Marxism, while Tomes builds on histories of consumption and political identity. Tomes shows how notions of cleanliness and dirt that derived from pre-germ-theory ideas about contagion were deployed by middle-class American homeowners and their wives to try to make their homes healthful. Manufacturers of all kinds of products, from soap to drains, capitalized on this preoccupation. The advent of germ theory did not, as one might expect, radically alter a perception of disease causation based largely on smell. Instead, germ theory was incorporated into older ideas about dirt and disease, and often manufacturers treated germicidal properties as add-ons or improvements to their cleansing products rather than incommensurables. As germ theory was appropriated and folded into an older set of ideas and practices, it was transformed in the process.

Tomes's essay also points toward an area of cultural history that has not yet been much explored by historians of medicine, namely, material culture. Before turning to my third theme of rhetoric and storytelling, I want to make a plea for more attention to the material world. In the past two decades, historians have come to think about the "world of goods" or the history of consumption. At the heart of many of these studies are questions about the ways in which people shape their identities through patterns of consumption. The 1990s, the decade that gave us the paradox of green consumerism, a decade of economic boom for many, saw historians remaking the eighteenth century into a moment of getting and spending, what Roy Porter called "pudding time."[39] Historians of the more recent past have explored the history of advertising and its complex cultural roles.[40]

There are not many histories of the material culture of medicine, but a few examples can suggest the potential for such work. In a fascinating book, Jacqueline Musacchio explores the material goods associated with childbirth that well-to-do Italians of the Renaissance consumed.[41] She gives us a kind of material culture of childbirth, from painted bowls and trays to paintings showing women wearing items of clothing with various significances for fertility.[42] Recently, John Styles has explored the ways in which the patent medicine Turlington's Balsam was packaged and repackaged in the eighteenth century, how innovation and novelty worked in an object as humble as a patent medicine.[43]

Many of these histories of consumption look toward the moment of production rather than the moments of use. It is as if the end of the story is the purchase of an item. Other histories of material objects tell us that the moment of acquisition is only the midpoint. For example, in a fascinating essay, Diane Hughes traced the shifting meanings of earrings in northern Italian cities in the later Middle Ages.[44] First, they connoted a kind of exoticism, but they then became more tightly linked with Jewish women. In some cities, Jewish women became legally bound to wear earrings, just as their husbands wore cloth circles or some other sign of difference. However, these associations eventually faded too, often as Jews were segregated in ghettos or banished, in effect marked in other ways. As women's fashions came to signify their husbands' wealth rather than their own female folly, well-to-do women began to wear jewels in their ears. In this story, the production of earrings is largely irrelevant. What matters are the shifting meanings ascribed to them.

An essay by Sara Pennell shows us how production and consumption might be linked to provide a richer account of material objects.[45] She notes that the overseers of the poor in an Oxfordshire parish redeemed a bastard-bearer's skillet from pawn in 1727. In this moment, the overseers both sought to return Mary Bassell to economic self-sufficiency and to underwrite the household as the unit of moral authority within the parish. A complex set of meanings could be ascribed to a commonplace tool like a skillet. But values were also inscribed literally on cooking pots. Pennell gives us an illustration of a pot whose handle read "y wages of sin is death" and notes another that said "pity the poor." It was cooking pots that were beaten and rattled in skimmingtons, community shaming rituals directed toward inadequate husbands who failed to rule hearth, home, and wife. Thus, a relatively simple material object can be understood as the bearer of multiple meanings, some built into the very substance of the item, others ascribed to it or performed with it. Such an analysis might be extended to any number of material objects employed in health and healing.

Earrings and skillets may seem very far removed from medicine, but for the early modern period they are not so distant. Ways of marking the body or the technologies of diet and regimen can point us toward thinking about material things. Medicine was practiced not just through narrative but with tools. For example, I have long wanted to read a good analysis of the illustrations of instruments in early modern surgical texts. There they are, these fierce-looking weapons that a surgeon must apply to the fragile human body. It is easy to understand that the images might serve as patterns for a local smith or metal-worker, but I suspect that they function in multiple ways. Lucia Dacome is beginning a study of wax anatomical models and their makers in eighteenth-century Bologna, building on Ludmilla Jordanova's pioneering study.[46] Much more could be explored in the material culture of medical education—the kymographs and other lab equipment of the early twentieth century, or skeletons so often hung in nineteenth-century American doctors' offices, mementoes of student days.

My plea for a medical history that pays attention to things is part of a larger desire to see what I call meaning and materiality addressed together. Too often, cultural history can seem to float free of any mooring in economic or social aspects of the past. A sophisticated and engaging reading of a text excites my imagination, but so too does an early modern wooden bed carved with elaborate figures (are they Adam and Eve? Who else would be depicted naked?). Living in our "world of goods," when forgetting to put out the recycling creates a mountain of tin cans and glass bottles, it is hard to imagine an early modern house in which almost all the material objects were deeply familiar, items of scarcity that might bear many meanings over their lifetimes of use.

Rhetorical Form: Listening to Stories after the Linguistic Turn

While things have been largely ignored in the history of medicine, words have not. One of the strongest themes in the cultural history of medicine is a new attention to rhetorical form, to the medium as well as the message. In part, an attention to narrative derives from historians such as Natalie Davis, who thought about the ways in which early modern French men and women framed their appeals in a court of law.[47] Attention to rhetoric is also a post-linguistic-turn habit of mind. Once we start thinking that language might be constitutive rather than reflective, we are drawn to consider how meanings are being conveyed through form as well as content. One of the most subtle practitioners of this art is Steven Stowe, who analyzed how ordinary doctors, the ones who were not the leaders of their profession, told stories about sickness and healing that enabled them to make their work, often bloody and unsuccessful, into something meaningful and even redemptive.[48] He showed how articles in medical journals related to other forms of autobiographical writing, such as letters and diaries. In the article on plague by Colin Jones that I discussed above, Jones noted how the style of writing about plague shifted as the threat of plague began to vanish. Writers became more "literary." Following an argument of Thomas Laqueur, Jones suggested that when suffering bodies were no longer dying of the plague and being dragged off to mass graves, they began to be invoked in texts as a way of evoking feelings—both horror and sympathy—in the reader, who might then engage in some form of humanitarian action.

Like my other two themes—making meaning and process over pattern—the recent attention to rhetoric has been shaped by cultural history's roots in both social and intellectual history. The habits of close reading and attention to modes of persuasion derives in part from the history of ideas, while social history's commitment to history from below has drawn attention to a much larger set of stories and storytellers. These shifts can be seen especially clearly in oral history. Originally, oral history was a part of social history, a practice intended to give voice to the people history had left out. A recent essay by Kate Fisher demonstrates the value of this kind of social history.[49] Fisher interviewed elderly working-class women and men about their contraceptive practices in the interwar period. Surprisingly, the demographic transition seems to have owed much to male rather than female behavior. The women Fisher interviewed often stated that they left all that sort of thing to their husbands. Most strikingly, a woman who had deployed contraceptive knowledge during her first marriage played dumb in her second marriage, feeling that it was not seemly for her to be more knowing than her husband. Most historians and public policy makers have assumed that since women are the ones who get pregnant, they are the most important players in contraceptive use. This assumption, Fisher suggested, needs questioning.

At one point, Fisher acknowledged that she was being told tales shaped by her sources' assumptions about what they should say about this sensitive matter, narratives whose tropes tell us more than just the facts of who bought the condoms. Fisher draws on a school of Italian historians that has developed a rich account of the process of oral history that shows how memory and meaning are intertwined. Allessandro Portelli, Luisa Passerini, and others have inverted the usual quest of oral historians.[50] Rather than looking for the facts, they look for the slippages, the misrememberings, the displacements—what I am calling the marginal. Their claim is that these moments tell us the meanings people ascribed to the events they witnessed and experienced. Their emphasis on the comments that other historians would have discarded as wrong is reminiscent both of psychoanalytic inquiry and of Ginzburg's analysis of Menocchio, the Friulian miller. Menocchio's claim that the world came into being just as worms are bred from decaying cheese was thought just as bizarre in his own day as it is in our own. But in this moment of misunderstanding, Ginzburg sees a creative process of appropriation and adaptation which tells us much about the possibilities of peasants' mental worlds.

An essay by Lynn Marie Pohl employs this new oral history to explore hospital desegregation in the American South. Pohl shows how her informants shaped their tales to save various phenomena.[51] For example, white doctors who cared for black patients emphasized their blindness to color in individual cases but did not acknowledge the profound inequities in the health care institutions in which they worked. Pohl has much to tell us, and often her analysis of the ways in which her sources' memories are constituted seems too brief. For example, many of her sources conflated two very different changes: hospital desegregation and the advent of what her informants called "high-tech" medicine. Pohl suggests that this trope of pretechnological "caring" medicine owes something to the myth of humane and caring plantation life often popular with white Southerners. I wish she had built an entire article around this insight and explored what it means to make segregation equal to "caring" and "humane" medicine. Pohl's work notwithstanding, the Italian-inspired mode of oral history has not been much adopted by historians of medicine despite some interest in the processes of memory.[52]

For some historians of Europe or North America, issues in oral history can feel somewhat peripheral unless one works on the recent past. Not so for Africanists. Historians of African cultures without written records have developed sophisticated methodologies for understanding and interpreting oral tradition, designed in part to grant oral testimony the gravitas usually assigned to written records. Recently, Luise White has turned these concepts upside down in her study of vampire stories in east Africa.[53] She takes material other historians would have discarded as clearly fictional and asks what it can tell us about a culture. Beliefs that forest rangers and firemen were actually vampires who fed on human blood suggest to White the ways in which some Africans experienced and transformed the traumas of colonial regimes.

Concluding Thoughts, or Wrestling with My Regrets

Throughout the above discussion of three aspects of cultural history—making meaning, from pattern to process, and attention to rhetorical form—I have pointed to a shift from social to cultural histories. Now I want to return to the anecdote I told at the beginning of this essay, my a-ha moment about the Old Poor Law, because it may shed light on the occasional hostility between social and cultural historians. All of a sudden, I had to think about process, about a kind of struggle between an overseer and a person seeking help. In social history terms, that struggle would have taken the form of a person seeking to demonstrate that he or she was "deserving." The category "deserving" was fairly stable and fixed at any one historical moment. From a cultural history perspective, both parties are negotiating, bringing different categories to bear, invoking a variety of scripts in this encounter. One of the pluses of a cultural history perspective is that we can see how social categories are always being renegotiated as they are enacted: the overseer is maneuvering his ideas about families and dependency just as much as his supplicant is invoking his or her sense of an appropriately "deserving" narrative. Both parties are reproducing a model of the family even as they struggle to compensate for the moments when the model does not seem to be working very well.

Without care, however, the fundamental inequalities between overseer and supplicant, the profound economic injustices inherent in this encounter can be sidelined. My guess is that some of the tensions between social and cultural history come from this sense that something important is being forgotten. Understanding dependency and the category of "deserving" gives us the feeling that we can analyze social injustice and that the analysis might be a step toward challenging injustice. Cultural history does not offer that kind of promise. Its analyses are at once too big and too small. Large-scale cultural structures, like ideas about the family, about gender relations, about work, come into play in unexpected configurations in specific instances. In a sense, the political is deeply personal in relation to cultural history, rather being a more general call to action.

This shift is, of course, a much more general one. I don't know if we are really "bowling alone" in the apt phrase of Robert Putnam—that is, if our communities have become increasingly atomized and privatized as the possibilities of communal action have diminished.[54] But I think we all have experienced a kind of shrinking of horizons of possibility. At some points in the recent past, radical rearrangements of the social world seemed possible, maybe even immanent. Collective action engaged our imaginations and sometimes our feet. Not surprisingly, some of the history that was written then embodied those hopes and commitments. But I don't want to cede all moral authority away from the enterprise of cultural history, to portray it as "Commitment Lite." It does, however, imply a different kind of connection between our work and our lives.

For me, understanding how early modern English women's reproductive bodies came to be sites of political and social contest is a pursuit very closely connected to contemporary concerns. This, my current project, has deep personal significances that I can't quite see yet, although it doesn't take a deep analysis to posit a connection between my writing a book on constructions of the maternal body with my own reckonings with motherhood and eventual adoption of a son. But it is not solely the personal. The past two decades have seen a curious valorization of "the child" in our culture, accompanied by various policings of women. The balance of social interest in an unborn child versus a pregnant woman's rights has shifted toward the child. (And no, I don't think it's a good thing for pregnant women to do cocaine, but I object to the various American legal remedies that have been employed to jail them as dealers to their unborn children.) All reproductive politics have been confounded by complexities of the legacies of *Roe v. Wade* (the 1973 U.S. Supreme Court case that legalized abortion on demand). All of this may seem quite remote from my arguments about how the Reformation reshaped women's bodies or how the English Civil War problematized women's knowledge of their own bodies, but like most historians, the processes I study in the past are analogous to and suggestive of patterns today. What this kind of cultural history does not do, however, is point us toward collective action. Instead, I see a different way in which historians might matter, such as when a friend of mine gave his twelve-year-old daughter a copy of Joan Jacob Brumberg's excellent book about body image in nineteenth- and twentieth-century America, and she saw herself and her friends in a new way.[55]

What's missing, at least for me, is a kind of indignation or desire for justice. I'd like to conclude by thinking about two exemplary works, one social, the other cultural: Ruth Richardson's *Death, Dissection, and the Destitute* (1987), and Mike Sappol's *A Traffic in Bodies* (2002).[56] Both works address the complex relationships between anatomists, grave-robbers, and the people who feared that they would end up on the dissector's table. Both are grounded in a mastery of rich and detailed knowledge of the nineteenth century. Richardson's book is a social history of anatomy and the impact of the 1834 Anatomy Act in England. She shows, first, how profoundly the emergent culture of dissection violated ordinary people's ideas about appropriate attitudes toward the dead, especially the newly dead. She analyzes the politics of the act, showing how a punishment formerly reserved for the worst felons was now extended to all who had the misfortune to die a pauper in a state institution. She explores forms of resistance to the act and their connections to other political movements, such as Chartism. It is a tour-de-force, taking what had formerly been a triumphalist moment in the history of medicine, when rationality triumphed over superstition and sentiment, and showing that this was a struggle deeply embedded in English politics and class structures.

There are many historical specificities that may account for the difference in Sappol's and Richardson's accounts: perhaps class relations in the United States differed from those in Britain so that the meanings of state-mandated dissection of the poor varied; perhaps the role of popular anatomy in a new republic differed from that in Britain; perhaps the florid array and large numbers of medical schools in the United States compared to Britain made anatomical dissection a different kind of political hot potato. These speculations aside, what makes the two books different is their positioning in relation to the stories they tell.

Like Richardson, Sappol has an ear for a good tale, but he steps back and shows us a story's cultural purchase in a way that Richardson does not. At the very end of his book, Sappol notes briefly that, in 1886, an old woman was murdered in Baltimore and her body sold to the University of Maryland medical school for fifteen dollars. Although widely reported, this story never achieved the "folkloric notoriety" of Burke and Hare, the Edinburgh murderers.[57] By 1886, anatomists were able to turn tales such as these to their own advantage, lobbying state legislatures to pass anatomy acts that would grant medical schools the unclaimed bodies of those who died in state and local institutions. Sappol emphasizes the retellings of this story rather than the story itself.

When I read Richardson, I am drawn into the stories of body-snatching and violation she tells; I am made indignant, angry, distressed. Richardson leads the reader to identify with the poor who were at risk of dissection, both before and after the Anatomy Act. Before the act, it was the bodies of the poor that were likeliest to be "resurrected" from graveyards and it was the poorer class of criminal who was sentenced to dissection; after the act, any pauper dying in a state institution was liable to dissection. Richardson opens the book with a compelling chapter on popular beliefs about the body and death that makes us understand why a criminal might be resigned to execution but plead desperately to avoid dissection. She enlists our sympathies, and those sympathies position us to regard the Anatomy Act with horror.

Sappol's prose is cooler and more detached, and he looks at the effects of the kinds of stories that Richardson embeds her readers within. Sappol can therefore ask fascinating questions about the relationship between the construction of a bourgeois professional self and the role of an anatomical way of thinking. It is not that Sappol is blind to the unjust class politics that granted a poor person's body to a physician who did not risk dissection himself, just that he takes a step back from these stories and looks at how they worked in a variety of realms.

Each book ends on a very different note. Sappol, writing in the post-postmodern moment, leaves us with an open-ended conclusion. On the one hand, things had changed by the 1880s: anatomy's central role in medical self-definition was beginning to cede to physiology, microbiology, and other "laboratory" sciences, while the newer social sciences borrowed anatomical metaphors and the authority with which they were imbued. On the other hand, as Sappol cheerfully acknowledges, the anatomical body is utterly dominant in Western culture: "The maps we carry of our innards are anatomical and will remain so for the foreseeable future."[58] Thus, newer techniques of self-making, such as the gland craze of the 1920s, often relied on an anatomical foundation.

Richardson's ending is far more poignant. She closes her book with two countervailing moments. She recalls interviewing an elderly man in 1978 who had contributed to a collection so that one of the regulars at his pub could avoid the ignominy of a pauper burial. She connects this deep working-class antipathy to pauper funerals to the violations sanctioned by the Anatomy Act that made what had been a dreadful sentence inflicted on executed felons into a routine deep indignity the state visited upon the poor. She goes on to suggest that three levels of silence have veiled the Anatomy Act from contemporary consciousness: that cultivated by the state bureaucracy in charge of the procurement of bodies for dissection, that of the larger society that does not want to know about the scapegoating of the poor, and last, a deep silence on the part of the working classes themselves. Richardson quotes George Steiner, "What is felt may occur at some level anterior to language, or outside it." She suggests that the profound repugnance for a pauper funeral derives from "the fact that the misfortune of poverty could qualify a person for dismemberment after death became too intensely painful for contemplation."[59] Turn the page, and Richardson implicitly invites her readers to atone for the sins of the past: "Bequests of bodies now ensure that the social injustice it [the Anatomy Act] represented before the Welfare State no longer operates. Should any reader wish to bequeath their body to anatomy . . ." and Richardson gives the address of the Inspectorate of Anatomy.[60]

In her attention to the ways in which the horrors of dissection were kept at bay both literally and figuratively by the desperate attempts of the poor to provide decent funerals for themselves, Richardson weaves together the social and the cultural for a moment. By inviting her readers to consider donating their own bodies for dissection, she urges them to imagine the weight of injustice that she has been describing. For me, the endings of both books suggest what we have gained and what we have lost. I miss both the emotional engagement produced by Richardson's account and the ways in which her account is profoundly a moral tale. We might think that history as an explicitly improving kind of writing had waned long ago, but rereading Richardson and other social historians reminds me that it was only yesterday. Sappol's story excites me intellectually, even as I am often frustrated by his unwillingness to tell his stories in depth, to engage my emotions as well as my mind by means of narrative detail and structure. Perhaps I am especially liable to persuasion on this point, but he convinces me fully that anatomy and the body imagined anatomically were central to many nineteenth-century Americans' understandings of themselves. Nor do I want to say that we have sacrificed moral purpose for intellectual complexity. Instead, the trade-off is something deeper to do with the possibility of writing stories that do not question themselves by means of their openness. Nostalgia is cheap, and perhaps my wish for the kinds of moral engagement implied by social history not much better. After reading Richardson, am I going to leave my body "to science," as the delicately obscure and high-minded phrase has it? Hell, no.

NOTES

I am very grateful to Lucia Dacome, Mary Henninger-Voss, Harry Marks, and the editors of this volume for their thoughtful comments on earlier drafts of this chapter.

1. This chapter is intended as a think-piece rather than a literature review. Consequently, the works cited below are only a sampling of recent work in the cultural history of medicine, a sampling shaped by my own early modern interests and eclectic tastes.

2. Donald Kelley, "The Old Cultural History," *History of the Human Sciences* 9 (1996): 101–126.

3. Carlo Ginzburg, *The Cheese and the Worms: The Cosmos of a Sixteenth-Century Miller,* John and Anne Tedeschi, trans. (Baltimore: Johns Hopkins University Press, 1980).

4. Mary E. Fissell, *Patients, Power, and the Poor in Eighteenth-Century Bristol* (Cambridge: Cambridge University Press, 1991).

5. Leonore Davidoff, "The Separation of Home and Work? Landladies and Lodgers in Nineteenth- and Twentieth-Century England," in *Fit Work for Women,* Sandra Burman, ed., (London: Croom Helm, 1979), 64–97, 64.

6. Margaret Pelling, "Medical Practice in Early Modern England: Trade or Profession?" in *The Professions in Early Modern England,* Wilfred Prest, ed. (London: Croom Helm, 1987), 90–128, and "Compromised by Gender: the Role of the Male Medical Practitioner in Early Modern England," in *The Task of Healing. Medicine, Religion and Gender in England and the Netherlands 1450–1800,* Margaret Pelling and Hilary Marland, eds. (Rotterdam: Erasmus Publishing, 1996), 101–134.

7. Clifford Geertz, *The Interpretation of Cultures* (New York: Basic Books, 1973).

8. Douglas Hay et al., *Albion's Fatal Tree: Crime and Society in Eighteenth-century England* (New York: Pantheon Books, 1975).

9. See, for example, *Microhistory and the Lost Peoples of Europe,* Edward Muir and Guido Ruggiero, eds., Eren Branch, trans. (Baltimore: Johns Hopkins University Press, 1991), and *Jeux d'échelles: La micro-analyse à l'expérience,* Jacques Revel, ed. (Paris: Gallimard Seuil, 1996).

10. In formulating this overview of cultural history, I have drawn on the following:

Explorations in Cultural History, T. G. Ashplant and Gerry Smyth, eds. (London: Pluto Press, 2001); *The Postmodern History Reader,* Keith Jenkins, ed. (London: Routledge, 1997); and *Beyond the Cultural Turn: New Directions in the Study of Society and Culture,* Victoria E. Bonnell and Lynn Hunt, eds. (Berkeley: University of California Press, 1999).

11. *The New Cultural History,* Lynn Hunt, ed. (Berkeley: University of California Press, 1984).

12. *Natural Order: Historical Studies of Scientific Culture,* Barry Barnes and Steven Shapin, eds. (Beverly Hills, Calif.: Sage, 1979); Roger Cooter, *The Cultural Meaning of Popular Science: Phrenology and the Organization of Consent in Nineteenth-Century Britain* (Cambridge: Cambridge University Press, 1985); Adrian Desmond, "Artisan Resistance and Evolution in Britain, 1819–1848," *Osiris* 3 (1987): 77–110.

13. Steven Shapin, "Phrenological Knowledge and the Social Structure of Early 19th-Century Edinburgh," *Annals of Science* 32 (1975): 219–243; "Homo Phrenologicus: Anthropological Perspectives on an Historical Problem," in Barnes and Shapin, *Natural Order,* 41–71; and "The Politics of Observation: Cerebral Anatomy and Social Interests in the Edinburgh Phrenology Disputes," in *On the Margins of Science: The Social Construction of Rejected Knowledge,* Roy Wallis, ed., *Sociological Review* Monograph no. 27 (1979): 139–178.

14. Barbara Sicherman, "The Uses of a Diagnosis: Doctors, Patients, and Neurasthenia," *Journal of the History of Medicine and Allied Sciences* 32 (1977): 33–54; H. Tristram Engelhardt, "The Disease of Masturbation: Values and the Concept of Disease," *Bulletin of the History of Medicine* 48 (1974): 234–248.

15. *The Problem of Medical Knowledge: Examining the Social Construction of Medicine,* Peter Wright and Andrew Treacher, eds. (Edinburgh: Edinburgh University Press, 1982).

16. *Framing Disease: Studies in Cultural History,* Charles E. Rosenberg and Janet Golden, eds. (New Brunswick, N.J.: Rutgers University Press, 1992).

17. Ibid., xiv.

18. Ibid., xv.

19. Ibid., xxii–xxiii. The works to which Rosenberg refers are Owsei Temkin, "The Scientific Approach to Disease: Specific Entity and Individual Sickness," in *Scientific Change: Historical Studies in the Intellectual, Social, and Technical Conditions for Scientific Discovery and Technical Invention from Antiquity to the Present,* A. C. Crombie, ed. (New York: Basic Books, 1963), 629–647, and Arthur Kleinman, *The Illness Narratives: Suffering, Healing, and the Human Condition* (New York: Basic Books, 1988).

20. This combination of social and cultural histories is characteristic of much of Rosenberg's work. His first book, for example, weaves together changes in ideas about cholera with the social particulars of nineteenth-century New York City: Charles E. Rosenberg, *The Cholera Years, the United States in 1832, 1849, and 1866* (Chicago: University of Chicago Press, 1962).

21. Colin Jones, "Plague and Its Metaphors in Early Modern France," *Representations* 53 (1996): 97–127. See also Giulia Calvi, *Histories of a Plague Year: The Social and the Imaginary in Baroque Florence* (Berkeley: University of California Press, 1989), which blends social analysis of the patterns of plague transmission with a cultural analysis of the meanings associated with a nun who was proposed for canonization during the 1633 plague.

22. Jones, "Plague," 116.

23. Jonathan Sawday, *The Body Emblazoned: Dissection and the Human Body in Renaissance Culture* (London: Routledge, 1995); Caroline Walker Bynum, *Holy Feast and Holy Fast* (Berkeley: University of California Press, 1985). The immensely influential collection by Michel

Feher on the history of the body had much medical content but not many historians of medicine among its authors, *Fragments for a History of the Human Body,* Michel Feher, ed. (New York: Zone, 1989).

24. Thomas W. Laqueur, *Religion and Respectability: Sunday Schools and Working Class Culture, 1780–1850* (New Haven: Yale University Press, 1976); "Bodies, Death, and Pauper Funerals," *Representations* 1 (1983): 109–131; and *Making Sex: Body and Gender from the Greeks to Freud* (Cambridge, Mass.: Harvard University Press, 1990).

25. Ludmilla Jordanova, *Sexual Visions: Images of Gender in Science and Medicine between the Eighteenth and Twentieth Centuries* (Madison: University of Wisconsin Press, 1989). Katharine Park, "The Life of the Corpse: Division and Dissection in Late Medieval Europe," *Journal of the History of Medicine and Allied Sciences* 50 (1995): 111–132; "The Criminal and the Saintly Body: Autopsy and Dissection in Renaissance Italy," *Renaissance Quarterly* 47 (1994): 1–33; and "The Rediscovery of the Clitoris. French Medicine and the Tribade, 1570–1620," in *The Body in Parts: Fantasies of Corporeality in Early Modern Europe,* David Hillman and Carla Mazzio, eds. (London: Routledge, 1997), 171–194.

26. Barbara Duden, *The Woman beneath the Skin* (Cambridge, Mass.: Harvard University Press, 1991).

27. Harold J. Cook, *Trials of an Ordinary Doctor: Johannes Groenevelt in Seventeenth-Century London* (Baltimore: Johns Hopkins University Press, 1994).

28. Ibid, xvii.

29. Paul Seaver, *Wallington's World: A Puritan Artisan in Seventeenth-Century London* (Stanford: Stanford University Press, 1985).

30. Guido Ruggiero, "The Strange Death of Margarita Marcellini: *Male,* Signs, and the Everyday World of Pre-modern Medicine," *American Historical Review* 106 (2001): 1141–1158.

31. Robert Darnton, "Workers Revolt: The Great Cat Massacre of the Rue Saint-Séverin," in *The Great Cat Massacre and Other Episodes in French History* (New York: Vintage Books, 1985), 75–106.

32. *Writing Culture: the Poetics and Politics of Ethnography,* James Clifford and George E. Marcus, eds. (Berkeley: University of California Press, 1986).

33. Terri Kapsalis, *Public Privates: Performing Gynecology from Both Ends of the Spectrum* (Durham, N.C.: Duke University Press, 1997). Thanks to Christine Ruggere for this reference.

34. Judith Butler, *Gender Trouble: Feminism and the Subversion of Identity* (London: Routledge, 1990); Laura Gowing, *Domestic Dangers: Women, Words, and Sex in Early Modern London* (Oxford: Oxford University Press, 1996).

35. See, for example, Gareth Stedman Jones, *Languages of Class: Studies in English Working-Class History, 1832–1982* (Cambridge: Cambridge University Press, 1983), and Patrick Joyce, *Visions of the People: Industrial England and the Question of Class, 1848–1914* (Cambridge: Cambridge University Press, 1991) and *Democratic Subjects: The Self and the Social in Nineteenth-Century England* (Cambridge: Cambridge University Press, 1994).

36. Roger Chartier, "Culture as Appropriation: Popular Cultural Uses in Early Modern France" in Steven L. Kaplan, ed., *Understanding Popular Culture: Europe from the Middle Ages to the Nineteenth Century* (Berlin: Mouton Publishers, 1984), 175–191.

37. Carolyn Steedman, *Strange Dislocations: Childhood and the Idea of Human Interiority, 1780–1930* (London: Virago, 1995).

38. Nancy Tomes, "The Private Side of Public Health," *Bulletin of the History of Medicine* 64

(1990): 509–539. Tomes incorporated this discussion in her subsequent book, *The Gospel of Germs: Men, Women, and the Microbe in American Life* (Cambridge, Mass.: Harvard University Press, 1998).

39. Lorna Weatherill, *Consumer Behaviour and Material Culture in Britain, 1660–1760* (London: Routledge, 1988); Carole Shammas, *The Pre-industrial Consumer in England and America* (Oxford: Oxford University Press, 1991); *Consumption and the World of Goods,* John Brewer and Roy Porter, eds. (London: Routledge, 1993).

40. T. J. Jackson Lears, *Fables of Abundance: A Cultural History of Advertising in America* (New York: Basic Books, 1994); Thomas Richards, *The Commodity Culture of Victorian England* (Stanford: Stanford University Press, 1990); Timothy Burke, *Lifebuoy Men, Lux Women: Commodification, Consumption, and Cleanliness in Modern Zimbabwe* (Durham, N.C.: Duke University Press, 1996).

41. Jacqueline Musacchio, *The Art and Ritual of Childbirth in Renaissance Italy* (New Haven: Yale University Press, 1999).

42. See also Jacqueline Musacchio, "Weasels and Pregnancy in Renaissance Italy," *Renaissance Studies* 15 (2001): 172–187.

43. John Styles, "Product Innovation in Early Modern London," *Past & Present* 168 (2000): 124–169.

44. Diane Owen Hughes, "Distinguishing Signs: Ear-rings, Jews, and Franciscan Rhetoric in the Italian Renaissance City," *Past & Present* 112 (1986): 3–59.

45. Sara Pennell, "'Pots and Pan History': The Material Culture of the Kitchen in Early Modern England," *Journal of Design History* 11 (1998): 201–226.

46. Jordanova, *Sexual Visions*.

47. Natalie Zemon Davis, *Fiction in the Archives: Pardon Tales and Tellers in Sixteenth-Century France* (Stanford: Stanford University Press, 1988).

48. Steven Stowe, "Seeing Themselves at Work: Physicians and the Case Narrative in the Mid-Nineteenth-Century American South," *American Historical Review* 101 (1996): 41–79. See also his reflections on the process of reading letters in "Singleton's Tooth: Thoughts on the Form and Meaning of Antebellum Southern Family Correspondence," *Southern Review* 25 (1989): 323–333.

49. Kate Fisher, "'She Was Quite Satisfied with the Arrangements I Made': Gender and Birth Control in Britain 1920–1950," *Past & Present* 169 (2000), 161–193.

50. See, for example, Alessandro Portelli, *The Death of Luigi Trastulli and Other Stories: Form and Meaning in Oral History* (Albany: State University of New York Press, 1991) and *The Battle of Valle Giulia: Oral History and the Art of Dialogue* (Madison: University of Wisconsin Press, 1997); Luisa Passerini, "A Memory for Women's History: Problems of Method and Interpretation," *Social Science History* 16 (1992): 669–692.

51. Lynn Marie Pohl, "Long Waits, Small Spaces, and Compassionate Care: Memories of Race and Medicine in a Mid-Twentieth-Century Southern Community," *Bulletin of the History of Medicine* 74 (2000): 107–137.

52. John Harley Warner, "Remembering Paris: Memory and the American Disciples of French Medicine in the Nineteenth Century," *Bulletin of the History of Medicine* 65 (1991): 301–325.

53. Luise White, *Speaking with Vampires: Rumor and History in Colonial Africa* (Berkeley: University of California Press, 2000).

54. Robert D. Putnam, *Bowling Alone: The Collapse and Revival of American Community* (New York: Simon & Schuster, 2000).

55. Joan Jacobs Brumberg, *The Body Project: An Intimate History of American Girls* (New York: Random House, 1997).

56. Ruth Richardson, *Death, Dissection, and the Destitute* (London: Routledge & Kegan Paul, 1987); Michael Sappol, *A Traffic of Dead Bodies: Anatomy and Embodied Social Identity in Nineteenth-Century America* (Princeton: Princeton University Press, 2002).

57. Sappol, *Traffic,* 318.

58. Ibid., 327.

59. Richardson, *Death,* 280, 281.

60. Ibid, 283.

CHAPTER SEVENTEEN

Cultural History and Social Activism
Scholarship, Identities, and the Intersex Rights Movement

Alice Domurat Dreger

A couple of years ago, I found myself whining in an e-mail message to my friend and colleague Tod Chambers, a bioethicist at Northwestern University, that I was running so many fundraisers for the Intersex Society that I didn't have time to do much new historical scholarship. "Ah, yes," Tod wrote back, "those who remember history are condemned to fundraise." I caught the reference to George Santayana's famous admonition that "Those who cannot remember the past are condemned to repeat it." Yet, much as Tod's response made me laugh—and feel chagrined to realize that all in all I had very little to be complaining about, the work of the Intersex Society being so important—I found myself wondering why I seemed to be a latter-day anomaly, a humanist whose scholarship had led to a life of activism wrapped around that scholarship.

I have been asked for this volume to explore this personal history and to share my thoughts on whether and how histories of medicine, particularly *cultural* histories of medicine, can shape the present and the future of medicine's constituencies. While this chapter cannot be anything like a definitive account of the relationship of medical history and reform movements, I hope that it will provide some guidance to scholars interested in using medical history to effect social change. I also hope that it may provide some inspiration in that vein. While the events of the last decade that I relate here have been exhausting, physically and emotionally, they also make up one of the most rewarding and meaningful parts of my life. It seems to me a great gift to be able to do history to alter the future.

So where to start? If, dear reader, you are like most people I meet, when I tell you I study intersex and that I recently completed three years as founding chair of the board of directors of the Intersex Society of North America, you will wonder whether the reason I am so involved in this arena might be that I am intersexed. In person, the question is usually couched as, "Now, why are you interested in *that?*" I notice this question is rarely asked of a historian interested in, say, hospitals. When I became visibly pregnant the question changed to, "I've always wondered, can intersexed people get pregnant?" I guess in spite of the relative success of our activist work, you still can't ask people bluntly, "Were you born with funny-looking genitals?" Maybe that's for the better.

Although I've come to believe that everyone is born with funny-looking genitals aesthetically speaking (in the same sense that there's no such thing as an elegant knee), I'm not intersexed. And as long as I'm confessing a certain lack of expertise, let me note also that I was not really trained as a historian of medicine. My Ph.D. is in history and philosophy of science, from a department that mostly saw medicine as something else entirely. Let me go back a bit, just to make clear this is no Whiggish history, to suggest that the roads to history-cum-activism are varied and open.

Bored with institutionalized education in 1985 after a year of college at Georgetown University, I dropped out of school and became a mortgage broker. It paid, and my boss was willing to hire a nineteen-year-old. Bored with that too after a few years (and seeing the savings and loan crisis coming straight at us, since we in the mortgage business were helping to cause it), I decided to go back to school around my work schedule and finish my degree at the cheapest, closest place, SUNY Old Westbury, thinking I'd probably then go on to do real estate law. Around the time I was wrapping up my B.A. in comparative humanities, I found out a school friend of mine had gotten a fellowship to go to Indiana University to do a graduate degree the Program in the History and Philosophy of Science (HPS). "What's a fellowship?" I asked my adviser.

"They pay you to read," he answered.

That sounded very good, very different from refinancing already-over-financed suburban dwellings and fighting with title companies. Perhaps not coincidentally, by that time the therapist I had been working with for a couple of years declared me healed and stated his opinion that I was likely to stay in remission if I got out of New York promptly. So I applied in April 1990 to Indiana University's HPS program, because I had always loved the humanities and the sciences and because my undergraduate adviser had gotten his degree there and thought he might be able to get me in so far after the deadline. They admitted me with the understanding I would have to pay my own way for at least a year. One of the secretaries later told me they actually admitted me because they lost my application and were embarrassed to ask for another, and I find that the only logical explanation why they would admit a mortgage broker with a fairly unimpressive degree to such a fine program. In any case, at twenty-four going on fifty, I cashed in my mortgage-brokering retirement accounts to pay for tuition and housing, asked the automobile club to tell me where the state of Indiana was, sold most of my belongings, and left.

That was the fall of 1990. The historian of biology Fred Churchill taught my first graduate course in historiography and that semester he helped me hone in on the topic of hermaphroditism as a possible dissertation topic. He knew I was interested in gender (feminism especially), and he told me that he thought hermaphroditism was a ripe but ignored topic. (Years later I discovered that Fred's father, a famous surgeon, had trained some of the very surgeons whose present-day practices I would come to challenge as an activist.)[1] Hermaphroditism, eh? I figured I could look at Darwin's barnacles, comparative embryology, that sort of thing. But Fred kept saying, "Look at medicine." I couldn't understand why, since I had never heard of human hermaphrodites. A few years later in a course outside our department on the social history of medicine, the historian of Victorian medicine M. Jeanne Peterson also told me to "look at medicine." This time I did and was astonished to discover, in the first series of the *Index-Catalogue of the Library of the Surgeon General,* something like 300 articles in English and French, the languages I could read. I ordered them all through interlibrary loan and realized there was plenty here ripe for feminist analysis, just as Fred had suspected.

Activism surely has as much to do with personality (in my case a persistent bitter optimism and resilient egotism) as circumstances, so I should note that in my first years in graduate school I was already something of an activist, and was supported as such by a few professors at Indiana, most notably the philosopher of science Stephen Kellert. I worked with my peers to improve the circumstances of graduate students and chose to teach a freshman-level course I named, after the Science for the People volume, "Biology as a Social Weapon."[2] I railed against the Human Genome Project, writing a critique of the specious metaphors used by James Watson and company to sell the project to Congress over the objections of many well-established biologists. In fact, I was raised to be an activist; my parents were committed members of Right-to-Life, and I have many childhood memories of picketing abortion clinics on the weekends. Though she wouldn't use the term today to describe herself, my mother was a vocal feminist who argued all the time with anyone she could for women's rights (she did not and does not see abortion as a right). I remember after I went off to graduate school my mother bemoaning my choosing history because she felt it was dead work. "Don't you want to write something new, do something new?" she asked, not recognizing the study of the past as being something in the present. So I always had the notion that you had to be doing something in the now, and a certain guilt would set in if I became too self-absorbed in the life of the mind.

Three years into my graduate degree, still very enamored with the ever-so-cohesive intellectual history of science I had been learning at Indiana, but fascinated also by the possibility of studying something with relevance to ongoing feminist political struggles, I had decided to focus my dissertation on the British and French treatment of hermaphroditism in the late nineteenth century. Fred Churchill and the historian of medicine Ann Carmichael (based in the history department at Indiana) would be my codirectors, a perfect team for what I wanted to study, the scientific-theoretical treatment of hermaphroditism as well as the medical-clinical. I believed this work had the potential to benefit living women, and later, with the help of historian-physician Vernon Rosario, I realized it could also benefit living gay and lesbian people. It didn't occur to me at all then it might benefit very many people born with intersex conditions because I still believed, in spite of all the evidence in front of me, that intersex was incredibly rare.[3]

Partly because of its relevance to gender politics at large, my dissertation topic caused consternation among some of my professors. My graduate department, highly traditional—never having strayed too much from its origin when it was known as the Department of the History and Logic of Science—found itself with a sudden influx of social-minded students, students who did not see the fall of communism and socialism as reason to give up on leftist critiques. So in the early 1990s our department found itself living a Freudian version of the science wars, where many of the students rebelled to do what we called "externalist" work, work criticized by many of the faculty as lacking rigor and purpose. As I recall it, I was informed that according to an obscure rule my dissertation proposal had to be approved by every member of the department before I could become a doctoral candidate, a problem since at least one member of the department (though not on my committee) found my chosen topic inappropriate. This was Noretta Koertge, vocal critic of women's studies departments and feminist studies of science.

Noretta was justifiably afraid I was slipping into sloppy social constructivism, arguing that sex is "just a social construction." I so distinctly remember one day standing in the hallway of Goodbody Hall in Bloomington and Noretta asking me, "Is this what you want kids in junior high health classes to learn? Don't you think it's important that a girl knows *she* can get pregnant and her boyfriend can't? Don't you think it's important she knows that *she* can get cervical cancer and he can't?"

I remember that conversation so well because the questions were troublesome. If I talked about sex being a social construction, would people really think boys could get pregnant and get cervical cancer? And wait—if they couldn't, then *was* it a social construction? Her questions were important . . . and *exciting:* Would people really think this work that important that it could affect how standard-bodied teenagers understood their bodies? Could it hold the power to liberate them, in the way that *Our Bodies, Ourselves,* first published in 1971, had so many women? I knew from previous conversations that Noretta and I were both interested in liberation, though we had very different ideas about which story would liberate. It reminded me of the history of Christianity I had learned at Old Westbury. The constructivists were essentially doing higher biblical criticism on science, historicizing, tracing the steps out as *human* steps. The true believers saw this as heresy, while to me it held the *only* promise of a liberation theology of science.

With all of us locked in a fabulously passionate debate about the meaning and purpose and nature of science, in spite of not having official doctoral candidate status (because my proposal was not approved by all) I plowed ahead. I joked to my friends, "They can approve my dissertation proposal when my book comes out." Ironically, it was institutionalized support of activism that got my proposal and my doctoral candidacy approved. Indiana University maintains the John Edwards Fellowship which is targeted specifically at graduate students of any discipline with a strong sense of social responsibility. I won the Edwards, and in part because of that important line on my c.v., the university wanted to nominate me for a National Endowment for the Humanities (NEH) fellowship, but couldn't do so without my being a doctoral candidate. So I was advanced, in spite of any objections that may have persisted. I did not win the NEH, but went on to apply for and win a Charlotte Newcombe Doctoral Dissertation Fellowship, funding targeted at those studying religious or ethical issues. Between the Edwards and the Newcombe, it was clear to me that at least some of the powers-that-be wanted to see socially oriented work, even activism, so I felt something of a duty—not to mention funding pressure—to make my work openly feminist (i.e., critical of oppressive gender norms) intellectually and politically.

In spite of this, even as late as 1995, while I finished my dissertation, I did not expect to get very involved, in terms of scholarship or activism, with the treatment of intersex today. I saw myself now as a real historian, and I had been given the conservative notion of history—that you don't study anything in your lifetime. I even had a sense you should downplay any connections to the present, that *not* to do so was disciplinarily gauche. Suzanne Kessler, the social psychologist already well known for her groundbreaking work on the social construction of gender, had published an excellent critique of the present-day treatment of intersex in SIGNS in 1990, and Anne Fausto-Sterling, the well-known biologist and feminist critic of science, had published an equally intelligent critique of the myth of two sexes in *The Sciences* and the *New York Times* in 1993.[4] (I remember Jeanne Peterson giving me the *Times* piece to photocopy, and the secretary in the graduate school office seeing the title, "How Many Sexes Are There?" asking me, "Aren't there two?" I raised my eyebrows but didn't answer; I wasn't sure at that point.) However interesting the question of today might be, it was not a matter for a real historian, and I wanted to be seen as a real historian.

So in 1995, using parts of my dissertation, I worked up my first two all-history articles, one for *Victorian Studies,* and one for a volume Vernon Rosario would edit based on the first major History of Science Society (HSS) sessions on science and homosexuality in 1994.[5] Probably because of what happened next, these would turn out to be the only explicitly all-history articles I have ever written. The historian of science Stephanie Kenen and I had talked at the HSS sessions about the history of intersex and afterward I think she gave my name and e-mail address to some people who had been born with intersex conditions. So my name entered that e-mail circle, and one day I opened my e-mail account to find a message from a woman who had read my *Victorian Studies* piece and said she had been born with a form of "pseudohermaphroditism." She wanted to know why I was interested in this issue, and what on earth I meant when I referred to intersex as a birth deformity. The tone of her message inspired a knee-jerk reaction: Here come the politically correct police to attack my work. While I was beginning to realize that intersex need not be constructed as a pathology, picking on my use of the word "deformity" seemed excessive. As I recall, I sent back a noncommittal message and laid low.

By early 1996, I had my Ph.D. in hand and was visiting assistant professor in the Program for the History of Science and Technology at the University of Minnesota. I had been invited to Berkeley and Stanford to give talks on my work in March, and, worried that the Intersex Society members based in San Francisco might crash my talks and disrupt them, in February I decided to send an olive branch. I wrote to the general address, info@isna.org, and alerted them to my upcoming presentations, adding: "I hope the historical work I am doing will be of interest & use to your members, so I want to make it as accessible as possible. Please let me know if there is a way I can facilitate getting info to your interested members."[6] Cheryl Chase, who had founded the Intersex Society of North America (ISNA) in 1993, wrote back in part to say, "Your work is indeed of interest to us. I believe that many hermaphrodites made out pretty well before medical intervention. Medical authorities today insist that life for a hermaphrodite w/o medical 'correction' would be unthinkable, without value. Though I haven't yet had a chance to look at what you are writing, I suspect it helps us undermine the medical claim to authority."[7]

Cheryl, a voracious learner, a brilliant woman seemingly willing to do anything to achieve her mission, soon started asking me explicitly about what the history did tell us about the days before massive concealment practices. She told me she didn't believe currently practicing doctors who thought all hermaphrodites must have killed themselves, as Herculine Barbin did, before the hypermedicalization of intersex started.[8] I assured her,

> Yes, I believe you are right—many hermaphrodites did well before/without medical intervention. Several of the people I studied seemed not to know anything was unusual, so that the medical men were the literal causes of their problems ("But my good woman, you are a man!" to which the woman responded, essentially, "You're nuts, thanks, what do I owe you?"). Others knew there was something unusual, and some even knew pretty well what was up, but it didn't cause massive trauma. Of course, others, as you may know, were rejected by family, lovers, and community. The cases differ dramatically.[9]

Cheryl wanted to know more: "Do you have a hypothesis about what factors determined differing outcomes?"[10] I answered:

> Nationality, ethnicity, class, family, age, marital status, partners' attitude(s), community size/ethos, time period, personality/status of involved medical men, and so on. Which is one reason I am not sure we can draw too many conclusions about what the *best* situation would be, but we do seem to be able to draw some conclusions about what *doesn't* seem to work positively. (Like trying to define biologically exactly what it means to be male, female, or "truly" both or neither, or insisting that someone take on an identity that s/he cannot identify with personally.)[11]

I was grateful that Cheryl was a patient teacher and someone who did not take my work and oversimplify it. When I told her the history was complicated—that it didn't offer a simple answer to "what should happen now?"—she neither disagreed with my expertise nor jettisoned the history for its inability to give her an easy or "right" answer.

Eventually, Cheryl moved on to the inevitable question, "How did you come to this work? Who are you?"[12] (code for "Are you one of us?"). In spite of my explanation, I know from later conversations she figured I must be a closeted intersexual or lesbian or *something,* so unsuccessful had been her attempts to communicate with other academic feminists who otherwise claimed to care about clitorectomies. Though Kessler and Fausto-Sterling and also Natalie Angier of the *New York Times* were great allies to Cheryl's work, they were notable exceptions. Others Cheryl contacted insisted clitorectomies only happened in Africa or that people with intersex conditions were really transsexuals and therefore undeserving of feminists' help.

I find it interesting that our e-mail of this time records my explanation to Cheryl about why I had been shy to contact ISNA earlier because it shows how I started to think about the work very differently now that real people with intersex conditions were talking with me:

> Several times I considered contacting you folks, but resisted for 2 main reasons: (1) primarily I had a hell of a time trying to get my work done at Indiana, and I feared—justifiably I think—that if they knew I was in contact [with] you, the conservatives could use it as a way to claim my "objectivty" [*sic*] as an historian had been compromised. Of course, it is true that no one thinks that, if you are studying biologists from 100 years ago and you talk to a biologist today that that somehow makes you unqualified to go on with the work.... (2) I felt that you, as a support group, deserved privacy and did not need people who were not born intersexed to go prying into your conversations and experiences. If I had learned anything from my work it was that the problems that happened arose primarily out of a lack of respect for others' privacy and right to define their communities/relationships. Now, in hindsight, I see that you are a lot more than a support group, and I see also that my work necessarily invaded your privacy in some ways, since it deals with the history of why you would need/want a support group. I always figured that once my work shaped up it would be of interest & use to you. I'm glad that appears to be the case.[13]

Cheryl pressed me to learn more about the present-day system, and with my partner, Aron Sousa (whom I had met in 1994), helping me I did. As a fourth-year medical student, Aron had the books and knowledge I needed to start this work. Though I originally figured Cheryl and other people with intersex conditions must be exaggerating how harmful the present-day system is, my still-tentative forays into the current literature proved them, Suzanne Kessler, and Anne Fausto-

Sterling right: the whole system was designed to act as if intersex didn't exist outside the clinic, to cover it up rhetorically, surgically, hormonally, thereby (however unintentionally) shaming and silencing the child and family, and leaving genitals of questionable function and appearance. The literature was also quite openly sexist and heterosexist, equating female sexuality with a penetrable hole, ignoring or even eliminating the ability to orgasm in the quest of "normality," holding reproductivity as the key to womanhood but the phallus as the key to manhood.

Soon, with the encouragement of Cheryl and other people with intersex conditions she introduced me to, I became convinced that my work could actually help improve the social and medical treatment of people with intersex conditions, and this became much more important to me than what it might do for nonintersexed straight women, gay men, and lesbians. I wrote to Cheryl, "Perhaps I could help get some of this stuff into actual medical discussions. I have some ways to get into those—my Ph.D. in history of medicine & science opens a lot of doors that were previously closed to me."[14]

I found myself determined to see my book, based on my dissertation, have an important impact. Still not having had the chance to meet her, I wrote to Cheryl in March of 1996, "I am considering placing my book w/ Routledge. Any thoughts on that? The folks at [one academic press] offered me a contract, but they were too into theory and didn't at all understand how important it is that this book be CLEAR in its language and messages. I am also considering going directly to a trade press if it takes that to make the message more widely heard."[15] I am sure that motivation was true, but I suspect part of my concern to write a "good book" (a readable, accessible book) also came from the fact that I thought of myself as a closeted writer who found academia a comfortable closet. My mother had been a writer, though she hid some of that from her children and had loved the English language so much she would read us Strunk and White's *Elements of Style* at the breakfast table. I found myself loving writing early. In fourth grade, given the essay topic of "what I want to be when I grow up," I wrote about becoming a writer. Unfortunately my teacher, Mrs. Kuzmier, let me know "that isn't a real job" and told me to rewrite the paper, so I had spent much of my life in approach-avoidance with professional writing, wanting like all good up-and-coming feminists eventually to have "a real job." Later I fell in love with Aron in part for our shared love of words. His undergraduate degree had been in chemistry, but he had almost enough credits for an English degree as well, and I remember well how, exhausted after a day on the wards, he would collapse on the couch with something good to read, by Toni Morrison or Mark Twain or Jane Austen. He soon became my home editor, and he read for style as much as accuracy and originality. The selfish desire to be recognized as a good writer by those close and far would have been on my mind as I decided how to write and where to publish, just as the knowledge that I wasn't especially good at postmodern theory would have made me shy to work with editors who wanted a lot of Foucault-like gender theory from me.

I had started, with Cheryl's and Aron's encouragement and help, to broach the subject of the present-day medical treatment in all the talks I gave and made sure my talks were advertised in a way that clinicians treating intersex would come. Surgeons were especially apt to show up, loving medical history as they do, so much more as a group than other specialties. Indeed, I noticed that at one university, when I was invited to join a new history of medicine reading group, though it was a group meant to include anyone interested, it was nearly all surgeons. I've come to assume this disproportionate interest exists because surgery is still largely taught by apprenticeship, and so surgeons find themselves far more aware of lineages, and therefore histories, than their counterparts, far more likely to believe and know that who came before you matters. Whatever the reason, it was a great boon to the cause of the Intersex Society, now *my* cause, because surgeons were the people doing the most irreversible damage.

But however much they enjoyed my histories—and some didn't because they saw it more as sociology than the great-man histories they were used to—they found my critiques of the present-day system bizarre, blindly politically correct, unsupportable, queer, wrong. Cheryl, when she did manage to get in to talk to an intersex clinician, found the same. I wrote to her,

> I've been trying to figure out how to defuse the rage we encounter w/ our critiques on the part of the American med. establishment. Really, why do they get so angry? I mean, I know why, but WHY? I have been trying to figure out how to help these people see how they can get from rage to productive thinking to change, so that they [can] change, and as long as they are at the rage stage, they can't move on. If you have ideas, let me know. I try to talk to them about how I appreciate that their goal is helping the patient and that we need to think deeply about what that means—i.e., letting them stay on that idealized, "heroic" plane but redirect that desire to be heroic into standing up for needed change. Reconfirugre [sic] what it means to be a "heroic surgeon," you know? I think that might work—maybe—as long as they don't feel like they have to be alone, or that they have to all at once deal w/ what has happened. I am sure the rage comes largely from a profound desire not to have to face what has happened.[16]

But people other than the clinicians were *very* interested in the critiques of the present system, and this kept us going: "I gave a talk here [Minnesota] yesterday which was 'an historical and ethical analysis of the biomedical treatment of human hermaphroditism,' and there was a marvelously mixed audience which seemed generally sympathetic. I continue to find the same everywhere: when it is explained carefully and given lots of supporting evidence, people find ISNA's critique eminently reasonable and urgent."[17]

How useful was the history, though? Not that much. The history drew in the surgeons, but the surgeons were stone-faced when I offered my critiques of their current practices. History was supposed to be about something else—progress, entertainment, pride, role models. Indeed, when I suggested to that mostly surgeons medical history reading group that we read something other than great-man histories, perhaps some new social history, one had exploded, "That's sociology, not history!" In response to my genuine bafflement, he explained he wanted history to get him through his day, with his stressed staff, his unreasonable insurers, his demanding patients, his unsympathetic chair. He needed history to show him how Lister persevered in spite of it all. If he was at all representative, and I tended to find that he was, new social histories were not going to do anything for him or his relationships with his patients.

Meanwhile, the gynecologists and the pediatric endocrinologists were more inclined to hear the critiques of present-day practice, to take them seriously, but they didn't really care all that much about the history; they sat through the historical parts of my talks interested but came more to hear about what it all meant to us now. Though still in love with history, I found myself inclined to agree with them, that what we needed was a good ethical analysis and some quality follow-up studies of the standard of care more than a knowledge of what happened to one Louise Bavet in Paris in the 1800s.[18]

But there was one group that did find the history life-changing, and that was intersexed people themselves. My work showed how the experience of intersex was contingent—contingent to time, place, nationality—all of those things I had named for Cheryl at the start of our conversation, and more. This meant that the profound shame felt by people with intersex conditions, the sense of being freaks, of having deserved to be mutilated by "reconstructive" surgeries were *not necessary* but rather constructed. Nature did not make them monsters; their cultures did. I could show them how they were undergoing a historical progression that had been traveled successfully before them by women, by queer people, by blacks—the progression from freak, to pathology, to variation, to full person. They would win—*we would win*. Cheryl and I would talk about this, about the history of successful civil rights movements, trying to figure out how to make it happen for intersex. I remember her schooling herself, reading Ronald Bayer's *Homosexuality and American Psychiatry: The Politics of Diagnosis* and the autobiography of Martin Luther King Jr.[19] The latter gave us comfort as our rights movement went through what they all have: spats between insiders, attempts by other groups (for example, the anti-male-circumcision groups) to equate their issues and demands with ours.

Sometimes the history helped people with intersex conditions just because it stopped them from feeling so alone. I could tell them story after story about people who came before them, and they realized they were not rare, exotic, bizarre—they were people, people who had shown up in every culture in every time. In the same way learning the story of how women got the vote had been so therapeutic for me as a girl, this history was explicitly therapeutic for many of them.

While it was rewarding to get letters of thanks from people with intersex conditions, hearing from more of them made me feel strange about speaking on their behalf. I wrote to Cheryl in April 1996,

> It has been really important to me that I not usurp the voices of people who have intersex experiences—i.e., that I do more directing to those voices than that I try to speak for them. But from all the censorship, from seeing you folks write VERY reasonable things which are suppressed and seeing Anne [Fausto-Sterling], her colleague from Brown, and me all get a fair hearing comparatively . . . it is clear to me that at least for some time those of us who are (god damned) "experts"—i.e., not ISNA members per se but academic and medical friends of ISNA—are going to have to do a lot of the talking since we seem to be given a lot more of a hearing than you do. It frustrates the shit out of me.[20]

Cheryl wasn't ready to let me off the hook. She answered, "I can't speak for everyone, but as far as I'm concerned, we could use all the help that you can provide, not only because of the great resistance to listen to what intersexuals have to say in the first person, but also because not all intersexuals are skilled writers/historians/social analysts, nor are able to make a career of those activities. And oftentimes those with skills are impeded by various sorts of emotional damage from speaking out."[21] So I became determined to make my career work for this cause, and now, in spite of my lack of training in medical ethics, to write a strong ethical critique that would demand that people with intersex conditions be listened to.

In 1996, as I was preparing to move from the University of Minnesota to a permanent position at Michigan State University, I proposed a book based on my dissertation to Harvard University Press, and to my delight they agreed to review the manuscript, though they were skeptical about including the ethical critique.

Indeed, when I sent them the manuscript, they found the final chapter, the ethical critique, wanting, and suggested leaving it out since it didn't seem to fit and lacked adequate support. The former argument (fit) was somewhat true, although I didn't care, but the latter (strength) was unavoidably true; in my haste to get the manuscript to them on time, the last chapter went to them thin and weak. (Nevertheless, I was dismayed to find out that one of the negative reviews of the critique chapter came from a clinician treating intersex whose work I had criticized by name in the chapter. It all fed our conspiracy theories.) With Aron's and Cheryl's help, I rewrote the chapter to be much stronger and fully supported, and we were amazed to hear that the clinician-reviewer now wanted to see my sources, being inclined to find my argument rather persuasive. The press did decide that the critique should be labeled an "epilogue," and my what-this-has-to-do-with-today introduction a "prologue."

As I wrote the book, Aron went into a tough editorship mode, reading as a general reader and also as a physician (he was now a resident in internal medicine at Michigan State). He rejected the first seventeen versions of my opening paragraphs for the book, reminding me (from memory) of the best opening lines of literature, warning me frequently not to employ the jargon of "social construction," which might cause the science wars conservatives to ignore my work, the science wars radicals to abuse it, and the physicians to put the book down for a better read. My editor, Michael Fisher, and manuscript editor at Harvard, Elizabeth Gretz, also showed great dedication to the project, Elizabeth in particular talking with me constantly via e-mail about how best to make my arguments heard. The epilogue was spun off as an article for the *Hastings Center Report,* and medical ethicists stopped telling me intersex wasn't an ethical issue because it just involved the medical treatment of a well-understood pathology.

The book, *Hermaphrodites and the Medical Invention of Sex* (1998), was generally well received, and much to the credit of Harvard University Press, was reviewed in major medical journals including *The New England Journal of Medicine* and *JAMA—* which is not to say some of the reviews were not very curious. One clinician writing in a medical journal claimed my epilogue was already outdated because clinicians don't lie to intersexed patients anymore (this at the exact time we had correspondence from a mother who was actively being lied to at that clinician's own institution). Another complained that I had failed to talk about Darwin's barnacles—a complaint I found amusing given my original intentions for this topic. (Cheryl assured me the intersex cause was not being delayed by my failure to analyze Darwin's take on hermaphroditic barnacles.) But people with intersex conditions and their parents seemed to appreciate the book unequivocally, and the word was out, louder than ever (and now with "Harvard University Press" attached) that the system was flawed. Now people with intersex conditions were coming out of the closet, thanks largely to the work of Cheryl and her organization of the academic and media allies. I started doing media appearances as a talking head and also wrote a piece published in the *New York Times* under the headline "When Medicine Goes Too Far in the Pursuit of Normality."[22] Cheryl and I went on the road together to make joint appearances, and more than once, on encountering a surgeon who insisted the standard practice was fine and would never change, I found myself quoting Winston Churchill: "History *will* vindicate me, and *I* will write the history." That seemed to get surgeons' attention.

At some level this had stopped feeling like scholarship and had pretty clearly become activism. My activism was supported by scholarship, of course, but the texts I read and the questions I was asked were all filtered through the mission: to build a world free of shame, secrecy, and unwanted genital surgeries for people born with atypical sex anatomies. Cheryl had become one of my best friends, and Aron and I would spend time with her plotting, drinking, fantasizing about the future. I had no time, as I now complained to Tod Chambers, to do anything new in terms of scholarship. But new scholarship hardly seemed important when I would be on the road with Cheryl and hear her tell her story again. Each time it was a little different, a few more details of her history emerged, and the parts had been assembling in my brain. I remember well one day in April 1999, when Tod and David Ozar, bioethicist at Loyola, had lined up four gigs in a row in Chicago, and I had to listen to Cheryl tell her own story four times in close succession. I found myself on the trip home having to pull off Interstate 94 from crying so hard about her, having a huge sense of helplessness to rewrite her history in any way that mattered.

By then, doctors had finally started to turn. At Michigan State, after meeting and talking with me, pediatric endocrinologist Bruce Wilson came down firmly on the side of ISNA, warning his colleagues in grand rounds around Michigan that they'd better get on the right side of "the paradigm shift." (I slapped my head to realize we should have been using that rhetoric all along! But I can't really do it, being historian-shy of the term.) Before the Chicago trip, we had been scared about our appearance at Children's Memorial Hospital, knowing we were going to show a video in which Angela Moreno openly blamed that very hospital for the loss of her clitoris in 1985.[23] We did what we came to do, however, and Jorge Daaboul, a pediatric endocrinologist who had taken an ethics class with Tod turned to his colleagues and told them they must listen to what we were saying. Jorge held up my book in his office after the presentation and asked me to sign it, and, so grateful for his kindness and willingness to fight the establishment with us, I quoted Galileo: "And yet, it moves." At Loyola, where Dave Ozar had arranged to have us face off against a pediatric surgeon, David Hatch, I sat shaking as Hatch began his response to our presentation. It took me several minutes to hear what Cheryl was hearing, and I finally understood why she had grabbed my wrist so hard she was stopping the blood flow to my hand: he was saying we were right.

Jorge was, and is, one of the few doctors for whom reading the history of intersex has made a difference. In a paper he wrote for the American Association for the History of Medicine 2000 meeting in Bethesda, Maryland, for a session I organized entitled "Does the Study of History Affect Clinical Practice? Intersex as a Case Study," Jorge opined,

> Until the nineteenth century medicine had a strong sense of history. Medicine has since lost its reverence for historical precedent. It is my contention that this has had a pernicious effect on the practice of medicine and has been detrimental to patient care. While the technological advances of the last one hundred years have led to increasing precision in diagnosis and therapy, many of the insights and knowledge acquired by our predecessors are still valid. Our headlong rush into technological excellence, however, has obscured those insights and today we find ourselves relearning what our predecessors knew about human reactions to diseases and conditions.[24]

Jorge recounted how his own training about intersex was like virtually all his colleagues', that it descended from the teachings of Lawson Wilkins and John Money at Johns Hopkins—the place intersex activists had come to call "the Death Star." Jorge and his colleagues were taught that gender was constructed after birth, that children with intersex conditions could (and should) for the most part be made surgically into convincing-looking girls. Intersex was to be concealed, surgically, rhetorically, whatever the cost. Jorge remembered,

> All of us trained in the tradition of Money/Wilkins followed their recommendations. We lied to the parents of intersexed children and we lied to the children and we subjected intersexed individuals to "corrective" surgery. I admit that I followed the party line, albeit with vague uneasiness. Three years ago, I had the good fortune to do an ethics fellowship at the University of Chicago and my vague uneasiness turned to horror as I realized that my specialty was willfully violating norms of informed consent and respect for patient autonomy. I brought this up to my colleagues and in effect two arguments were used in defense of the current practice.

First, Jorge's colleagues responded that so many smart forefathers couldn't possibly be wrong. Second, people with intersex conditions *had* to be "fixed";

they couldn't be treated like other people because they weren't like other people. In short, "intersexed individuals could not possibly live normal lives as intersexed individuals . . . the only chance they had for happiness and psychological well being was the establishment of a secure male or female gender identity. There simply was no precedent for intersexed individuals living as normal people in our society. This argument proved difficult to refute." That was, as Jorge notes, until he read *Hermaphrodites and the Medical Invention of Sex* and realized that there was indeed *historical* evidence that intersexed people had fared pretty well before the age of surgery. Jorge realized,

> Nineteenth-century medicine "discovered" hard science and tried to reduce all human functions to mechanistic processes to be dissected and analyzed. Physical normals were established and anything that differed from the norm was deemed abnormal. Obviously, other sociocultural factors such as homophobia were at work in the case of intersex, but it is my belief that the impetus for classifying intersexed individuals as diseased or abnormal sprang from nineteenth-century medicine's obsession with establishing statistical normals. Once normality was defined in this artificial, statistical way, anything which deviated from the norm was by definition classified as abnormal.

For Jorge, the history was revelatory: "For me the study of history proved invaluable in my formulating an approach to intersex. The moment I realized that there was a historical precedent for individuals with intersex leading happy, normal productive lives I revised my approach to intersex and have become a strong advocate of minimal intervention. The study of the history of intersex gave me the knowledge to improve and refine my approach to this condition. Consequently, I am a better doctor to my patients." But Jorge is unusual in being moved this way; when Jorge puts this history to his colleagues, they respond that it is irrelevant to current practice, that they needed *current* evidence, not "ancient history."

In fact, as I've come to learn from my work on intersex, from Aron's growing interest in evidence-based medicine, and from my friend and colleague Libby Bogdan-Lovis's study of (and activism against) the medicalization of childbirth, much of medicine as it is practiced today is not about any kind of rigorous notion of scientific evidence. It's about an insidious kind of oral history, an amorphous set of beliefs that are passed down from attending physician to resident to student about what is "true." So, in intersex, the apocryphal (and so far untraceable) myth that one "untreated" (surgically unaltered) adolescent with intersex killed himself; in childbirth, the myth that hospitals are much safer than homes. Working against more scientific evidence are the worst-case scenarios feared by every practicing physician. Howard Brody and James Thompson have articulated this as the source of the "maximin" strategy of medical practice, in which physicians maximize intervention in the hopes of minimizing the odds that the worst possible outcome will happen. In obstetrics, at least, "this approach has tended to underestimate the risks of intervention and to overestimate the utility of a maximin strategy."[25] The little horror stories that underlie "maximin" behavior work much more powerfully than scholarly histories to shape what happens in the clinic. So, when in 2001 Tod found himself brought in to speak with a clinical group about a case of intersex, he found the surgeon on the case persuaded that he had to avoid having the nurses in the pediatric intensive care unit become squeamish about "a girl with a penis" as they had in a previous case. He had every intention of cutting down the new patient's clitoris as he had the previous one's. (The result of this encounter was not entirely fruitless; completely outraged by this kind of reasoning, reasoning he had seen in no other realm of medicine, Tod ended up donating a large sum to ISNA and letting me document the story in the ISNA newsletter.)[26]

To my mind, the deep historical stories of what happened to people with intersex conditions before the age of surgery can't really assure us that people with intersex conditions raised without surgery today would be better off. First, we don't know enough about what happened to draw any real conclusions about well-being back then. Second, the cultural settings are so different we can't use one as a predictor of the other. Finally, Cheryl and I don't advocate "doing nothing" as happened sometimes in the nineteenth century but rather argue that, in cases of childhood intersex, gender as boy or girl be assigned, that competent mental health professionals be involved with the family from the start, that medically indicated procedures be done to ensure physical health, that cosmetic surgeries be available when (but only when) those old enough to consent for themselves wish to use them. It might be worth noting that, to Cheryl's mind, scholarly histories won't change doctors' practices because, as she told the AAHM, "doctors don't read," no matter how well we write.[27]

While I do not think the history can be positively helpful—that is, it cannot tell us what we *should* do—I do think it can be negatively helpful, in telling us what to stop doing. So, for example, examining the history of the treatment of intersex can help physicians see that the nosology still used today of the "three types" of hermaphroditism ("true hermaphroditism," "male pseudohermaphroditism," and "female pseudohermaphroditism") is a silly, useless, even harmful classification system that came to be over 125 years ago for social rather than scientific reasons. Indeed, I have made some inroads with the North American Task Force on Intersex (a group headed by pediatric urologists and pediatric endocrinologists) in terms of getting them to work on changing the nomenclature of intersex to something more precise and patient-centered.

The work of those of us in the Intersex Society of North America to improve the social and medical treatment of people with intersex conditions is hardly over. Most practices still appear to be using the Lawson Wilkins and John Money system of concealment, and few rigorous longitudinal studies of this treatment are underway. But we have made some progress. *The Lancet* in July 2001 published a research letter that openly acknowledged "There are few, if any data on the long-term outcome of feminising genital surgery for children with ambiguous genitalia." In view of the authors' own retrospective study of forty-four cases, they conclude "This information [on poor outcomes] must be available to parents and clinicians planning such surgery. Cosmetic genital surgery in infancy needs to be reassessed in the light of these results."[28] In December 2001, the *British Medical Journal* published an editorial arguing that "most vaginal surgery in childhood should be deferred," citing ISNA's Website in the opening paragraph. The conclusion is something that, by 1998, Cheryl and I thought we might never see:

> We [clinicians treating intersex] need to rethink our approach to the management of intersex conditions. We must abandon policies of non-disclosure and manage patients within a multidisciplinary team. Long term follow up studies of adults with intersex conditions are crucial. However, such studies can be done only with the equal involvement of people with these conditions and of peer support groups and the cooperation of all clinicians managing intersex. It is time to create a major intersex research partnership to begin tackling these questions and move forwards towards enlightened and patient empowered care.[29]

The medical history and the historical advice I have provided to the intersex rights movement forms only a small piece of the foundation of this shift, a shift I still sometimes think we will never see through to completion, so great are the institutional powers working against it and us. But I do think that my historical work has lent important support to Cheryl, to Jorge, and to others along the way. Certainly it has offered comfort to many people with intersex conditions, and if that were all it had achieved, I would be satisfied that the work had been worth it. In retrospect, probably the most useful thing I did was to take Cheryl seriously and to use doctors' curiosity about history to get Cheryl and others in the door to tell their own personal stories. Nothing has turned out to be more powerful than getting patients to testify to doctors about their own lives. Even Jorge, who claims to have changed his practice because of my book, also tells another conversion story: A patient for whom he had cared for a long time found out accidentally what he had refused for many years to tell her, that she had a Y chromosome, and came back to slap him (literally) and berate him for having lied to her.

Are people who remember the past condemned to a life of activism and fundraising? Probably not often enough. This, for me, has been a worthwhile, even joyous fate. Santayana was right that those who cannot remember the past are condemned to repeat it. But I now know that those who *only* remember the past are in some ways condemned to the past, unable to change the future.

NOTES

1. The book Fred Churchill lent me that led me to realize this was G. Wayne Miller's *The Work of Human Hands: Hardy Hendren and Surgical Wonder at Children's Hospital* (New York: Random House, 1993).

2. Ann Arbor Science for the People Editorial Collective, *Biology as a Social Weapon* (Minneapolis: Burgess Publishing Company, 1977).

3. On the issue of the frequency of intersex, see Alice Domurat Dreger, *Hermaphrodites and the Medical Invention of Sex* (Cambridge, Mass.: Harvard University Press, 1998), 40–43.

4. See Suzanne Kessler, "The Medical Construction of Gender: Case Management of Intersexual Infants," SIGNS: *Journal of Women in Culture and Society* 16 (1990): 3–26; Anne Fausto-Sterling, "The Five Sexes: Why Male and Female Are Not Enough," *The Sciences* 33 (1993): 20–25; and Anne Fausto-Sterling, "The Five Sexes: Why Male and Female Are Not Enough," *New York Times,* 12 March 1993.

5. Alice Domurat Dreger, "Doubtful Sex: The Fate of the Hermaphrodite in Victorian Medicine," *Victorian Studies* 38 (1995): 335–369, and "Hermaphrodites in Love: The Truth of the Gonads," in *Science and Homosexualities,* Vernon A. Rosario, ed. (New York: Routledge, 1997), 46–66.

6. Alice Dreger, personal communication to ISNA, 21 February 1996.

7. Cheryl Chase, personal communication to Alice Dreger, 21 February 1996. For Cheryl Chase's account of the founding of ISNA and the start of the intersex rights movement, see Cheryl Chase, "Hermaphrodites with Attitude: Mapping the Emergence of Intersex Political Activism," *GLQ: A Journal of Gay and Lesbian Studies* 4 (1998): 189–211.

8. On Herculine Barbin, see Michel Foucault, *Herculine Barbin, Being the Recently Discovered Memoirs of a Nineteenth-Century Hermaphrodite,* R. McDougall, trans. (New York: Colophon, 1980).

9. Alice Dreger, personal communication to Cheryl Chase, 22 February 1996.

10. Cheryl Chase, personal communication to Alice Dreger, 22 February 1996.

11. Alice Dreger, personal communication to Cheryl Chase, 23 February 1996.

12. Cheryl Chase, personal communication to Alice Dreger, 24 February 1996.

13. Alice Dreger, personal communication to Cheryl Chase, 24 February 1996.

14. Ibid., 23 February 1996.

15. Ibid., 20 March 1996.

16. Ibid., 7 April 1996.

17. Ibid., 16 April 1996.

18. For what did happen to Louise Bavet, see Dreger, *Hermaphrodites,* 114.

19. Martin Luther King Jr., *Autobiography of Martin Luther King, Jr.,* Clayborne Carson, ed. (New York: Warner Books, 1998); Ronald Bayer, *Homosexuality and American Psychiatry: The Politics of Diagnosis* (Princeton: Princeton University Press, 1987).

20. Alice Dreger, personal communication to Cheryl Chase, 17 April 1996.

21. Cheryl Chase, personal communication to Alice Dreger, 17 April 1996.

22. Alice Dreger, "When Medicine Goes Too Far in the Pursuit of Normality," *New York Times,* 28 July 1998, B10. This article is reprinted in *Health Ethics Today* 10 (August, 1999): 2–5.

23. Intersex Society of North America, "Hermaphrodites Speak!" (color video, 26 minutes). Copies may be obtained via the ISNA Website, www.isna.org.

24. Jorge Daaboul, "Does the Study of History Affect Clinical Practice? Intersex as a Case Study: The Physician's View," presented at the American Association for the History of Medicine, Bethesda, Md., 19 May 2000.

25. Howard Brody and James R. Thompson, "The Maximin Strategy in Modern Obstetrics," *Journal of Family Practice* 12 (1981): 977–986, 977. I am grateful to Libby Bogdan-Lovis for this reference.

26. See Alice Dreger, "Why Do We Need ISNA," *ISNA News* (May 2001): 2–5.

27. Cheryl Chase, "Does the Study of History Affect Clinical Practice? Intersex as a Case Study: The Patient-Advocate's View," presented at the American Association for the History of Medicine, Bethesda, Md., 19 May 2000.

28. Sarah M. Creighton, Catherine L. Minto, Stuart J. Steele, "Objective Cosmetic and Anatomical Outcomes at Adolescence of Feminising Surgery for Ambiguous Genitalia Done in Childhood," *The Lancet* 345 (14 July 2001): 124–125.

29. Sarah Creighton and Catherine Minto, "Managing Intersex," *British Medical Journal* 323 (1 December 2001): 1264–1265.

CHAPTER EIGHTEEN

Transcending the Two Cultures in Biomedicine
The History *of* Medicine and History *in* Medicine

Alfons Labisch

Scientific knowledge always aims at generalized conclusions labeled as "laws of nature." What matters in medicine is to meet the requirements of an individual human being who is particularly in need in the case of illness. The dilemma of science-oriented medicine can be characterized as follows: in the immediate encounter with patients, doctors have to put their generalized scientific knowledge into action to address the individual conditions and needs of the patient. Thus, in the doctor-patient encounter the scientific object-orientation of medical science turns into the subject-orientation of medical practice.

At the beginning of the twenty-first century, the more advanced variants of the life sciences do not strive for a more or less ultimate truth: "laws of nature," valid across time and context. Accordingly, the distinction between natural sciences and the humanities begins to blur, as the categories of time and of individuality are introduced as objects of science. Time and individuality are essential categories of historical analysis as well. The physicist and biologist Max Delbrück (1906–1981), Nobel laureate in 1969, commented on these fundamental developments. "The complex accomplishment of any one living cell is part and parcel of the first-mentioned feature, that any one cell represents more a historical than a physical event," Delbrück insisted in 1966. "These complex things do not rise every day by spontaneous generation from the nonliving matter—if they did, they would really be reproducible and timeless phenomena, comparable to the crystallization of a solution, and would belong to the subject matter of physics proper. No, any living cell carries with it the experiences of a billion years of experimentation by its ancestors. You cannot expect to explain so wise an old bird in a few simple words."[1]

As a consequence, the changing scientific and technological appropriation of the world and the interrelated appropriations of the general social world and the different inner worlds of individuals are more and more being viewed not as opposites but as part of a comprehensive cultural process. Neuroscience, to give just one example, is a modern field in which not only physical, chemical, and biological issues are addressed in a transdisciplinary context but questions relating to the social sciences and humanities are routinely raised as well. In the interaction of neuroanatomy, neurophysiology, neurology, modern visualizing procedures, behavioral research, philosophy of cognition, anthropology, and theology and religion, the neurosciences have grown into a conglomerate of questions, methods, and results. This compound mixture renders a strictly defined classification of "natural sciences vs. the arts and humanities" absurd. Such a disintegration of traditional scientific entities, traditional subjects, and traditional disciplines should not be considered as a problem but as an opportunity; like economics, the sciences have to rearrange themselves in a globalizing world.

An integrated way of looking at these new developments is already well advanced in the Anglo-American concept of the humanities. In a pragmatic manner, it is oriented partly toward technical, partly toward economical, and partly toward social and political phenomena and decisions. Such a culturally specific pragmatism can hardly suffice in the traditions of continental Europe. What matters is to arrive at a complex and integrated view of a globally industrialized form of culture. In the first place, it has to accommodate the more and more differentiating forms of knowledge, techniques, and action. The starting point for sorting out how this matters for rethinking medicine must be to develop new orientations, perhaps new interpretations, but at the very least to raise the proper questions. In such a scheme, the humanities do not merely offer a complementary function to science and technique, by simply commenting on developments that already happened. Rather, they can be indispensable ingredients in a common approach to those compound phenomena of today that demand explanation.

The main purpose of this paper is to outline the specific character of medicine as a science and as an acting discipline (*Handlungswissenschaft*) in the special perspective of its historicity. Before going into detail, the editors of this volume urged me to put the story into perspective by giving a few autobiographical thoughts on acting as a doctor among historians and as a historian among doctors respectively. After studying philosophy, classics, history, social sciences, and medicine in Aachen and Cologne, I taught health policy and medical sociology at a faculty of social sciences and social security in Kassel beginning in 1979. In addition, I worked part-time as a doctor in general practice and in public health. So I had the opportunity to experience a doctor's life while at the same time teaching the machinery of medical care in a highly developed social system. Researching the sociology and history mainly of public health, as well as the interrelationship of health and society, led to some pragmatic, practical endeavors in public health policy on the national level and to initiatives in primary health care policy on the international level.[2]

What I had found to be a nearly perfect match of history, sociology, and medical practice began to unravel, when in 1991 I shifted to teaching medical history at a medical faculty. In Germany, medical faculties usually are large academic bodies reaching from medical theory to clinical practice, sometimes made up of as many as fifty or sixty independent institutions for basic medical sciences like anatomy, physiology and biochemistry; for specialized clinical basics like pathology, microbiology or radiology; and for medical clinics of every specialty. Being also co-opted as a member of the faculty of arts and humanities quite soon thereafter, I once again gained the opportunity to look at medicine both from the inside and from the outside.

I continued my research in the field of historical sociology of public health and the interdependence of medicine, health, and society, but it became obvious that this work had begun to fall apart. Neither students nor colleagues at the faculty of arts and humanities appeared to understand what history as applied to medicine meant. At the same time, neither students nor colleagues at the medical faculty appeared to understand what history could mean to their everyday business—except for opening talks and historical lectures at festive occasions. What about those medical students who are able to reproduce masses of scientific basics in any exam but are utterly unable to get three sentences of a historical plot into a sensible line? What about those students of history who get absorbed in their reading but lose any sense of its larger meaning outside the circle of their pet teachers and colleagues, not to mention its possible application to any relevant, pragmatic question? Serving as vice-dean of the medical faculty starting in 1998, and as dean since 2002, the more I got into the machinery of the university clinic, the more these worlds drifted apart. How to meet those fascinating, but totally different personalities in arts and humanities and in medicine on a workaday level? What would it mean for a combination of medicine and history to become a genuine scientific approach and valid tool for history in general, but especially a means to address the obvious problems of modern medicine?

My impression today is that students make basic decisions how to approach science, research, and professional action before they begin their university education. Medical students primarily want to help ill people—and they want to do this on a qualified level of scientific knowledge. So most of them are interested in analyzing scientific problems as technical questions, but, above all, they are intrigued by *acting*. Quite rarely are they interested either in asking theoretical questions or in researching theoretical problems; quite often they are interested in analyzing situations as a doctor would. Students of history, by contrast, are from the first intrigued by analyzing events and processes of former times. The more they dive into the subject, the more they get interested in the methodological arsenal of the humanities. Most of them, however, are hardly able to get into action, especially when speed and on-the-spot decisions are required. So whereas a medical student gets lost in a library, a history student gets lost in an intensive care unit in a very similar way. These basic decisions, more or less bound by character, are met by the different educational cultures and requirements in the humanities and in medicine.

Yet the more we get into the ongoing molecular transition of medicine, the more we have to face that the boundaries of medicine in people's lives are being profoundly reshaped. Public understanding of medical theory and medical practice is tested and challenged again in fundamental questions regarding birth (e.g. in-vitro fertilization, preimplantation diagnostics, abortion) and death (e.g. euthanasia), and in technical developments that raise other fundamental questions, as in the Human Genome Project or regarding stem cells or gene therapy. Contributions from the humanities are essential, both in the individual encounter of doctor and patient and in sorting out the social implications of modern medicine.

In the course of the nineteenth century, a series of fundamental debates brought to light that medicine is not a pure science—a *Naturwissenschaft*—and never will be. This raises the question as to the meaning of those "sciences" within medicine that deal with aspects of human action. These are the *Geisteswissenschaften*, the humanities in the widest sense. The following discussion seeks to extract from the manifold subjects and methods of the humanities the meaning of historical thinking *in* and *for* medicine. Leading questions include: What is the fundamental meaning of historical thinking in medicine? Which types of historical thinking can be distinguished? What is their range and function?

Shining Examples from the Past

Medicine is a discipline that aims at application and action. The patient as the point of reference is a constant reminder of the need to act. In the classic phrase *ut aliquid fiat* (something needs to be done), this urge to action is coined in unique precision. Because of its reference to individual human subjects, medicine requires specific knowledge individually applied to the individual human being in a particular situation of need. Consequently, medicine is in an ambiguous situation with regard to the theory and philosophy of science. For—as scientists have learned since Aristotle—singular statements and individual cases can never generate scientific knowledge: *de singularibus non est scientia* (singularities do not constitute science).

Ever since there has been a scientific approach to medicine, there has also been a version of history of medicine that is an appeal to former times. From ancient until early modern times medicine was grounded in Hippocratic-Galenic humoral pathology. In such a comprehensive, all-embracing concept of physiology, pathology, and therapy, the history of medicine was not a specialized mindset but an integral part of thinking. Different dogmas or therapies were freely discussed with authors of many centuries. In humoral pathology, the human being, the patient as a person (not as a mere "case"), was the center of attention. Thus the discussion was especially pertinent for the individuated history of the sick person. This kind of—in our sense—*a*historical discussion, spanning many centuries, was illustrated by individual histories of patients. The gynecologist and medical historian Paul Diepgen (1878–1966) expressed this kind of historical argumentation as follows: "A patient's history as told by Hippocrates still had the same value as one told by Boerhaave."[3]

Due to medicine's orientation toward cases, historical approaches and historical justifications remained common forms of argumentation until the nineteenth century, even when medicine entered its "materialistic phase." This was also true for scientists who are considered heroes of a conscious renunciation of romantic forms of medicine, which were dismissed as "philosophically speculative." In clinical medicine, this can be seen with Carl August Wunderlich (1815–1877) and his circle of friends; in theoretical medicine, the pathologist Rudolf Virchow (1820–1902) and his companions may serve as example. For the Leipzig physician Wunderlich, who introduced systematic thermometry into the clinic, the historical approach was an essential means of acquiring knowledge: it was all about understanding both medicine and oneself—in medical theory and clinical practice—as being historically conditioned and able to extract new insights from history in critical reflection.[4] The famous circle of the *Archiv für physiologische Heilkunde*—which included Wunderlich, the Marburg surgeon Wilhelm Roser (1817–1888), and the psychiatrist and neurologist Wilhelm Griesinger (1817–1868)—criticized the modern "ontological" scientific theory of illness because it drew attention away from the patient. Griesinger, for example, resisted the ontological model, and drew the "multidimensional picture of the history of individuality" of the patient.[5]

Virchow, for his part, insisted nearly all his life that medicine needs historical knowledge more than any other science.[6] Virchow and his circle—among them the epidemiologist and "poor-doctor" (a physician paid by the state to attend the poor for free) Salomon Neumann (1819–1908)—repeatedly emphasized the social context and the social responsibility of medicine. So in his many historical studies—which, by the way, in coverage and quality match the lifetime oeuvre of many professional medical historians—Virchow dealt both with socially significant diseases and with socially significant medical institutions. Inherent in medicine, in this view, was a medical historiography that deals explicitly with the particular problems of medicine, and thus should be called "pragmatic." This holistic view might be chiefly what led Virchow to plead *against* the introduction of medical history as a separate academic discipline in the late nineteenth century.

A few decades later, a modern (that is, specialized) kind of clinical medical history served to display the knowledge of a specific field of scientific research. A master of this new kind of medical history was Emil von Behring (1854–1917), the founder of modern immunology and serum therapy, and in 1901 the first Nobel laureate for medicine and physiology. Behring, who had a comprehensive command of the humanities and the sciences, developed his concept of immunity from a profound knowledge and understanding of medical history. In his *Geschichte der Diphtherie unter besonderer Berücksichtigung der Immunitätslehre,* he meticulously analyzed the history of epidemiology, the clinical history, and pathology of diphtheria to develop his ideas about immunity, immunization, and serum therapy.[7] In addition, Behring used this history of medicine to explain his research to his colleagues,[8] particularly to real or assumed scientific opponents like Virchow, who then was esteemed as the "Pope of Medicine."[9]

This kind of specialized clinical history initially served to open the door for a new *method* in the course of scientific progress. To give an example, Adolf Gottstein (1857–1912), a bacteriologist, then epidemiologist, and finally leading health administrator of Prussia and Germany, analyzed Behring's serum therapy historically and epidemiologically. A public debate ensued—stirred up by different medical approaches, clinic vs. public health; undermined by national and racial prejudice, conservative vs. democratic and German vs. Jewish. The argument finally led to a strict supervision of the production of serum on the clinical side, by none other than Paul Ehrlich (1854–1915), and to an improvement of mother and child care as well.[10]

Modern experimentation brought about the end of this kind of reliance on history in medical science and research. One among many German examples is Robert Koch (1843–1910), Nobel laureate in 1905, and his paradigmatic school of experimental bacteriology. Although Koch did not explicitly turn against history as a means of producing knowledge, the experimental method became de facto the only acceptable source of evidence for new knowledge. So, in the last decades of the nineteenth century, the historical mode of argumentation simply disappeared from the production of knowledge in medicine. Medical history vanished as a way of reasoning for medical researchers. The history of medicine developed into a stock of traditions of antiquarian interest, living more and more apart from the medical mainstream. But nineteenth-century clinical history of medicine also inevitably contained *legitimating* functions. It later more or less degenerated into a means to place a medical author, invention, or discovery at the end of the historical process, thus locating the author at the summit of progress. In medical life, therefore, medical history survived with slightly different aims: to legitimize and to set the scientific scene for a specific medical author; for a specific medical procedure, concept, or therapy; for specific medical schools; and, finally, to display an author or speaker's learning and cultural refinement.[11]

However, the total scientification of medicine, desired by many leading doctors in the mid-nineteenth century, did have unwanted repercussions, and the high hopes raised by the ambitious period of scientifically based therapy in the 1890s were disappointed. There was the scandalous failure of Robert Koch's tuberculin as a remedy against tuberculosis, and later the publicly disputed clinical effects of the new serum therapy against diphtheria. These disappointments brought about sustained reactions. Patients, expressing a typically German turn to "nature" and critical of contemporary civilization, turned to the natural healing movement, which took root and prospered.[12] In medicine, clinical practice, as opposed to the mere production and application of scientific knowledge in the laboratory, was again pushed to the center of attention. New action-oriented concepts were developed, among them the constitutional theory (*Konstitutionstheorie, Konstitutionshygiene*) in a now natural-scientifically, clinically, and epidemiologically supported form.

In the course of this movement, medicine's indebtedness to the humanities came back into focus and underwent reevaluation. In 1861, the generally unpopular *tentamen philosophicum* (exam in philosophy) was abolished in favor of the *tentamen physicum* (exam in science). At the close of the nineteenth century, however, with the revision of the medical examination system, a movement of leading professors and administrators emerged to make history of medicine a compulsory subject. This requirement did not find its way into the examination regulations promulgated on 28 May1901, but as a compensatory gesture, it was decreed that medical examinations should consider the history of each respective subject.[13] Thus, while the history of medicine did not become a compulsory subject of medical teaching, it did become a compulsory part of medical examination. History of medicine gradually achieved a new, independent status and meaning in and for medicine: historical education was meant to bring a purely scientifically designed medicine to practical reason.[14]

The outstanding protagonists of a historical- and philological-oriented history of medicine at the end of the nineteenth century were the psychiatrist and medical historian Theodor Puschmann (1844–1899)—who became associate professor in Vienna in 1879 and full professor in 1888—and Julius Pagel (1851–1912) in Berlin, who, without formal academic affiliation, earned his living as poor- and panel-doctor.[15] Puschmann, Pagel, and the neurologist and medical historian Max Neuburger (1868–1955), Puschmann's successor in Vienna, represented a history of medicine that met the historical and philological standards of the time. The practitioners themselves often came from the broad field of medical practice, and epitomized an approach to history of medicine that strove to make historical thinking and professional analyses of use to contemporary medicine.[16] "True knowledge and skill," Puschmann insisted in 1879, was only granted to those who were able to recognize the development of their science and art from its beginnings to its present state. Those who wished to understand a fact "truly and completely, have to study the history of its origins."[17] Only such a historical mode of understanding could inform a reliable and independent evaluation of medical innovations. More than this, it was the history of medicine that safeguarded the universal character of medicine by maintaining its connection to other disciplines: the history of medicine represented part of the general history of civilization. The connection of medicine as a special branch of science—with all its limitations—to the humanitarian and cultural endeavors was thereby maintained.

Julius Pagel, as his son, the pathologist Walter Pagel (1898–1983), later recounted, saw the special status of history and its mission in medicine in a similar fashion.[18] Medical history embraces the whole of modern medicine. Medical history reveals the deep cultural and social implications of medicine. It is only history that will enable the modern worker to assess his own time in its proper perspective and to understand its tendencies and position. History should include an attempt to reexamine the methods and results—right or wrong—used by the great historical figures, and thereby inspire us in our own medical work. So, at least, history teaches modesty.[19] Thus, the mission assigned to the history of medicine was to view the current problems of medicine within the context of the whole, to evaluate them, and to draw conclusions about the practical performance of doctors.

The debate on the character of medicine as a natural or an applied science continued into the early twentieth century. In these discussions, the history of medicine was always considered to represent the dominating part of the humanities in medicine.[20] Richard Koch (1882–1949), philosopher, historian, and physician in Frankfurt, drew new attention to *action* in medicine. Being both a medical practitioner and a medical theorist, he marked out the edge when medical science turns into a doctor's action, and at that point he introduced the historical argument.[21] Koch criticized the modern history of medicine: it had strayed from its original scope and task, aiming instead at scholarly sophistication and the production of new historical knowledge. An awareness of the whole, and therefore of the essentials, was lost, just as had happened in the natural sciences. Instead, the history of medicine should be history from doctors for doctors. This history should not refer to progress, thereby legitimizing the current state of medical science; rather, it should refer to the task of the doctor helping the ill—a task to be emotionally reexperienced through history. The essence of a doctor's performance, Koch maintained, is not scientific knowledge, which is only a vehicle, but rather acting and helping. Consequently, medical research can also be philosophical, historical, or philological: "That is the first true point of reference between medicine and the history of medicine: the urge to break through the loneliness of the acting doctor," Koch emphasized in 1921. "The one and only true point of reference between the history of medicine and medicine lies in the peculiarity of medicine as an action. Further points of reference are determined by the kind of object of this activity. The object of a doctor is the ill patient, from whom the doctor has to derive two different groups of insights: general and individual ones."[22] Thus, Koch introduced a kind of psychological a priori, which forces doctors in their lonesome and always doubt-stricken decisions to seek refuge in historical contemplation.

At this point, Karl Sudhoff (1853–1938) should be mentioned. At the turn of the century, Sudhoff was the leading figure, if not the godfather of historico-philological-based medical history in the German-speaking world. Although he could look back at a long and successful career as medical practitioner, he forcefully threw out any pragmatic aspect of professional medical history.[23] When publishing the second edition of Julius Pagel's famous *Einführung in die Geschichte der Medicin* (Introduction to the history of medicine) in 1915, Sudhoff left out Pagel's first chapter on the meaning and the value of medical history.[24] Later in his career, Sudhoff obviously had to notice that something was missing in professional historico-philological medical history at medical faculties. This may have been the reason why, in the early 1920s, he promoted the academic career of Richard Koch in Frankfurt, although Koch obviously was much more a medical philosopher than a medical historian.

During the interwar period, a time of a deeply felt "crisis of medicine,"[25] even renowned clinicians supported the idea of using historical reflection as a kind of regulative means. In 1926, for example, the famous surgeon Ferdinand Sauerbruch (1875–1951) argued in favor of teaching in university "that medical science is culturally bound and hovers up and down alongside the trends of time.... Therefore: History of medicine [is] not as an end in itself, but in connection with the living art of healing."[26]

Historical thinking *in* medicine, then, can be an applied mode of thinking in terms of the humanities, dealing with problems and questions of medical knowledge and practice in their dimension of time. This important realization often has come under threat. Both the scientific sophistication of medicine and the scholarly sophistication of medical historiography have at times contributed to the dismissal of the historical argument from medical research and medical practice. The historical-critical history of medicine became a specialized discipline, initially practiced by medical specialists primarily outside medical institutions. For its part, medicine came to be considered as a pure science. Together, this led to a great loss, the estrangement between history and medicine. Medical history can be saved for medicine by making a distinction between a history *of* medicine and a history *in* medicine.

Historicity of Medicine: Two Styles of Medical History

Where, in his or her practical routine, does the doctor need historical knowledge? At which point in medical decisionmaking does he or she feel drawn to history? This obviously does not happen quite often today, but the problem of history comes to the fore whenever customary routines of thinking and treatment are losing validity. In such moments, the dimension of time becomes evident in the guidelines of medical action. At this point, further analysis of the interrelation of medicine and history begins. Such an analysis starts from an altering experience of time and temporality, but as soon as new ways of looking at things, new routines of action, have gained broader acceptance as being evident and in a colloquial sense "natural," both historicity and uncertainty about how to act will again recede into the background. This becomes immediately evident in the contrast between debates over molecular medicine and medical ethics, and the now forgotten debates about heart transplantation in the 1970s and 1980s. But, as all of us experience today, changes in science are accelerating drastically, and our common and accepted landmarks of orientation and action are also changing. This obviously is a stinging stimulus of uncertainty.

Despite its scientific frame of reference, medicine is and always will be an applied science primarily focused on individual subjects. This emphasis on a subject-oriented, individual performance centered locally and temporally on a person in need of help involves personal and social aspects. So action-oriented, and thus pragmatic, approaches of the humanities are and always will be constituents of medical knowledge and its practical performance. The various approaches to medical thought that are derived from the humanities with their specific objects and methods (medical theory/theory of science, ethics/moral theory/theory of values, psychology/individual human interaction, sociology/social interaction, psychosomatics/intrapersonal action, general medicine/doctor and patient interaction) can be neither detailed nor discussed in this chapter. Among these different disciplines of the humanities, the category of knowledge and action in time becomes a specific object of historical analysis in medicine. In the context of changes in science and society, in a doctor's knowledge-oriented practical behavior, temporality, historicity, and coincidence come prominently to the fore.

Historiography as a scientific analysis is primarily directed at action in time, and especially at actions that are finished. History deals with the question of what was the case in a given time. To understand this special case, historians have to analyze context, contingency, possible options, and—if possible—the reasons for decisions and results. They distinguish their work from fiction or myth in their methodologically skilled way of searching and interpreting so-called sources as remains of the past. The historical context and the contingency, anchored by the historical events, offer possibilities for distant comparisons and examples as a sort of collective memory and experience. Genuine objects of a history of medicine are found in internal medical and external social actions, as well as areas of action referred to medicine that are finished as cases. The main focus is those forms of action whose features are comparable to issues raised in the historical observer's own times. This relationship over time implies historical relevance.

The question of the range of medical history is immediately related to the question of how far medicine ranges itself. Due to the natural-biological foundation of human existence, medicine is integrated into society.[27] The notion of health mediates—in an interpretation appropriate to a given civilization—between the societal and the natural demands of human beings. The interpretation and influence of the notion of health determines the place of medicine and the range of medical practice within a society. Notions of disease, in turn, prompt individual and public medical intervention. What bundles together all these interpretations and actions is the human body, conceived as an individual or as a public body. In the course of modern rationalization, this body becomes scientifically designed. The transgression of science into everyday life and the (theoretically desired) rational organization of all aspects of life are familiar to everyone today.

Being subject to continuity and change—and therefore to historicity—the aforementioned processes of rationalizing the body keep going on. If change is more rapid, and if the present seems to be "shrinking," it will become evident in people's everyday lives, and history will be called on.[28] The human range of experience and the scope of expectations are reflected in a conscious turn toward history. The underlying rationale of this reflection is the experience of time—perceived as inherent in any point of reference and as historicity, as each person is a subject to continuity and change.[29] This is where history develops its pragmatic effect: by piecing together the confusing events of the day in their temporal dynamics and thus integrating humans into a coherent picture. Of course, this takes place in all people's lives—for example, under the heading of remembrance and experience. It also takes place in societies relying on rationality in a special way, namely by professional historiography.

At the start of the twenty-first century, we are observing two processes simultaneously. First, medicine is developing into an institution of everyday life that is about to penetrate every part of society. Second, a worldwide change is recognizable in which classical meanings, such as faith, and classical institutions that give a meaning to life, such as religion and churches, are losing importance. At this point, a boom of historical literature appears, which can be observed almost everywhere in Western culture. Also at this point, in its societal realm of social influence and the inner developments of medicine, lies the specific responsibility of medical history. It has to be repeated that in view of the continually increasing influence of medicine on society, more and more questions will be posed to medicine from the outside. The mission of medical history thus turns just as much toward society as it does toward medicine. This means that history in general and medical history in particular have become a necessity of life in a secularizing world.

"And what about criticism?" is the question that by now must be expected to arise. It is a question that may be categorized as historical because it belongs to a special period of writing medical history in the 1970s and 1980s. A first answer may read: a criterion of professional historical research is neither to make "everything historically suitable" nor to be mainly or even exclusively critical. To study and offer answers to such relevant questions with appropriate historiographical methods produces clarity—especially toward the self-stylization of everyday life and the historical myths of collective action. This means that professional historical research is a matter of criticism in itself.

The question as to the critical function of historical research in medicine still deserves another answer, however, especially in view of (medical) ethics. Historical analysis creates a distance. The object of historiography is always action that has already taken place. This means that it is no task of historiography to deal with immediate evaluation of action yet to take place. This specific relationship between proximity and distance toward human action is peculiar. In the case of medicine, it is even the advantage of historical analysis and therefore has to be undertaken with deliberation. Here, appropriate questions and methods can serve to release the productive potential of different views and approaches that may help to solve current problems in the light of potential ranges of action. To say it more philosophically, historical analysis clarifies possible areas of contingency, without being under the immediate pressure of decisionmaking. Thus, historiography keeps the necessary distance from impending problems. When I speak of distance in the following, it is exactly this genuinely critical function to which I am referring.

Above, I stated that history in general and medical history in particular becomes a necessity of everyday life. As proof of this analysis, we can see today that the history of medicine gains ever more weight within general history. The number of books published annually on medical history increases, for the most part from general history, especially in Britain and North America. The history of medicine, once established in medical faculties, starts losing this formerly monopolized field largely to the faculties of arts and humanities. This is just another symptom of the ever-growing importance of medicine in modern societies. So an increasing space is opening up for *history of medicine as an aspect of general history*. The general criteria for this branch of medical history will be quite narrowly associated with methods, questions, and approaches of general historical research.

The traditional and still major field of medical history is the *history of medicine as an institution:* this is the case for the *external* history *of* medicine, as a self-reflective view on the development of medicine as an institution and its organizations, as well as for the *internal* history *of* medicine, as a self-reflective view on the internal developments of medical knowledge and practice. From an external perspective, the form and scope of medical services come into focus, such as the history of public health or the history of hospitals. The same is true for the effects of medical procedures on everyday life, as in transplantation medicine, perinatal medicine, the various fertilization methods, and the emerging questions concerning the beginning and end of human life. The internal perspective requires a sound knowledge of scientific and practical problems of medicine. Beyond the quite legitimate maintenance of tradition, it is essential to clarify the long-term perspective in which new medical developments will have to be considered.

It is in the dialectics between medical science and medical practice that we recognize the object and necessity of *history in medicine,* connecting general knowledge with the individual case. Yet neither case-oriented practical performance nor highly specialized medical knowledge is able to analyze the field of action that "medicine" constitutes. This view of specific aspects of the area of action called "medicine" in its temporal dynamics is imperative both for the internal (self-)assessment and for the external perception of medicine. To the extent to which medicine becomes an ever more influential institution within people's individual and social lives, it will become equally imperative to reflect the external perspective from an internal point of view. The historical argument becomes an instrument of conveying an understanding of the emerging medicine to the public. More than this, it is the task of history *in* medicine to work both internally and externally and to analyze in well-defined questions the changing conditions for medical knowledge and medical practice.

In a paper of such brevity and density one has to argue with abstract concepts: "society," "culture," "time," "medicine," and finally "scientific knowledge" and "medical practice." In historical work, these concrete, "objectified" and abstract notions—seeming to act on their own—include not only individual agents differentiated by gender, class, and ethnicity but also many occupational groups, who represent medicine in science and practice. This includes especially all those who are engaged less in curing than in caring. As the entire deduction of the notions of science and action in medicine as applied in this essay shows, the patient in his or her world is the very individual subject who directs every thought and action in medicine. Medicine steadily and inevitably has oscillated between science and action since its scientific offspring in antiquity.[30] So it is the individual patient, in his situation of urgent need, who unifies these different branches of knowledge and experience in a unique action. Without a patient, medicine would be just another science or just another technique. This means, in short, that in medicine, the dialectics of science and action are directed toward, as well as aligned by, the patient. Thus, the patient constitutes medicine. No one else has described this peculiar foundation of medicine in so short and comprehensive a way as Hippocrates in his first Aphorism. The patient has to be an outstanding object of medical-historical analysis.

"History *in* medicine" denotes the self-reflection of the conditions of medical knowledge and practical medical action in the special light of their change over time.[31] History *in* medicine signifies that historical forms of analysis are immediately focused on helping solve current medical problems.[32] Unlike a clinically oriented medical ethics, it takes place in a more distant form: it is either about the exploration of historical fields of action and their possible analogy to current fields of action or about the—admittedly very distant—analogy of singular actions in past and present. History *in* medicine is indebted to the methodological standards of historiography; its subject matter, however, is derived from the internal problems of the present development of medicine. History in medicine is therefore always *applied* and, in that sense, *pragmatic history of medicine*.

My differentiation of a history *of* and a history *in* medicine refers mainly to the scope of possible objects and problems. This distinction includes adequate methods and different areas of application. From another point of view, we can also differentiate between distinct modes or manners of historiography. We may distinguish a *contemplative kind of medical history* (history *of* medicine), which is the domain of historians, be they general historians or medical historians, from an *"acting" kind of medical history* (history *in* medicine), which is the domain of physicians or of historians who are familiar with medical problems. This characterization may get labeled even more clearly by saying there is *medicine in context* (history *of* medicine), using a retrospective perspective to analyze historical context and contingency as an aim in itself, and *medicine in action* (history *in* medicine), using the analysis of historical context and contingency in a prospective perspective related to medical problems.

Epilogue

A taxonomy has been presented here of different variants of the historiography of medicine. It is rooted in particular assumptions about the character of medicine and in some philosophical premises about contextuality, historicity, and contingency. The future will show whether this classification and the derivative tasks will be of any use to the current discussion of the history and historiography of medicine. At a time when biomedicine is facing multidimensional problems, medical history can provide an important conceptual tool that goes far beyond those still-ruminated self-justifications of medical history within medicine we have known for many decades. However, in order to make a difference and really contribute to medicine, it is imperative that medical history be integrated into the medical curriculum.

As I have pointed out, at the close of the nineteenth century, a movement of leading medical professors tried to include history of medicine in the medical curriculum as an obligatory subject. Although the history of medicine did not become a compulsory teaching subject with the examination regulations issued in 1901, the stipulation that examinations should consider the history of each subject gave medical history a new, independent place and meaning in and for medicine.

Only in 1939, when in Germany a new medical curriculum was presented, did the history of medicine become a compulsory teaching subject. The decisive motivation was the National Socialists' idea of using history in order to teach racist thinking, to distinguish a certain "German character" of medicine, and to spread a new medical ethics. The history of medicine authorized a particular medicine of a "genetically healthy and racially pure Aryan 'Volk.'" George Rosen (1910–1977), himself educated in medical history in early Nazi Germany as a Jewish medical student, stated at a conference of U.S. medical historians in the 1960s (by which time he was professor of health education at the School of Public Health and Administrative Medicine at Columbia University in New York): "I think you are aware that medical history was being taught in practically every university in Germany before World War II and apparently had no effect on the medical students at that time."[33] This reminder is necessary to stress that—in medical studies or elsewhere—a contemplative stance in the spirit of the humanities is not in itself any guarantee of its moral-ethical tendency, let alone its quality. After 1945, the history of medicine had to pay the same price as every other German science: the dramatic loss of leading figures never to be compensated for (e.g. Henry Sigerist, Ludwig Edelstein, Owsei Temkin, Walter Pagel, Erwin H. Ackerknecht, who all emigrated to the United States to found a new era of medical history on an international level, and Richard Koch who emigrated to the USSR to survive as a spa doctor on the Crimea).

The decisions that have enabled the institutional life of medical history in Germany at the beginning of the twenty-first century, however, did not arise from within but were made outside the history of medicine. In the early 1960s, the

German "Scientific Council" (*Wissenschaftsrat*) recommended that the subject of medical history should be established at all faculties of medicine. In the late 1960s, it was the new Licensing Order for Physicians (*Approbationsordnung für Ärzte*) that made the "cultural and social fundaments in the history of medical thinking, knowledge and performance; change of ideas about health and disease" objects of examination. With that, the history of medicine achieved decisive institutional significance in the former Federal Republic of Germany—yet without the pressure of a "permanent accountability of relevance."[34] Since then, a chair of medical history or a comparable institute has belonged to every faculty of medicine in Germany, which total more than thirty academic institutions with their respective scientific and nonscientific staff.[35] All in all, the number of professional medical historians (in this context, those working at institutes of medical history) in Germany roughly adds up to more than 150 people (including those funded by external resources). The professional qualification of the full-time medical historians is divided into medical, scientific, and philosophical qualifications. Approximately a third of the leading scholars hold doctorates both from faculties of medicine and from faculties of arts and humanities.

The most dangerous threat to the future existence of medical history today emerges from the subject itself and also from other developments within the humanities. The history of medicine finds itself once more at the brink of commanding an appropriate level of methodology referred to as historiography, while being in danger of losing the attention of its reference discipline of medicine. Therefore, medical history as an integral part of medical education is in danger of drifting away from the faculties of medicine. The sense-giving and action-oriented component of medicine that owes much to the humanities is more and more taken charge of by medical ethics. But, like practical medicine, medical ethics is essentially case-oriented. Therefore, only in exceptional cases is medical ethics able to appropriately consider continuity and change as a temporal dimension of medical knowledge and practical performance. This includes its own historicity—up to the point that there also had been a special medical ethic in Nazi medicine.

But help sometimes comes from outside. The new Licensing Order for Physicians passed the German Bundesrat on 26 April 2002, after more than five years of heated debates. It declared medical history, theory, and ethics, grouped together, to be an obligatory subject in the clinical phase of medical education. In the long run, this will lead to more institutes for the history, theory, and ethics of medicine—which were already common, as for instance in Hannover since 1988. History, theory, and ethics of medicine are definitely single academic subjects on their own. So it will be left to the future how history will survive in this challenging environment.

In that perspective, an adequate history of medicine promptly adopts a key position of deploying a transdisciplinary synopsis. Its eye is turned toward the whole spectrum of scientific approaches: from that of the life sciences, via approaches directed to an individual person, to those of the social sciences and the humanities. The field of responsibility of medical history will further expand both toward the general public and toward medicine. This development is already evident in a boom of medical-historical publications, topics, and questions issued—mainly in the field of genetics and molecular medicine but also in the history of the body, gender studies, science studies, history of health and disease, and many other areas. A huge field is opening up for the history of medicine. Medical history has a treasure of examples of how to act, which have endured thousands of years across diverse cultures. The analysis of historical cases can understand the contingency of present-day medical problems, helping us in turn to solve them.

A historiography of medicine—one that is oriented toward the problems of medicine—could (from its distance both toward the object and toward the immediate individual decision) generate those heuristics and hermeneutics that are being called for.

The field, however, has yet to be conquered by medical historians.

NOTES

For help and advice I am grateful to Ulrich Hadding, Düsseldorf, as well as to Werner F. Kümmel, Mainz. I am also indebted to the team of DESK (Deutschland-und Europa-Studien in Komaba), University of Tokyo, and the socio-historical forum of Osamu Kawagoe at Doshisha University, Kyoto. Special thanks go to Yasutaka Ichinokawa, Tokyo. Moreover, I thank Andreas Frewer, Göttingen, and Volker Roelcke, Lübeck, for correcting an earlier version of this essay. In Düsseldorf, the German manuscript was translated by Tanja Ziesemer, and the translation was supervised by Marita Bruijns-Pötschke, Ulrich Koppitz, and Sandra Lessmann. I am very obliged to all of them. Finally, I am especially indebted to Frank Huisman, Maastricht, and John Harley Warner, New Haven, for their encouraging criticisms, which changed my text drastically for the better. Still, I am ambivalent about their advice to get involved personally in this history of medical history. Of course concept, style, and remaining errors are all mine.

1. Max Delbrück, "A Physicist Looks at Biology," in *Phage and the Origins of Molecular Biology,* J. Cairns et al., eds. (Cold Spring Harbor: CSHLabPress, 1966), 9–22, 10ff.

2. See, e.g., Alfons Labisch, "Die 'gemeinschaftliche Gesundheitssicherung' (Primary Health Care) in der Bundesrepublik Deutschland und in der europäischen Gemeinschaft. Bericht über die Ergebnisse einer internationalen Arbeitsgruppe der EG-Kommission/Brüssel," *Öffentliches Gesundheitswesen* 43 (1981): 500–506; "The Public Health Service in the Federal Republic of Germany," in *Public Health and Industrialized Countries*, P. Duplessis et al., eds. (Quebec: Les publications de Québec 1989), 113–131; "Emergence de la profession médicale et de l'assurance maladie en Allemagne (1883–1931): Un compris institutionalisé au fondement de la société industrialisée," *Espace social européen, dossier spécial no. 4: Les politiques de santé en France et en Allemagne*, Bruno Jobert, Monika Steffen, eds. (Paris: Observatoire européen de la protection sociale, 1994), 23–39; "La salud y la medicina en la época moderna. Características y condiciones de la actividad en la modernidad," in *La medicalización de la sociedad*, J. P. Barran et al., eds. (Montevideo: Nordan, 1993), 229–251; Alfons Labisch with Yasutaka Ichinokawa, "Bunmeika no Katei ni okeru Kenkou-Gainen to Iryou," *Shiso* 878 no. 8 (1997): 121–154.

3. Paul Diepgen, "Aufgaben und Bedeutung der Medizingeschichte," *Geistige Arbeit. Zeitung aus der wissenschaftlichen Welt* 1, no. 14 (1934): 1.

4. See Carl August Wunderlich, *Wien und Paris. Ein Beitrag zur Geschichte und Beurtheilung der gegenwärtigen Heilkunde in Deutschland und Frankreich* (Stuttgart: Ebner und Seubert, 1841); "Die medicinische Journalistik," *Archiv für physiologische Heilkunde* 1 (1842): 1–42; and, for a widely known example, *Geschichte der Medicin. Vorlesungen gehalten zu Leipzig im Sommersemester 1858* (Stuttgart: Ebner und Seubert, 1859). On Wunderlich, see Volker Hess, *Der wohltemperierte Mensch. Wissenschaft und Alltag des Fiebermessens (1850–1900)* (Frankfurt: Campus, 2000).

5. Wilhelm Griesinger, *Die Pathologie und Therapie der psychischen Krankheiten. Für Aerzte und Studirende*, 2d ed. (Stuttgart: Krabbe, 1861), 133.

6. See, among many other examples, Rudolf Virchow, "Die naturwissenschaftliche Methode und die Standpunkte in der Therapie," *Archiv für pathologische Anatomie und Physiologie und für klinische Medicin* 2 (1849): 3–37; and "Ueber die Standpunkte in der wissenschaftlichen Medicin," *Archiv für pathologische Anatomie und Physiologie und für klinische Medicin* 70 (1877): 1–10.

7. Emil Behring, *Die Geschichte der Diphtherie. Mit besonderer Berücksichtigung der Immunitätslehre* (Leipzig: Thieme, 1893).

8. Ibid.

9. Emil Behring, *Gesammelte Abhandlungen zur ätiologischen Therapie von ansteckenden Krankheiten* (Leipzig: Thieme, 1893).

10. Adolf Gottstein, *Geschichte der Hygiene im neunzehnten Jahrhundert* (Berlin: Schneider, 1901); *Das Heilwesen der Gegenwart. Gesundheitslehre und Gesundheitspolitik* (Berlin: Deutsche Buch-Gemeinschaft, 1924); *Die Lehre von den Epidemien* (Berlin: Springer, 1929). On Adolf Gottstein, see Gottstein, *Erlebnisse und Erkenntnisse. Nachlass 1939/1940. Autobiographische und biographische Materialien*, Ulrich Koppitz and Alfons Labisch, eds. (Berlin: Springer, 1999).

11. A strange form of this maneuver can still be observed today in many scientific treatises of medicine, especially in lectures. Two, maybe three sentences are considered sufficient to deal with thousands of years of medical history. First and obligatory, Hippocrates is to be mentioned, then Galen; in Germany, maybe Paracelsus (1493–1541) or, in the UK and the United States, Thomas Sydenham (1624–1689), then probably one of the modern "leading authorities"—and finally the author himself. This is the order in which the

speaker wants to be viewed, and that is how "precursors" are created. See the lucid essay of Marc Andresen, "Die Konstruktion von Vorläufern in der Wissenschaftsgeschichtsschreibung. Bemerkungen zu einer Fiktion," *Berichte zur Wissenschaftsgeschichte* 22 (1999): 1–8.

12. For further reading, see Robert Jütte, "The Historiography of Nonconventional Medicine in Germany: A Concise Overview," *Medical History* 43 (1999): 342–358.

13. Hans-Heinz Eulner, *Die Entwicklung der medizinischen Spezialfächer an den Universitäten des deutschen Sprachgebietes,* Studien zur Medizingeschichte des neunzehnten Jahrhunderts, vol. 4 (Stuttgart: Enke, 1970), 434.

14. *Die Institutionalisierung der Medizinhistoriographie. Entwicklungslinien vom 19. ins 20. Jahrhundert,* Andreas Frewer and Volker Roelcke, eds. (Stuttgart: Steiner, 2001).

15. Theodor Puschmann, *Geschichte des medicinischen Unterrichts von den ältesten Zeiten bis zur Gegenwart* (Leipzig: Veit, 1889). Julius Pagel, *Einführung in die Geschichte der Medicin. Fünfundzwanzig akademische Vorlesungen,* Geschichte der Medicin, vol. 1 (Berlin: Karger, 1898); *Biographisches Lexikon hervorragender Ärzte des neunzehnten Jahrhunderts. Mit einer historischen Einleitung* (Berlin: Urban und Schwarzenberg, 1901); *Grundriss eines Systems der Medizinischen Kulturgeschichte. Nach Vorlesungen an der Berliner Universität (Wintersemester 1904/05)* (Berlin: Karger, 1905).

16. A wonderful example of combining scientific historiography with a pragmatic approach, focused on the special task of medicine, is Neuburger's "Einleitung," in *Handbuch der Geschichte der Medizin, begründet von Theodor Puschmann,* vol. 2: *Die neuere Zeit,* Max Neuburger and Julius Pagel, eds. (Jena: Fischer, 1903), 3–154.

17. Theodor Puschmann, "Die Geschichte der Medizin als akademischer Lehrgegenstand," *Wiener Medizinische Blätter* 44 (1879): 1069–1072; 45 (1879): 1093–1096; cf. "Die Bedeutung der Geschichte für die Medizin und die Naturwissenschaften," *Deutsche Medizinische Wochenschrift* 15 (1889): 817–820.

18. Walter Pagel, "Julius Pagel and the Significance of Medical History for Medicine," *Bulletin of the History of Medicine* 25 (1951): 207–225.

19. Ibid., 212ff.

20. For a concise summary of arguments on the part of the history of medicine, see Werner F. Kümmel, "Vom Nutzen eines 'nicht notwendigen Faches': Karl Sudhoff, Paul Diepgen und Henry E. Sigerist vor der Frage 'Wozu Medizingeschichte?'" in *Geschichte und Ethik in der Medizin. Von den Schwierigkeiten einer Kooperation,* Medizin-Ethik, vol. 10, Richard Toellner and Urban Wiesing, eds. (Stuttgart: Fischer, 1997), 5–16; "'Dem Arzt nötig oder nützlich'? Legitimierungsstrategien der Medizingeschichte im 19. Jahrhundert," in Frewer and Roelcke, *Die Institutionalisierung der Medizinhistoriographie,* 75–89; and "'Ein Instrument medizinischen Lebens': Henry E. Sigerist und die Frage 'Wozu Medizingeschichte?'" *Gesnerus* 58 (2001): 201–214.

21. See Richard Koch, "Die Bedeutung der Geschichte der Medizin für den Arzt," *Fortschritte der Medizin* 38 (1921): 217–225; cf. "Die Geschichte der Medizin im Universitätsunterricht," *Klinische Wochenschrift* 6 (1927): 2342–2344, and "Die Geschichte der Medizin im Universitätsunterricht," *Archiv für Geschichte der Medizin (Sudhoffs Archiv)* 20 (1928): 1–16. On Richard Koch, see Karl E. Rothschuh, "Richard Hermann Koch (1882–1949). Arzt, Medizinhistoriker, Medizinphilosoph (Biographisches, Ergographisches)," *Medizinhistorisches Journal* 15 (1980): 16–43, 223–243; and *Richard Koch und die ärztliche Diagnose,* Frankfurter Beiträge zur Geschichte, Theorie und Ethik der Medizin, vol. 1, Gert Preiser, ed. (Hilde-

sheim; Olms 1988). The personal archives of Richard Koch are being prepared for publication by Urban Wiesing and Frank Töpfner, Tübingen.

22. See Koch, "Bedeutung," 223f; on Richard Koch and medical history, see Urban Wiesing, "Die Einsamkeit des Arztes und der 'lebendige Drang nach Geschichte'. Zum historischen Selbstverständnis der Medizin bei Richard Koch," *Gesnerus* 54 (1997): 219–241.

23. This is apparent in an argument about the new chair of history of medicine at Leipzig; see N. N. [J. Hermann Baas], "Die Puschmann-Stiftung für Geschichte der Medizin," *Münchener Medizinische Wochenschrift* 51 (1904): 884ff; Max Seiffert, "Aufgabe und Stellung der Geschichte im medizinischen Unterricht," *Münchener Medizinische Wochenschrift* 51 (1904): 1159–1161; and Karl Sudhoff, "Zur Förderung wissenschaftlicher Arbeiten auf dem Gebiete der Geschiche der Medizin," *Münchener Medizinische Wochenschrift* 51 (1904): 1350–1353. See, finally, Karl Sudhoff, "Theodor Puschmann und die Aufgaben der Geschichte der Medizin. Eine akademische Antrittsvorlesung," *Münchener Medizinische Wochenschrift* 53 (1906): 1669–1673.

24. Julius Pagel, "Erste Vorlesung: Einleitung, Begriff, Werth, Object, Quellen und Eintheilung der medicinischen Geschichte," in *Einführung in die Geschichte der Medicin* (Berlin: Karger, 1898), 1–22. For the second edition, see Karl Sudhoff, *J. L. Pagels Einführung in die Geschichte der Medizin in 25 akademischen Vorlesungen, durchgesehen, teilweise umgearbeitet und auf den heutigen Stand gebracht von . . .* , 2d ed, (Berlin: Karger, 1915). According to Owsei Temkin, Pagel's first chapter was "one of the most comprehensive outlines of the pragmatic point of view" ("An Essay on the Usefulness of Medical History for Medicine," *Bulletin of the History of Medicine* 19 (1946): 9–47, 37).

25. Eva-Maria Klasen, "Die Diskussion um eine 'Krise' der Medizin in Deutschland zwischen 1925 und 1935" (M.D. thesis, Mainz, 1984); *Medizinkritische Bewegungen im Deutschen Reich (ca. 1870–ca. 1933), Medizin, Gesellschaft und Geschichte,* suppl. 9, Martin Dinges, ed. (Stuttgart: Steiner, 1996).

26. Ferdinand Sauerbruch, "Heilkunst und Naturwissenschaft," *Die Naturwissenschaften* 14, no. 48–49 (1926): 1081–1090, 1090.

27. See esp. Alfons Labisch and Norbert Paul, "Medizin, 1. Zum Problemstand," in *Lexikon der Bioethik,* Wilhelm Korff et al., eds. (Gütersloh: Gütersloher Verlagshaus, 1998), vol. 2:631–642. In general, see Alfons Labisch, *Homo hygienicus. Gesundheit und Medizin in der Neuzeit* (Frankfurt: Campus 1992).

28. Hermann Lübbe examined the phenomenon of "the shrinking of the present"; see his "Erfahrungsverluste und Kompensationen. Zum philosophischen Problem der Erfahrung in der gegenwärtigen Welt," in Lübbe et al., *Der Mensch als Orientierungswaise? Ein interdisziplinärer Erkundungsgang,* Alber-Broschur Philosophie (Freiburg: Alber, 1982), 145–168.

29. Reinhart Koselleck, *Vergangene Zukunft. Zur Semantik geschichtlicher Zeiten* (Frankfurt: Suhrkamp, 1979); *Zeitschichten. Studien zur Historik* (Frankfurt: Suhrkamp, 2000).

30. See, for example, Albert Jori, "Wissenschaft, Technik oder Kunst? Verschiedene Auffassungen der Medizin im Corpus Hippocraticum," *Historizität. Erfahrung und Handeln in Geschichte der Medizin, Naturwissenschaft und Technik (Sudhoffs Archiv, Beiheft),* Ulrich Koppitz and Alfons Labisch, eds. (Stuttgart: Steiner, 2003).

31. With regard to a theoretically and methodologically appropriate history in medicine, there are exceptional works of orientation, Karl E. Rotschuh (1908–1984) in particular; see his *Konzepte der Medizin in Vergangenheit und Gegenwart* (Stuttgart: Hippokrates, 1978).

From the comprehensive oeuvre of Fritz Hartmann (b. 1920), I refer only to a key essay, "Krankheitsgeschichte und Krankengeschichte (naturhistorische und personale Krankheitsauffassung)," *Sitzungsberichte der Gesellschaft zur Beförderung der gesamten Naturwissenschaften zu Marburg* 87, no. 2 (1966): 17–32. Eduard Seidler (b. 1929) deserves to be mentioned, too; see, in particular, "Gedanken zur Funktion der historischen Methode in der Medizin," *Medizinische Klinik* 70 (1975): 726–731.

32. For the skeptics, there are some current examples. In the late 1980s the World Health Organization (WHO) commissioned a historical analysis of the species sanitation of malaria in order to make this forgotten knowledge available again after it had been buried in the era of eradication; see Willem Takken et al., *Environmental Measures for Malaria Control in Indonesia. An Historical Review on Species Sanitation,* Wageningen Agricultural University papers, 90-7 (reprint, Wageningen: Agricultural University, 1991).

33. *Education in the History of Medicine: Report of a Macy Conference, Bethesda, Maryland, June 22–24, 1966,* John B. Blake, ed. (New York: Hafner, 1968), 49ff.

34. Cf. Hermann Lübbe, *Geschichtsbegriff und Geschichtsinteresse. Analytik und Pragmatik der Historie* (Basel: Schwabe, 1977).

35. On the recent developments and current situation of medical history in Germany, the *Nachrichtenblatt der Deutschen Gesellschaft für Geschichte der Medizin, Naturwissenschaft und Technik* gives a survey of the situation of the respective institutes in the first issue of each year (the latest issue being 52, no. 1 [2002]). These reports are also available on the homepage of the DGGMNT, at www.mpiwg-berlin.mpg.de/DGGMNT/. An international survey is presented by Christoph Meinel, "Geschichte der Naturwissenschaft, der Technik, und der Medizin in Deutschland, 1997–2000/History of Science, Technology, and Medicine in Germany, 1997–2000," *Berichte zur Wissenschaftsgeschichte* 24, no. 2 (2001): 77–146.

CHAPTER NINETEEN

A Hippocratic Triangle
History, Clinician-Historians, and Future Doctors

Jacalyn Duffin

> Declare the past, diagnose the present, foretell the future; practise these acts. As to diseases, make a habit of two things—to help, or at least to do no harm. *The art has three factors, the disease, the patient, the physician.* The physician is the servant of the art. The patient must cooperate with the physician in combatting the disease.
> HIPPOCRATES, *Epidemics I*, XI.

STORY: **The Walk**

Every Thursday afternoon, I put on my lab coat, checking the right pocket for my stethoscope (red for blood) and the left for my prescription pad (almost never used). Without an overcoat, even at −20° C, I walk the length of Stuart Street to the hematology clinic. The short trajectory takes me past the medical school, its library and labs, under the windows of deans, across the old and new entrances to the Kingston General Hospital (a National Historic Site), near two main lecture halls, and between several stately homes converted into offices for clinicians and patient-support groups.

It is a useful walk. Corner consults with faculty expedite administrative matters. More effective, however, are the encounters with medical students. They smile or nod hello; a few stop to tell tales of their courses, love affairs, exams, and residency aspirations. Some, especially the newcomers, do a double take when they recognize me in this costume. At the next history class, they will listen more closely. Practicing hematology does not make me a better medical historian, but it helps to draw the attention of my audience of future physicians.

As a clinician-historian working in a medical center, I have been asked by the editors to consider history written and read by clinicians, how it might have changed in the last two decades, and how it could be used in teaching. What stories from the past should we bring to future doctors? If they listen readily to a historian who is also a clinician, could it be that doctor-written history would be best?

In this unapologetically subjective paper, I explore the analogy between historical practice and medical practice, both methodologically and conceptually. The Hippocratic triangle of doctor, patient, and disease finds an analogy in the relationship between a historian, her sources, her audience, and the histories that they build together. Then, I examine the so-called genre of clinician history over the last few decades to conclude that there really is no such thing: an M.D. degree is not a predictor for subject, method, or style; history is written by individuals. Finally, I turn to the role of audience, particularly medical students. Not only do my medical-student listeners influence the history that I write—they also influence me as a individual at least as much as I hope to influence them. I deny that clinician history is a genre, but my audience reminds me that being (perceived as) a clinician is part of my identity as a historian.

Hippocratic Triangles: History as an Analogy for Medicine

STORY: Historical Practice and Medical Practice

I harbor doubts about my clinical abilities, especially on Thursday at noon. Will this be the day that I hurt someone? Yet the nursing station at the end of the corridor and the warm greetings of the outpatient staff put me at ease. It feels like home. Here, I understand the task, the banter, the humor, the resident-paging fascination of a big spleen with a bruit in room #12, and the importance of completing the requisitions with care (although I always need an update on their location and format, which seem to be altered every week).

In the consulting room, I love talking to the patient, eliciting her story—"elicit" not "illicit," but enticing enough to imply a connection. When it is done well, method disappears into pleasant conversation. A differential diagnosis of the possible explanations (diseases) emerges by the end of our chat.

A physical examination is essential to the consultation. No relevant sign must be left untried. I contemplate the centuries of wisdom that molded the intimate rite of those formal maneuvers of touch, and I adore the elegance of discernment, draping, and display. I think of Laennec every time I auscultate a chest. Then, further investigations are to be ordered: blood tests, X rays, scans, and biopsies. A vocabulary of signs that will point to the theory (disease/diagnosis) that will allow me to explain the problem on the patient's return. Then I get to turn her story through retelling into a patient history, which I compose, as evidence for the diagnostic process. When the clinic is over, I retreat to my other dwelling safe in the past, and the doubts slowly rise again.

The patient history is created by *shaping* a story—its content, rhythm, and sequence—and then *juxtaposing* it to the canon of known diseases, which are, in essence, theories about suffering.[1] The physical examination is merely an exercise in evidence-gathering, sifting through the document of the organism for signs that will help to eliminate or confirm the diagnostic impression derived from the story. The laboratory investigations exploit an archive of less accessible parts.

Like the doing of history, clinical practice uses experience, training, and method to collect stories and invest them with meaning. It demands convincing evidence and because medical thinking is subject to change, it attends to current scholarship. Far from being a rote exercise, the choice of tests, interpretations, and treatments requires imagination and erudition. The diagnosis must be explained to the patient. Should the findings be unusual, it should be communicated to the profession as a whole: each individual diagnosis could have an epistemic impact on the disease concept itself. In other words, a diagnosis can become a "secondary source" in a historiographic sense.

For many years now, historians have been well aware of the social construction of disease, in which the Hippocratic triangle—disease, patient, and doctor—plays a decisive role. The opinions of both doctors and patients about sickness and each other combine to make disease a socially and temporally contingent idea about suffering. This famous triad also provides a good analogy for scholarly history. In the same way, history is built by its authors and their listeners in a manner that reflects their desires and needs.

Carlo Ginzburg also drew a similar analogy between history and medicine, describing them both as semiotic disciplines that interpret signs in the world around us as evidence pointing to metaphysical concepts.[2] I would push this semiotic analogy further into the matter of audience and voice. In clinical practice and historical practice, two people or groups of people (doctors/historians and patients/sources) embark on a mutually agreed upon relationship, organized around a metaphysical concept (history or medicine). Their ultimate task is to derive from the particular (illness/past event) a meaning that can be generalized (diagnosis/a history). At least two steps occur in this process for the doctor and for the historian: the first is receptive (listening and seeing); the second is active (interpretation and communication). Doctors and patients take turns in the roles of listener and voice. Historians and their sources/audience do too (see fig. 19.1).

1. Clinical Practice

Disease:	illness		a diagnosis (idea)
	↓		↑
Patient:	voice	↔	listener
	↓		↑
Doctor:	listener	↔	voice

2. Historical Practice

History:	past episode		a history (idea)
	↓		↑
Sources:	voice	↔	listener
	↓		↑
Historian:	listener	↔	voice

Figure 19.1. Hippocratic triangles

To be effective as practitioners and as scholars, historians must also be attentive to the signs of the past, and they must explain them. Research through primary and secondary sources combines observation and reasoning, akin to listening and deciphering signs. Writing, or telling, or teaching history actively reverses that role. It is a privilege to occupy two worlds that complement each other so well.

What Is Clinician History?

STORY: Attacked by Historians

It is December 1985 and I am about to take my children tobogganing. The mail arrives—a thick envelope from a prestigious journal. In great excitement, I make everyone sit down and wait, melting in their snowsuits, while I rip open the package. So green was I then, that I failed to take its fatness as a warning. "I am sorry to say, . . ." began the editor's letter, gently rejecting my first-ever submitted article. That paper had been carefully reworked from what my mentor had called "the best" chapter of my thesis on Laennec, the nineteenth-century inventor of the stethoscope.[3]

Readers' reports were enclosed. I picked up the first. It was only a few lines long. The words blurred: "internalist," "presentist," "Whiggish," and "of interest only to Laennec enthusiasts"—obviously a low form of life. If my best work was hopeless, what of the rest?

We went tobogganing.

The analogy between history and medicine drawn in the first section of this paper is not widely accepted. Indeed, in certain historical circles, an M.D. degree is something to expiate. "Yes, I have medical training," the clinician must say, "but I can write history in spite of it." Scorn of doctor-written history is a problem of stereotyping, semantics, and class.

This lesson took me years to learn, and it began with that rejected paper. I did not look at it again for many weeks; assimilating the sense of the reader's powerful words came first. My postdoctoral supervisor, Toby Gelfand, tried to explain what was meant by "presentist," "Whiggish," "internalist." These words were new—I could not remember hearing them during the doctoral program at the Sorbonne. I soon discovered that they are often applied to the history written by doctors who are said to judge the past with arrogance and naïveté, as if it exists in a series of milestones culminating in a glorious present that is so much better than the past, as if unseen danger does not lurk in what we do now. This helped. The attack was not personal, then; it was an attack on a physician who dares to write history.

Whiggism might be understood as an extreme or "egocentric" form of presentism; usually, however, "Whiggish" histories are simply bad history, marred by skewed or absent evidence and wishful interpretation. Presentism, on the other hand, did not seem incompatible with good history. I admitted that doctors did tend to write in that manner because they write about the history of what they know and do. Nevertheless, once I had sorted out the meanings, I eventually concluded that accusing all doctor-written history of either Whiggism or presentism was false, and not only because it traded on stereotype. At least two other reasons sprang to mind.

First, many doctors are indeed arrogant, and yes, they like to explain their present-day practices; however, medics are always looking to improve on and replace that present, especially the academics. Citing articles that are more than five years old is scarcely permitted. If clever doctors doubt the merits of their present so much, how can they be described as seeing it as a culmination? Certainly those of us engaged in chemotherapy are committed to ensuring that it is not; this dissatisfaction with the present provides an impetus to examine the past. In the eyes of historians, this reason may seem weak because it does not challenge the naïve belief in "progress" through time. But without belief in the possibility of progress, medical inquiry would vanish. Would we skeptical historians want medical researchers not to think that what they do today is better than what was done yesterday? Does it mean that a committed doctor cannot also be a good historian?

Second, and perhaps stronger, good questions in any kind of history about any period or place emerge from the present—from what is in vogue in our culture and from preoccupations in the literature produced by colleagues. So-called social historians lob the presentist criticism at doctors, especially when they do not admire the topic or understand the method of analysis and if the prose is overly couched in "medical-ese." Doctors do not make this criticism of the so-called social historians; lacking the vocabulary of "historical-ese," they fail to perceive, much less label the problem. But they could, especially in terms of politics and analytic framework. For example, would we have those elegant histories of public health or the incisive criticisms of the asylum movement, if we academics had not found it appropriate to flirt with leftist ideas in the aftermath of World War II? Would we have the rise of nursing history and the outstanding works on women as physicians and as patients, had we not experienced the feminism of the 1970s? Would we be able to admire the intriguing new work on the health of aboriginal peoples if those peoples had not stood up for their rights and been heard? Histories of these topics that are poorly done, with selective evidence or narrow interpretations, might even be considered Whiggish. And what if doctors did criticize these works for being presentist? It would be a noncriticism. Good history *is* presentist, just as it is original, well researched, well argued, well contextualized, and well written with passion and care. Bad history is just plain bad.

I now had a face-saving explanation for my first rejection. The critical reader of my paper could tell that I was a clinician from clues in the text; he or she then assumed that my work was necessarily presentist, Whiggish, internalist—and dismissable. The paper *was* internalist, in that it was a thought experiment about a medical idea and my evidence lay within published and unpublished texts. But it was no more "Whiggish" or "presentist" than works published by historians who dwell on externals.

Consoled by finding myself part of a newly hypothesized collective, I began to wonder if other doctors who write history were "internalist" too. Were clinician-historians more comfortable with clinical texts, manuscripts, and problems? Was I a new member of an epistemic community? I needed to learn more.

Clinician-Historians: A Survey

STORY: Are All Clinicians Internalists?

By the spring of 1986, at my first meeting of the American Association of the History of Medicine in Rochester, N.Y., I found it instructive to walk up to strangers at the receptions, and say, "Hi! I'm an internalist! What do you do?" I still remember the bemused reactions. To my surprise, and disappointment, some physician-historians whom I met there argued vigorously that external factors were as much, if not more, important. When I tried the line again, as I was introduced to an eminent nonphysician historian of medicine, he replied, "Good for you! The rest of us need you; we rely on you." I took it as a compliment, but later wondered if he might have meant it ironically, as "for fodder."

I went away thinking that being internalist was not a product of my M.D.; it was merely a consequence of my current interests. By the time I finally published my first book, I had left the so-called internalist fold for something else closer to social history.[4]

Is clinician history different? Does it have distinctive qualities, topics, and methods? Has it changed over the last generation? My answer to the editors' questions is "no." This conclusion opposes the claims of a small body of literature. Both physicians and sympathetic historians have lamented the trend to "medical history without medicine"; they imply that the historical view of doctors, if not better, is at least worthy for being different.[5] Others, including a few physicians, deplore or mock history produced by doctors—a key descriptor used in this debate is "amateur."[6] Both these opinions, whether they praise or denigrate doctor-written history, approach the issue as if doctor-written history is a genre. Thereby, they *create* a genre, and each new member of the species is made to fit, for better or for worse.

STORY: If We Say That It Looks Like a Duck,
Quacks Like a Duck, . . .

In 1988, when I applied for the position I hold now, the search committee consisted of two historians and five doctors. The historians insisted on a private interview and told me that, "as a physician," I would be unable to write good history because I could not be "objective" about my own discipline. That and their other complaints about me ("intellectual history," French degree, female) led them to rank me last of the seven candidates. When I was offered the job and wished to establish a cross-appointment in history, I went to the head of that department. Acting on advice from the colleagues who had interviewed me, he replied: "We viewed our entire role in this matter to help the medical school to avoid making the mistake of appointing a rank amateur. Since they failed to heed our advice, we wash our hands of the entire matter."

Much later I learned, to my immense chagrin, that the more numerous doctors on the committee had passed a motion stating that an M.D. degree was essential to the teaching of medical history. Their strategy probably gave me my job. But it also

A Hippocratic Triangle   439

helped to explain why I would be shut out of the history department for seven biblical years. When a genre is widely accepted as an entity, it is difficult to see that it might actually be a granfalloon.[7]

Clinician-written history is not a genre. I sought evidence by constructing a bibliography of good history written by clinicians. Some critics might consider that I have "cheated" by attending only to good history, but so much of what gets labeled as "doctor-written history" is not history at all.[8] Some clinicians can and do write good history, and those are the ones I describe.

I approached the bibliography in two ways: first by author, then by topic. I constructed a list of clinician-historian authors using three criteria: they are clinicians; they have published books with academic presses or papers in scholarly journals devoted to history of medicine or science in English, French, or Spanish;[9] and I have met them personally.

The third criterion was easy; I used it to define the time frame, and to increase confidence in the credentials of the authors. It meant excluding some big names (e.g., Singer, Sigerist, Canguilhem, Rosen), but it also meant that I did not need to look for lesser lights to more fully represent their periods.

The second criterion, publishing in history journals, was delicate. A historical article in a medical journal may be "good history," but one cannot assume it to have been peer-reviewed by historians. I decided to use this narrower criterion to avoid disputes over whether or not the clinicians' history could be considered "good." If it passed peer review for a history journal, at least some historians had found it acceptable. As a result, the list is necessarily a subset of all the names I could have used. I excluded fifteen or twenty clinician-historians who have never published in history journals, despite having excellent work appear in leading medical journals. Some of the excluded are good friends and valued colleagues, and I would not like them to think that I do not admire their work. For this reason, I do not provide the list of names.

The first criterion, however, posed a more difficult problem. I wondered how to handle the numerous categories of clinician-historian and toyed with the idea of subsets. Should all who hold an M.D. be lumped together in one group? What if the people with M.D.s had never practiced—would they still be considered clinicians? Should I use current clinical income as a criterion? What of the people who have only an M.D. degree and yet work exclusively as historians? Should I include those who once practiced with those who still do? Should additional credentials, such as a doctorate or a master's degree in history, be handled as special cases? So many categories! The only solution was to accept the M.D. credential alone

Table 19.1 Numbers and Attributes of Clinician-Historians by Generation

	Total	Men	Women	Ph.D.	M.A.	Nations Represented	Deceased
1	34	31	3	10	2	5	13
2	27	24	3	14	1	5	0
3	18	12	6	13	2	6	0

Note: Generation 1 = retired or deceased; 2 = born ≤ 1950; 3 = born > 1950

because it was a sign that the author could claim, at a minimum, to have had the clinical experience of every newly minted doctor.

I have a list of seventy-nine names of clinician-historians from ten different countries; sixty-six are still living (Table 19.1). Their work spans every period and a wide variety of topics from prehistory to classical antiquity to the recent past, from "biographies" of people, diseases, and institutions to "body history" and to politically motivated social criticism. I grouped them into three "generations": senior (retired or deceased), middle (born up to 1950), and younger (born after 1950). Twelve are women, all of whom are still living. Reflecting trends in both medicine and history, the proportion of women increased from 9 percent, to 11 percent, to 33 percent across the three generations.

The second approach to evidence gathering was to reexamine the bibliography of clinician-written works by topic. I began with subjects that have been addressed in the hundreds of publications by these physician-authors; then I looked for absence or presence of an equivalent treatment by one or more historians who do not hold M.D. degrees. For example, I wondered if clinician-historians concentrate on questions close to their own specialty; for every physician who did, I could find another who did not—and plenty of other historians who had examined the same topic. I wondered if clinicians are generally more interested in the history of disease, or in biography. Some are, some are not, and for every clinician who writes on these topics, I could find a match among historians without clinical training. I wondered if clinicians have a proclivity to focus on the discoveries that inform current practice. Again, some do and some do not, and Ph.D.-historian matches were even easier to find. I wondered if clinicians could be identified by the topics that they might avoid, such as scrutiny of medical failures, medical atrocities, alternative medicines, and collectivities. Clinicians have written on all these topics too, from a wide variety of perspectives—descriptive, analytic, laudatory, and accusatory.

As for the editors' second question—has clinician history changed recently?—because I could find no "hallmark" to identify the genre, the answer, logically, must be "no." The shifts that can be discerned within this group across the years simply reflect the trends in the entire discipline of history writ large, as it moved from a focus on great personages, practices, politics, and change, to one emphasizing numbers, social groups, economics, culture, and continuity. The contributions of the senior generation display philological and philosophical erudition, and a readiness to take on a greater variety and more distant periods (for example, Mirko D. Grmek, Saul Jarcho, Lester S. King, L. J. Rather, Owsei Temkin). Until the mid-twentieth century, medical schools often required Latin and Greek as prerequisites for admission. Those skills once common to members of the medical profession made medicine of the distant past more accessible to would-be historians than it is now; however, representatives in the middle and younger generations have similar erudition and interests. The much-touted "social history" of the last twenty years, with its associated activism and transcultural sensitivities, is not peculiar to the younger generation, when the work of certain senior and middle generation scholars is considered.

In contrast, the proportion of publishing clinicians who hold doctoral degrees in history has increased from approximately 35 percent (ten of thirty-four) in the senior generation, to 50 percent (fourteen of twenty-seven) in the middle generation, to 72 percent (thirteen of eighteen) in the younger generation. The trend is more obvious if men only are considered. Additional credentials in history were held by all but one of the women in the three generations (ten Ph.D.s, one M.A.)—inviting the old feminist observation that women have had to be "better" to be "equal."[10]

This apparent increase over time in history credentials at the doctoral level among clinician-historians may be real or factitious. It could be a product of age and an artifact of my "generations": if a doctor takes up history late in life, pursuit of a degree might be irrelevant; therefore, the senior generation would have fewer degrees. But most senior and middle generation authors on my list began publishing history early in their careers.

Other explanations for the apparent increase in graduate history training can be considered. Degree programs in medical history are relatively new and more numerous than they were in the past, making the option of graduate work both more attractive to clinicians and more available to everyone.[11] Perhaps more people in general (including doctors) are pursuing graduate degrees in medical history than before. Another reason for the apparent rise in graduate degrees may be a gatekeeper effect in publishing through credentialism, which now affects many fields other than history. More submissions than space would mean that journal editors (and their readers) are influenced favorably by the additional credentials or personal familiarity with the authors' instructors; they may be influenced negatively by the stereotyping described above. As a result, it may now be more difficult for a clinician-historian who holds only an M.D. to publish in a history journal (and to meet my second criterion). We do not know how many clinician-historians without Ph.D. degrees apply for and are denied publication in history journals. Nor do we know how many of the clinicians who are excluded from my list for having published in medical journals may have had the same work rejected by history journals. Impressive exceptions whose names *do* appear on the list caution us not to place too much weight on the apparent value of the double doctorate.

Clinician-historians, like all other historians, are first and foremost individuals. Their identities participate in their creations. Those whose books and articles achieve scholarly recognition produce solid work, in which their medical training may have been used. But there is nothing secret about that training; it is described in books. More important are the encounters of ordinary life which pique curiosity, foster tastes, and sharpen skills of perception and inquiry. The beauty of history is the infinite variety of individuals that constitute it, the originality of their questions, and unique ways that they go about answering them. I hold with those who see history as a humanities discipline, a writerly art.

If history is an art created by individuals and not a science, why do some scholars insist on telling us what we should do? Programmatic statements can be intriguing, but less often are they useful, and sometimes they denigrate individual spark. They come in many forms: the exalted "call" for a new kind of history (often invited as an intended honor and awkward to refuse); the essay review that surveys the literature on a topic and finds all contenders wanting; the conclusion of a research paper that otherwise would never end; descriptive lists of our "needs," "failures," or "missed opportunities." Splenic berating of our collective inadequacies provides us with what those individuals perceive to be gaps in the literature—gaps that we can accept as a challenge or food for thought. They are informed by energetic familiarity with an intimidating mass of literature that I can barely manage to skim, let alone criticize, and they are delivered with confident aplomb.

I used to feel inadequate and frightened when I encountered such tirades. My work was far too much fun to be serious, noble, or useful enough to meet these standards, and I did not want the heavy responsibility of filling a "need." Now the demands bore me. I think: "Why don't you go and do that history, if you think it is so important? We can learn by your example, and emulate it, or not." Or I wonder, "If you are already doing that history, then why would you want me to do it too? Are you that lonely? Egotistical? Intolerant?" Musing about why we have not followed one or another avenue may be worthwhile, but I am more interested in learning the why about what has been and is being written. Historians and doctors, alike, are not lemmings to be goaded into doing something simply because someone else says it is a good idea. The best reason for addressing a topic is a passionate and personal desire to answer the question.

Medical Students: As Audience

Many people have written on the value of teaching of history to medical students;[12] others describe methods.[13] The challenge lies in making a presentation of the past mesh with the demands of scientific medical training. Some historians contend that trying to teach history to medical students is a waste of time; better to reach them before training, or after. Their reasons are familiar: students are not interested; faculty provides no curriculum time; medicine is so anchored in actual science that the past seems irrelevant; trying to communicate the subtleties of good scholarship is unsatisfying, even frustrating, because future doctors will never become real historians. These colleagues readily concede that it would be A GOOD THING if some doctors knew at least a little history. But they envisage a long uphill battle, dejection, and ultimate failure in attempts to deliver it. They fear having to compensate for the privilege of "air-time" by somehow compromising their standards in order to keep the sessions palatable.

I disagree on all these points. I like to teach history to medical students because I believe it will make them better doctors—not because of the stories that I may (or may not) select, not because I think it will make them more cultivated or humane, but because of the analogy described above. History draws attention to organized reasoning. Learning a second language always enhances understanding of the first. I see no reason to skimp on content or on scholarship. All doctors will become historians of their own patients.

Medical students may not be historians yet, but they are intelligent, enjoy controversy, and bore easily. They can assimilate a historical problem quickly if the historian is willing to travel at their speed and use visual resources. Medical students have no desire to believe that doctors were always right or that medicine never harmed; nor do they need a final resolution of the issues (especially if both sides are presented well). Of course, it helps to connect the past with the present—but not a scholar among us fails to do that already (see above), if not as clinicians, then as researchers and citizens of the world.

It does not really matter at all what stories are taught to medical students, or who composed them, as long as they are good history with engaging questions, review of the literature, credible evidence, and a robust interpretation that does not seek to hide its own weaknesses. The stories we should tell, then, come from scholarly research—our own, admired work of others, or in answer to the questions that students ask. It is not necessary to follow a chronological sequence or to cover an entire period, or a nation, or a theme. A homeopathic dose of history on a topic that resonates with their current concerns works wonders. Less is more.

At Queen's University, I have been permitted to "infiltrate" the core curriculum with small presentations of history on topics of current study: history of anatomy during anatomy; history of obstetrics in reproductive medicine, and so forth. Each session may have an overriding chronological progression: for example, I show our copy of Vesalius before I talk about "body history." But I make no attempt to be exhaustive; rather, I strive to offer interesting examples of epistemic shifts, both "good" and "bad." I emphasize how historical discoveries have been made and dwell on current research dilemmas. Dignifying its place in the curriculum, history enjoys at least one question on every examination.[14]

At first, I refused to teach history as an option, believing that marginality would diminish respect and effectiveness. But some students asked for more and we now have electives in medical history: Hollywood films, directed research projects, annual field trips, and a course using literature that is taught in conjunction with the law school.[15] It makes for an erratic schedule.

But I am selfish. My audience serves me well in return. Probably without realizing it, they readily slip from the role of listener into the stimulating role of source. Bright, warm, enthusiastic, and usually completely ignorant of history, medical students ask questions for which I have no answer, excellent questions that I would never have found by myself, questions that sometimes become passions of my own. What I choose to teach or investigate, then, is not only what I *like* to do—which, as I said above, is the best reason for approaching a historical topic—but it can also be what my students teach me to appreciate. Solely because of my medical students I have reviewed the politically charged language of cell differentiation; built a problem-based learning module for medical history; investigated aspects of the history of women in medicine; developed lectures on health and human rights; investigated the history of medical tuition fees; written a textbook; and explored the socioeconomic status of medical students.

Most of these projects resulted in some form of public dissemination, including peer-reviewed conference papers or publications. Some continue as long-standing favorite projects. In other words, my audience has had a profound influence on me as a scholar and as a person; it inevitably influences the histories that I write.

Medical Students as Source: How the Audience Constructs the Historian

STORY: Attacked by Doctors

It is 9:30 A.M. on a crisp autumn day in 1996, and I am in a small, dim amphitheatre, wrapping up my history of psychiatry lecture to the second-year class in the first week of their psychiatry block. We are running about five minutes late. Psychiatrist Dr. X arrives to give the next lecture. I wave him in, and he takes a seat. I finish presenting Ewan Cameron's CIA-funded, brainwashing of human subjects in Montreal and move on to discuss the successive editions of the DSM, explaining it as an ingenious solution to the thorny problem of shifting nosology. I use the example of homosexuality. In closing, I point out that psychiatry, more than other disciplines, is open to the relativism of diagnosis and its potential to harm. The lights come up. We agree the class is to have a short break; half the students head to the coffee machine, while the rest mill about.

The psychiatrist—a charismatic fellow with whom I have had pleasant committee contact—stalks to the front with a grim look on his face. "You historians are all the same! You only talk about the bad stuff and the easy stuff! Cameron is such an outrageous exception! And homosexuality—some of my best friends are homosexuals!" [sic]

I dislike confrontations and was gathering up the books, slides, and overheads, hoping to make an escape. So as not to ignore him, I said "Well, it was an example taken right out of the DSM."

He grew louder. "The DSM! The DSM! What does that matter? We don't need that!" Many students were listening now, and others were drifting back. "Why don't you pick something more subtle, something really complicated for a change? Why don't you pick pedophilia? There's a good one for you! Do you want to say it is not a disease? Do you want to say that Paul Bernardo is not sick?" [This query was a reference to a notorious rapist and murderer.] "Bernardo, last I checked, was incarcerated for rape and murder," I retorted, feeling recklessly annoyed. Undaunted, he pressed on: "Why don't you talk about the good stuff? Why don't you talk about phenothiazines?"

At that point, I lost it. Most of the students were standing in stunned silence. I fixed them and blared: "Dr. X thinks that I have been unfair to psychiatry in this lecture. Did I talk about phenothiazines?" They all nodded, yes. "Well then," he said, "Lithium." I raised my eyebrows. They nodded again. "O.K., electroshock.

There's a good one, and I bet you didn't mention that!" They nodded some more. Banging my books and boxes together, I said, "Next time, come for the entire lecture, or don't come at all!"

It was a good exit—but the effect did not last long. On the street two minutes later, I realized I had left my purse, wallet, and keys in the amphitheatre, and had to slink back, cringing along the wall to the place where they lay. I left quickly again.

By noon, the entire school knew what had happened. Students in the class reported what Dr. X had said after I left (it was not flattering). Others in upper years dropped in wondering if I needed to be consoled. A gadfly in the first year proposed our names for an annual faculty debate. The purse notwithstanding, I was generally seen as the victor but was ashamed of having lost my temper with a colleague in front of students. Our clash became legendary. A year later, the next second-year class showed up in force and high spirits, hoping that Dr. X would be in attendance again (he was not). By 2000, Dr. X had left Queen's and the cohort of students concerned had graduated. I looked forward to an end, but somehow the newcomers in the class of 2004 found out. I realize that the story will just hang in the atmosphere until I retire or die.

Every medical historian has had a painful encounter with an irate clinician who rejects a historical interpretation because it is excessively hard on medicine. It may come as a surprise to the nonmedical members of our discipline, that clinician-historians experience these unpleasant run-ins too. We are attacked on both sides. Clinically trained or not, historians report what they believe to be valid interpretations, sometimes painful truths.

Practitioners identify with their predecessors, and they feel defensive. I've been there too. When a patient tells me that it is good to have a woman physician because "only a women can understand," I am irritated on behalf of the kind men who taught me the skills now being praised. When anxious family members tell me how oncologists suppress alternative cures for cancer because they like to make money, I burn with the memory of the young people who died of malignancies while in my care. What would I not have given to have saved them? When I read observations that women physicians are lax feminists, I prickle with defensiveness as I pass it on to my students.

Practitioners complain that historians write unfairly biased distortions of the past that trade on tales of greed, sensationalism, and deliberate harm to helpless patient victims. They say that historians make the subject needlessly boring by speaking in generalities and by omitting the medicine, the people, and the pictures. When clinicians assume that I must agree with these criticisms, I feel defensive once again, but now it is on behalf of the historians. I am often accused of these faults within my own institution, and my students know it.

History is unique to the people who write it, but the individual historian is shaped by her audience and her task. The influence of history on my medical students is little related to what I say. It stems from who I am—or more precisely, who and what *they* invite, allow, or require me to represent. They are curious about the past, prepared to deplore as well as to celebrate; they are optimistic but apprehensive about the future, hoping never to do harm and knowing harm has been done. The person who teaches them history acknowledges the anxiety and validates it, while providing some comfort. This is the ultimate privilege, then, to be *perceived* by my students as a lonely, therapeutic humanist in the midst of hundreds of scientific medics, a solo historian waging a salutary campaign of attempted truth against a menacing and self-congratulatory establishment that they nevertheless admire and seek to join. I could never have constructed that identity all by myself, even if I had tried! Nor would I have imagined it. I am a clinician after all.

When I surrender (or lose) my hospital privileges, I intend to don a white coat every Thursday afternoon and stroll down Stuart Street so that I can feel my stock rise.

NOTES

Epigraph: Hippocrates, "Epidemics I," in *Hippocrates,* W. H. S. Jones, ed. and trans., 20 vols. (London: W. Heinemann; New York: G. P. Putnam's Sons, 1923), vol. 1:147–211, 165 (my emphasis).

1. For more on this process, see Kathryn Montgomery Hunter, *Doctors' Stories: The Narrative Structure of Medical Knowledge* (Princeton: Princeton University Press, 1991).

2. Carlo Ginzburg, "Clues: Roots of an Evidential Paradigm," in Carlo Ginzburg, *Clues, Myths, and the Historical Method,* John and Anne C. Tedeschi, trans. (Baltimore: Johns Hopkins University Press, 1989), 96–125, esp. 105–108). Ginzburg claimed that this process, being qualitative and individualizing, distinguished both history and medicine from science. While I agree with the semiotic analogy, I am not convinced of the distinction he makes between medicine or history and science.

3. Jacalyn Duffin, "Laennec: Entre la pathologie et la clinique" (Thèse de doctorat de 3e cycle, Université de Paris I-Sorbonne, 1985). This thesis was much later revised and published as *To See with a Better Eye: A Life of R. T. H. Laennec* (Princeton: Princeton University Press, 1998).

4. Jacalyn Duffin, *Langstaff: A Nineteenth-Century Medical Life* (Toronto: University of

Toronto Press, 1993) is a social history of an isolated rural practice based on a computer-assisted analysis of forty years of medical daybooks.

5. See, for example, [Leonard G. Wilson], "Medical History without Medicine," *Journal of the History of Medicine and Allied Sciences* 35 (1980): 5–7. William B. Spaulding, "How Can University Presses Publish Canadian Medical History," *Canadian Bulletin of Medical History* 7 (1990): 5–7; Sherwin B. Nuland, "Doctors and Historians," *Journal of the History of Medicine*, 43 (1988): 137–140; Genevieve Miller, "The Fielding H. Garrison Lecture. In Praise of Amateurs: Medical History in America before Garrison," *Bulletin of the History of Medicine* 47 (1973): 586–615.

6. See, for example, Thomas Schlich, "How Gods and Saints Became Transplant Surgeons: The Scientific Article as a Model for the Writing of History," *History of Science* 33 (1995): 311–331. S. E. D. Shortt, "Antiquarians and Amateurs: Reflections on the Writing of Medical History in Canada," in *Medicine in Canadian Society: Historical Perspectives*, S. E. D. Shortt, ed. (Montreal: McGill Queen's University Press, 1981), 1–17; Owsei Temkin, "Who Should Teach the History of Medicine?" in *Education in the History of Medicine*, John B. Blake, ed. (New York: Hafner, 1968), 53–60.

7. Kurt Vonnegut introduced the once-popular term "karrass," defined as a "team" that is organized to do God's work "without ever discovering what they are doing," while his term "granfalloon" was a "false karass," like "the communist Party, the Daughters of the American Revolution, the General Electric company . . . and any nation, anytime, anywhere." Kurt Vonnegut, *Cat's Cradle* (1963; reprint, New York: Dell, 1988), 11, 67.

8. For example, Schlich's engaging article "Gods and Saints" uses hilarious quotations to examine the kind of "history" that is written by surgeons; his purpose is not to mock but to ask "why intelligent and educated people, such as transplant surgeons, engage in producing a sort of history that for many historians looks little short of absurd" (311). Limiting his cohort to surgeons, further limited to transplant surgeons, further limited to transplant surgeons of the pancreas who publish in scientific literature, he found some delightfully bad examples of writing history and concluded that it was not history "in the sense a historian would understand it" (326). Drawing close to Ginzburg's "Clues" (cited above) in finding differences between the scientific method and the doing of history, he places the blame on the scientific model used appropriately by his sample doctors in their medical publishing, and inappropriately in their writing of history. He is right; those examples are not good history, and because they are not history, he did not address the genre of "clinician history" at all.

9. The journals of medical history accepted in this bibliography included but were not confined to the *Boletin de la Sociedad Mexicana de historia y filosofia de la medicina, Bulletin of the History of Medicine, Canadian Bulletin of the History of Medicine, Histoire des sciences médicales, History and Philosophy of the Life Sciences, History of Science, Isis, Journal of the History of Medicine and Allied Sciences, Milbank Memorial Fund Quarterly, Social History of Medicine*, and *Social Science and Medicine*.

10. "Whatever women must do they must do twice as well as men to be thought half as good. Luckily, this is not difficult"; Charlotte Whitton (1896–1975), who in 1951 became the first woman to be elected mayor of Ottawa.

11. In the mid-1980s this trend was sufficiently novel to merit description: Christopher Lawrence, "Graduate Education in the History of Medicine: Great Britain," *Bulletin of the History of Medicine*, 61 (1987): 247–252; Ann G. Carmichael and Ronald L. Numbers, "Grad-

uate Education in the History of Medicine: North America," *Bulletin of the History of Medicine,* 60 (1986): 88–97.

12. *On the Utility of Medical History,* Iago Galston, ed. (New York: International Universities Press, 1957); Saul Jarcho, "Some Observations and Opinions on the Present State of American Medical Historiography," *Journal of the History of Medicine and Allied Sciences,* 44 (1989): 288–290; Howard Markel, "History Matters: Why History Is of Importance to Academic Pediatricians in the 21st Century," *Journal of Pediatrics* 139 (2001): 471–472; Plinio Prioreschi, "Does History of Medicine Teach Useful Lessons," *Perspectives in Biology and Medicine* 35 (1991): 97–104; Plinio Prioreschi, "Physicians, Historians, and the History of Medicine," *Medical Hypotheses* 38 (1992): 97–101; Jack Pressman and Guenter B. Risse, "Is History Relevant to Medical Education Today?" *UCSF History of Health Science Newsletter* 6, no. 1 (1995): 4–5; Guenter B. Risse, "The Role of Medical History in the Education of the Humanist Physician," *Journal of Medical Education* 50 (1975): 458–465; George Rosen, "The Place of History in Medical Education," *Bulletin of the History of Medicine* 22 (1948): 594–627; S. E. D. Shortt, "History in the Medical Curriculum: A Clinical Perspective," *Journal of the American Medical Association* 248 (1982): 79–81.

13. Donald G. Bates, "History of Medicine at McGill," *Canadian Medical Association Journal* 144 (1991): 412–413; Jacques Bernier, Theodore M. Brown, J. T. H. Connor, John K. Crellin, Julian Martin, T. Jock Murray, and Meryn Stuart, "Applied Medical History and the Changing Medical Curriculum" [the Glenerin accord], *ACMC Forum* 30 (1997): 1–2, 18; *Education in the History of Medicine: Report of a Macy Conference,* John B. Blake, ed. (New York: Hafner, 1968); *Teaching the History of Medicine at a Medical Center,* Jerome J. Bylebyl, ed. (Baltimore: Johns Hopkins University Press, 1982); "Correspondence," *Journal of the History of Medicine and Allied Sciences* 45 (1990): 99–100; "Correspondence," *Canadian Bulletin of Medical History* 7 (1990): 121–130; Barron H. Lerner, "From Laennec to Lobotomy: Teaching Medical History at Academic Medical Centers," *American Journal of the Medical Sciences* 319 (2000): 279–284; Erich H. Loewy, "Teaching the History of Medicine to Medical Students," *Journal of Medical Education* 60 (1985): 692–695; "Special Issue: Viewpoints in the Teaching of Medical History," *Clio Medica* 10 (1975): 129–165.

14. On this approach, see Jacalyn Duffin, "Infiltrating the Curriculum: An Integrative Approach to History for Medical Students," *Journal of the Medical Humanities* 16 (1995): 155–174. See also Jacalyn Duffin, *History of Medicine: A Scandalously Short Introduction* (Toronto: University of Toronto Press, 1999; Macmillan, 2000).

15. Mark Weisberg and Jacalyn Duffin "Evoking the Moral Imagination: Using Stories to Teach Ethics and Professionalism to Nursing, Medical, and Law Students," *Journal of the Medical Humanities,* 16 (1995): 247–263.

CHAPTER TWENTY

Medical History for the General Reader

Sherwin B. Nuland

> The story of medicine is vital and inspiring, no matter from what angle you approach it. It is closely interwoven with the story of peoples, of civilizations, and of the human mind. It deals with great men and small men—with philosophers and scientists, with monarchs and ecclesiastics, with scoundrels and humbugs. On the one hand, it springs from folkways, legends, credulity, and superstition; on the other from intelligence, culture, labor, valor, and truth. And always it seems to reflect the character and progress of the people with whom for the time it is lodged—be they reactionary or be they progressive. Whatever else it is, the history of medicine is never dull.
>
> JAMES GREGORY MUMFORD, 1913

There was a time when purchasing a copy of Fielding Garrison's classic, *An Introduction to the History of Medicine*,[1] was among the first steps any fledgling devotee of the field would take as he embarked on his explorations. In those days, such acolytes included far more physicians than embryonic professional historians, of whom there were still very few.

Garrison's was not a book to be read from cover to cover. Most of its owners and borrowers probably used it as I did, which was to look up a thumb-nailed biography in a particular field or era and read the author's entertaining and often idiosyncratic comments on his career and contributions—and not infrequently his personality. It was a spotty way to learn medical history, but the passages had the unique quality of piquing further interest, largely by the way its author was able, in a paragraph or two, to convey the individuality of each historical figure and his own enthusiasm for the topic. In time, the spots would begin to enlarge and finally coalesce, both those gleaned from Garrison and those many additional ones assembled from sources to which his attractive sentences led their readers. It was possible to acquire a workmanlike overview of the panorama of medical history in this way, just by starting with a copy of *An Introduction to the History of Medicine,* a good light, and a comfortable chair. Like so many others did, between the publication in 1913 of the first of its four editions and the gradual decline in the book's popularity after the early 1980s, I spent many an evening hour browsing in my Garrison.

My well-worn copy of the 1929 final edition—which I bought brand-new, half a century after its publication—contains the famous preface familiar to several generations of readers like me, who first approached the field of medical history by using its pages as a guide. It is here that the epigraph to the present essay is to be found,[2] quoted from a reviewer's comments at the time of the book's original appearance in 1913. This particular passage has always appealed to me, for several reasons: it was written by a clinical doctor, Mumford, without formal credentials as a historian; the clinical doctor was a surgeon at the Harvard medical school; the clinical doctor was himself the author of books and articles on the history of medicine meant for the general reader and physicians; and in one of his own earlier works, the clinical doctor had stated that his intent in writing the book was to "show to laymen as well as to physicians something of the meaning of medicine and of the life of its votaries."[3] Except for the name of the school, all of these words might just as well be said of me.

I might also have made a declaration that sounded remarkably like the one in the epigraph by that clinical doctor, James Gregory Mumford. Of course, Mumford's might be a statement just as well made of *any* variety of historiography, and that in itself tells a tale with a lesson. The lesson is that all history is intrinsically colorful, eminently readable, incomparably edifying, and has lessons vastly important for the general public to know—in a word, fascinating; in another word, instructive. Perhaps even more directly, the study of medical history is great fun.

Medical history is knowledge to be shared. In contemplating the spectrum of formats in which it can be presented, one should not be misled into the error of thinking that historiography consists only of the meticulously documented studies of today's academic historians writing for one another, using the specialized language in which their contributions are often couched. Professional journals and meticulously detailed books are not the only place or the only way to disseminate the results of one's research. Examples abound, from other fields related to science. Are Steven Rosenberg's laboratory findings any less true when they appear in *The Transformed Cell*[4] (his popular 1992 memoir of a career spent investigating immunotherapy for cancer) than they are as originally published in *Science* in 1984? Similar comments might be made concerning the work of Candace Pert[5] in neuroscience and of many other investigators who not only elucidate their findings for the general reader but use the medium of the trade press to do so. The outstanding contemporary example of such things is Stephen Hawking.[6] None of these authors has lost stature in the academic community, simply by writing for the public. A popular book does not in itself make of its writer a Deepak Chopra.[7]

Such contributions are being brought forth not only by the researchers themselves but by science writers and other observers of their work. Many a general reader has learned a great deal about biology from the books and articles of Natalie Angier,[8] infectious disease and public health from Laurie Garrett,[9] and general medicine and epidemiology from James LeFanu.[10] That these writers have not themselves done the research they describe does not at all lessen the accuracy of the information they provide. They, the scientists, and the clinicians who write for the public have done something else as well. Among their accomplishments has been to explicate the ways in which clinical and investigative discoveries affect the general society and the individual. In these efforts, they have been interpreters, guides, and, in a sense, philosophers of the disciplines they study, to the great benefit of large numbers of readers. Thus, the light of biomedicine does not remain hidden under the ivory bushel in which academicians seem to feel most comfortable.

The end result of such work is to add to the pool of knowledge to which every man or woman should have access. In other words, the Hawkings and Rosenbergs and Angiers have taken information from its sometimes arcane sources, intellectually edited it, and then brought it in a useful form to the general reader. By presenting their offerings attractively, they have expanded the number of those who may become interested in them. They have even created some aficionados.

And these authors have gone even beyond that: In this era of inadequate science education in secondary schools—at least in the United States—their works have served to draw youth who might otherwise be uninterested into the seductive web of scientific inquiry. In my generation, far (far, far) more young people were drawn to medical school by reading *Microbe Hunters* (1926) and *Arrowsmith* (1925) than by the *New England Journal of Medicine;* far (far, far) more young people were drawn to physics by reading Eve Curie's 1937 biography of her mother[11] than by *Comptes rendu de l'Académie des sciences;* and it can also probably be safely said that far (far, far) more young people were drawn to the study of medical history by reading Sigerist's *The Great Doctors*[12] than by the *Bulletin of the History of Medicine.*

It is not only potential participants who need drawing. Truth be told, medical history units are too often considered tangential to the real work of a medical school, or seen as secondary sections of a university's history department. So inaccurate a perception can only be corrected by fostering an appreciation of the significance of the contributions such units make. But that formidable transformation of sensibility requires the help of a supportive group of colleagues from other sections or departments who comprehend the value of these studies, to their own disciplines and to the life of the institution. What can best bring fellow faculty to such an understanding is not the specialized literature in journals they will never read but the popular writings that bring the subject to life, the very writings with which they may curl up on those pleasure-filled evenings. For reasons that surpass understanding, however, the great majority of academic historians have convinced themselves that to produce such works is to sell out.

Of course, medical historians already have a built-in claque, if only they would make use of it. Granting myself the luxury of quoting my own words, I'll repeat a few sentences from an editorial of more than a dozen years ago, in the *Journal of the History of Medicine and Allied Sciences:* "From wherever else it may come, support for the scholarly study of medical history must ultimately find its strongest source among the doctors. Whether it is academic support from fellow faculty members to strengthen a department, or the financial support that comes from buying the publications of university presses, the medical community is our natural constituency."[13]

There is a certain irony in this statement. The very group of those to whom in 1988 I was exhorting academic readers to appeal is the selfsame group by whom birth was given to the selfsame discipline whose members I was exhorting. The meaning of the complex sentence you have just attempted to read is really quite simple: The field of study we call medical history was begun by clinical physicians, whether we refer to the scattered historical writings of the Hippocratics or the later, more formalized, endeavors of William Osler and Theodor Billroth—and yes, James Gregory Mumford, among many others. In the days when the discipline was becoming professionalized in the hands of the John Fultons and Arturo Castiglionis and Erwin Ackerknechts, it continued to look to the working physicians for intellectual sustenance, contributions to the literature, and support of various sorts. But theirs is a constituency now scattered; much of its partnership has become lost to us.

The reasons are several. First and foremost is the increased emphasis on the science of medical practice and its consequent blurring of the importance of humanism. A corollary of this is the great increase in the number of specialties and specialists—super-specialists, in fact—which has led to concentration on ever smaller aspects of human biology and therefore bred a breed of ever less well rounded practitioners whose interests are ever less worldly and ever more circumscribed.

Then there is the mood of the time—the Zeitgeist, if you will—in which pragmatic concerns are dominant over thoughtful considerations of the past of one's profession or its philosophical aspects. In the competitive rush forward to new heights of attainment, the contemplative person is a rara avis. In the view of the average clinician or biomedical researcher, the past of their technological field began when they entered it, and twenty years later he or she cares to look back no further than twenty years.

But individuals and groups never solve problems by studying factors of *tempora* and *mores*. There is little that they can do to alter such ingredients of the mix. It is only when attention is paid to their own contributions to an undesired situation that the possibility of change may present itself. For the case here being addressed, much can be done in this regard.

The first step is to recognize that the academization of medical history has not been all to the good. It brings with it a panoply of problems. Like all university studies, the discipline has become split into more than a few specialized areas, the number of which seems to increase by the year; many of these are of serious interest only to their participants; social history in various denominations is all the rage, and clinical history is peripheralized; the vast majority of medical historians are primarily just that: historians, bearing a Ph.D. They rarely have any training, knowledge, or interest in the clinic or the laboratory, and many of them do not hesitate to signify their contempt for the literary efforts of the few bedside doctors who essay to do historical research, publish their findings in clinical journals, or—seemingly most irksome of all—are among the few who somehow have their papers accepted for presentation at national meetings. Over the past three decades, the field has become, as the title of an editorial by Leonard Wilson long ago called it, "Medical History without Medicine."[14]

There are several ways to address these self-inflicted wounds, without taking an iota from the great transformation into professionalism that has so elevated the levels of research, discovery, and interpretation. Obviously, one of them is to provide training in historical methods for those clinicians whose interest is so high that they would be willing to end, fracture, or temporarily disrupt a career in order to take advantage of such an opportunity. Those few individuals would reap a gratifying reward, and they might also bring a refreshed clinical orientation to historical inquiry, which has been in short supply for so many years. But the success of such programs presupposes the interest of a sufficient number of practicing doctors and laboratory researchers that the possibility of some significant usefulness is real. Because today's medical historical literature is so academic and has such a social orientation, it is rarely of any deep interest to clinicians or bench scientists. It does not possess the sort of lure that attracts a busy practicing doctor or molecular biologist, for example, toward an awakening to the beguilements of a medical history of the sort described by Dr. James Gregory Mumford a century ago. The way to fashion that lure is by writing popular history. Even the noncontemplative person of a few paragraphs back may be hooked by it, though he or she will rarely bite on anything that has a highly specialized academic flavor. Having discovered the fascination of medicine's past when it is recounted as a vivid saga, they may then join the natural constituency of which I wrote in 1988, the constituency that appreciates the work of medical historians and contributes to its nurturing.

I have discovered, however, that physicians represent only a small fraction of those attracted to tales of medical discovery and the lives of its participants. As Fielding Garrison himself wrote, in the first edition of his book, "The history of medicine is, in fact, the history of humanity itself, with its ups and downs, its brave aspirations after truth and finality, its pathetic failures."[15] Who can resist stories that tell of such things? Who can resist them, that is, when the clear stream of narrative has been allowed to course unhindered and undiverted by the necessary rocks and eddies of scholarly detail that impede the flow of an academic essay? Historians have wonderful tales to tell, and very few of them are known by the general reader. Far more people than can be imagined by most professors and graduate students are captivated—or would be if they were but made acquainted with it in an engrossing way—by our portion of "the history of humanity itself."

All historians know that, ultimately, they are telling a story. Of late, the bedside doctors have come to understand such things as well. The recent emphasis on narrative in clinical writing is only an expression of the realization that unfolding a story of illness as seen through the experience of the patient holds readers' interest and expands their imaginations. It also draws them to read similar stories, once it is realized that they do not need to be deathly dull. Certainly, being "interwoven with the story of peoples, of civilizations, and of the human mind," the history of medicine is the unfolding of a saga that speaks to the heart of every person who comes into contact with it. Whether focused on science or society, medical history is the Bildungsroman of the art of healing. It tells of people's evolving understanding of their bodies, but it tells also of the development of cultures, nationalities, and individuals. It tells of ideas, as they came to be formulated by innovators of genius and those of a more pedestrian sort; it tells of conflicts, intrigues, wrongheadedness, blind alleys, and bravery; it tells of the human condition. Both before and beyond its virtues as an academic discipline, it is a great story.

Of course, telling that story to the general reader is not everyone's cup of tea. Whether by disinclination or disability, many scholars of medical history will not find a comfortable place in the vanguard of the popularizers. In fact, that very word, popularizer, is anathema to some. The French call the process *vulgarization,* and that sounds even worse, though its meaning is actually not at all like its sound to an Anglophone sensibility. But it has been my experience since trying my own hand at it, that within the hearts of far more scholars than I had previously imagined, there is to be found the soul of a man or woman who yearns for a far wider audience than the constricted one to which they have been addressing themselves. I'm sure motives vary widely, but it hardly makes a difference. Whatever the reason a medical historian takes up the pen in the cause of popularization, the result is sure to be a widening interest in our field. That can only benefit everyone, professor and public alike—and in the long run gain support for our more academic endeavors.

Perhaps some comments on my own experience are appropriate at this juncture. I wrote my first historical essay for my fellow physicians, in 1976,[16] and it appeared in a clinical journal. The process went into high gear in the early 1980s, when I published some fifteen such papers over a period of about three years. Up to this point, they were all biographical in nature. After all, we do take seriously the oft-quoted aphorisms of Carlyle and Emerson, who assured us that not only was biography not a mere refuge for those unqualified for more serious analysis, but it is in fact the very marrow of history. Furthermore, I have the word of our very own Sir William himself, who was convinced that though a recounting of the events may interest readers, the real reason we study history is because we seek what he called "the silent influence of character on character."[17] Any experienced popularizer would agree.

In 1988, a trade press published my first book for the general reader.[18] Using as a structural basis the biographies of fourteen physicians, I traced the course of medical history from the Hippocratic period until the second half of the twentieth century. I had expected a limited audience and was astonished to find myself the object of extensive reviews in the lay press and even a small notoriety as the author of that month's Alternate Selection of the Book of the Month Club, in the days when that group chose for its list only one book and an alternate. One after another, invitations to speak came pouring in. They originated from physicians' organizations, medical schools, and the public. Mirabile dictu, several were even from groups whose membership was composed of professional and amateur medical historians. As tenuous as it may have been before, my associations with the history community were actually strengthened. Always having been identified primarily as a surgeon, I was gratified to find that this book written for the laity caused my scholarly colleagues to consider me one of their own far more than they had in the past. At the 1988 meeting of the American Association for the History of Medicine, more than a few approached me for suggestions to help them with their quest to do what I had done. So much for the fear of selling out.

Whether my experience can be transposed to what might be encountered by a man or woman whose full-time career has been as a scholar of medical history—festooned with a long list of dense publications and a Ph.D.—is impossible to know. But I would guess that were such a person to choose a path similar to mine, he or she would not lose a bit of academic prestige. In the field of general history, I can point to such distinguished historians in my own institution as Peter Gay, Paul Kennedy and Jonathan Spence, whose trade publications have only enhanced their authority among colleagues.[19]

My 1988 book has been reprinted again and again. Whatever else it may have contributed to the advancement of medical history studies, it continues to be used in a number of courses for undergraduates and medical students. It has not had the effect of an *Arrowsmith* or a *Microbe Hunters*, but college and medical students continue to send letters, saying that it influenced their choice of a career. I have heard from many ordinary people telling me how fascinating they had discovered medical history to be. They want more.

The book did not make me rich by any means, but after fourteen years I still receive a royalty check every six months. And this brings up the question of money. I have never understood why my academic colleagues profess an aversion to adding something, no matter how small, to their income. If pecuniary distaste is so profound a factor in decision-making about going public, or if faculty members or a dean are concerned lest pocketing the profits distract from more academic pursuits, monetary returns can always be assigned to one's department or used for charitable purposes. I suppose there is some possibility that one or two authors will be made wealthy by some future book of medical history, but that seems unlikely. Financial considerations, pro and con, should not be a factor in deciding whether to be a popularizer.

Since that 1988 publication, I have written four more books for the general reader. Though none has been restricted to history, I have woven it into the narratives in various ways. Almost always, the critical reviews and the letters sent by readers have contained comments alluding to the interest to be found in passages or chapters that concern the history of medicine. This in itself, is a lesson for those not yet convinced of the inherent fascination of stories telling of mankind's millennia of success and failure in understanding the human body.

Again and again in this essay, I have used the term "trade press." This has been done to make my meaning clear: Books written for the general reader should not be submitted to university presses. As draconian a statement as that would seem, I have over the years come to certain conclusions that support it, and which I am convinced are valid. The typical university press functions under financial constraints that inhibit its ability to compete in the making and sale of publications. Its volumes are priced at a figure substantially higher than an equivalent trade book; its choice of editors is limited; its publicity budget is small; its support staff is sufficiently underpaid that the most qualified personnel are not attracted to it. In short, a trade press will with rare exceptions do better by a book and its author. In those two areas so crucial to sales as jacket price and publicizing, there is no contest.

Everyone wants his or her book to be read by as many people as possible. The best way to bring the products of a writer's dedication and talent to the largest possible audience is to have it published by a trade press. Misplaced loyalty to the concept of the intellectual and marketplace purity of the university press has stood in the way of too many authors who have a message that deserves far wider circulation than it has received. Observation has taught me that the business practices of the publisher on your campus are no better or worse than those originating from a multinational conglomerate based in a New York skyscraper.

I have been keeping company with medical historians for the better part of thirty years. If I know nothing else about them, I have learned that, with only rare exceptions, they are fascinated with their work and love to talk about it; they love to share their knowledge with colleagues, and in fact with anyone else who will sit still long enough to listen. Also, with only rare exceptions, each of them believes himself to be in the midst of uncovering knowledge and making interpretations that have significance far beyond the narrow confines of the area he has chosen to explore, and very often beyond the narrow confines of the academy. Many of them are absolutely correct in these assumptions. There is a wide audience that would agree, were they only given intelligible access to the burgeoning discoveries being made and the burgeoning philosophies being formulated.

Writing for the public is not a monologue. It is a dialogue that may add to a scholar's understanding of the work that he has previously presented only to his scholarly colleagues. Readers respond, and their letters not infrequently illuminate darkened corners or even darknesses not previously recognized. Writing for the public is a huge seminar, in which many are enlightened and a few make enlightening contributions that can add immeasurably to the meaning of the studies being discussed. In this, it is hardly a one-way street.

For such reasons, and for others that I have outlined in previous paragraphs, my partnership with the general reader has brought an abundance of rewards. It is a different way of teaching, and it is a different way of learning. Come on in, the water's fine.

NOTES

1. Fielding H. Garrison, *An Introduction to the History of Medicine: With Medical Chronology, Suggestions for Study, and Bibliographic Data* (Philadelphia: W. B. Saunders Company, 1929).

2. Ibid., 10.

3. James G. Mumford, *A Narrative of Medicine in America* (Philadelphia: J. B. Lippincott Company, 1903).

4. Steven A. Rosenberg, *The Transformed Cell: Unlocking the Mysteries of Cancer* (New York: G. P. Putnam's Sons, 1992).

5. For example, see Candace Pert, *Molecules of Emotion: Why You Feel the Way You Feel* (New York: Scribner, 1997).

6. For example, see Stephen Hawking, *A Brief History of Time: From Big Bang to Black Holes* (New York: Bantam Books, 1988).

7. For example, see Deepak Chopra, *Quantum Healing: Exploring the Frontiers of Mind/Body Medicine* (New York: Bantam Books, 1989).

8. For example, see Natalie Angier, *The Beauty of the Beastly: New Views of the Nature of Life* (New York: Houghton Mifflin, 1995).

9. For example, see Laurie Garrett, *The Coming Plague: Newly Emerging Diseases in a World out of Balance* (New York: Farrarr, Straus & Giroux, 1994).

10. For example, see James Le Fanu, *The Rise and Fall of Modern Medicine* (London: Little, Brown, 1999).

11. Eve Curie, *Madame Curie* (Garden City, N.Y.: Doubleday & Doran, 1937).

12. Henry Sigerist, *The Great Doctors: A Biographical History of Medicine* (Garden City, N.Y.: Doubleday Anchor Books, 1958).

13. Sherwin B. Nuland, "Doctors and Historians," *Journal of the History of Medicine and Allied Sciences* 43 (1988): 137–140.

14. [Leonard G. Wilson], "Medical History without Medicine," *Journal of the History of Medicine and Allied Sciences* 35 (1980): 5–7.

15. Fielding H. Garrison, *An Introduction to the History of Medicine,* 1st ed. (Philadelphia: W. B. Saunders, 1913), 10.

16. Sherwin B. Nuland, "Astley Cooper of Guy's Hospital," *Connecticut Medicine* 40 (1976): 190–192.

17. William Osler, "Books and Men," *Aequanimitas, with Other Addresses to Medical Students, Nurses, and Practitioners of Medicine,* 3d ed. (Philadelphia: P. Blakiston's Sons, 1932), 213.

18. Sherwin B. Nuland, *Doctors: The Biography of Medicine* (New York: Knopf, 1988).

19. For example, see Peter Gay, *The Cultivation of Hatred* (New York: Norton, 1993).

CHAPTER TWENTY-ONE

From Analysis to Advocacy
Crossing Boundaries as a Historian of Health Policy

Allan M. Brandt

My doctoral dissertation, on which my first book was based, offered a historical assessment of the considerable social, cultural, and political obstacles to the successful treatment of sexually transmitted diseases. At the conclusion of my defense of the dissertation at Columbia University in 1982, historian David Rothman asked pointedly, "So what would you *do?*" At the time, I was completely nonplussed by the question. What would I do? What difference would it make what I would do? I had just completed narrating and analyzing the complex history of sexually transmitted diseases over more than a century. Certainly, as a historian I need not focus on "what to do." At the time, I considered it something of a non sequitur; my work was dedicated to illuminating the reasons why STDs persisted in the face of effective treatments. Many of these factors were deeply structural, cultural, and persistent, and it wasn't the role of the historian—so distant from the levers of power—to propose policy approaches. But the question has continued, nonetheless, to reverberate throughout my professional work. And Rothman certainly had a point. If in fact I now understood why disease persisted as a result of powerful historical forces, might I have some applicable insights about policy approaches in the present and the future? Hadn't fundamental concerns about contemporary society and policy motivated my work in the first place? In the fall of 1982, in the first years of the AIDS epidemic, could I find some utility in my historical work for the looming here and now?[1]

In the intervening years I have reflected often on this episode and on the relationship between historical inquiry and policy-making. There has been a long-standing engagement between scholarship, politics, and advocacy that is a central element of the emergence of the field of social medicine. The policy process is an unavoidable aspect of studying science and medicine. Perhaps it is the universal aspects of health, disease, and its treatment that have continued to draw recent historians to contemporary policy and advocacy. Ultimately, I believe that many of us sought careers as historians because of our desire to connect the past with the present, our desire to discover approaches to contemporary social problems in a sophisticated recovery of the past. This may not be the only—or even the most important—motivation for historical investigation, but it certainly is one, especially in the instance of the history of medicine, where profound moral and material questions remain so fundamentally unresolved.[2]

In this chapter, I will identify some of the problems and prospects of defining the policy arena as one of the constituencies for modern historical research and teaching in the history of medicine and public health. I propose that policy-directed history falls roughly into three categories: the historian as author of policy-relevant history; the historian as policy participant/consultant; and the historian as policy advocate. Each of these three approaches raises a set of specific questions and problems, and together they help to define a continuum. After all, virtually all historical work—especially in a field such as history of medicine—possesses some "relevance" for contemporary policy questions. Moving along this gradient, some historians have avoided any overt connections, while others have sought to make them explicit. Finally, as I will illustrate, some recent historians have sought to take historical insights explicitly into the realm of social and political advocacy.

Policy-Relevant History

It is difficult to pick up a book in the field of twentieth-century medical history without finding a blurb on the jacket announcing that this is a book that should be read by all policy-makers. If only all policy-makers had time to read! The central premise of such studies, of course, is *not* to influence public policy directly but rather to explicate and analyze a series of critical historical developments; few historians working in this mode set out to write histories that have clear and immediate implications for the formation of public policy. Rather than histories being constructed to affect public policy, most of these policy-relevant historical studies are *generated by* contemporary policy debates, without any clear or overt desire to influence policy. There seems little doubt that current events are more likely to drive historical study, than historical study to drive current events.[3] Historians inevitably discover a wide range of critically important questions for scholarly attention in contemporary policy contests surrounding medicine and public health. By disposition, we are eager to go back to find roots, antecedents and precedents, continuities and discontinuities.

In fact, the questions that historians take up for serious analytic scrutiny often have their origins in the vicissitudes of modern public life. In this respect, there is a powerful dialectical relationship between historical studies and the conflicts of contemporary society and politics. For example, if the hospital had not become such a fundamentally problematic and contested institution within contemporary American medicine, it seems unlikely that we would have seen the efflorescence of historical scholarship seeking to understand its multiple roles and functions within American culture—and the significance of policy in shaping its particular nature. During the 1980s, as a result, a series of important and sophisticated books opened up the hospital to historical scrutiny. Without question, each of these studies—implicitly or explicitly—raised a wide range of policy-oriented questions about the character of bureaucracy, expertise, and the allocation of medical resources, all questions of continuing importance in contemporary institutions.[4]

Indeed, the persistent problem in the United States of securing access to health services in the last century has generated a significant body of scholarship evaluating past attempts to shape the delivery system. Repeated efforts to develop comprehensive insurance schemes—from World War I to the Clinton administration—all resulted in calamitous failure.[5] Understanding these cyclic failures to legislate comprehensive access to health services offers no simple solution to a century-long policy conundrum. But to not take advantage of what can be learned from these experiences almost certainly leads to future debacles à la the Clinton reforms. Could a competent historian have saved the Clinton health reform? Probably not! but, nonetheless, some of their many political and policy gaffes might have been avoided.[6]

Also, the very data historians use may well have their origins in specific political and/or judicial processes. Historians of the last hundred years (and earlier) often have found themselves immersed in the primary documents produced by the processes of policy-making, both at the institutional and state levels. In this way, we become part of the complex, post-hoc process of policy assessment and evaluation. Whether working in the archives of a local hospital or the National Archives, we get a sharp picture of the internal dynamics of policy and politics as it has shaped issues such as the delivery of care or public health campaigns. The production and availability of primary source materials themselves are one element of the social processes we investigate.

To recognize that contemporary events influence the selection and framing of historical problems is not to suggest that such studies need be presentist and unduly influenced by contemporary values and ideologies. Sophisticated analogical reasoning may help us to discover important historical problems, in addition to considering their relevance for understanding contemporary policy options. Indeed, this very process of evaluating how the present is similar to the past—and how and why it is different—is one of the most basic and important elements of historical reasoning. This approach forces a careful consideration of questions of change, continuity, and causality critical to the historical sensibility.

Further, there is a growing recognition that there is much that is relevant in the past for our consideration in the present, that policy-relevant studies offer the possibility of a *usable* history that goes beyond the simple aphorism of "remembering the past." First, such policy studies often ask how did we get here (or, more typically, how did we get into this mess)? Second, they explore the full range of social, cultural, political, and economic forces influencing the organization of policy and its outcomes. Third, policy-relevant history may provide a sense of options and alternatives, as well as an appreciation for the significance of unanticipated consequences. Such studies draw attention to interests and forces in the policy-making process that typically may be outside public awareness.[7] Finally, such histories fracture a sense of the inevitable and provide an antidote to powerful reductionist tendencies in the world of policy-making.[8]

The AIDS epidemic generated a whole series of critical historical questions about disease and public health that has fundamentally involved issues of public policy. As the dimensions and significance of the AIDS epidemic became overwhelmingly clear in the early to mid-1980s, I found myself returning to Rothman's question. There were, of course, no simple answers to HIV, but many people now looked back to other epidemics as a means of understanding the contemporary dynamics of AIDS, as well as with the more concrete design of historically guiding contemporary policy. Suddenly, the history of civil liberties and the state, public health and prevention, were no longer of "merely" academic interest. Especially since there were no easy answers, history offered one mechanism for "locating" the epidemic: How was it similar to, and different from, other historical episodes of epidemic disease? How might historical approaches be embraced or discarded given contemporary needs, expectations, and conditions?[9] Questions relating to the physician's response in times of epidemic, the nature of experimentation and the regulation of new drugs, and the role of the state in relation to public health all posed important questions for historical research. The answers might hold important policy implications, but the questions came to be understood as historically significant in their own right.[10]

In this sense, I believe there were potential advantages for policy-makers in "seeing" the epidemic through a historical lens. In 1988, I published a brief but explicit assessment of four such lessons. Although these observations in some ways seemed mundane, if not obvious, I wanted to be direct about what a historical perspective might bring to a series of contested policy debates. This article, which appeared in the *American Journal of Public Health,* drew from my earlier research on sexually transmitted disease. First, I called attention to how fear—both rational and irrational—would shape the context of decision-making for the current epidemic. Second, I cautioned that education, as important as it would be, could not be relied on to stem the spread of HIV. Third, I warned that reliance on compulsory measures was also likely to fail. Finally, I suggested that public-policy-makers should not expect a biomedical solution to the epidemic; even effective treatments and vaccines would only offer new and critically difficult policy dilemmas. Although there was no specific policy advice in the paper, I had outlined what I believed could be a more relevant context for the policy debate that would follow. In retrospect, I believe I failed to assess adequately how the reduction of fear and the implementation of effective treatments in the developed world would lead to complacency and routinization as the global epidemic emerged.[11] But although my four lessons were not comprehensive, they did provide an outline of issues that have persistently proven to be obstacles to effective and equitable AIDS policies. These lessons seem fairly obvious in retrospect, yet in the heat of the early HIV battles, many seriously proposed oppressive, mandatory approaches despite considerable historical data to suspect their utility.[12]

My paper was certainly not the only attempt by historians to use the past to comprehend and confront the AIDS crisis. Volumes like the widely cited and influential anthologies edited by Daniel Fox and Elizabeth Fee provided numerous thoughtful and probing essays that considered many aspects of AIDS. The historians generally avoided explicit policy recommendations, in favor of constructing a sophisticated historical context for contemporary policy debates. Not only did these essays speak clearly to policy-makers, but they also influenced historiography within medical history by directing attention to critical questions concerning epidemics and public health.[13]

Policy-relevant histories like these have important implications for the organization and implementation of contemporary policy. For instance, in the early years of the HIV epidemic, historians often sought to demonstrate why traditional approaches to disease control—especially compulsory measures like quarantines—were likely to fail. But finding contemporary debates in the past, or their antecedents, does not necessarily provide guidance about their contemporary resolution. Understanding, for example, how states dealt with healthy carriers of disease in the past does not make clear how they should be dealt with now in the context of AIDS in different societies and cultures. It is also critical to recognize that historians will have limited influence on how such analogues are constructed and used in the process of policy-making.

Of course, a little analogical thinking may be quite a dangerous thing. The history of public policy is replete with the misuse of historical analogues. If history can be effectively *used*, it can also be radically misused as a rationale for policy. In 1974, for example, the national swine flu immunization program was largely based on the assumption that we faced a pathogen similar to that which caused such dramatic mortality in 1918. In the end, the epidemic never materialized, and the vaccine inflicted iatrogenic harms.[14] Such occurrences remind us that the very nature of analogy is the critical evaluation of similarities *as well as* differences, an analytic problem to which historians may bring specific skills. Historian Daniel Fox has reminded historians of medicine numerous times of the dangers of oversimplification. "Order" in historical writing, he has warned, can come "at the sacrifice of complexity."[15] In other words, although many histories may have "policy relevance," how such scholarship will be *used* in the public sphere remains far from clear. Historical complexity might expose potential obstacles to effective policy that a simplified recommendation might not. Understandably, historians rigorously and systematically studying these historical questions have typically been circumspect in attempting to make any policy implication of their work explicit.

Still, historians may offer important insights regarding opportunities for reducing unintended consequences while avoiding explicit policy recommendations. Additionally they may offer an important antidote to oversimplifications of complex issues. History—with its emphasis on context and contingency—may remind the world of politics and policy of the temporality of the status quo.

Within the history of medicine and public health policy, we see significant data to remind us that "it wasn't always this way." As a result, those opportunities for change, and its sources, may be illuminated.

The Historian as Policy Consultant/Participant

It is, of course, one thing to suggest that a carefully drawn history has significance for contemporary public policy and quite another actually to attempt to detail what those implications are. There are occasions when historians are invited into the policy-making process. In such instances, defining the role of historians and their potential contributions has certain problems. Because we recognize the contingent nature of historical analysis and interpretation, being asked by a group of avid policy-makers what the history of a given institution or issue *is,* puts most historians into a state of anxiety. The very presumption that there is *a* particular, explicit, and objective history that simply can be produced at will for policy purposes distorts both historical methodology and analysis. Typically, historians are asked to write brief prefaces in volumes that present a policy proposal that has little or no substantive relationship to their preface. History provides a bit of interesting background and some academic legitimacy to the hardball politics that are to follow. In such contexts, is it possible for the historian to transcend the mode of academic Muzak? Are there any opportunities for sustained historical inquiry?

In recent years, I have participated in two conferences sponsored by the Macy Foundation, an organization whose mandate focuses on the improvement of medical education. In each instance, the organizers were eager to set a firm historical foundation for developing explicit policy-level proposals. The first conference focused attention on strategies for ameliorating the long-standing antagonisms between medicine and public health and integrating population-based knowledge and methods into medical education. The second conference addressed opportunities for improving relationships between doctors and nurses. In both instances, I came to be impressed by the considerable obstacles to attending systematically to historical knowledge while addressing contemporary questions.

The basic premise in both of the papers I wrote for these conferences was one quite familiar to historians: only by carefully addressing deep historical tensions and antipathies could current policy overcome the constraints on building genuine collaborations. The historical antipathies between public health and medicine, for example, centered on a series of dichotomies relating to notions of clinical fidelity to the patient versus commitments to the health of populations. As a result, I argued, no simple synthesis was likely to reconfigure and ameliorate such tensions, which also reflected fundamental political and economic interests. Similarly, in my historical assessment of nurse-physician relationships, I drew attention to the emergence of professional nursing in an age of strict gendered hierarchies in clinical care.[16]

Despite my admonition to develop contemporary approaches taking seriously these powerful historical forces, the discussions that followed my presentations rarely, if ever, addressed these historical and cultural contexts. In this sense, the past truly was but prologue to more concrete approaches to immediate programmatic concerns. Although I continue to believe in the value of sophisticated and complex contextualization of contemporary and persistent problems in medicine, health care delivery, and public health, in these instances my efforts may well have been perceived as only opening old wounds. Historians are all too familiar with past failure. My policy-oriented colleagues, respectful of the past, seemed eager to move ahead.

John Demos, a noted historian of Colonial America, has offered an important observation about how historians might avoid the fate of only providing cultural window-dressing for policy-makers. In an essay about his experience as a member of the Carnegie Commission on Children during the early 1970s, Demos perceptively wrote, "I suspect that history's relevance to policy is most easily appreciated along the route that reverses chronology—that is, from present to past. . . . One needs questions to put to history, and questions arise only as policy itself receives shape and substantive definition." This approach leads to a number of important rationales for the consideration of a historical perspective. As Demos noted, "What elements in the culture facilitate change? What forces block change or inhibit reform? What traditions can be marshaled on the side of policy initiatives; and which serve merely to maintain the status quo?"[17] In these questions, Demos captures some of the analytic perspectives that historians may offer policy-makers. Historians seeking to account for both stability and change—purposeful and inadvertent—center attention on those aspects of policy often peripheral to the worldview of the policy-making process. In this mode, it is not so much that historians are called on to advise or make explicit recommendations; rather, a deeply historical sensibility may alert those with policy-level responsibilities to unintended consequences, legitimate complexities, and the precise timing and contexts of both potentials and obstacles to reform.

Rather than provide policy-makers with a set of historical facts, the historian might therefore bring to the policy process a particular analytic approach, what might be called "historical thinking." This analytic approach would necessarily focus attention on the historical context of policy reform. Take, for example, the ongoing acrimonious policy debate about testing and regulation of new drugs. This conflict, I would argue, may be understood clearly only in its historical context. The HIV epidemic suddenly and dramatically fractured the risk-aversive ethic of human experimentation that had evolved since World War II. It created a constituency that found the protections of the past unduly restrictive in the present and who were *eager* to take experimental risks with unproven therapies. In the midst of the AIDS epidemic, when access to clinical trials would become a bitterly contested question, it is worth considering that only a short time prior the focus on policy discussion had been the protection of research subjects from potentially dangerous protocols. What the regulatory process failed to recognize was that in certain specific contexts individuals might aggressively seek access to experimental drugs even if their safety and efficacy had yet to be proven by orthodox scientific criteria.[18]

Since then, the regulatory process has allowed some drugs onto the market that later were recalled for safety reasons, creating public outcry.[19] Risk therefore can be defined only in a very specific historical and social context. In this respect, an appropriate "margin of safety" for any set of clinical trials cannot be uniformly set. After all, it is a distinctly *historical* question to ask why is the problem framed in this particular way at this particular point in time? How has it differed at other historical moments? For what reasons? The critical question in this light is how the historical context had changed the political calculus of drug regulation. It is perhaps for these reasons that historians have been so uncomfortable with principle-based bioethics where context is often overlooked.

In some cases, particular incidents in the past generate a need for investigation so that policy-makers can discern what happened and why and learn what policy improvements are necessary. Historians, I would suggest, may play a significant role in such instances. In 1994, at the request of the U.S. Secretary of Health and Human Services, the Institute of Medicine (IOM) began an investigation of the contamination of blood and blood products by HIV during the period from 1982 to 1985. A large number of patients came to be infected through contaminated transfusion in these three years after AIDS was first revealed. Further, patients with hemophilia were especially vulnerable to infection since they were dependent on the use of antihemophilic factor (Factor VIII), a blood-clotting medication derived from human plasma. During this period, in which the risks of infection were uncertain, almost one half of the nearly 20,000 hemophiliacs in the United States became HIV infected. These infections marked one of the most serious iatrogenic epidemics in the history of modern medicine, with thousands of deaths. The IOM was charged with assessing the decisions associated with protecting the blood supply and assuring the safety of blood products during this early phase of the HIV epidemic.[20]

The IOM study group had a particularly strong interest in utilizing historical methods and techniques in their investigation. In the summer of 1994, I agreed to be the historian on the committee. The committee was specifically interested in a meticulous reconstruction of the early history of the AIDS epidemic and the process of recognition of blood as a vector. It is for precisely this reason that many policy review committees have turned to historians for advice and consultation regarding archival research, records recovery, narrative reconstruction, and interpretation. The IOM report also produced and secured primary source materials that will be of considerable value to historians in the future, including a useful cache of oral histories.

In a sharp historical perspective (a.k.a. 20/20 hindsight) the failures of the federal government to protect the blood supply and blood products in the early years following the identification of HIV stood in bold relief. By carefully reviewing a wide range of archival materials and interviewing public officials, industry executives, patients, and physicians, we were able to reconstruct a highly articulated narrative of events and actions that helped to account for the tragic vulnerability of blood and its products.

Certainly, the committee was not so naïve as to believe that all risks to vulnerable resources like blood can be eliminated. But the HIV epidemic had manifested the dramatic implications of the failures of federal vigilance. At the core of the IOM study and the HIV blood crisis stood the conundrum of risk and fear so central in many public health policy dilemmas. Officials, activists, patients, and doctors had struggled in the early years of the HIV epidemic to weigh the risks of disease against potential risks to the system of blood collection and allocation. Among the most important recommendations of the committee was the establishment of a Blood Supply Council to continuously monitor and assess potential threats to the blood supply and of no-fault compensation for individuals who—in spite of aggressive efforts to reduce blood-related harms—suffered severely adverse consequences from the utilization of blood products in the treatment of disease. Although limited by our charge to make prospective recommendations, our advocacy of a compensation program made explicit our commitment to those who had been harmed in the course of the HIV epidemic.

Later, critics of the IOM report centered their attention on the limited policy questions that framed the investigation: what regulatory and procedural steps could be taken to limit the contamination of blood or blood products *in the future?*

The failure of the IOM report to identify individual and institutional fault in the past was declared by these critics as the bureaucratic limitation of the policy arena. According to one analysis, the IOM report "led to an exculpatory solution that obfuscated the moral dimensions of suffering." But, as a member of the committee, I was certain that our report was never intended to be "exculpatory." To the contrary, the committee had worked steadfastly to represent the ongoing concerns and needs of the hemophilic community, despite the constraints on our charge and investigation. The general reception of the report within this community was tremendously positive. Indeed, in my own post-hoc assessment of the IOM blood study, I believe that its bureaucratic, rhetorical form, its emphasis on objective and disinterested evaluation of the decision-making process, was ultimately of significant value. Not only did the report help to generate reforms in blood policy, but it affirmed the character of this iatrogenic catastrophe, identifying a series of federal, industrial, and clinical missteps that together led to tragedy. In this sense, there was always a direct line between the heart-breaking testimony of HIV-infected hemophiliacs, blood recipients, and their surviving family and friends and the often spare, but direct prose of the report. Further, litigants now had the report of a blue ribbon, *independent* investigation, detailing egregious lacunae in the nation's ability to protect a community resource and the structured shortcomings of regulatory processes in which powerful commercial interests had become deeply entrenched and influential.[21]

Similar questions arose on the president's Advisory Committee on Human Radiation Experiments (ACHRE) in 1995. This committee, appointed by President Clinton to review abuses in the use of human subjects by the federal government and others in the 1940s and 1950s, centered its attention on a series of critically important historical questions. Historian Susan Lederer, author of a respected volume exploring human experimentation before World War II, was a member of the committee, alongside lawyers and policy-makers. The ACHRE created an "Ethics Oral History Project" to help establish the harms and practices of physicians and researchers at the time. Here again is another example of contemporary concerns and debates generating new historical data and research. The ACHRE was careful to avoid the Scylla and Charybdis of blame and exoneration. On the one hand, its members did not want to assess these human experiments exposing individuals to radiation on the basis of contemporary expectations of research ethics and informed consent (which had subsequently been articulated in law and policy); on the other hand, it would have been equally inappropriate simply to exonerate such research by claiming ethical standards had yet to be articulated and embedded in the research processes. As the members explained: "Moral consistency requires the advisory committee to conclude that, if the use of healthy subjects without consent was understood to be wrong at the time, then the use of patients without consent in nontherapeutic experiments should also have been discerned as wrong at the time, no matter how widespread the practice." Clinton issued an apology to survivors of these researches based on the careful historical assessment offered by the committee in its 906-page report.[22]

Historians working within the public policy process offer a potential to explore more deeply the full set of variables shaping policy definition, development, implementation, and outcomes. Further, historians of science, medicine, and technology may bring a specific body of knowledge and analytic frameworks to this assessment. This type of historical thought might be contrasted with traditional, cost-benefit, politicocentric thinking. I do not expect this perspective to have immediate or wide cache in policy circles. Nor do I believe it likely that historians will obtain unfettered access to the corridors of power. Nevertheless, there is reason to believe that historians will continue to play important but limited roles in certain circumstances. Unlike policy-relevant history, historians participating in the world of policy-making will consistently be asked to articulate clearly the immediate implications of their historical analyses for a set of contemporary dilemmas. There remains an ongoing debate within our profession about the rectitude and implications of such involvement and its impact on the historian's craft.

The Historian as Policy Advocate

These questions are more starkly framed in those instances in which historians have actively crossed the boundary between scholarship and political advocacy. If historians as policy consultants offer new opportunities to bring complexity to debate, as advocates, complexity is often given short shrift. When the divide is crossed, a series of critical tensions and professional dilemmas are revealed. I am not suggesting by any means that the boundary is sacrosanct and should not be crossed; rather, it becomes even more important to articulate the process of and reasons for taking on advocacy roles in our professional identities as historians. There is, of course, a critical difference between advocacy as a citizen and advocacy in which claims are based on our expertise and training. To what degree does our particular advocacy position rest on our historical method and analysis? If I advocate, for example, against compulsory screening for HIV, do I do so as a medical historian or simply as a citizen? Certainly, op-eds and opinion pieces in which we draw on our scholarship and that of others to take a position on a contemporary social or political problem have become increasingly common among historians.[23]

These questions become explicit when examining some of the issues that arose when a group of professional historians submitted an amicus brief in the *Webster* case heard by the Supreme Court in 1989, a process in which I participated. *Webster* tested the constitutionality of a Missouri antiabortion statute. The brief, written by lawyers and a committee of historians, was eventually signed by over 400 professional historians. The fact that the Constitution is to be interpreted in light of "our history and traditions" generated the idea of seeking the perspective of professional historians, as did the solicitor general's assertion that nothing in our history or traditions supports a right to abortion. This assertion was based on a book written by historian James Mohr, who disagreed with their selective—and in his view inaccurate—use of his work.[24]

The historians' brief reached several important conclusions about the history of abortion deemed relevant to the Court's deliberations. First, the brief argued that for much of the nation's history, including the time of the founding of the Constitution, abortion was not illegal. Second, moves to restrict abortion, especially in the mid- to late nineteenth century, arose from a series of specific historical forces: developments in the efforts to define the medical profession; beliefs about women's appropriate social roles; and general attacks on the control of reproduction, especially as they related to broader—often racist and nativist—eugenicist interests. The brief concluded that these particular rationales for restricting abortion clearly would be considered discriminatory and unconstitutional and could not be justified as a basis for restricting abortion now; therefore, the brief concluded, there is no legitimate *historical* tradition for restrictions on a woman's right to choose.

Historians working with legal counsel in preparing the brief soon found that the judicial process left little room for ambiguity, uncertainty, or contingency. The lawyers wanted the best, most fully substantiated historical statement that could be used for the explicit purpose of making the legal case for a woman's right to choose. As Sylvia Law, one of the lawyers who wrote the brief, later observed, "The document was constructed to make an argumentative point rather than to tell the truth." James Mohr, one of the historians involved in writing the brief, expressed certain misgivings about historians moving into the realm of genuine advocacy. He explained, "For a lawyer the best history is that version of a complex story that best serves a predetermined end." To cite but one example, the fact that most late-nineteenth-century feminists supported restrictions on abortion was raised as the brief was being written. While this finding did not necessarily contradict the argument made in the brief, its full exploration was viewed as possibly compromising the position of the brief. Discussion of this issue was left on the cutting room floor.[25]

Some of the restrictive qualities of history as advocacy created concern on the part of historians who signed the brief. And some historians refused to sign, even while proclaiming their political commitment to the pro-choice position. In the aftermath of the brief, a number of historians, including Mohr, suggested the possibility that historians taking on advocacy positions *as historians* might ultimately compromise their legitimacy and authority. The brief, some argued, was partisan and not objective. Some historians stated their preference for historical arguments that stated all sides, celebrated complexity, and avoided overt political claims. History as advocacy, they concluded, may conflict with essential professional ideals, norms, and values.

Nevertheless, while the brief might have made scholarly compromises, many historians still felt it was important to sign. As historian Alice Kessler-Harris, one of the signers explained, "Historians are actors in the debate whether we like it or not . . . so we should be conscious actors."[26] With the history of abortion invoked as justification for *Webster,* the historians felt a need to respond in order to correct this assessment. James Mohr reflected later that while signing the brief was somewhat problematic, not signing it would have been far more so: "If asked whether that document comports more fully with my understanding of the past than the historical arguments mounted on the other side, some of which also cite my work, I would state strongly that it does. Consequently, I had no hesitation signing [Sylvia] Law's statement to the Court."[27]

Both public health professionals and lawyers stressed how even though historians might see the brief as simplified and selective, it added a layer of complexity to the courtroom's understanding of abortion and its legacy. As physician and reproductive-choice advocate Wendy Chavkin explained: "Because of the way the debate about abortion has been framed in the contemporary United States, it is eye-opening for the American public and judiciary to learn that current views about the fetus differ from those of the past." She went on to assert, "The historical perspective offers tools and facts for correcting misconceptions."[28] Sylvia Law discussed how the historians' brief was, in fact, used to counteract the simplified version of abortion history presented by the government. "Thus," she asserted, "the historians' core objective was negative. We contended [in the brief] that history is complex and not determinative. . . . [H]istorians should take pride in having precluded the Court from using history in a shallow and deterministic way."[29] Historical knowledge and interpretation had provided valuable information and perspective to the Court.

Of course, the problem of historical complexity and simplification is not limited to the area of advocacy. Historians are always and inevitably balancing notions of legitimate complexity, generalization, and argument. Sensitive to the problems of audience, for example, textbook authors seek to write history that is "correct" from a scholarly standpoint but accessible to novice students.

Concerns about contemporary advocacy of course are not limited to medical history. Similar issues about the role of advocacy among historians were raised during the Clinton impeachment in 1998. Some 400-plus historians signed an open letter warning that Clinton's impeachment would critically damage the modern presidency, leaving it "at the mercy as never before of the caprices of any Congress."[30] While the substance of the letter is itself of considerable interest, tellingly it generated a debate among historians about advocacy and professional responsibility. Eric Foner, a Columbia University professor and expert on the impeachment of Andrew Jackson, declined to sign: "I did not in any way support the impeachment of Clinton, but I did not think history tells us that Clinton ought not to be impeached, or that there is a single meaning of the impeachment clause in the Constitution." Pauline Maier, who did sign, later regretted the decision. "I agreed with the constitutional position that it expressed, but we went far beyond the area where we had professional expertise." Many of the historians who signed lacked what she believed was the necessary expertise to consider what the framers of the Constitution meant. For these historians, whatever their contemporary political concerns and dispositions, their historical knowledge did not speak clearly to the contemporary contest.[31]

There is, unfortunately, little room for ambiguity and nuance in the world of advocacy. Given these concerns—the loss of ambiguity, the loss of intellectual autonomy, the potential for the loss of professional authority and legitimacy— why take on the role of historian advocate? This is a question that has preoccupied me at those times I have sought to straddle the boundary between scholarship and its potential influence in public health policy. Expert witness work constitutes an important aspect of advocacy in that historians—as experts—come to participate in adversarial processes with sometimes enormous implications for health policy.

If expert testimony may limit the historical complexity, nuance, and ambiguity that we prize in our academic work, why become a participant? There are really two answers. The first centers on compensation; many historians today offer their professional research and analytic skills to litigants. In recent years, historians have become frequent participants in litigation as expert witnesses. Expert testimony may be well-compensated, and there are now a number of agencies offering historical services in this area. In such instances, the historian-expert may have no sense of commitment to one of the litigants but is simply deploying her or his historical methods and skills for a paying client.[32] The second reason comes to bear in cases where our scholarship aligns with a litigant whose cause we share. Of course, at times, it may be difficult to discern these two rationales. But certainly the potential to utilize our scholarship on behalf of a plaintiff constitutes an important and increasingly significant aspect of public history and historical advocacy.

My work on the history of the cigarette has brought these issues into sharp relief. For some time I have been working on a social history of cigarette smoking in the United States. Among several goals for the work was to attempt to demonstrate that one could only understand the contemporary public health dilemmas associated with smoking and its harm if one were to locate smoking as a behavior—a biopsychosocial phenomenon—in a deep cultural and historical context. But I was also eager to define the particular social policy questions that the history of the behavior explicitly engages. My hope has been that understanding the history of cigarette smoking and its health risks might offer insights into strategies for reducing the use of tobacco and its tremendous impact on morbidity and mortality.

Over the last decade, as I continued to work on this project, legal suits against the tobacco industry began to be brought—first a slow trickle and then a more continuous flow. Early on, I was approached by lawyers working both sides of this legal battle about my availability as an expert witness in tobacco litigation. In these cases, smokers or groups of smokers alleged that the tobacco industry should be held liable for the serious harms to health, especially lung cancer, which these individuals have endured. In recent years, such cases have been augmented by suits brought by the states and insurers to recover health care costs incurred by smokers. Although my research clearly indicated the malfeasance of the industry in obscuring and denying the harms of smoking, as well their aggressive efforts to promote their product in the face of these risks, I was skeptical about participating in these growing legal battles. I decided that the courtroom was unlikely to be an effective forum for me to present my work. Obviously, the courtroom is not the ideal place for historians to present original scholarly research and findings. The court has a specific set of questions it seeks to adjudicate and resolve; historical data and analysis may well be brought to bear on these questions, but they are utilized in an explicitly adversarial context.

As a graduate student, I had carefully followed the bruising, sometimes vicious conflict that had arisen when historian Rosalind Rosenberg had agreed to serve as an expert witness on behalf of Sears Roebuck in a noted sex discrimination case brought by the Federal Equal Opportunity Commission. In her testimony, Rosenberg contended that women had typically refrained from taking on higher-paying commissioned jobs. Women's and labor historian Alice Kessler-Harris testified on behalf of the government. Sears won the case, but Rosenberg found herself facing recrimination from the women's history community.

In a subsequent interview, Rosenberg explained, "A lot of people feel I have abandoned feminism for the sake of scholarship. Others believe I've abandoned both feminism and scholarship. I don't feel I've done either. I did what I thought was my intellectual duty." Later she noted, "The most depressing part of all this is that no one seems to believe me when I say I decided to testify because it seemed like the right thing to do." It might be an academic's duty to report one's findings and interpretations in new scholarship, but was it Rosenberg's duty to work on behalf of Sears? Strikingly, Rosenberg seemed unaware of the boundary she had crossed.[33]

Feminists did not necessarily object to her scholarship, but they did question her judgment in participating in a sex discrimination case. Historian Kathryn Kish Sklar described the common reaction among feminist historians: "The feeling people have now is just one of betrayal. I find it tremendously disturbing to think that our special knowledge as scholars might be used to reverse years of struggle by a social movement that improved the lives of countless women—us among them." At the annual meeting of the American Historical Association in 1985, the Coordinating Committee of Women in the Historical Profession and the Conference Group on Women's History passed a series of resolutions in response to the case. In the third resolution, they stated, "We believe as feminist scholars we have a responsibility," the third resolution read, "not to allow our scholarship to be used against the interests of women struggling for equity in our society."[34]

The Rosenberg/Kessler-Harris legal combat galvanized professional attention. The issues at stake in the Rosenberg/Kessler-Harris face-off were substantive and important, and the judicial process had proven a poor venue for them to be effectively debated. Moreover, the sociopolitics of the case overwhelmed any substantive historiographic questions.[35]

The Sears case drew attention among historians to the dilemmas of courtroom work. On the one hand, the case demonstrated the contemporary policy significance of historical understanding and scholarship. On the other hand, it made historical scholarship vulnerable to charges of bias and ideology. Moreover, it seemed more than apparent that the subtlety, nuance, and complexity that we prize as historians had been lost in the courtroom and the vituperative professional battles that followed. In the historical aftermath of the Sears case, I had come to the conclusion that serving as an expert witness had little appeal.

Nonetheless, I watched the emergence of a new wave of tobacco litigation with considerable interest. These cases directed attention to issues at the center of my own research and writing: the character of changing knowledge and epistemology in modern medicine and science, as well as assessments of risk and responsibility for disease. These questions had been critical in my earlier work on STDs and HIV, and I was now eager to apply them to the epidemic of disease associated with smoking. For my work, the cases marked a sensitive indicator in the historical evaluation of responsibility for harm. Notably, too, much of my recent work on the history of tobacco and cigarette smoking rests on industry documents only made available as a result of antitobacco litigation. Prior to this litigation, these critically important historical materials were unavailable for research. In this instance, the litigation process has brought to light an unprecedented volume of internal materials.

During the 1990s, the policy significance of this litigation rose substantially. Given the failures to enact significant legislation in Congress to regulate and control the use of tobacco, and the failure of the Food and Drug Administration to gain regulatory authority over cigarettes, the courts proved to be among the most potent potential devices for public health.[36] By the mid-1990s, the release through legal discovery of impressive incriminating documents from the tobacco industry's internal records made these cases increasingly viable, demonstrating that the companies well understood the harms of their product and took decided steps to deceive the public. In my earlier assessments, I had assumed that cultural propensities for victim-blaming in the United States would lead to the legal vindication of the industry. But with evidence now mounting that the industry *intended* to promote cigarettes (often to children) while *knowing* of their well-documented carcinogenic and addictive qualities, the potential for industry culpability and liability grew. The importance of the cases—and their focus on historical questions of medical and scientific knowledge—began to lead me to rethink my earlier qualms about testimony. Further, my clear historical assessment that the industry had capriciously abandoned any regard for public well-being, publicly denying the harmfulness of their product through relentless promotion and deceptive public relations, spurred me to reconsider.

Early in 2002, I agreed to serve as an expert rebuttal witness for the Department of Justice in their historic case against the tobacco industry. Although my concerns about taking on this role remained, having already written and published on cigarettes, I now concluded that this would certainly not be the last or exclusive forum to describe my historical findings and analysis. Further, I was motivated to agree by the fact that the tobacco industry had secured several noted historians of medicine who were prepared to serve as experts for the defense. These historians had fashioned arguments that were broadly consistent with the traditional industry defense: that considerable scientific uncertainty about the harms of smoking persisted late into the twentieth century and, moreover, that the public was well aware that smoking *might* be harmful and should have taken necessary precautions. My own reading of the historical evidence on the history of causality, as well as my research in the thousands of industry documents, sharply contradicted these claims. I now concluded—despite my misgivings about expert testimony—that such industry claims, now offered by well-regarded historians, must be powerfully and effectively rebutted. Further, I had become impressed that the courts now offered perhaps the most effective approach to the regulation of this deadly product. So I embarked on this new advocacy role.

Expert testimony is, without question, a clear form of advocacy. Experts are sought precisely because it is understood that their expertise (professional standing and knowledge) will be used on behalf of a particular client's position. It thus seems disingenuous to me to defend expert witness testimony as an expression of academic freedom. Certainly a historian has a right to testify on either side of a litigation, but no one is compelled to offer arguments on behalf of a particular litigant, be they Sears Roebuck, the Department of Justice, or Philip Morris. Historians in the court have consciously aligned their work with the particular interests of their clients; the historian's testimony is subsumed to these goals.[37]

For a witness to purport to simply be offering statements of historical "truth" is to deny the context in which these historical arguments are made. In an interview about his testimony in an individual smoker's suit against Philip Morris in Portland, Oregon, historian Kenneth Ludmerer asserted, "The testimony was purely historical. When did we first become suspicious? When did a consensus arrive? It was purely historical testimony." He went on to explain, "I'm firmly on record that tobacco is a great threat to public health. I am not an apologist for the tobacco industry."[38] No one is required to become an expert witness; to claim that one is doing so simply to serve historical truth would be to overlook the fact that historical truths are best served in our scholarly writings where we exercise considerable autonomy in the exercise of nuance, complexity, interpretation, and argument. In the courtroom, experts work on behalf of advocates; in some instances there may well be an alignment of interest. I am eager for my work as an expert witness to serve the interests of regulating and controlling tobacco use. Historians working for the tobacco industry must recognize that their work serves the interests of the industry rather than the traditional academic goals of inquiry and understanding.

I am convinced that I can serve as an advocate in this context and that my work will still be persuasive within the realm of scholarship, that the many substantive questions that I continue to have about the history of medicine and public health will not be shaped or altered by my work for the government. There are, of course, risks in taking on the role of advocate. My own sense is that it is best to be open and explicit about this work and its implications for professional identity. So far—and I have yet to be deposed—I am pleased to be working on behalf of the government's case in this potentially landmark litigation.

Although traditional notions of historical objectivity and truth have come under intense discussion in the last decades, claims of expertise, authority, and objectivity are often viewed in the courtroom in the light of the powerful interests at stake. When historians take on the role of "hired guns" can the rectitude of their claims be sustained? I believe that ultimately they can, especially when augmented and balanced by our work presented in other, more neutral forums. In this sense, as historians, we would do well to give more critical thought to the production and character of historical knowledge (much as we do for science and medicine). We need to assess more carefully the context of this knowledge and how it is interpreted and utilized in politics and policy-making.

It is, of course, not just what a historian argues, but where—in what context—it is argued that matters. In our scholarly work there is no greater ambition than the truth, although we know new interpretations and revisions emerge with every generation. The world of the academy is structured to provide unfettered access of inquiry and argument. The courts, while committed to the truth, are structured as adversarial and procedural devices for adjudicating competing claims. As a result, historical findings and interpretations are fundamentally shaped by this context in which they are presented.[39]

A critical question looming across issues of advocacy is its impact on the authority and objectivity of historical analysis. When historians cross the boundary into the realm of explicit policy and political advocacy it is sometimes viewed as compromising the integrity of their work. According to some observers, history in the service of advocacy becomes simply advocacy. Of course, it is possible that historians may forfeit some professional legitimacy and authority when they act as advocates. But in recent decades it has become clear that historical authority rests less on traditional notions of facticity and objectivity than was true in earlier historiographies. There has been a growing recognition of the inevitable influence of values and context in shaping historical work itself. Rather than weakening or discrediting historical argument, it merely becomes one more aspect of our assessment of credibility and effectiveness of interpretation.

The Historian in the Public Realm

As historian James Kloppenberg has recently written: "History always involves interpretation; it never reveals simple truths." I am convinced that historians have something important to say in the world of advocacy and politics. The knowledge and methods of inquiry and interpretation may well have an important impact in the adjudication of political and social conflicts. To insist on insularity and objectivity at the cost of such social engagement is to deny among our most important constituencies and significant functions as historians.[40]

Despite the potential dangers of advocacy-oriented history, few of us, I think, would be willing for historians to forsake the ideal that historical work may speak clearly to a set of contemporary—and sometimes contentious—political issues. There are historical moments in which historians necessarily will be asked—or feel compelled—to develop explicit and overtly political arguments about the contemporary implications of their substantive work. To say simply that the historical record speaks for itself—especially given the potential for our work to be misread or misunderstood in the public realm—and to let our work be used as others see fit, is, I think, to abdicate a critically important professional responsibility, what historian Michael Grossberg termed "the [historians'] responsibilities of their craft to inform the present."[41] Moreover, these occasions when we may be engaged in political action serve as an important reminder that historical work has genuine moral and political currency. It helps us to link the past to the present, the personal to the professional.[42]

In instances when I become involved in policy, it has seemed critical to me to carefully and critically align my historical work with my "interests" in policy and politics. In such contexts, historians are using their methods, skills, and professional legitimacy on behalf of an advocacy position. Depending on our particular sensibilities and dispositions, our relative eagerness or reticence to engage the powerful values and ideologies implicit in the construction of public policy, historians in the years ahead will nonetheless recognize new and critical constituencies for our work outside the academy. While the problems of crossing this border are both clear and significant, I would suggest that there are potential gains for our scholarship, our discipline, as well as the tenor of political debate. In this respect, policy-related history is a chance worth taking.

Of course, the option of historical work is not to see academic scholarship and public history as a dichotomy but rather as a continuum. There will be times when moving into an advocacy role makes good political and personal sense. There will also be times when it is crucial that our work conform to the "internal" professional dictates of academic presentation and discourse. But these realms are neither separate nor distinct. In the end, we must work to assure that they are interrelated in critical and self-conscious ways. Historians—at their best—have often and effectively played this crucial role in contemporary societies and culture. For me, these opportunities for engagement continue to be essential to my motivation and identity as a historian.[43]

Echoing philosopher George Santayana's famous aphorism, many have asked are we "condemned to repeat the past"? Can history intelligently and effectively guide contemporary efforts in public policy and political advocacy? The answer to these questions remains far from certain. Nonetheless, it seems more than evident that historians do possess some professional moral responsibility to produce work of value to contemporary questions *and* to attempt to assure that this work is used in thoughtful and sophisticated ways. This commitment moves along a gradient from relevance to advocacy, but inevitably we find our work along this line. In its connection to such important, ongoing, and contemporary questions of health, disease, and social justice we may well find important new ways to sustain our commitments to our scholarship and historical investigation.

NOTES

1. Allan M. Brandt, *No Magic Bullet: A Social History of Venereal Disease in the United States since 1880,* expanded ed. (New York: Oxford University Press, 1987).

2. Elizabeth Fee and Theodore M. Brown, "Intellectual Legacy and Political Quest: The Shaping of a Historical Ambition," in *Making Medical History: The Life and Times of Henry E. Sigerist,* Elizabeth Fee and Theodore M. Brown, eds. (Baltimore: Johns Hopkins University Press, 1997), 179–193.

3. Examples include David J. Rothman, *Beginnings Count: The Technological Imperative in American Health Care* (New York: Oxford University Press, 1997); Daniel M. Fox, *Power and Illness: The Failure and Future of American Health Policy* (Berkeley: University of California Press, 1993); and Keith Wailoo, *Dying in the City of the Blues: Sickle Cell Anemia and the Politics of Race and Health* (Chapel Hill: University of North Carolina Press, 2001).

4. Charles E. Rosenberg, *The Care of Strangers: The Rise of America's Hospital System* (New York: Basic Books, 1987); David Rosner, *A Once Charitable Enterprise: Hospitals and Health Care in Brooklyn and New York, 1885–1915* (New York: Cambridge University Press, 1982); Rosemary Stevens, *In Sickness and in Wealth: American Hospitals in the Twentieth Century* (New York: Basic Books, 1989); Morris J. Vogel, *The Invention of the Modern Hospital, Boston, 1870–1930* (Chicago: University of Chicago Press, 1980).

5. See Ronald L. Numbers, *Almost Persuaded: American Physicians and Compulsory Health Insurance, 1912–1920* (Baltimore: Johns Hopkins University Press, 1978); Daniel S. Hirshfield, *The Lost Reform: The Campaign for Compulsory Health Insurance in the United States from 1932–1943* (Cambridge, Mass.: Harvard University Press, 1970); Beatrix Hoffman, *The Wages of Sickness: The Politics of Health Insurance in Progressive America* (Chapel Hill: University of North Carolina Press, 2001); and Paul Starr, *The Social Transformation of American Medicine* (New York: Basic Books, 1982).

6. See Haynes Johnson and David S. Broder, *The System: The American Way of Politics at the Breaking Point* (Boston: Little, Brown, 1996), and Theda Skocpol, *Boomerang: Clinton's Health Security Effort and the Turn against Government in U.S. Politics* (New York: Norton, 1996).

7. See, for example, Wendy B. Young, "Unintended Consequences of Health Policy Programs and Policies: Workshop Summary," Robert Wood Johnson Health Foundation, Institute of Medicine, Washington, D.C., 2001.

8. See Richard E. Neustadt and Ernest R. May, *Thinking in Time: The Uses of History for Decision-Makers* (New York: The Free Press, 1986).

9. See Charles E. Rosenberg, "What Is an Epidemic? AIDS in Historical Perspective," in *Explaining Epidemics and Other Studies in the History of Medicine,* Charles E. Rosenberg, ed. (New York: Cambridge University Press, 1992), 278–292.

10. See Allan M. Brandt, "Emerging Themes in the History of Medicine," *Milbank Quarterly* 69, no. 2 (1991): 199–213.

11. Allan M. Brandt, "AIDS: From Public History to Public Policy," in *AIDS and the Public Debate,* Caroline Hannaway, Victoria A. Harden, and John Parascandola, eds. (Washington, D.C.: IOS Press, 1995), 124–131.

12. Allan M. Brandt, "AIDS in Historical Perspective: Four Lessons from the History of Sexually Transmitted Disease," *American Journal of Public Health* 78 (April 1988): 367–371.

13. *AIDS: The Burden of History,* Elizabeth Fee and Daniel M. Fox, eds. (Berkeley: University of California Press, 1988); *AIDS: The Making of a Chronic Disease,* Elizabeth Fee and Daniel M. Fox, eds. (Berkeley: University of California Press, 1992).

14. Richard E. Neustadt and Harvey V. Fineberg, *The Epidemic That Never Was: Policy-Making and the Swine Flu Scare* (New York: Random House, 1982).

15. Daniel M. Fox, "Historians and Analysts and Advocates: Disability Policy," *Reviews in American History* 17, no. 1 (1989): 123.

16. *Education for More Synergistic Practice of Medicine and Public Health: Proceedings of a Conference Chaired by Stuart Bondurant,* Mary Hager, ed. (New York: Josiah Macy Jr. Foundation, 1999); *Enhancing Interactions between Nursing and Medicine: Opportunities in Health Professional Education; Proceedings of a Conference Chaired by Sheila A. Ryan,* Mary Hager, ed. (New York: Josiah Macy Jr. Foundation, 2001). See also Allan M. Brandt and Martha Gardner, "Antagonism and Accommodation: Interpreting the Relationship between Public

Health and Medicine in the United States during the 20th Century," *American Journal of Public Health* 90, no. 5 (2000): 707–715.

17. John Demos, *Past, Present, and Personal: The Family and the Life Course in American History* (New York: Oxford University Press, 1986), 210.

18. Harold Edgar and David J. Rothman, "New Rules for New Drugs: The Challenge of AIDS to the Regulatory Process," *Milbank Quarterly* 68 (1990): 111.

19. Jennifer Washburn, "Undue Influence: How the Drug Industry's Power Goes Unchecked and Why the Problem Is Likely to Get Worse," *American Prospect* 12, no. 14 (2001): 16–22.

20. Institute of Medicine, Committee to Study HIV Transmission, *HIV and the Blood Supply: An Analysis of Critical Decisionmaking* (Washington, D.C.: National Academy Press, 1995); see also *Blood Feuds: AIDS, Blood, and the Politics of Medical Disaster,* Ronald Bayer and Eric A. Feldman, eds. (New York: Oxford University Press, 1999).

21. For criticism and analysis of the report, see Salmaan Keshavjee, Sheri Weiser, and Arthur Kleinman, "Medicine Betrayed: Hemophilia Patients and HIV in the US," *Social Science & Medicine* 53, no. 8 (2001): 1081–1094; Ronald Bayer, "Review: HIV and the Blood Supply: An Analysis of Critical Decisionmaking," *American Journal of Public Health* 87, no. 3 (1997): 474–476; and Abigail Trafford, "Bad Blood," *Washington Post,* 19 September 1995, Z6.

22. Advisory Committee on Human Radiation Experiments, *The Human Radiation Experiments: Final Report of the President's Advisory Committee* (New York: Oxford University Press, 1996); Ruth R. Faden, Susan E. Lederer, and Jonathan D. Moreno, "Us Medical Researchers, the Nuremberg Doctors Trial, and the Nuremberg Code: A Review of Findings of the Advisory Committee on Human Radiation Experiments," *Journal of the American Medical Association* (hereafter, *JAMA*) 276 (1997): 1667–1671; Susan Reverby, "Everyday Evil: Book Review of 'Subjected to Science,'" *Hastings Center Report* 26, no. 5 (1996): 38.

23. See Gerald Markowitz and David Rosner, "'Cater to the Children': The Role of the Lead Industry in a Public Health Tragedy," *American Journal of Public Health* 90, no. 1 (2000): 36–46, "Pollute the Poor," *The Nation,* 6 July 1998, 8–9, and "The Reawakening of National Concern about Silicosis," *Public Health Reports* 13 (July/August 1998): 302–311; and David Rosner and Gerald Markowitz, "Battle for Breath: Industry Lobbyists, Government Watchdogs, and the Silicosis Crisis," *Dissent* 45 (Spring 1998): 44–48. Allan M. Brandt and Julius B. Richmond, "Settling Short on Tobacco; Let the Trials Begin," *JAMA* 278 (1997): 1028.

24. Laura Flanders, "Abortion: The Usable Past," *The Nation,* 7–14 August 1989, 175–177; Clyde Spillinger, Jane E. Larson, and Sylvia A. Law, "Brief of 281 American Historians as Amici Curiae Supporting Appellees," in *Webster v. Reproductive Health Services,* 492 U.S. 490 (1989).

25. Sylvia A. Law, "Conversations between Historians and the Constitution," *Public Historian* 12, no. 3 (1990): 11–17, 16; James C. Mohr, "Historically Based Legal Briefs: Observations of a Participant in the *Webster* Process," ibid., 19–26, 20.

26. Flanders, "Abortion," 175.

27. Mohr, "Historically Based Legal Briefs," 25.

28. Wendy Chavkin, "*Webster,* Health, and History," *Public Historian* 12, no. 3 (1990): 53–56, 53.

29. Law, "Conversations," 12.

30. "Historians in Defense of the Constitution," *New York Times,* 30 October 1998, A15.

31. Delia M. Rios, "Historians Ponder Role in Impeachment; Letter Sharpens Existing Divisions," *New Orleans Times-Picayune,* 16 February 2001, 6.

32. See, for example, the twenty-year history of an organization of historians who provide their professional services to businesses and the government: Sarah A. Leavitt, "The Best Company in History: History Associates Incorporated, 1981–2001," History Associates Incorporated, Rockville, Md., 2001.

33. Carol Sternhill, "Life in the Mainstream: What Happens When Feminists Turn Up on Both Sides of the Courtroom?" *Ms. Magazine,* July 1986, 48–51 and 86–91, 48, 88.

34. Ibid., 48, 87.

35. Thomas Haskell and Sanford Levinson, "Academic Freedom and Expert Witnessing: Historians and the *Sears* Case," *Texas Law Review* 66 (June 1988): 1629–1659, and "On Academic Freedom and Hypothetical Pools: A Reply to Alice Kessler-Harris," *Texas Law Review* 67 (June 1989): 1591–1603; Alice Kessler-Harris, "Academic Freedom and Expert Witnessing: A Response to Haskell and Levinson," *Texas Law Review* 67 (December 1988): 429–440; Ruth Milkman, "Women's History and the Sears Case," *Feminist Studies* 12 (1986): 375–400; Peter Novick, *That Noble Dream: The "Objectivity Question" and the American Historical Profession* (New York: Cambridge University Press, 1988), 502–510.

36. Larry O. Gostin, Allan M. Brandt, and Paul D. Cleary, "Tobacco Liability and Public Health Policy," *JAMA* 266 (1991): 3178–3182.

37. See Laura Maggi, "Bearing Witness for Tobacco," *American Prospect* 11, no. 10 (2000): 23–25. Historians who have testified for the industry include Stephen Ambrose, Lacey Ford, Kenneth M. Ludmerer, Peter English, and Theodore Wilson. Robert N. Proctor is also expert witness for the Department of Justice.

38. Ibid., 25.

39. Daniel A. Farber, "Adjudication of Things Past: Reflections on History as Evidence," *Hastings Law Journal* 49 (1998): 1009–1036; Reuel E. Schiller, "The Strawhorsemen of the Apocalypse: Relativism and the Historian as Expert Witness," *Hastings Law Journal* 49 (1998): 1169–1177; Wendie Ellen Schneider, "Case Note: Past Imperfect," *Yale Law Journal* 110 (2001): 1531–1545. Schneider works to establish a "conscientious historian" standard, taking into account the historians' specialization, quality of research for testimony, and analysis, 1541.

40. James T. Kloppenberg, "Why History Matters to Political Theory," in *Scientific Authority and Twentieth-Century America,* Ronald G. Walters, ed. (Baltimore: Johns Hopkins University Press, 1997).

41. Michael Grossberg, "The *Webster* Brief: History as Advocacy, or Would You Sign It?" *Public Historian* 12, no. 3 (1990): 45–52, 52.

42. See John Demos, "Using Self, Using History . . . ," *Journal of American History* 89, no. 1 (2002): 37–42.

43. Alice Kessler-Harris, "Historians Should Take Stands to Stay Relevant," *Chronicle of Higher Education,* 21 September 2001, B11–B13, William E. Leuchtenburg, "The Historian and the Public Realm," *American Historical Review* 97, no. 1 (1992): 1–18.

Notes on Contributors

Olga Amsterdamska teaches science studies and history of medicine at the University of Amsterdam. She has published articles on the history of the biomedical sciences, especially on the relations between the laboratory and the clinic in microbiology and biochemistry. Her current research centers on the history of British and American epidemiology and the changing notions of what makes epidemiology a science.

Warwick Anderson is the Robert Turell Professor of Medical History and Population Health and chair of the Department of Medical History and Bioethics at the University of Wisconsin, Madison. When he wrote this essay he was director of the history of the health sciences program at the University of California at San Francisco. In 2003, Basic Books published his study of race science in Australia, *The Cultivation of Whiteness: Science, Health, and Racial Destiny*. He is currently working on what he hopes is a postcolonial study of kuru investigations in the highlands of New Guinea and in Bethesda, Maryland.

Allan M. Brandt is the Kass Professor of the History of Medicine at Harvard Medical School, where he directs the Program in the History of Medicine. He holds a joint appointment in the Department of the History of Science, where he is currently chair. His work focuses on the social history of medicine, disease, and public health policy in the twentieth-century United States. He is the author of *No Magic Bullet: A Social History of Venereal Disease in the United States since 1880* (1987) and editor of *Morality and Health* (1997).

Theodore M. Brown is a professor and chair of the Department of History and Professor of Community and Preventive Medicine at the University of Rochester. He earned his Ph.D. in History of Science from Princeton University and was a postdoctoral fellow at the Johns Hopkins Institute of the History of Medicine. His research currently focuses on the history of psychosomatic medicine and on American public health and health policy. He is a contributing editor for the *American Journal of Public Health* and, with Elizabeth Fee, co-edited *Making Medical History: The Life and Times of Henry E. Sigerist* (1997).

Roger Cooter is a Wellcome Professorial Fellow at the Wellcome Trust Centre for the History of Medicine at University College London. The author of *The Cultural Meaning of Popular Science* (1984) and *Surgery and Society in Peace and War* (1993), he has also edited volumes on child health, alternative medicine, accidents, war and medicine, and most recently, with John Pickstone, *Medicine in the Twentieth Century* (2000).

Martin Dinges is deputy director of the Institut für Geschichte der Medizin der Robert Bosch Stiftung, Stuttgart, Germany, and adjunct professor of modern history at the University of Mannheim, Stuttgart, Germany. The volumes he has edited include *Neue Wege in der Seuchengeschichte* (1995, with Thomas Schlich), *Weltgeschichte der Homöopathie, Länder—Schulen—Heilkundige* (1996), *Homöopathie. Patienten, Heilkundige und Institutionen. Von den Anfängen bis heute* (1996), *Medizinkritische Bewegungen im Deutschen Reich (ca. 1870–ca. 1933)* (1996), and *Patients in the History of Homoeopathy* (2002). He has also written on the social history of medicine during the Enlightenment and on the reception of the work of Michel Foucault.

Alice Domurat Dreger is an associate professor of science and technology studies in the Lyman Briggs School at Michigan State University and associate faculty at the university's Center for Ethics and Humanities in the Life Sciences. She served for three years as the founding chair of the board of directors of the Intersex Society of North America, a nonprofit policy and advocacy group. Her research and outreach focus on the biomedical treatment of people born with unusual anatomies and the relation between anatomy and identity. Her books include *Hermaphrodites and the Medical Invention of Sex* (1998) and *Intersex in the Age of Ethics* (1999).

Jacalyn Duffin is a hematologist and historian who teaches medicine, history, and philosophy from the Hannah Chair at Queen's University in Kingston, Ontario, Canada. She is author of *Langstaff: A Nineteenth-Century Medical Life* (1993), *To See with a Better Eye: A Life of R. T. H. Laennec* (1998), and *History of Medicine: A Scandalously Short Introduction* (1999). Her current research interests are in disease concepts, diagnostic semeiology, and medical saints.

Elizabeth Fee is chief of the History of Medicine Division of the National Library of Medicine, National Institutes of Health, in Bethesda, Maryland, and adjunct professor of history and health policy at the Johns Hopkins University in Baltimore, Maryland. She is author of *Disease and Discovery: A History of the Johns Hopkins School of Hygiene and Public Health, 1919–1939* (1987), and co-editor, with Theodore M. Brown, of *Making Medical History: The Life and Times of Henry E. Sigerist* (1997).

Mary E. Fissell teaches in the Department of the History of Science, Medicine, and Technology at the Johns Hopkins University in Baltimore, Maryland. She is completing a book about the politics of reproduction in early-modern England. Her earlier work includes essays on the history of medical ethics, patients' narrative, hospital and welfare patients, popular medical books, and *Patients, Power, and the Poor* (1991), a social history of patients in eighteenth-century Bristol.

Danielle Gourevitch is directeur d'études at the École Pratique des Hautes Études, in Paris. She teaches history of medicine, especially Greek and Roman medicine and medical erudition in the nineteenth century. Her main works about this period are *La mission de Charles Daremberg en Italie (1849–1850)* (1994) and *Médecins érudits, de Coray à Sigerist* (1995). She has also published a collective manual, *Histoire de la médecine. Leçons méthodologiques* (1995). She is currently working on a book about Galenic *pathocenosis*.

Anja Hiddinga teaches science studies and sociology of medicine at the University of Amsterdam. Her research focuses on clinical medicine, especially obstetrics and maternal health. She is currently studying cultural differences and historical changes in the treatment and experience of pain during delivery. She is also interested in the cultural meanings of deafness and has recently produced a film on this subject.

Frank Huisman teaches in the Department of History of Universiteit Maastricht. He is the author of *Stadsbelang en standsbesef: Gezondheidszorg en medisch beroep in Groningen, 1500–1730* (1992), and co-editor with Catrien Santing of *Medische geschiedenis in regionaal perspectief: Groningen, 1500–1900* (1997), both local case studies of early modern Dutch health care. He has published on medical historiography, quackery, and the cultural authority of medicine. Currently, he is working on a book exploring the transformation of the Dutch health care system between 1880 and 1940.

Ludmilla Jordanova joined the University of East Anglia in 1996 having previously worked at the Universities of Essex and York. She is a trustee of the National Portrait Gallery, London, for whom she organized the exhibition of scientific and medical portraits, *Defining Features*, in 2000. Her recent publications include *Nature Displayed: Gender, Science, and Medicine, 1760–1820* (1999) and *History in Practice* (2000). Her current research is on self-portraiture and artistic practice in Britain from the sixteenth century to now and art in fiction and poetry.

Alfons Labisch, historian, social scientist, and physician, is a professor of the history of medicine and head of the Institute for the History of Medicine at Heinrich Heine Universität, Düsseldorf, Germany. His main areas of research include social history of health and medicine (especially of preventive medical disciplines and institutions), social history of the hospital, and science and practice in medicine. His recent publications include *"Einen Jedem Kranken in einem Hospitale sein Eigenes Bett." Zur Sozialgeschichte des Allgemeinen Krankenhauses in Deutschland im 19. Jahrhundert* (1996, co-edited with Reinhard Spree), *Adolf Gottstein: Erlebnisse und Erkenntnisse. Nachlass 1939/1940. Autobiographische und biographische Materialien* (1999, co-edited with Ulrich Koppitz), and *Krankenhaus-Report 19. Jahrhundert. Krankenhausträger, Krankenhausfinanzierung, Krankenhauspatienten* (2001, co-edited with Reinhard Spree).

Hans-Uwe Lammel is Privatdozent for medical history and head of the Department for the History of Medicine at the University of Rostock (Mecklenburg/Vorpommern). He is author of *Nosologische und Therapeutische Konzeptionen in der Romantischen Medizin* (1990) and editor of *Kranksein in der Zeit* (1995) and *Medizingeschichte und Gesellschaftskritik* (1997, co-edited with Michael Hubenstorf, Ragnhild Münch, Sabine Schleiermacher, Heinz-Peter Schmiedebach, and Sigrid Stöckel). His special fields are early modern historiography and epidemics in early modern times.

Sherwin B. Nuland is clinical professor of surgery and a fellow of the Institution for Social and Policy Studies at Yale University. He is a member of the executive committee of Yale's Whitney Humanities Center and of the university's Interdisciplinary Bioethics Forum. His

1994 book, *How We Die,* won the National Book Award and was a finalist for the Pulitzer Prize and the Book Critics Circle Award. He has written on medicine, medical history, and ethics for the *New York Times, New York Review of Books, Wall Street Journal, New Republic, Time, Life, National Geographic, New Yorker, American Scholar,* and other periodicals. All eight of his books were written for the general reader, including *Doctors: The Biography of Medicine* and biographies of Leonardo da Vinci and Ignac Semmelweis.

Vivian Nutton is a professor of the history of medicine at University College London. He has edited several ancient medical texts, including Galen's *On Prognosis* (1979) and *On My Own Opinions* (1999), and has published many books and articles on classical medicine and its transmission down to the twentieth century. He is at present writing a large history of medicine in classical antiquity.

Roy Porter, before he retired, was a professor of the social history of medicine at University College London. Among his many publications are *Mind-Forg'd Manacles* (1987), *In Sickness and in Health: the British Experience, 1650–1850* (1988, co-authored with Dorothy Porter), *Health for Sale: Quackery in England, 1660–1850* (1989), *Reassessing Foucault: Power, Medicine, and the Body* (1994, co-edited with Colin Jones), *The Greatest Benefit to Mankind: A Medical History of Humanity from Antiquity to the Present* (1997), and *Enlightenment: Britain and the Creation of the Modern World* (2000). He died in the spring of 2002.

Susan M. Reverby is a professor of women's studies at Wellesley College. Her work in the history of health care has focused on gender, class, and race issues. Her prize-winning book, *Ordered to Care: The Dilemma of American Nursing* (1987), is an overview on the history of American nursing. Her latest book is an edited collection, *Tuskegee's Truths: Rethinking the Tuskegee Syphilis Study* (2000). She is currently completing a manuscript exploring the differing ways the stories of the Tuskegee study are told.

David Rosner is a professor of history and public health at Columbia University and director of the Center for the History of Public Health at Columbia's Mailman School of Public Health. He is author of *A Once Charitable Enterprise* (1982; 1987) and editor of *Hives of Sickness, Epidemics and Public Health in New York City* (1995) and *Health Care in America: Essays in Social History* (1979, with Susan Reverby). In addition, he has co-authored and edited with Gerald Markowitz numerous books and articles, including *Deadly Dust: Silicosis and the Politics of Occupational Disease in Twentieth Century America* (1991; 1994), *Children, Race, and Power: Kenneth and Mamie Clark's Northside Center* (1996), and *Deceit and Denial: The Deadly Politics of Industrial Pollution* (2002). Recently he has been awarded the Viseltear Prize for Outstanding Work in the History of Public Health from the APHA.

Thomas Rütten, holder of a Wellcome Trust University Award in the History of Medicine at the University of Newcastle upon Tyne, holds a Ph.D. from the University of Münster and has over the last twelve years been teaching history of medicine at various universities, both in Germany (Münster; Witten/Herdecke) and in France (Paris VII). He is the author of *Demokrit-Lachender Philosoph und Sanguinischer Melancholiker* (1992), *Geschichten vom Hippokratischen Eid* (Habilitation, 1995), and numerous articles on classical, medieval, and early modern history, ethics, and iconography of medicine, as well as on the history of his discipline.

Heinz-Peter Schmiedebach is a professor in the history of medicine and has been head of the Institut für Geschichte der Medizin, Universität Greifswald, Germany, since 1993. Among his publications are *Psychiatrie und Psychologie im Widerstreit* (1986), *Medizin und Krieg: Vom Dilemma der Heilberufe 1865 bis 1985* (1987, co-edited with Johanna Bleker), and *Robert Remak (1815–1865): Ein Jüdischer Arzt im Spannungsfeld von Wissenschaft und Politik* (1995). His research interests include the history of psychiatry, development of public health, the history of deontology, and the history of medical theories and disciplines in the nineteenth and twentieth centuries.

Christiane Sinding is a senior researcher at the Institut National de la Santé et de la Recherche Médicale (INSERM). She was first trained as a physician, then as a historian of science. She is the author of *Le clinicien et le chercheur. Des grandes maladies de carence à la médecine moléculaire* (1991) and of numerous articles on the history of endocrinology. She is currently completing a manuscript on the history of therapeutical innovation in diabetes mellitus. She is co-editor of the *Dictionnaire d'histoire et philosophie de la médecine* (2004).

John Harley Warner is a professor and chair of the Section of the History of Medicine at the Yale University School of Medicine, and at Yale is also a professor of history and of American studies and chair of the Program in the History of Medicine and Science. His books include *The Therapeutic Perspective: Medical Practice, Knowledge, and Identity in America, 1820–1885* (1986; 1997), *Against the Spirit of System: The French Impulse in Nineteenth-Century American Medicine* (1998; 2003), and (co-edited with Janet Tighe) *Major Problems in the History of American Medicine and Public Health* (2001). He is now working on a book exploring clinical narrative and the grounding of modern medicine.

Index

Abel, John J., 145
abortion, 393, 472–74
Académie des inscriptions et belles lettres, 57, 60
Académie Française, 61
Académie Royale des Sciences, 55, 57, 58, 62, 275
Achelis, Johann Daniel, 103, 112n55
Ackerknecht, Erwin, 159, 170, 252–53, 255, 276, 290, 320, 425, 453
Ackermann, Christian Gottlieb, 42
activism (by historians), 22, 23, 142, 159, 169, 170, 177, 327, 390–409, 460–84
Adams, Francis, 116
Adelung, Johann Christoph, 42, 43
Advisory Committee on Human Radiation Experiments (ACHRE), 470–71
African Americans, 179, 183, 184, 189
African cultures, 380
agency: for change, 181, 182, 296, 317, 322, 327; as differentiated view, 227; patient's, 182; women's, 179, 180, 191n23
Agnew, Jean-Christophe, 172
AIDS epidemic: as activist impetus, 179, 461, 463–65, 468–70; as biological reality, 321; as challenge to myth of progress, 195, 287; and history of epidemics, 216, 287, 463–65. *See also* sexually transmitted diseases
Allbutt, Sir Clifford, 120
Altertumswissenschaft, 10, 118, 121, 128
Althusser, Louis, 267, 268
amateurism, 13, 29n56, 175, 438, 456. *See also* professionalism
American Association for the History of Nursing, 191n25
American Association of the History of Medicine (AAHM): Charleston meeting, 188; membership, 257; as platform, 175, 404, 406, 457; schism in, 140, 174, 176–77, 188, 437–38; traditional, 190n11. *See also* Deutsche Gesellschaft für Geschichte der Medizin, der Naturwissenschaften und der Technik; medical history, schism in
American Historical Association, 476
American Historical Review, 180
American Journal of Philology, 128
American Journal of Public Health, 464
Amsterdamska, Olga, 283n75
Amundsen, Darrel, 137n91
anatomy: anatomists, 382–85; discipline, 202, 326, 382–85; models, 378; theatres, 202
Anatomy Act, 382–85
ancient medicine, 10, 20, 115–38
Angier, Natalie, 397, 452
Annales school, 199, 212
Annals of Medical History, 144
Anonymus Londinensis, 120
Anonymus Parisinus, 119
anthropology: Canguilhem on, 267; impact of, 287, 291, 295, 342, 365, 374; medical, 278, 292
antiquarianism, 77, 85, 89, 116, 129, 185, 238, 416. *See also* internalism; presentism; Whiggism
anti-Semitism, 101, 104, 109n32, 109n33, 109n34, 127
antiwar movement, 169, 174, 190n7
Appleby, Joyce, 316
appropriation, 374–78, 380
archaeology, 98, 116
archaeology of knowledge, 274
Archiv für die Gesammte Medizin, 9
Archiv für Geschichte der Medizin, 13, 75, 96
Archiv für physiologische Heilkunde, 415
Aristotle, 121, 124
Armstrong, David, 278, 334n63
Arnold, David, 252, 288, 294, 296, 297, 298
Artelt, Walter, 102
art history, 341

Index

Aufklärungshistorie, 36, 38, 43, 46
autobiographies, 216, 218

Baas, Johann, 12–13, 14
Bachelard, Gaston, 263
bacteriology, 146, 213, 288, 416
Barbin, Herculine, 396
Barnes, Barry, 254, 341
Bataille, Georges, 332n37
Baudrillard, Jean, 317, 318
Baxandall, Michael, 350
Bayer, Ronald, 186, 400–401
Bayle, A. L. J., 269
Beck, Ulrich, 314
Behring, Emil von, 415
Bell, Heather, 290
Benassar, Bartolomé, 212
Benedetti, Vincenzo di, 138n105
Bennett, James Risdon, 59, 60
Benoist, Alain de, 332n37
Berge, Ann La, 276–77
Berger, Peter, 341
Bergsträsser, Gotthelf, 123
Berlin, University of, 5, 42, 118. *See also* Humboldtian university
Berlin Academy of Sciences, 34, 37, 45, 46, 118, 121
Berliner, Howard, 175
Bernard, Claude, 67, 264, 266
Berrios, German, 204
Beveridge, William, 156; report, 272
Bevoise, Ken de, 293
Bibliothèque des médecins Grecs et Latin, 57, 59
Bibliothèque Mazarine (Paris), 55, 56, 63
Bichat, Xavier, 269
Bildung: as educational ideal, 6, 11, 15, 16, 46, 74–94, 117, 417–18; as social strategy, 34, 39, 74–94. *See also* humanism; Humboldtian university; Two cultures, the
Bildungsbürgertum, 27n30
Billings, John Shaw, 14, 15, 28n49, 143
Billroth, Theodor, 453
biographical approach/biography, 6, 33–52, 139–64, 144, 170, 243, 245, 249, 291, 349, 456. *See also* role model
biology, 85, 263–67
biopower, 271, 272, 283n74
Biraben, Jean-Noël, 212
birth rate, 197, 200. *See also* death rate; demographic transition; fertility; historical demography; mortality

Bismarck, Otto von, 156–57
Blackmar, Elizabeth, 172
Blake, John, 173
Bleker, Johanna, 30n62, 215
Bliss, Michael, 141, 145
Bloch, Iwan, 84
blood (products), 468–69
Blood Supply Council, 469
Bloor, David, 341
Bodleian Library (Oxford), 56, 57, 149
body: discourse on, 225, 296, 317, 318, 390–409; embodiment, 181, 296, 305n46; female, 178, 179, 366, 381; hermaphrodite, 390–409; history, 177, 198, 202–3, 211, 334n63, 369, 387n23; as medical object, 269, 421; policing of, 314, 327; relation to cultural history, 371–72; studies, 319, 320, 326. *See also* mind; self
Boerhaave, Herman, 36
Boletín de la Sociedad Mexicana de historia y filosofía de la medicina, 448n9
Bonnechose, Cardinal de, 64
Bose, Girindrasekhar, 297
botany, 40
Bourdelais, Patrice, 212
Bourdieu, Pierre, 267, 278
Bowker, Geof, 278, 284n84
Bowman, Isaiah, 158
Brandenburg-Prussia (the state of), 5, 34, 37, 38, 42, 45, 46, 47
Brandt, Allan, 189
breast cancer, 179
Brieger, Gert, 176
British Journal of the History of Science, 254
Brockliss, Lawrence, 194
Brody, Howard, 406
Broeckx, Cornelius, 54, 64, 66, 67
Broglie, family de, 55, 62
Broman, Thomas, 34
Broussais, François, 264, 270, 271
Brown, Kathleen, 180
Brown, Richard E., 175; *Rockefeller Medicine Men,* 175
Brown, Theodore, 176
Browne, Richard, 373
Browne, Sir Thomas, 142
Bruckner, Pascal, 332n37
Brumberg, Joan Jacob, 382
Brunn, Walter von, 96, 100, 101, 104, 109n33, 109n36, 110n37, 111n49, 112n51, 112n52
Bryder, Lynda, 260

Bueltzingsloewen, Isabelle von, 214–15, 220
Bulletin of the (Institute of the) History of Medicine, 128, 152, 172, 174, 175, 177, 241–61, 448n9, 452
Burton, Antoinette, 325–26
Burton, Robert, 203
Bussemaker, Ulco, 56, 58, 59, 63
Butler, Judith, 319, 375
Butterfield, Herbert, 341; *Whig Interpretation of History,* 341
Bynum, Caroline, 371
Bynum, William F., 252, 255

Cambridge Group for the History of Population and Social Structure, 196, 199
Camus, Albert, 332n36
Canadian Bulletin of the History of Medicine, 448n9
Canguilhem, Georges, 21, 262–84; *Normal et le pathologique,* 263, 264, 271, 274
capitalism, 268, 292, 317, 318, 324, 332n35
care of the self (le souci de soi), 273
Carlyle, Thomas, 326
Carmichael, Ann, 393
Cartesian philosophy, 267
case history, 40, 144–45
Casey, Catherine, 326
Cassedy, James, 173, 252–53
Castellani, Aldo, 288
Castiglioni, Arturo, 453
celebration, 140, 173, 289, 341
Celsus, Cornelius, 116
Chakrabarty, Dipesh, 299, 300
Chalmers, Alfred J., 288
Chambers, Tod, 390, 403
charisma, 96–97, 141, 158. *See also* icon; personality cult
Chartier, Roger, 375
Chase, Cheryl, 396–407
Chavkin, Wendy, 473
chemical pollutants, 186–87
Cheney, James, 171
Chickering, Roger, 85
childbirth, 222–23, 376, 405
childhood, 375–76, 382
cholera, 213, 216, 287, 294
Choulant, Ludwig, 9, 59
Churchill, Fred, 392, 393
cigarette, history of the, 475, 477–79
citizenship, 296, 297

civic ideal (society), 39, 44, 46, 213
civil rights movement, 169, 174, 190n7, 400
Clarke, Adele, 255, 278
class: evaluation of a category, 181–83; historicity of, 318, 375; and medicalization, 223; as new analytic category, 170, 173, 177, 239, 340, 344. *See also* colonial, the; frame, the; gender; place; race
Classical Association, 148, 149
classics, 68, 115, 117–18, 124
clinician: as historian, 432–49; history of, 432–49, 450–59; as teacher, 159, 451–52, 458–59. *See also* Osler, William
Clinton, Bill, 462, 470, 471, 474
Clio Medica, 254
clitorectomy, 397
collective memory, 20, 33, 35, 43, 46, 420. *See also* icon; role model
Collège de France, 63, 67
College of Physicians, 368, 372–73
Collins, Harry, 254, 255
colonial, the, 288. *See also* class; frame, the; gender; place; race
colonialism, 22, 289, 290, 291, 292, 293, 300; and discipline, 294, 296; history of, 287, 298; intervention of, 294, 295; legacy of, 298. *See also* decolonization
colonial medicine, 253, 291, 293–94, 295, 298, 299; history of, 287, 292, 295, 296, 299
commercialism, 80, 201
community history, 182
comparative method, 355
Comte, Auguste, 20, 53, 54, 61, 62, 264
Condillac, Étienne de, 269, 281n28
Condrau, Flurin, 217–18
constitutional theory, 416
consumer society, 201, 376, 377
contagion, 326
contingency, 316, 400, 420, 422, 434, 465
continuity, 4, 263, 273–74, 281n45, 463–64. *See also* discontinuity; finalism; history, periodization in
Cook, Harold, 372–73
Cooter, Roger, 2, 261n28
Corpus Inscriptionum Latinarum, 121
Corpus Medicorum (Graecorum, Latinorum), 10, 57, 121, 123, 128
court, 313, 475–80
Crell, Christopher, 373
Croce, Benedetto, 17

Croix, Alain, 212
Crosby, Alfred W., 292
cultural, the, 316, 325, 326, 328, 331n32, 384. See also economic, the; political, the; social, the
cultural history of medicine: and activism, 390–409; and body history, 202–3, 390–409; evaluation of a program, 364–89; and the history of madness, 204–5; (lack of) integration into cultural history, 240, 259; as a new program, 23, 237, 347; and William Osler, 149; pragmatic German version of, 14, 20, 42, 43, 74–94, 102; and Charles Rosenberg, 321; and social constructionism, 340. See also social history of medicine
culture: criticism of, 315; model of, 370; popular, 367, 369, 375; practices of, 320; space for, 34, 44–47; studies of, 22, 180, 181, 184, 277, 317, 318, 319, 325, 326, 327; turn in, 3, 227, 296. See also discourse, analysis of; linguistic turn; literature, studies of; social studies of science
culture wars, 311. See also history wars; science wars; theory wars
curriculum, 46, 67, 152, 186
Curschmann, Geheimrat, 97–98
Curtin, Philip, 293, 295
Cushing, Harvey, 144, 151, 168

Daaboul, Jorge, 403–5, 407
Dacome, Lucia, 378
Dagognet, François, 275, 276
Damerow, Heinrich, 6, 42
Daremberg, Charles (father of Georges), 13, 20, 53–73, 80, 117; *Histoire des sciences médicales*, 63, 64–65, 68
Daremberg, Georges (son of Charles), 62
Darnton, Robert, 318, 374
Davidoff, Leonore, 367
Davis, Natalie, 378
death, 269, 271
death rate, 200, 272. See also birth rate; demographic transition; fertility; historical demography; mortality
decolonization, 289, 299. See also colonialism
deconstruction, linguistic, 178, 317, 321, 331n35, 370, 375
Deichgräber, Karl, 123, 126, 134n59
deinstitutionalization, 17–18, 30n61, 30n62
Delaporte, François, 277, 282n51
Delbrück, Max, 410–11

Deleuze, Giles, 317
Delumeau, Jean, 216
demographic transition, 197, 200, 201, 210, 211, 213–18, 379. See also birth rate; death rate; fertility; historical demography; mortality
Demos, John, 467
dentists, 219
Derickson, Alan, 182
Derrida, Jacques, 317, 320, 323, 332n37
Descartes, René, 203, 267
Dessertine, Dominique, 215
Deutsche Gesellschaft für Geschichte der Medizin, der Naturwissenschaften und der Technik (DGGMNT): as general platform, 12, 79, 86, 100, 107n11, 112n57; Gleichschaltung of, 102, 112n55; and Karl Sudhoff, 96, 97, 99, 100, 108n27. See also American Association of the History of Medicine
deviance, 364
diagnostics, 215, 433
Dictionary of Greek and Roman Biography and Mythology, 66
Dictionnaire encyclopédique des sciences médicales, 66
didactics, 35, 46, 78, 159
Diels, Hermann, 10, 118, 120, 122, 123
Diepgen, Paul, 75, 89, 102, 103, 104, 111n44, 113n58, 127, 414
différance, 34, 47n2, 318, 332n35
Digby, Anne, 204
Diller, Hans, 126
Diocles of Carystos, 119, 124
diphtheria, 214
discontinuity, 9, 180, 273, 319, 463–64. See also continuity; history, periodization in
discourse, 277, 312, 317–18, 320, 323, 325, 347; analysis of, 180, 187, 204, 326–27, 331n35, 347; formation of, 224, 225, 227. See also culture, turn in; linguistic turn
discrimination, 179
disease, 262, 264–66, 271, 326; causation of, 289; digestive, 214; ecology of, 291; framing of, 320, 370–71; geographical distribution of, 285–306; history of, 249, 285–306; infectious, 213; ontological conception of, 7, 81, 269–70, 370, 415; physiological conception of, 370–71; responsibility for, 477; study of, 292. See also health; illness; norms; occupational disease/health; pathology
dissection, 202–3, 269, 382–85
Dobson, Mary, 199–200

Dock, George, 144
Donzelot, Jacques, 326
Dörner, Klaus, 211
Douglas, Mary, 325, 342
doxographic method, 5, 6
Drabkin, I. E., 127, 137n92
Drees, Anette, 211
Dreger, Alice Domurat, 402, 405
Dreyfus, Hubert, 278
Driscoll, Lawrence, 327
Duden, Barbara, 211, 225, 372
Durkheim, Émile, 317

Eagleton, Terry, 331n35
eco-activists, 315
ecological approach, 291, 303n25
ecology: of disease, 291; impact of colonialism on, 292
economic, the, 317. *See also* cultural, the; political, the; social, the
Edelstein, Ludwig, 15, 16, 101, 108n31, 111n44, 112n54, 115, 116, 126, 127, 128, 425; *Ancient Medicine,* 128
Edinburgh school, 341
Ehrenreich, Barbara, 191n23
Ehrlich, Paul, 416
Eley, Geoff, 316
Eliot, Martha May, 156
Ellerkamp, Marlene, 216
empiricism, 40, 41, 66
endocrinologists, 400
engineers, 214
Englert, Ludwig, 102
English, Deirdre, 191n23
English, Peter, 484n37
Enlightenment, 76, 270
entomology, 288
epidemics, 212, 216–17, 272, 292, 465
epidemiological transition. *See* demographic transition
epidemiology, 197–98, 265; colonial, 293
épistème, 224, 274, 277, 347, 348
epistemology, 40, 67, 238; break in, 239, 269, 444; status of, 340
Epstein, Jim, 326
Eribon, Didier, 276
errors of metabolism, 266, 280n13. *See also* philosophy, of error
essentialism, 178, 180, 325. *See also* social constructivism

Estes, J. Worth, 252
ethics, 262, 266, 273; critique of , 401–2; practice of, 268; and standards, 470–71. *See also* medical, ethics
ethnicity, 185, 239. *See also* race
ethnography, 372–74
etiology, 7, 183
eugenics, 109n34, 183, 198, 344
Evans, Richard, 210, 316
evolutionism, biological, 265
experience: collective, 271; of pain, 225; of patient, 271. *See also* illness, experience of
expropriation of health, 202
externalism, 185, 290, 302n14, 346, 350, 356, 393, 437. *See also* internalism

Faltin, Thomas, 219–20
family reconstitution, 197
Fanon, Frantz, 295
Farley, John, 291
Faure, Olivier, 214, 215, 219
Fausto-Sterling, Anne, 395, 397, 401
Fee, Elizabeth, 176, 464–65
Feher, Michel, 386n23
femininity, 298, 343
feminism: academic, 397; as antiprofessionalism, 315, 342–43, 345; contribution to cultural history, 374–76, 437; second-wave, 191n23, 198, 392, 476; as social activism, 393, 394
feminist: interpretation, 222–23; studies, 277, 343
Ferngren, Gary, 137n91
fertility, 197. *See also* birth rate; death rate; demographic transition; historical demography; mortality
Feyerabend, Paul, 322
finalism, 350–51, 428n11. *See also* continuity; presentism
Fisher, Kate, 379
Fissell, Mary, 201–2
Flexner, Abraham, 29n52
Floud, Roderick, 198
Foner, Eric, 474
Ford, Lacey, 484n37
Foucault, Michel, 21, 186, 262–84; *Archéologie du savoir,* 347; on biopower, 312, 314, 315; and colonial medicine, 295–96, 304n40; on discourse, 225, 314, 326; *Folie et déraison,* 268, 275, 276, 347; *Histoire de la folie,* 275; *Histoire de la sexualité,* 317; on the hospital, 214; on

Foucault, Michel (*cont.*)
  insanity, 204; as inspiration to Anglo-Saxon historians, 252, 275–78; as inspiration to French historians, 212, 214, 275; *Mots et les choses*, 263, 273, 275, 347; *Naissance de la clinique*, 186, 214, 215, 268–72, 273, 276, 277, 347; as (pre-)postmodernist, 317, 319, 322, 323, 326, 332n37, 332n39, 333n43, 334n57; on sexuality, 198; and social constructivism, 346–47, 348; *Surveiller et punir*, 270, 276, 279, 317
founding father, 35, 42, 45, 96, 117. *See also* Hippocrates; icon; Sprengel, Kurt; Sudhoff, Karl
Fourier, Charles, 53
Fox, Daniel, 464–65
frame, the, 309–10, 321, 322, 323, 324, 325. *See also* class; colonial, the; gender; place; race
Freeman, Hugh, 204
free market: ideology, 312, 315; forces, 316
French Revolution, 268, 276
Freud, Sigmund, 203, 322
Frevert, Ute, 211–12
Frey, Michael, 224–25
Friedländer, Hermann, 42
Friedrich II (king of Prussia), 37, 45
Fuchs, Konrad Heinrich, 26n12
Fuchs, Robert, 119
Fukuyama, Francis, 319
Fulton, John, 169, 453

Galen of Pergamum, 5, 55, 68, 116, 122, 145–46
Garden, Maurice, 212
Garrett, Laurie, 452
Garrison, Fielding H., 15–16, 285–88, 300, 450–51, 455; *Introduction to the History of Medicine*, 450–51
Garrod, Archibald, 147, 266
Gay, Peter, 457
gaze, clinical, 269–70, 271
Geertz, Clifford, 368, 374; *Interpretation of Cultures*, 368
Geisteswissenschaften, 125
Gelehrtengeschichte, 36
Gelfand, Toby, 436
Gélis, Jacques, 222
gender: as determinant for morbidity, 216; evaluation of a category, 178–81; medical framing of, 296, 298, 345, 390–409; and medicalization, 222–23, 368; as new analytic category, 170, 173, 177, 237, 253, 318; as performance,

366, 375. *See also* class; colonial, the; frame, the; place; race
genitals, 391, 398
gentleman-physician, 14–15. *See also* Billings, John Shaw; Osler, William
German model: of higher education, 29n52; of laboratory-based research, 145; of medical history, 15–16, 117–22, 145, 151. *See also* Institute of the History of Medicine, in Leipzig; Johns Hopkins
Gesellschaft Deutscher Naturforscher und Ärzte, 10–11, 97
Gierl, Martin, 39
Gilhaus, Ulrike, 214
Gilman, Sander, 319
Ginzburg, Carlo, 365, 367, 372, 379–80, 434, 447n2, 448n8; *Formaggio e i vermi*, 365, 372. *See also* microhistory
Gleichschaltung, 102, 104
globalization, 287, 299
Glucksman, André, 332n37
Göckenjan, Gert, 211, 224
Goffman, Erving, 320, 322
Goodman, Andy, 171
Gossen, Hans, 132n27
Göttingen: medical school of, 87; University of, 5, 35, 36, 43, 45, 46. *See also* Humboldtian university
Gottstein, Adolf, 415–16
Goubert, Jean-Pierre, 199, 212
Gramsci, Antonio, 294, 324
grand narrative. *See* master narrative
grave-robbers, 382
Greenhill, William Alexander, 54, 57, 59, 62, 66, 117
Gregg, Alan, 158
Gregory, Annabel, 198
Griesinger, Wilhelm, 415
Grimm, Johann Friedrich, 58
Grmek, Mirko, 35, 441
Grob, Gerald, 167, 173
Groenevelt, Johannes, 372–73
Gross, Grace Linzee Revere, 142, 147
Grossberg, Michael, 480
Grosz, Elizabeth, 319
Gruner, Christian Gottfried, 42
Guillaume, Pierre, 212, 217
Gutting, Gary, 274
gymnasium, 118
gynecologists, 400

Győry, Tibor, 97, 98, 100, 101, 104, 108n27, 109n34, 109n36, 111n50

Haberling, Wilhelm, 104, 110n39
Haeser, Heinrich, 9–10, 13, 26n12, 54, 65–66; *Lehrbuch der Geschichte der Medizin und der Volkskrankheiten,* 9, 54, 66
Hahnemann, Samuel, 81
Hähner-Rombach, Sylvelyn, 217
Hall, Stuart, 299, 323
Halle, University of, 5, 34, 35, 40, 46. *See also* Humboldtian university
Haller, Albrecht von, 270
Hammonds, Evelynn M., 180, 184, 191n28
Handbuch tradition, 6
Hannaway, Caroline, 276–77
Haraway, Donna, 318, 327
Hardy, Anne, 198, 252–53
Harig, Georg, 80
Harnack, Adolf von, 121, 133n42
Harrison, Gordon, 291
Harrison, Mark, 290
Harvey, William, 146, 202, 203
Hatch, David, 404
Hawking, Stephen, 452
healer, popular concept of, 209, 219–20, 239
healing power of nature, 90
health, 262; beliefs, 292; consumers, 187, 313; discourse on, 224; education, 296; history of, 225, 227, 288; improvement, 214; legislation, 214; measures, 272; political economy of, 292. *See also* disease; norms; political, economy
health care: commodification of, 312; configurations of, 270; economic dimension of, 209, 221; history of, 4, 167, 180; popular, 201
health policy, 140, 168, 211, 212, 213, 221, 243, 244; activist, 170; analyst, 169, 181; program, 176. *See also* public health
Hecker, Justus Friedrich Karl, 7, 8, 9, 26n12, 42, 59
Hegel, Georg W. F., 6, 53, 66, 87, 267, 332n37
Heiberg, J. L., 121
Heller, Geneviève, 224
helminthology, 288
Helmreich, Georg, 132n27
Henry, Louis, 197
Henschel, August, 9, 10, 54, 117
hermaphroditism, 390–409
Herodotus, 42, 315
heroism, 399, 455. *See also* role model

Hesse, Mary, 341
Heteren, Godelieve van, 329n3
heuristics, 344, 371
Hiddinga, Anja, 283n75
Hill, Owen Berkeley, 297
Hindenburg, Paul von, 103
Hippocrates: and Galen, 145, 146; Littré edition of, 54, 67; Littré image of, 60; as professional role model, 27n27, 34, 35, 40, 46, 124, 424; studied by philologists, 117, 122. *See also* founding father; role model
Hippocratic corpus, 78, 116, 126–27
Hippocratism, 125. *See also* neo-Hippocratism
Hippolyte, Jean, 267
Hirsch, August, 7–8, 290
Hirschfeld, Ernst, 101, 121
Histoire des sciences médicales, 448n9
Historia literaria, 38, 41, 42, 43, 44, 45, 76
historian: of health policy, 460–84; as policy advocate, 471–80; as policy consultant, 466–71; professional identity of, 460–84; professional responsibility of, 474, 480–81; in the public realm, 480–81
historical consciousness, 5, 36
historical demography, 196, 199, 200, 211, 212. *See also* birth rate; death rate; demographic transition; fertility; mortality
Historical Studies in the Physical and Biological Sciences, 254
historicism, 10, 11, 13, 85, 99, 116, 117, 239. *See also* New Historicism; Ranke, Leopold von; Sudhoff, Karl
Historische Zeitschrift, 77
history (historiography): analogies in, 465; as analogy for medicine, 433–35; as an art, 442; audience for, 433, 474; boundaries in, 182; causality in, 42, 44, 76; chronology in, 43; of concepts, 273–74; continuity in, 148; credibility of, 43, 76; as a discipline, 315–16; falsification in, 267; from below, 181–82, 198, 204, 294, 366; of ideas, 20, 124–27, 145; interpretation in, 36; as invention, 315; narrativity in, 42, 43, 76, 315, 369, 378–80, 383–85, 455; oral, 379, 380; pattern in, 369, 374–78; periodization in, 7, 38, 41, 43, 315; pluralistic approach to, 322; popular, 455; as profession, 174; relevance of, 316, 461–66; research in, 42; as semiotic discipline, 434, 447n2; sources of, 43, 54, 80, 99, 106n8, 119, 204, 216–18, 352, 433; of the state, 37; teleology in, 44, 141,

498   Index

history (historiography) (*cont.*)
    291, 299, 345, 350–51; truth in, 42, 315–16, 325; universal, 35, 42, 44, 86; utility of, 44, 74–75, 76, 78, 79, 80, 81, 88, 140, 147, 159, 185; as window-dressing, 467; writing, 42, 311. *See also* colonial medicine, history of; community history; cultural history of medicine; institutional approach/history; labor history; medical history; women, history of
History and Philosophy of the Life Sciences, 448n9
history of medicine. *See* medical history
history of science, 86, 212, 239, 240, 254, 259, 351–54. *See also* journals
History of Science, 254, 448n9
history of technology, 352
history wars, 311. *See also* culture wars; science wars; theory wars
Hitchcock, Tim, 198–99
Hitler, Adolf, 100, 101, 102, 103, 109n34, 110n37, 110n39
HIV epidemic. *See* AIDS epidemic
Hobsbawm, Eric, 294, 316
Holy Roman Empire, 46
homeopathy, 81, 223
Homer, 124
homosexuality, 393, 395, 398
Horn, David, 278
Horton, Robin, 342
hospitals: Foucault on, 268–70, 271, 272, 276, 347; history of, 182, 214–16, 223; and hospitalization, 215, 216; and medical knowledge, 239; and racial segregation, 380; reformatory function of, 201. *See also* pathology; protoclinic
Howell, William, 145
Hubscher, Ronald, 220
Hudemann-Simon, Calixte, 221–22
Huerkamp, Claudia, 211
Hughes, Diane, 377
human experimentation, 470
humanism: as antidote to medical specialization, 128, 141, 149, 150, 453; and Aufklärungshistorie, 36; as educational ideal, 128; and William Osler, 141, 149; and George Sarton, 149; and Henry Sigerist, 150; and Kurt Sprengel, 44. *See also* Bildung; Two cultures, the
humanities, 118, 272, 411, 413, 417, 418, 420
Humboldt, Wilhelm von, 6, 42, 117
Humboldtian university, 5–6, 11, 46, 117–18, 128. *See also* Berlin, University of; Göttingen, University of; Halle, University of; Sprengel, Kurt
humoral pathology, 225, 414
Hunt, Lynn, 316, 319, 325, 369
Hunter, John, 146
hygiene: discourses of, 296; personal, 214; social, 213

iatrogenic disease, 468–71
icon, 140, 159–60, 278. *See also* collective memory; founding father
idealism, 346. *See also* realism
identity, 3, 4, 180, 318, 375–76, 390–409; colonial, 295–98. *See also* professional identity
ideology: end of, 317, 318–19, 323, 332n36; as research topic, 210, 224–26; shaping medicine, 312, 372, 480
Ilberg, Johannes, 79–80, 132n27
Illich, Ivan, 202, 273, 310
illness: experience of, 213–18; pattern of, 286; representation of, 319; risks, 216; social construction of, 183, 210, 216, 243, 264–66, 285–306, 354, 370–71, 434. *See also* disease; experience, of pain
Imhof, Arthur, 211
immunity, 415
imperialism, 290, 292, 293, 340. *See also* state, colonial
incarceration, 268
independent scholar, 141
Indian Psychoanalytical Association, 297
industrialization, 182, 196, 198, 201
industry liability, 475, 477–80
infant care, 222–23
influenza pandemic, 287
Ingram, Allan, 204–5
inoculation, 220. *See also* smallpox
insanity, 204, 262, 267–68, 296–97; and discipline, 268. *See also* norms; reason
inspiration, 139, 142, 168, 289, 390, 400, 418, 450, 451
Institute of the History of Medicine: in Berlin, 102, 103, 127; in Leipzig, 16, 17, 95, 96, 98, 99, 100, 107n22, 150. *See also* German model; Johns Hopkins; Puschmann Stiftung; Sudhoff, Karl; Wellcome, Units for the History of Medicine
Institute of Medicine (IOM)-report, 468–70
institutional approach/history, 243, 245, 249, 289

institutional epistemology, 275
insurance: companies, 314; health, 81, 84, 152, 155, 211, 217, 219, 220, 221, 462; social, 81, 156
interests, 317, 343, 348–50, 463, 479
internalism: in traditional medical history, 2, 369, 435–37, 438; transcended, 185, 188, 346, 350, 356. *See also* antiquarianism; externalism; presentism; Whiggism
International Congress of Tropical Medicine, 286
International Union of Academies, 121
intersex rights movement, 390–409
Intersex Society of North America (ISNA), 390, 391, 395, 396, 397, 401, 406, 407
intersexuals, 23, 390–409
Isensee, Emil, 6–7, 42
Isis, 149, 153, 254, 448n9

Jackson, Andrew, 474
Jacob, Jim, 349
Jacob, Margaret, 316, 349
Jaeger, Werner, 124, 126, 128, 135n65
Jameson, Fredric, 323
Jankrift, Kay-Peter, 216
Janus, 8, 9, 117, 254
Jarcho, Saul, 441
Jenner, Edward, 146
Jewson, Nicholas, 342
Johns Hopkins: Hospital, 15, 141, 212; Hospital Historical Club, 14, 139, 143; Institute for the History of Medicine, 16, 29n52, 96, 116, 127–29, 139, 142, 151, 160n2, 174; *Johns Hopkins Hospital Bulletin*, 143; School of Medicine, 28n49; University, 15, 139, 145. *See also* German model
Jones, Colin, 194, 210, 275, 371, 378–79
Jones, Gareth Stedman, 316
Jones, Gordon, 175
Jones, W. H. S., 120
Jordanova, Ludmilla, 240, 241, 244, 259, 310, 312, 319, 324, 328n3, 329n6, 357, 372, 378
Jouanna, Jacques, 138n105
Journal of the American Medical Association, The (JAMA), 256
Journal of the History of Biology, 254
Journal of the History of Ideas, The, 127
Journal of the History of Medicine and Allied Sciences, 159, 174, 241–61, 448n9
journals: general history, 256–58; history of science, 254–58; medical history, 241–61; medicine, 256–58; social sciences, 257; social studies of science, 239, 254–58
Joyce, Patrick, 320

Kakar, Sudhir, 297
Kalbfleisch, Karl, 123
Kantorowitz, Ernst, 123
Kapsalis, Terri, 374
Keith, Arthur, 147
Kellert, Stephen, 392
Kelly, Donald, 365
Kelly, Howard, 143
Kenen, Stephanie, 395
Kennedy, Paul, 457
Kenyon, Sir Frederic, 120
Kessler, Suzanne, 395, 397
Kessler-Harris, Alice, 473, 476
Kilpatrick, Robert, 349
King, Lester S., 252, 441
Klebs, Arnold, 168
Klee, Ernst, 211
Kleinman, Arthur, 370
Kloppenberg, James, 480
Knorr, Karin, 254
knowledge, 254, 262; codes of, 269; object of, 272; positive, 268, 283n73; production of, 318, 342; provisionality of, 323. *See also* social constructivism
knowledge/power nexus, 267–72, 276, 277
Koch, Richard, 101, 102, 112n51, 112n52, 112n55, 418, 419, 425
Koch, Robert, 416
Koertge, Noretta, 393–94
Korman, Gerd, 169
Kraepelin, Emil, 203
Kruif, Paul de, 452, 457
Kudlien, Fridolf, 138n106
Kühlewein, Hugo, 132n27
Kühn, Karl Gottlob, 116
Kuhn, Thomas, 321–22, 340–41, 347; *Structure of Scientific Revolutions*, 321–22, 340–41
Kulturgeschichte. *See* cultural history of medicine
Kyklos, 16, 125

Labisch, Alfons, 211, 224
laboratory studies, 228
labor history, 179, 182
Labouvie, Eva, 222

Lacan, Jacques, 276
Lachmund, Jens, 216, 226
Laclau, Ernesto, 317
Laennec, René, 433, 435
Lammel, Hans-Uwe, 75
Lamprecht, Karl, 12, 77, 84–85, 86, 87, 88, 89
Lamprechtstreit. *See* Methodenstreit
Lancaster, Henry Carrington, 137n95
Laqueur, Thomas, 199, 371
Latour, Bruno, 254, 255, 278, 284n84, 341; *Science in Action,* 342
Laveran, Alphonse, 286
law(s): of contrary effects, 86–87; of history, 85, 86, 90; of nature, 82–83, 410
Law, Sylvia, 472, 473
Lawrence, Christopher, 252–53
lawsuit, 188; experts in, 193n49
lead poisoning, 171, 186–87, 188, 189
Leavitt, Judith, 175, 190n20, 252–53; *Sickness and Health in America* (with Ronald Numbers), 175
Lebrun, François, 212
LeClerc, Daniel, 40
Lederer, Susan, 470
LeFanu, James, 452
Leibniz, Gottfried, 39
Leipzig University, 95
Lejeune, Fritz, 104
Léonard, Jacques, 212, 219, 220
leprosy, 216, 291, 296
Lévy, Bernard-Henri, 332n37
Lewis, Jane, 252–53
Lewis, Milton, 298
Lewis, Sinclair, 452
life expectancy, 213
life sciences, 263–67
Lindemann, Mary, 210
linguistic turn, 181, 319, 332n35, 369, 375, 378–80. *See also* culture, turn in; discourse, analysis of
literature: and genre, 33, 34, 37; studies of, 319; and literary theory, 315, 317
litigation, 472–80
Littré, Émile, 20, 27n27, 53–73, 116, 117; *Conservation, révolution, et positivisme,* 61; *Dictionnaire historique de la langue française,* 54, 61; *Hippocrate,* 58
Lloyd, Geoffrey, 138
localism, 344, 352, 356
Locke, John, 281n28

Loetz, Francisca, 221
Long, Diana, 170, 173
Loudon, Irvine, 252
Loux, Françoise, 212
Luckman, Thomas, 341
Ludmerer, Kenneth, 252–53, 478, 484n37
Lunbeck, Elizabeth, 172
Lupton, Deborah, 334
Lyons, Maryinez, 291
Lyotard, Jean François, 317, 322

MacDonald, Michael, 275
Mackenzie, Charlotte, 204
MacKenzie, Donald, 341
MacLeod, Roy, 298–99
Macy Foundation, 466
Magendie, François, 67
Magnus, Hugo, 28n37
Maier, Pauline, 474
malaria, 200, 289, 291
Mall, Franklin P., 145
Malthusian model, 196
Man, Paul de, 331n35
Manderson, Lenore, 290, 293
Mann, Gunter, 211, 212
Manson, Patrick, 290
manuscripts (collecting, editing, translating of), 54, 55–60, 63, 68, 98, 117, 123, 125. *See also* positivism
Marcellini, Margarita, 373–74
Markley, Robert, 278
Markowitz, Gerald, 172, 186
Marks, Harry, 172, 190n20
Marland, Hilary, 252–53
Marwick, Arthur, 316, 331n35
Marx, Karl, 315, 322, 324, 332n37
Marxism, 154, 268, 317, 331n35, 343, 375. *See also* socialism
masculinity, 298
Massachusetts History Workshop, 170
master narrative, 174, 299, 317. *See also* postmodernism
material culture, 376, 377–78
materialism, 9, 11, 81
McCarthyism, 322
McCulloch, Jock, 297
McKeown, Thomas, 197, 198, 213, 252, 253, 310, 323
McNeill, William, 292
M.D./Ph.D. credentialism, 247–50, 436, 439,

441–42. *See also* medical historian, educational background of; medical history, schism in mechanism, 266–67, 274. *See also* vitalism

medical: corporations, 313; culture, 212, 339; dictionary, 60, 64; economics, 328; education, 78, 79, 147, 149; ethics, 78, 81, 313, 328, 422, 426, 468; hierarchy, 373; industrial complex, 169; institutions, 210, 237, 277; instruments, 99, 377–78; malpractice, 313; marketplace, 201, 207n23, 210, 218–24, 313; monism, 270; nomenclature, 407; police, 272; practice, 210, 212, 223, 226, 240, 243, 245–46, 249, 273, 314, 339, 433, 410–31; profession, 81, 243, 249; power, 276, 342, 345, 347; topography, 200. *See also* ethics; medical history, praxeological turn in; professionalization

medical historian: Canguilhem as, 273–74; educational background of, 242–43; as expert witness, 475–80; Foucault as, 273; professional background of, 248, 250, 439–42. *See also* M.D./Ph.D. credentialism

medical history: academization of, 454; aims/agenda of, 3, 17, 18, 19, 168, 175, 180; approach of, 18; audience for, 3, 18, 23, 256–58, 259–60, 347, 432, 450–59; boundaries of, 4, 168, 175, 205, 226; canonical literature of, 253, 255, 341; case studies in, 213, 239, 342, 346, 351; chair in, 67, 97, 151; citation patterns in, 241, 251–56, 258, 259; constituency of, 17, 390–409, 450–59, 480–81; dialog in, 2; as estranged from medicine, 416, 418–19, 426; fragmentation of, 258–60; from below, 218, 355, 356; functions of, 1, 24; for the general reader, 450–59; in Germany, 5–14; and history of science, 238; institutional base of, 6, 14, 15–16, 27n30, 74, 88, 90, 96, 97, 106n8, 127–29, 151–52, 195, 206n8, 211, 212, 275, 425–26; intellectual change in, 237–61; intellectual configuration of, 251–56; interdisciplinarity in, 202, 227, 356; internal organization, 258–60; isolated position of, 258–60; legitimacy, 30n62; in the medical curriculum, 417, 425–27; methodology, 3, 17, 18, 175, 209, 215, 365; national socialist version of, 425; pluralism in, 3, 4, 5, 18, 195, 204; popularization of, 453, 456; practice, 18, 202, 433; practitioners, 17, 175, 258–60; praxeological turn in, 213, 224–26, 227, 228n17; programmatic calls in, 442; reception of Canguilhem and Foucault, 274–79, 283n75; relevance of, 205, 412; research, 9, 13, 96; schism in, 172–77, 188, 435–37, 445–47, 454; styles of, 419–24; teaching of, 8, 443–47; top-down, 196, 221; trends in, 195, 237; in the United States, 14–17; without medicine, 438, 454. *See also* deinstitutionalization; journals; medical, practice; national styles; professionalization

Medical History, 241–61

medicalization: and activism, 405; and colonial medicine, 294; as crude history, 314; of the intersexed, 396; and medical biopower, 210, 218–24, 272, 276, 277, 278, 324, 345, 421; of the poor, 202. *See also* patient

medicine: as acting discipline, 410–31; alternative, 223; as analogy for history, 433–35; as an art, 264; boundaries of, 168, 315; of cases, 269; casuistic character of, 414, 420, 423; causality in, 270; clinical, 53, 271; cognitive dimensions of, 356; complimentary, 223; constituency, 390–409; disciplinary effects of, 273; as a discipline, 312–15; environmental, 183; evidence-based, 312, 313, 405; experimental, 88, 146; folk, 201; historicity of, 411, 419–24; history in (medicine in action), 410–31; history of (medicine in context), 410–31; as knowledge, 312; magical, 201; narrative-based, 313; of nosology, 269; occupational, 183; physiological, 82, 146, 265, 270, 271; pluralism, 201, 224; popular, 202; as practice, 312; public understanding of, 413; as semiotic discipline, 434, 447n2; social implications of, 413; visual experience in, 356

mentalités, history of, 365

Merleau-Ponty, Maurice, 267

Merton, Robert, 343

metaphor, 148, 203, 320, 322, 371, 392

Methodenstreit, 77, 84–85, 88

methodology, 76, 89, 97, 99, 102, 126; experimental, 53, 147, 149, 416; laboratory, 146, 147, 238

Mewaldt, Johannes, 122, 123

Meyer-Steineg, Theodor, 101, 109n36, 112n52, 119

microhistory, 369, 372–74. *See also* Ginzburg, Carlo

Middle Ages, 82, 99, 117

Middleton, Conyers, 116

midwives, 222–23

Milbank Memorial Fund Quarterly, 321, 448n9

Miller, Genevieve, 15

mind, 202–3. *See also* body; self

Mitteilungen für Geschichte der Medizin und der Naturwissenschaften, 97, 99
modernism, 325
modernity, 299, 314, 317, 318, 319, 320, 332n35
modernization, 199, 201, 210, 225, 345
Moehsen, Johann Carl Wilhelm, 20, 33–52; *Geschichte der Wissenschaften in der Mark Brandenburg,* 38
Mohr, James, 472, 473
Mommsen, Theodor, 10, 118
Money, John, 404, 407
Morantz-Sanchez, Regina, 180, 191n28
morbidity, 216, 253, 475
Morello, Paolo, 65
Moreno, Angela, 403
Morman, Edward, 177
Morris, J. N., 252
mortality, 197, 213, 216, 253, 475; change in, 213–14; patterns of, 197–98; rate of, 197, 215, 293. *See also* birth rate; death rate; demographic transition; fertility; historical demography
Mulkay, Michael, 255
Müller, Martin, 104
multiculturalism, 314, 329n17
Mumford, James Gregory, 450, 451, 453, 455
Murray, Gilbert, 123
Musacchio, Jacqueline, 376
museology, 316

Nandy, Ashis, 297
national health services, 312
nationalism, 291; and historians, 289; and optimism, 297
national socialism, 20, 100, 101, 110n39, 425
National Sozialistische Deutsche Arbeiter Partei (NSDAP), 96, 100, 101, 104, 108n29, 112n55, 114n72
national styles, 115, 127, 209–36, 299. *See also* medical history, styles of; zeitgeist
natural healing movement, 416
Naturphilosophie, 7
Naumann, Moritz, 26n12
Nazis, 100, 101, 108n31, 109n33, 128, 137n98, 211
Needham, Joseph, 154
neo-Hippocratism, 7. *See also* Hippocratism
neoliberalism, 22, 323
Neuburger, Max, 14, 20, 28n37, 74–94, 141, 417; *Handbuch der Geschichte der Medizin,* 88

Neumann, Salomon, 415
New Historicism, 319
Nicolson, Malcolm, 284n83, 348
Nietzsche, Friedrich, 317, 322, 332n37
normal, the, 266, 268, 312
normality, 179, 262, 266, 272, 398, 400, 403, 405; colonial, 297. *See also* health
normalization, 272, 273, 314
normativity, 224, 262, 263–67
norms, 199, 262–84; clinical, 267–72; as productive force, 262; as source of power, 262
North American Task Force on Intersex, 406–7
nosology, 406, 445
Novick, Peter, 321
Numbers, Ronald, 175, 190n20, 237–38, 240, 241, 242, 244, 246, 249, 252–53, 258
nursing, 170; history of, 179, 191n25, 437; practitioners of, 220, 239
nutrition, 171, 198

objectivity, 265–66, 321, 335n69
obstetricians, 222–23
occupational disease/health, 182–83, 187. *See also* disease
O'Connor, Erin, 326–27
Oppenheimer, Gerald, 186
Organization of American Historians, 193n48
orientalism, 318
Osiander, Friedrich B., 223
Osler, William, 14–16, 20, 139–64, 169, 453, 456
Osler Society, 21, 140, 159, 177
Our Bodies, Ourselves, 394
Oxford University, 147
Ozar, David, 403, 404

Packard, Randall M., 292, 293
Pagel, Julius, 13, 14, 20, 28n37, 74–94, 112n51, 141, 417–18, 419; *Geschichte der Medizin,* 83, 419; *Grundriss eines Systems der Medizinischen Kulturgeschichte,* 75, 83–84
Pagel, Walter, 75, 76, 101, 102, 417, 425
Palmer, Bryan, 316, 317
paludism, 286
Paracelsus, 95, 97, 98
paradigm, 347
Parchappe de Vinay, Jean-Baptiste, 67
Park, Katharine, 372
Paris: Faculté de médecine, 67; Muséum d'histoire naturelle, 67; "school" of medicine, 276, 282n63

Passerini, Luisa, 379
Passeron, Jean-Claude, 267
Pasteur, Louis, 148, 275
pathological, the, 264–66, 268, 312
pathology, 8, 146, 264, 266, 269–70, 395; colonial, 297; geographical, 286, 289; historical, 7–8, 26n12, 286, 289. *See also* disease; hospitals
patient: (lack of) agency of, 182, 201, 202, 217, 222, 223, 227, 271, 313, 399; and demand for medical services, 210, 224; and the hospital, 215, 269, 270, 271; and narrativity, 434; as new research object, 177, 209, 210, 218, 225, 243, 244, 246, 258, 313, 320; role of, 216; view of, 218, 355, 356. *See also* medicalization
patronage, 46, 223, 340, 343
Pauly, Félix-Alphonse, 62
Paulys Realenzyklopädie der klassischen Altertumswissenschaft, 120
Pelling, Margaret, 252–53, 355, 367–68
Pennell, Sara, 377
Pernick, Marty, 172, 190n20
personality cult, 96. *See also* charisma; icon
Pert, Candace, 451
Peterson, Jeanne, 392, 395
Petrequin, J. P. E., 116
Pfaff, Franz, 123, 132n27
pharmaceutical companies, 314
philhellenism, 141, 147–48
Phillipps, Thomas, 56
philology, 41, 80, 88, 89, 117–22, 127; as method, 10, 11, 14, 20, 42, 68, 77, 80, 85, 89, 99, 102, 117–22, 126, 141, 150, 152; and philologists, 58, 60, 67, 75, 78, 79, 117–22
philosophical approach, 14, 75, 76, 90, 102, 111n51, 123–27, 141, 150
philosophy: of error, 266; of knowledge, 263, 275; of medicine, 262–84; of remedy, 275; of science, 321, 340, 353, 414; of values and action, 275
phrenology, 370
Pickstone, John, 277, 311
Pinel, Philippe, 203, 269
place, 288. *See also* class; colonial, the; frame, the; gender; race
plague, 216, 371, 378–79
Plato, 124, 143
Pliny the Elder, 42, 116
Plutarch, 33
Pohl, Lynn Marie, 380
policy: debate over , 460–84; making of, 460–84; review committees and, 468–70

political, the, 317, 325, 327, 328, 381. *See also* cultural, the; economic, the; social, the
political: action, 318; correctness, 314, 329, 399; economy, 292, 295; engagement, 140, 141, 155, 384–85; philosophy, 323; power, 317; reform, 327
poor, construction of the deserving, 366–67, 381
Poor Law, 366–67
Poovey, Mary, 326
population growth, 196
population technology, 272
Portelli, Allessandro, 379
Porter, Dorothy, 319
Porter, Roy, 252, 255, 275, 277, 313, 355, 376
positivism: approach of, 77, 238; criticism of, 123, 263, 264, 273; and Charles Daremberg, 20, 53–73; and Émile Littré, 20, 53–73; and Max Neuburger, 75; and Julius Pagel, 75; and philology, 115, 124, 137n94; as scientific ideal, 121, 124, 137n94, 264; and Karl Sudhoff, 124; in traditional medical history, 2. *See also* manuscripts
postcolonial medicine, 278, 291; history of, 285–306. *See also* colonial medicine
postcolonial studies, 181, 184, 288, 295
postmodernism: and methodological relativism, 18, 22, 315, 321, 323, 332n39; and neoliberalism, 316, 319, 323, 331n35; (lack of) influence on the social history of medicine, 177, 253, 315, 323, 325, 326. *See also* master narrative
poststructuralism, 178, 180, 317, 374–78
power, 184, 209, 226, 341; colonial, 296; of norms, 272; spatial distribution of, 287
practice, clinical, 145, 434
pragmatism (in medical historiography): and cultural history, 14, 41–44, 74–94; and Charles Daremberg, 65, 66, 72n60; as didactic instrument, 6, 10, 11, 20, 41–44, 74–94, 116, 421, 424; and Max Neuburger, 74–94, 417–18; and William Osler, 141; and Julius Pagel, 13, 74–94, 417–18; versus philology, 116; and Theodor Puschmann, 11; and Henry Sigerist, 141, 150; and Kurt Sprengel, 6, 10, 41–44; and Karl Sudhoff, 13, 419; and Rudolf Virchow, 415
preindustrial societies, 196
presentism, 141, 238, 267, 435–37, 463. *See also* antiquarianism; externalism; finalism; history, teleology in; internalism; Whiggism

probability thinking, 269
Probst, Christian, 211
Probyn, Elspeth, 319
Proctor, Robert, 188, 484n37
professional identity: of medical historians, 89, 168, 176, 242; of physicians, 5, 8, 21, 33–52, 78, 139–64. *See also* medical historian
professionalism, 13, 173, 175, 456. *See also* amateurism
professionalization: of American medical history, 16, 140–41, 151, 175, 237–61, 453, 454; of British medical history, 195, 237–61; of German medical history, 14, 90, 119; of medicine, 210, 211, 218–24, 314, 324, 341, 345, 367–68. *See also* medical history, schism in; medical, profession; Sudhoff, Karl; Welch, William Henry
Proschaska, Georg, 267
protoclinic, 269. *See also* hospitals
providence (divine), 39, 43, 44, 47
Prussia. *See* Brandenburg-Prussia
psychiatry, 298; colonial, 297; history of, 203–4, 277–78, 445–46; movement against, 204
psychoanalysis, 297
psychology, 87, 268
psychotherapy, 298
public domain, 34, 39
public health, 152, 176, 245–46, 258, 376, 465; colonial, 293, 294; ethics and, 186; history of, 249; infrastructure of, 213, 214; policy, 186, 219, 460–84; school of, 186; system, 172
publishing: trade press, 452, 456, 458; university press, 401–2, 458
Puschmann, Marie (née Fälligen), 11–12
Puschmann, Theodor, 10–11, 12, 14, 117, 417
Puschmann Stiftung, 11–13, 17, 95, 97–98, 121, 123. *See also* Institute of the History of Medicine, in Leipzig
Putman, Robert, 315, 324, 381
Putnam, Hilary, 322

quackery, 80, 81
qualitative, the, 264–67
quantitative, the, 264–67
Quellenforschung. *See* history, sources of
quinine, 200

Rabinow, Paul, 274, 278
race, 173, 177, 183–85, 237, 288, 293, 297, 305n46, 375; biomedical framing of, 296; formation of, 295; and racism, 183–84; and segregation, 293, 304n36. *See also* class; colonial, the; ethnicity; frame, the; gender; place
Ramsey, Mathew, 211
Ranke, Leopold von, 10, 13, 99. *See also* historicism
Rapaport, Dr., 121
Rather, L. J., 441
rationalism, 321
rationality, 274, 278, 314, 317
Rattansi, Piyo, 342
Rawls, John, 313–14, 324
realism, 321, 341, 346. *See also* idealism
reason, 262, 263, 268, 317; diffusion of, 38, 44, 47; myth of, 270. *See also* insanity; norms
reductionism, 11, 14, 265; historiographic, 238
reflex, concept of, 264, 266, 267
relativism, 18, 22, 23, 321, 322, 336n84
Renzi, Salvatore de, 54
representation: as elusive concept, 180–81, 315, 318; of the female body, 178, 179; models of, 268–69, 312, 324, 340; of the past, 4, 315, 316; and politics, 187, 316
Representations, 319, 372
reproductive rights, 179
Republica literaria, 38
research funding, 316, 328n2, 394, 408
Reverby, Susan, 187, 192n29, 286; "Beyond the Great Doctors" (with Rosner), 167, 168, 185; *Health Care in America* (with Rosner), 167, 175
Revue d'histoire de science, 254
Richard, Philippe, 212
Richardson, Ruth, 202, 382–85; *Death, Dissection, and the Destitute,* 202, 382–85
Riddle, John M., 137n91
Riley, James C., 252
Risse, Guenter B., 252
Robin, Charles, 64
Robinson, David M., 137n95
Rockefeller Foundation, 151
Roe v. Wade, 382
Rohlfs, Heinrich, 84
role model, 27n27, 33, 142, 143, 144, 145, 400. *See also* collective memory; heroism; Hippocrates
Romantic period, 6–7, 44, 54, 66
Rosario, Vernon, 393, 395
Rose, Nikolas, 315, 334n63
Rosen, George, 159, 170, 252–53, 320, 425
Rosenbaum, Julius, 9, 26n12, 54, 59, 72n60

Rosenberg, Charles, 167, 170, 173, 252–53, 255, 261n28, 309, 310, 311, 320, 321–24, 366, 370, 371, 386n20; *Framing Disease* (with Golden), 320–24, 370, 371
Rosenberg, Rosalind, 476
Rosenberg, Steven, 451
Rosenkrantz, Barbara, 171, 172, 173
Rosenzweig, Roy, 172
Roser, Wilhelm, 415
Rosner, David, 286; *Deadly Dust* (with Markowitz), 186; *Deceit and Denial* (with Markowitz), 186–87
Ross, Ronald, 290
Roth, Moritz, 78, 79
Rothman, David, 173, 186, 460
Rotschuh, Karl, 430n31
Royal Society of London, 373
Royal Society of Medicine (Great Britain), 149
Ruggiero, Guido, 373–74
Russett, Cynthia, 345
Ryle, John A., 323

Sadowski, Jonathan, 297
Said, Edward, 295, 304n40, 318
Saint-Simon, Claude-Henri de, 53
Salomon-Bayet, Claire, 275
Samuel, Raphael, 316, 331n35
Sander, Sabine, 219
Santayana, George, 390, 408, 481
Sappol, Mike, 382–85
Sarasin, Philip, 225
Sarton, George, 149–50, 153
Sauerbruch, Ferdinand, 419
Savitt, Todd, 183
Sawday, Jonathan, 202–3, 319, 371
Scarborough, John, 137n91
Schaffer, Simon, 323
Schelling, Friedrich von, 6, 59, 72n59
Schiller, Johann von, 39
Schlich, Thomas, 448n8
Schofield, Roger, 196, 200–201
Schöne, Hermann, 121–22, 123
Schönlein, Johann Lukas, 26
Schulze, Johann Heinrich, 116
Schwerner, Mickey, 171
science: applied, 79; experimental, 74, 239; multiple meanings of, 238, 271, 274
Science and Culture, 254
science and technology studies (STS). *See* social studies of science

Science in Context, 254
sciences, rearrangement of the, 411
Science, Technology, and Human Values, 254, 255
science wars, 18, 321, 393, 402. *See also* culture wars; history wars; medical history, schism in; theory wars
scientometric approach, 237–61, 439–42
Scott, H. Harold, 288–89, 290
Scott, Joan Wallach, 180
Seaver, Paul, 373
secularization, 220, 421
Seidel, Hans-Christoph, 223
Seiffert, Max, 12–13, 14, 77, 78, 85
self, 202–3, 281n28. *See also* body; mind
Seligmann, Franz Romeo, 64–65
semiotics, 316, 317, 434. *See also* signs
sexism, 179
sexuality, 177, 178, 179, 198, 199, 277
sexually transmitted diseases, 460, 464. *See also* AIDS epidemic
Shapin, Steven, 341, 370
Short, Thomas, 200
Shorter, Edward, 199
Shryock, Richard, 173, 252–53, 286
Sigerist, Henry, 139–64; and Baltimore, 15–17, 127, 425; *Einführung in die Medizin*, 150, 152; and the geography of disease, 285–88, 300; *Great Doctors*, 452; *History of Medicine*, 124, 157, 159; as inspiration to Anglo-Saxon historians, 252–53; intellectual development of, 124–25; as a "Jew," 109n36; on Max Neuburger, 76; on Julius Pagel, 75; and professionalism, 13; in Saskatchewan, 157; and the social history of medicine, 168–69, 170, 173, 323; *Socialized Medicine in the Soviet Union*, 153–54; on Kurt Sprengel, 35; on Karl Sudhoff, 14, 101, 109n31; and the Sudhoff Medal, 103; and the utility of medical history, 20–21
Sigerist Circle, 21, 140, 177
signs, 433, 434, 435. *See also* semiotics
silicosis, 186, 189
Singer, Charles, 109n32, 149, 151
Singer, Dorothea, 149
Sklar, Kathryn Kish, 476
Sloterdijk, Peter, 325
smallpox, 216–17, 294. *See also* inoculation
Smith, David, 260n6
Smith, Wesley D., 137n91
social, the: as category of analysis, 310, 326, 327, 328, 366, 384; evaluation of a category, 316–

social, the (*cont.*)
21, 366, 384; replaced by the cultural/the frame, 309, 316, 322, 324, 325, 331n32. *See also* cultural, the; economic, the; political, the
social construction: of medical knowledge, 338–63; of social facts, 367; of womanhood, 178. *See also* illness
social constructivism, 22, 239, 240, 253, 260n9, 322, 323, 393. *See also* essentialism
social control, 272, 314, 317
social engagement, 141, 384–85. *See also* political, engagement
social history of medicine: definition of, 339; as estranged from medicine, 20, 454; failure of, 309–37; as generational break, 237; (lack of) integration into social history, 240, 259; as a new program, 2, 3, 21, 140, 167–93, 194–208, 209–36, 244, 249, 286, 287, 288, 441; transition of, 367, 380. *See also* cultural history of medicine
Social History of Medicine, 195, 241–61, 319, 339, 448n9
socialism, 150–59. *See also* Marxism
socialized medicine, 155, 175
social medicine, 310, 323
Social Science and Medicine, 448n9
social sciences, 240, 259, 411
social studies of science: audience for, 256; changes in, 239, 240; and gender, 187; lack of interest in, 278; (absence of) links with social history of medicine, 179, 181, 184, 254, 255, 259, 284n83. *See also* culture, studies of; journals; laboratory studies; literature, studies of
Social Studies of Science, 254, 255, 351
Society for the Social History of Medicine, 195, 319
Society for the Social Studies of Science, 239
sociology, 85, 155, 157, 278, 310; approach of, 152–53; categories of, 317; of disease, 369; of knowledge, 226, 316, 341, 342, 369; of science, 290, 321, 338, 353
Soviet medical system, 153
specialization, 11, 14, 53, 149, 151, 453
Spence, Jonathan, 457
Spengler, Oswald, 150
Spicker, Stuart F., 274
Spree, Reinhard, 211, 213
Sprengel, Kurt, 5–6, 7, 10, 11, 20, 33–52, 72n60, 75, 84, 89, 141; *Versuch einer pragmatischen Geschichte der Arzneikunde*, 6, 36, 41–44, 46. *See also* Humboldtian university; pragmatism
Staden, Heinrich von, 137n91
Stannard, Jerry, 137n92
Starr, Paul, 168
Starr, Susan Leigh, 255
state: colonial, 289, 292, 294; formation of, 295; power of, 296, 313
Steedman, Carolyn, 375–76
Steiner, George, 384
Stevenson, Lloyd, 172, 175, 190n19, 252
Stöckel, Sigrid, 213
Stollberg, Gunnar, 215, 216
Stone, Lawrence, 199
Stowe, Steven, 378
Stricker, Wilhelm, 84
Studies in the History and Philosophy of Biology and the Life Sciences, 254
Stumm, Ingrid von, 216
Styles, John, 376
subjectivity, 265–66, 297, 298, 316, 332n37
Sudhoff, Karl, 11–14, 15–16, 20, 28n37, 74, 75, 79, 80, 86, 88, 89, 95–114, 119, 124, 125, 141, 150, 170, 418–19. *See also* founding father; historicism; Institute of the History of Medicine, in Leipzig; professionalization
Sudhoffs Archiv. *See Archiv für Geschichte der Medizin*
surgeon, 399, 403
surgery: genital, 403, 404, 406, 407; transplant, 448n8
Swanson, M. W., 293
Sydenham, Thomas, 146, 269
Sydenham Society, 59
synthesis, 75, 143, 148, 157
Szreter, Simon, 252

Tamm, Ingo, 215
Technology and Culture, 352
Temkin, Owsei: in Baltimore, 127, 128, 152; *Galenism: The Rise and Decline of a Medical Philosophy*, 125, 128; as German exile, 15, 16, 425; and the German model, 151; as historian of ideas, 115, 116, 170, 320; as inspiration to Charles Rosenberg, 370; intellectual development, 125–26; as a Jew, 101, 110n36; on Julius Pagel, 82; on pragmatic history, 76
theory wars, 325. *See also* culture wars; history wars; science wars
therapeutics, 80, 81, 266, 373

Thompson, E. P., 294, 310, 319
Thompson, James, 406
Thucydides, 43, 315
tobacco industry, 188–89, 193n49, 475, 477–79
Tomes, Nancy, 376
Touati, François-Olivier, 216
travel books, 39, 41, 45, 46
Treacher, Andrew, 370
trial, clinical, 468
tropics, the, 285; disease in, 288–92; expertise in, 290; medicine in, 285–306
trypanosomiasis, 286, 291
tuberculosis, 216, 217, 218
Turner, Bryan, 334n63
Turner, Roy, 34
Turner, Trevor, 204
Tuskegee syphilis study, 187–88
Two cultures, the, 11, 78, 149, 150, 453, 410–31. *See also* Bildung; humanism

Unit for the History of Medicine. *See* Wellcome
U.S. Food and Drug Administration, 477
utilitarianism, 80, 83. *See also* history, utility of

Vaughan, Megan, 296, 297
Verbrugge, Martha, 172
Vesalius, Andreas, 202
veterinarians, 220
Veyne, Paul, 267
Vienna school of medicine, 81, 87
Virchow, Rudolf, 8, 414, 415
vitalism, 266–67, 274. *See also* mechanism
Vögele, Jörg, 213–14
Vretto, Athena, 319

Wachter, Kenneth, 198
Wagner, Gerhard, 112n55
Wailoo, Keith, 184
Wallington, Nehemiah, 373
Walzer, Richard, 123
Warburg Institut (Hamburg), 123
Warner, John Harley, 143, 238–39, 240, 241, 244, 252–53, 255, 259, 260n6
Warren, Christian, 186
Watson, James, 392
Wear, Andrew, 310
Weber, Max, 317
Webster, Charles, 252–53, 323, 333n54, 342, 349
Webster case, 472–74
Weimar, 123–27

Weindling, Paul, 252
Weisz, George, 282n63
Welch, William Henry, 16, 127, 139–40, 143, 151, 168. *See also* professionalization
welfare: medicine, 312; state, 272
Wellcome: Institute for the History of Medicine London, 195; Trust, 18, 195, 206n8; Units for the History of Medicine (Cambridge, East Anglia, Glasgow, Manchester, Oxford), 195, 206n8
Wellmann, Max, 120, 122
Wenkebach, Ernst, 132n27
Whiggism: belied, 204, 437; as dismissive category, 23, 199, 391, 435–37; rejection of, 239, 340, 341, 391. *See also* antiquarianism; internalism; presentism
White, Hayden, 318
White, Luise, 380
Wilamowitz-Moellendorff, Ulrich von, 10, 118–19, 121, 123
Wilkins, Lawson, 404, 407
Williams, Raymond, 347
Willis, Thomas, 267
Willowbrook decree, 171
Wilson, Bruce, 403
Wilson, Charles Morrow, 289
Wilson, Leonard, 174, 176, 454
Wilson, Theodore, 484n37
Winckelmann, Johann Joachim, 117
Wischhöfer, Bettina, 221
witchcraft, 342, 374
Wittwer, Philipp Ludwig, 42
Witzler, Beate, 213
Wolff, Eberhard, 216–17
Wölfflin, Heinrich, 150
womanhood, social construction of, 178, 179, 180
women: activism of, 187; bodies of, 343, 366; diseases affecting, 178; health movement of, 169, 174, 179; history of, 169–70, 178
women's studies, 176
Woods, R. I., 252
Worboys, Michael, 290
Wright, Peter, 370
Wrigley, Tony, 196, 197, 200–201
Wunderlich, Carl, 8–9, 12, 13, 72n59, 414–15

Yates, Frances, 342
yellow fever, 291

Zeiss, Heinz, 104
zeitgeist, 7, 150, 176. *See also* national style

R
133
.L62